ex libris

Energy Economics

Subhes C. Bhattacharyya

Energy Economics

Concepts, Issues, Markets and Governance

Dr. Subhes C. Bhattacharyya
Centre for Energy, Petroleum and Mineral
   Law and Policy
University of Dundee, UK
e-mail: S.C.Bhattacharyya@dundee.ac.uk;
         subhes_bhattacharyya@yahoo.com

ISBN 978-0-85729-267-4 e-ISBN 978-0-85729-268-1
DOI 10.1007/978-0-85729-268-1

Springer London Dordrecht Heidelberg New York

© Springer-Verlag London Limited 2011

Apart from any fair dealing for the purposes of research or private study, or criticism or review, as permitted under the Copyright, Designs and Patents Act 1988, this publication may only be reproduced, stored or transmitted, in any form or by any means, with the prior permission in writing of the publishers, or in the case of reprographic reproduction in accordance with the terms of licenses issued by the Copyright Licensing Agency. Enquiries concerning reproduction outside those terms should be sent to the publishers.

The use of registered names, trademarks, etc., in this publication does not imply, even in the absence of a specific statement, that such names are exempt from the relevant laws and regulations and therefore free for general use.

The publisher makes no representation, express or implied, with regard to the accuracy of the information contained in this book and cannot accept any legal responsibility or liability for any errors or omissions that may be made.

Cover design: eStudio Calamar, Berlin/Figueres

Printed on acid-free paper

Springer is part of Springer Science+Business Media (www.springer.com)

Preface

The idea for this book came about 4 years ago when I attended a workshop in Oxford on energy economics teaching in the U.K. organised under the auspices of the UK Energy Research Centre (UKERC). That was a time when oil prices started its upward journey and concerns about the security of energy supply were becoming a common man issue. It occurred to me that despite this great interest in understanding the common energy problems around us, there is a lack of critical appreciation of the problem and its inter-linkages with other issues. While the interest in the field of energy studies has seen a tremendous growth over the past decade, there is a serious gap in terms of a holistic understanding of the energy problems around us. That workshop clearly demonstrated that the economic concepts that are relevant to the energy industry are poorly understood by researchers of inter-disciplinary background. The main reason behind this state of affairs is the lack of a good, accessible reference book in energy economics that anyone interested in the subject can hold onto.

Luckily for me, this revelation came as a good opportunity to deliver such a book. Last year, 2010, marked the completion of 25 years of my continuous involvement with the energy field of study. I have been teaching the subject to students of inter-disciplinary backgrounds for quite sometime now. I have taught various aspects of energy economics and policies, and have provided training to senior staff. Moreover, having worked in the industry and in high level professional consulting, I understand the need for a balanced approach for such a book. In addition, my current research focuses on practical, applied problems where technology, engineering, economics, finance, regulation and the environment all feature in different proportions. This came handy while preparing for this book.

My desire to put a holistic picture by including various dimensions of the problem in the book has meant that the size has gone up. The feedback from my students has influenced the outline and the content of the book. While all of them want to gain some analytical skills and concepts so that they can analyse any given problem using simple economic logic, they have also shown great interests in understanding the environmental aspects related to energy use and the regulation and governance of the industry. I have complied with their desires and hope that this volume helps any reader to gain a wider and balanced understanding of the energy issues.

Most of the content of the book is accessible to persons of non-mathematical background. The economic concepts have also been explained in simple terms, often using graphical presentations. However, for those who cannot imagine an energy economics book without mathematics, I have added some materials and have provided references for further reading. Essentially, mathematics has been used as an aid and not for the sake of it.

I am grateful to my students who provided valuable feedback and encouraging comments on most of the materials of this book that have been tested in various classes. Their questions and reflections/ criticisms have always have helped me in improving my work. Although I have included additional materials based on my personal research activities or to reflect the changes taking place in the energy sector, I am very confident that other readers would find the content useful to them. I am also thankful to my colleagues with whom I have co-authored some of my academic publications that are included in this book under various chapters. However, I am only responsible for any errors and omissions that may still remain.

A book of this size always takes special personal efforts. Although I thought I would be able complete the work in a short period of time given the state of preparedness of the initial manuscripts, it proved too optimistic in the end. I am thankful to Ms Claire Protherough and Mr Anthony Doyle for their understanding and flexibility. Above all, I could not have realised this work without the support and sacrifice of my family members—my spouse Debjani and my daughter Saloni. The order in which your names appear in the print does not matter—you are always special and priceless to me.

Contents

1 **Introduction to Energy Economics** 1
   1.1 Introduction 1
   1.2 Energy and Multidimensional Interactions 2
   References 5

Part I Energy Demand Analysis and Forecasting

2 **Energy Data and Energy Balance** 9
   2.1 Introduction 9
   2.2 Energy Basics 9
      2.2.1 Energy Defined 9
      2.2.2 Alternative Classifications of Energy 10
   2.3 Introduction to the Energy System 11
   2.4 Energy Information 14
   2.5 Energy Accounting Framework 15
      2.5.1 Components of the Energy Account 16
      2.5.2 Commodity Accounts and Overall
            Energy Balance 18
      2.5.3 Units, Conversion Factors and Aggregation of
            Energy Flows 19
   2.6 Accounting of Traditional Energies 24
      2.6.1 Features of TEs 25
      2.6.2 Data Availability, Data Collection and Reporting 26
   2.7 Special Treatments of Some Entries in the
      Energy Balance 27
      2.7.1 Treatment of Primary Electricity Production 27
      2.7.2 Treatment of Electricity in Final Consumption 28
      2.7.3 Self Generation 28
   2.8 Analysis of Energy Balance Information 29
   2.9 Alternative Presentation of Energy Accounting Information 31

		2.9.1	Energy Flow Diagrams	31
		2.9.2	Reference Energy Systems (RES)	32
		2.9.3	Common Energy Data Issues	34
	2.10	Conclusion		35
	References			38
3	**Understanding and Analysing Energy Demand**			41
	3.1	Introduction		41
	3.2	Evolution of Demand Analysis		42
	3.3	Overview of Energy Demand Decisions		44
	3.4	Economic Foundations of Energy Demand		46
		3.4.1	Consumer Demand for Energy: Utility Maximization Problem	47
		3.4.2	Cost Minimization Problem of the Producer	50
	3.5	Alternative Approaches for Energy Demand Analysis		51
		3.5.1	Descriptive Analysis	51
	3.6	Factor (or Decomposition) Analysis		57
		3.6.1	Analysis of Change in Total Energy Demand	58
		3.6.2	Analysis of Changes in Energy Intensity	61
	3.7	Analysis Using Physical Indicators		64
	3.8	Energy Demand Analysis Using the Econometric Approach		65
	3.9	Conclusion		71
	References			74
4	**Energy Demand Analysis at a Disaggregated Level**			77
	4.1	Introduction		77
	4.2	Disaggregation of Demand		77
	4.3	Sectoral Energy Accounting		79
	4.4	Analysis at the Sectoral Level		81
		4.4.1	Industrial Energy Demand Analysis	81
		4.4.2	Energy Demand Analysis in the Transport Sector	93
		4.4.3	Energy of Energy Demand in the Residential and Commercial Sectors	101
	4.5	Conclusion		105
	References			105
5	**Energy Demand Forecasting**			107
	5.1	Introduction		107
		5.1.1	Simple Approaches	107
		5.1.2	Advanced or Sophisticated Techniques	112
		5.1.3	Econometric Approach to Energy Demand Forecasting	113
		5.1.4	End-Use Method of Forecasting	115

		5.1.5	Input–Output Model	116
		5.1.6	Scenario Approach	119
		5.1.7	Artificial Neural Networks	120
		5.1.8	Hybrid Approach	121
	5.2		Review of Some Common Energy Demand Analysis Models	122
		5.2.1	MAED Model	123
		5.2.2	LEAP Model	124
		5.2.3	Demand Module in NEMS (National Energy Modeling System)	125
		5.2.4	Demand Modelling in WEM (World Energy Model)	127
	5.3		Conclusion	128
	References			132
6	**Energy Demand Management**			**135**
	6.1		Introduction	135
	6.2		Energy Demand Management	136
		6.2.1	Definition	136
		6.2.2	Evolution of DSM	137
		6.2.3	Justification for DSM	138
	6.3		Load Management	139
		6.3.1	Direct Load Control Method	140
		6.3.2	Indirect Load Control	141
	6.4		Energy Efficiency Improvements and Energy Conservation	142
		6.4.1	What is Energy Efficiency?	142
		6.4.2	Opportunities for Energy Saving	144
		6.4.3	Economics of Energy Efficiency Improvements	146
	6.5		Analysing Cost Effectiveness of DSM Options	148
		6.5.1	Participant Test	149
		6.5.2	Ratepayer Impact Measure (RIM)	149
		6.5.3	Total Resource Cost Test	150
		6.5.4	Programme Administrator Cost or Utility Cost Test	150
	6.6		Energy Efficiency Debate	151
		6.6.1	Market Barriers and Intervention Debate	151
		6.6.2	What are the Market Barriers to Energy Efficiency?	152
		6.6.3	Government Intervention and Its Nature	155
		6.6.4	Energy Efficiency Versus Economic Efficiency Debate	156
		6.6.5	Rebound Effect	158
		6.6.6	Use of Market-Based Incentives for Energy Efficiency	159

	6.7	Conclusion	159
	References		159

Part II Economics of Energy Supply

7 Economic Analysis of Energy Investments 163
    7.1   Introduction 163
          7.1.1   Main Characteristics of Energy Projects 163
    7.2   Basics of the Economic Analysis of Projects 165
          7.2.1   Identification of Costs 166
          7.2.2   Identification of Benefits 168
          7.2.3   Valuation of Costs and Benefits 168
    7.3   Economic Versus Financial Investment Analysis 174
    7.4   Indicators of Cost-Benefit Comparison 175
          7.4.1   Methods Without Time Value 175
          7.4.2   Methods Employing Time Value 176
    7.5   Uncertainty and Risk in Projects 179
    7.6   Conclusion 182
    7.7   Example of a Project Evaluation Exercise 182
          7.7.1   Problem Statement 182
          7.7.2   Answer 183
    References 189

8 Economics of Fossil Fuel Supply 191
    8.1   Introduction 191
          8.1.1   Exploration 191
          8.1.2   Exploration Programme 193
          8.1.3   The Economics of Exploration Activities 195
          8.1.4   Investment Decision 196
          8.1.5   Risks in Exploration Projects 197
    8.2   Field Development 200
          8.2.1   Investment Decision 200
          8.2.2   Resource Classification 202
          8.2.3   Classification of Crude Oil, Natural Gas and Coal ... 204
    8.3   Production 205
          8.3.1   Oil Production 205
          8.3.2   Production Decline and Initial Production Rate 207
          8.3.3   Gas Production 208
          8.3.4   Coal Production 209
    8.4   Economics of Fossil Fuel Production 210
          8.4.1   Field Level Economics 210
          8.4.2   Industry Level Economics 210
    8.5   Resource Rent 212

	8.6	Supply Forecasting		215
		8.6.1	Relation Between Discoveries and Production	215
		8.6.2	Supply Forecasting Methods	216
	8.7	Conclusion		217
	References			218
9	**Economics of Non-Renewable Resource Supply**			**219**
	9.1	Introduction		219
	9.2	Depletion Dimension: Now or Later		219
	9.3	A Simple Model of Extraction of Exhaustible Resources		221
		9.3.1	Effect of Monopoly on Depletion	222
		9.3.2	Effect of Discount Rate on Depletion Path	224
	9.4	Conclusion		225
	References			225
10	**Economics of Electricity Supply**			**227**
	10.1	Introduction		227
	10.2	Basic Concepts Related to Electricity Systems		228
	10.3	Alternative Electricity Generation Options		231
		10.3.1	Generation Capacity Reserve	233
	10.4	Economic Dispatch		233
		10.4.1	Merit Order Dispatch	234
		10.4.2	Incremental Cost Method	234
	10.5	Unit Commitment		235
	10.6	Investment Decisions in the Power Sector		237
		10.6.1	Levelised Bus–Bar Cost	237
		10.6.2	Screening Curve Method	239
	10.7	Sophisticated Approaches to Electricity Resource Planning		242
	10.8	Conclusion		243
	References			246
11	**The Economics of Renewable Energy Supply**			**249**
	11.1	Introduction: Renewable and Alternative Energy Background		249
		11.1.1	Role at Present	249
	11.2	Renewable Energies for Electricity Generation		252
	11.3	Bio-Fuels		254
	11.4	Drivers of Renewable Energy		257
	11.5	The Economics of Renewable Energy Supply		258
		11.5.1	The Economics of Renewable Electricity Supply	258
	11.6	The Economics of Bio-fuels		268
		11.6.1	Bio-Ethanol Cost Features	268
		11.6.2	Bio-Diesel Costs	269

		11.6.3 Support Mechanisms	270
	11.7	Conclusion	271
	References		271

Part III Energy Markets

12 Energy Markets and Principles of Energy Pricing ... 277
    12.1 Introduction: Basic Competitive Market Model ... 277
    12.2 Extension of the Basic Model ... 280
        12.2.1 Indivisibility of Capital ... 281
        12.2.2 Depletion of Exhaustible Resources ... 283
        12.2.3 Asset Specificity and Capital Intensiveness ... 283
    12.3 Market Failures ... 285
        12.3.1 Monopoly Problems ... 285
        12.3.2 Natural Monopoly ... 287
        12.3.3 Existence of Rent ... 293
        12.3.4 Externality and Public Goods ... 293
    12.4 Government Intervention and Role of Government in the Sector ... 294
    12.5 Conclusion ... 296
    References ... 297

13 Energy Pricing and Taxation ... 299
    13.1 Introduction ... 299
        13.1.1 Basic Pricing Model ... 299
    13.2 Tradability of Energy Products and Opportunity Cost ... 301
    13.3 Peak and Off-Peak Pricing ... 304
        13.3.1 Peak Load Pricing Principle ... 305
        13.3.2 Short-Run Versus Long-Run Debate ... 308
    13.4 Energy Taxes and Subsidies ... 310
        13.4.1 Principles of Optimal Indirect Taxation ... 311
        13.4.2 Equity Considerations ... 314
        13.4.3 Issues Related to Numerical Determination of an Optimal Tax ... 315
        13.4.4 Energy Taxes in Nordic Countries: An Example ... 317
        13.4.5 Who Bears the Tax Burden? ... 318
        13.4.6 Subsidies ... 319
    13.5 Implications of Traditional Energies and Informal Sectors in Developing Economies for Energy Pricing ... 321
    13.6 Conclusion ... 322
    References ... 322

14 International Oil Market 325
14.1 Introduction 325
14.2 Developments in the Oil Industry 325
14.2.1 Pre-OPEC Era 325
14.2.2 OPEC Era 332
14.2.3 Commoditisation of Oil 339
14.3 Analysis of Changes in the Oil Market 340
14.3.1 Evolution of Oil Reserves, Oil Production and Oil Consumption 340
14.3.2 Constrained Majors 343
14.3.3 Analysis of the OPEC Behaviour 344
14.3.4 A Simple Analytical Framework of Oil Pricing 349
14.4 Conclusion 351
References 351

15 Markets for Natural Gas 353
15.1 Introduction 353
15.2 Specific Features of Natural Gas 354
15.2.1 Advantage Natural Gas 354
15.2.2 Gas Supply Chain 354
15.2.3 Specific Features 356
15.3 Status of the Natural Gas Market 357
15.3.1 Reserves 357
15.3.2 Production 358
15.3.3 Consumption 360
15.3.4 Gas Trade 362
15.4 Economics of Gas Transportation 366
15.4.1 Economics of Pipeline Transport of Gas 366
15.4.2 Economics of LNG Supply 369
15.4.3 LNG Versus Pipeline Gas Transport 371
15.5 Gas Pricing 372
15.5.1 Rules of Thumb 372
15.5.2 Parity and Net-Back Pricing 374
15.5.3 Spot Prices of Natural Gas 376
15.6 Natural Gas in the Context of Developing Countries 376
15.7 Conclusion 380
References 380

16 Developments in the Coal Market 383
16.1 Introduction 383
16.2 Coal Facts 383
16.3 Changes in the Coal Industry 388
16.4 Technological Advances and the Future of Coal 389

		16.5	Conclusion	390
			References	391

17 Integrated Analysis of Energy Systems — 393
- 17.1 Introduction — 393
- 17.2 Evolution of Energy Systems Models — 393
  - 17.2.1 Historical Account — 394
- 17.3 A Brief Review of Alternative Modelling Approaches — 397
  - 17.3.1 Bottom-up, Optimisation-Based Models — 397
  - 17.3.2 Bottom-up, Accounting Models — 402
  - 17.3.3 Top-down, Econometric Models — 403
  - 17.3.4 Hybrid Models — 404
  - 17.3.5 Some Observations on Energy System Modelling — 405
- 17.4 Energy Economy Interactions — 406
  - 17.4.1 Modelling Approaches — 408
- 17.5 Conclusion — 414
- References — 414

Part IV Issues Facing the Energy Sector

18 Overview of Global Energy Challenges — 419
- 18.1 Introduction — 419
- 18.2 Grand Energy Transitions — 420
- 18.3 Issues Facing Resource-Rich Countries — 424
  - 18.3.1 Co-Ordination of Global Influences — 424
  - 18.3.2 Resource Management Issues — 427
- 18.4 Issues Facing Resource-Poor Countries — 428
  - 18.4.1 Managing Global Influence — 429
  - 18.4.2 Issues Related to Supply Provision — 431
- 18.5 Other Sector Management Issues — 434
  - 18.5.1 Management of Environmental Issues of Energy use — 434
  - 18.5.2 Renewable Energies and the Management Challenge — 435
  - 18.5.3 Reform and Restructuring — 436
- 18.6 Conclusion — 437
- References — 438

19 Impact of High Energy Prices — 441
- 19.1 Introduction — 441
- 19.2 Recent Developments in Energy Prices — 441
- 19.3 Impacts of Energy Price Shocks: Case of Importing Countries — 443
  - 19.3.1 Consumer Reaction to Oil Price Increases — 443

	19.3.2	Transmission of Reactions to the Economy	445
	19.3.3	Linkage with the External Sector	446
19.4		Energy Price Shocks and Vulnerability of Importers	448
19.5		Impact of Higher Oil Prices: Case of Oil Exporting Countries	451
	19.5.1	Windfall Gains	451
	19.5.2	Effect of Windfall Gains	455
19.6		Conclusions	461
References			461

20 Energy Security Issues 463
- 20.1 Introduction 463
- 20.2 Energy Security: The Concept 463
  - 20.2.1 Simple Indicators of Energy Security 464
  - 20.2.2 Diversity of Electricity Generation in Selected European Countries 467
- 20.3 Economics of Energy Security 469
  - 20.3.1 External Costs of Oil Imports 470
- 20.4 Optimal Level of Energy Independence 472
- 20.5 Policy Options Relating to Import Dependence 473
  - 20.5.1 Restraints on Imports 473
  - 20.5.2 Import Diversification 475
  - 20.5.3 Diversification of Fuel Mix 476
  - 20.5.4 Energy Efficiency Improvements 476
- 20.6 Costs of Energy Supply Disruption 477
  - 20.6.1 Strategic Oil Reserves for Mitigating Supply Disruption 478
  - 20.6.2 International Policy Co-ordination 480
- 20.7 Trade-Off between Energy Security and Climate Change Protection 480
- 20.8 Conclusions 483
- References 483

21 Investment Issues in the Energy Sector 485
- 21.1 Problem Dimension 485
  - 21.1.1 Global Investment Needs 485
  - 21.1.2 Regional Distribution of Energy Investment Needs 487
  - 21.1.3 Uncertainty About the Estimates 488
- 21.2 Issues Related to Investments in the Energy Sector 492
  - 21.2.1 Resource Availability and Mobilisation 492
  - 21.2.2 Foreign Direct Investments 495
  - 21.2.3 Risks in Energy Investments 496
  - 21.2.4 Energy Pricing-Investment Link 497

	21.3 Developing Country Perspectives on Investment	498
	21.4 Reform and Investment	500
	21.5 Global Economic Crisis and the Energy Sector Investments	500
	21.6 Conclusions	501
	References	501
22	**Energy Access**	**503**
	22.1 Problem Dimension	503
	22.1.1 Current Situation	503
	22.1.2 Future Outlook	506
	22.2 Indicators of Energy Poverty	507
	22.3 Energy Ladder and Energy Use	509
	22.4 Diagnostic Analysis of Energy Demand by the Poor	511
	22.5 Evaluation of Existing Mechanisms for Enhancing Access	514
	22.6 Effectiveness of Electrification Programmes for Providing Access	516
	22.7 Renewable Energies and the Poor	517
	22.8 Alternative Solutions	520
	22.9 Conclusion	522
	References	522

Part V Economics of Energy–Environment Interactions

23	**The Economics of Environment Protection**	**527**
	23.1 Introduction	527
	23.2 Energy–Environment Interactions	527
	23.2.1 Energy–Environment Interaction at the Household Level	531
	23.2.2 Community Level Impacts	532
	23.2.3 Impacts at the Regional Level	533
	23.2.4 Global Level Problems: Climate Change	534
	23.3 Environmental Kuznets Curve	535
	23.4 Economics of the Environment Protection	537
	23.4.1 Externalities	537
	23.4.2 Spectrum of Goods	538
	23.4.3 Private Versus Social Costs	540
	23.5 Options to Address Energy-Related Environmental Problems	541
	23.5.1 Regulatory Approach to Environment Management	542
	23.5.2 Economic Instruments for Pollution Control	545
	23.5.3 Assessment and Selection of Instruments	553

	23.6	Effects of Market Imperfection	555
	23.7	Valuation of Externalities	557
	23.8	Government Failure	559
	23.9	Conclusion	560
	References		560

24 Pollution Control from Stationary Sources 563
- 24.1 Introduction 563
- 24.2 Direct Pollution Control Strategies 563
  - 24.2.1 Pollution Standards 565
  - 24.2.2 Emission Taxes and Charges 566
  - 24.2.3 Emissions Trading 566
- 24.3 Indirect Policies 570
  - 24.3.1 Pollution Control Technologies 570
  - 24.3.2 Options Related to Fuels and
    Conversion Processes 572
- 24.4 Indoor Air Pollution 574
- 24.5 Conclusion 576
- References 576

25 Pollution Control from Mobile Sources 579
- 25.1 Introduction 579
- 25.2 Special Characteristics of Mobile Pollution 580
- 25.3 Social Costs of Transport Use 581
  - 25.3.1 Infrastructure Usage Related Costs 582
  - 25.3.2 Environmental Pollution Costs 584
  - 25.3.3 Infrastructure-Related Costs 585
  - 25.3.4 Internalisation of Externalities 586
- 25.4 Mitigation Options 587
  - 25.4.1 Vehicle Emission Standards and Technologies 588
  - 25.4.2 Cleaner Fuels 589
  - 25.4.3 Traffic Management and Planning 593
- 25.5 Conclusion 594
- References 594

26 The Economics of Climate Change 597
- 26.1 Climate Change Background 597
  - 26.1.1 The Solar Energy Balance 597
  - 26.1.2 GHGs and Their Global Warming Potential 598
- 26.2 The Economics of Climate Change 603
  - 26.2.1 Problem Dimension 603
  - 26.2.2 Overview of GHG Emissions 604
- 26.3 Economic Approach to Control the Greenhouse Effect 608
  - 26.3.1 Integrated Assessment 609

	26.4	Alternative Options to Cope with Global Warming	610
		26.4.1 Generic Options	610
		26.4.2 National Policy Options	611
		26.4.3 Emissions Trading System (ETS) of the EU	615
		26.4.4 International Policy Options	617
	26.5	Climate Change Agreements	618
		26.5.1 UNFCCC	618
		26.5.2 The Kyoto Protocol	619
	26.6	Conclusion	620
	References		621
27	**The Clean Development Mechanism**		**623**
	27.1	Basics of the Clean Development Mechanism	623
		27.1.1 CDM Criteria	624
		27.1.2 Participation Requirement	624
		27.1.3 Eligible Projects	625
		27.1.4 CDM Entities/Institutional Arrangement	626
		27.1.5 CDM Project Cycle	629
		27.1.6 Additionality and Baseline	633
		27.1.7 Crediting Period	635
	27.2	Economics of CDM Projects	636
		27.2.1 Role of CDM in KP Target of GHG Reduction	636
		27.2.2 Difference Between a CDM Project and an Investment Project	637
		27.2.3 CDM Transaction Costs	637
		27.2.4 CER Supply and Demand	640
		27.2.5 Risks in a CDM Project	643
	27.3	Conclusions	644
	References		644

Part VI Regulation and Governance of the Energy Sector

28	**Regulation of Energy Industries**		**649**
	28.1	Introduction	649
	28.2	Traditional Regulation	650
		28.2.1 Rate Level Regulation	650
		28.2.2 Rate Structure Regulation	658
	28.3	Problems with Traditional Regulatory Approach	660
		28.3.1 Regulatory Alternatives	662
	28.4	Price-Cap Regulation	665
		28.4.1 Choice of Inflation Factor	667
		28.4.2 X Factor	668
		28.4.3 Z Factor	669
		28.4.4 Choice of Form	670

Contents

		28.4.5	Advantages and Disadvantages of Price Cap Regulation.	670
		28.4.6	Comparison of Price Cap and RoR Regulation	671
		28.4.7	Experience with Price Cap Regulation.	672
	28.5	Revenue Caps		673
	28.6	Yardstick Competition		674
	28.7	Performance Based Regulation.		676
		28.7.1	Base Revenue Requirement	678
		28.7.2	Sharing Mechanism.	678
		28.7.3	Quality Control.	679
	28.8	Conclusion		680
	References			680
29	**Reform of the Energy Industry**			**683**
	29.1	Introduction.		683
	29.2	Government Intervention in Energy Industries		683
	29.3	Rationale for Deregulation.		686
	29.4	Reform Process		689
		29.4.1	Changing the Rules Requires Stability of Rule Makers.	689
		29.4.2	Danger of Derailment at Every Stage of the Reform Process	690
		29.4.3	Importance of Overall Acceptance of Changed Rules.	692
		29.4.4	Adaptation to the New Environment	693
		29.4.5	Transition Management.	694
	29.5	Options for Introducing Competition.		694
		29.5.1	Competition for the Market	695
		29.5.2	Competition in the Market.	696
	29.6	Restructuring Options		698
		29.6.1	Vertically Integrated Monopoly Model (VIM)	699
		29.6.2	Entry of Independent Power Producers (IPP)	701
		29.6.3	Single Buyer Model	703
		29.6.4	Transitional Models	705
		29.6.5	Wholesale Competition: Price-Based Power Pool Model	707
		29.6.6	Wholesale Competition: Net Pool	709
		29.6.7	Wholesale Competition: Cost-based Pool.	711
		29.6.8	Wholesale Competition through Open Access	712
		29.6.9	Full Customer Choice: Retail Competition Model.	713
	29.7	Reform Sustainability: A Framework for Analysis		715
	29.8	Experience with Energy Sector Reform.		718
	29.9	Conclusions.		720
	References			720

Abbreviations

A

AAU	Assigned Allocation Units
AC	Average cost
ADB	Asian Development Bank
ANN	Artificial neural network
APERC	Asia Pacific Energy Research Centre
ARIMA	Integrated Auto regressive moving average
ARMA	Auto regressive moving average

B

BCM	Billion cubic metres
BF	Blast furnace
BP	British Petroleum

C

CAIR	Clean Air Interstate Rule
cal	Calories
CAPM	Capital asset pricing model
CBO	Congressional Budget Office (US)
CC	Combined cycle
CDD	Cooling degree days
CDM	Clean Development Mechanism
CEGB	Central Electricity Generation Board
CER	Certified emissions reductions
CERI	Canadian Energy Research Institute
CES	Constant elasticity of substitution
CF	Capacity factor
CFC	Chlorofluorocarbon

CFL	Compact fluorescent lamp
CGE	Computable General Equilibrium model
CHP	Combined heat and power
CIF	Cost insurance freight
CNG	Compressed natural gas
CO_2	Carbon-di-oxide
COP	Conference of Parties
COPD	Chronic pulmonary obstructive disease
CPI	Consumer price index
CRA	Charles River Associates
CRF	Capital recovery factor

D

DCF	Discounted cash flow
DECC	Department of Energy and Climate Change (UK)
DfID	Department for International Development
DNA	Designated National Authority
DOE	Designated Operational Entities (CDM)
	Department of Energy (US)
DR(I)	Direct reduction (of Iron in steel making)
DSM	Demand-side management
DTI	Department of Trade and Industry (UK)
DWL	Deadweight loss

E

EB	Executive Board (CDM)
EC	European Commission
ECA	Energy Commodity Account
ECM	Error correction model
EDI	Energy Development Index
EEA	European Environment Agency
EGEAS	Electric Generation Expansion Analysis System
EIA	Energy Information Administration (of the Department of Energy, USA)
EMV	Expected monetary value
EPA	Environment Protection Agency (US)
EPRI	Electric Power Research Institute
ESI	Electricity supply industry
ESP	Electrostatic precipitator
ETS	Emissions trading system
EU	European Union

F

FAO	Food and Agricultural Organisation
FGD	Flue gas desulphurisation
FOB	Free on board
FSU	Former Soviet Union (countries)

G

GCC	Gulf Co-operation Council
GDP	Gross domestic product
GGFR	Global gas flaring reduction
GHG	Greenhouse gas
GWh	Giga watt hour
GWP	Global Warming potential

H

HDD	Heating degree days
HH	Henry Hub (US)
HHI	Herfindahl Hirschman Index

I

IAEA	International Atomic Energy Agency
IEA	International Energy Agency
IGCC	Integrated Gasified combined cycle
IIASA	International Institute for Applied Systems Analysis
IMF	International Monetary Fund
IOC	International Oil companies
IPCC	Inter-Governmental Panel on Climate Change
IPP	Independent Power producers
IRR	Internal rate of return

J

JI	Joint Implementation (projects)
JODI	Joint Oil Data Initiative

K

kcal	Kilo calories
KP	Kyoto Protocol
kW	Kilo watt

L

LEAP	Long-range Energy Alternatives Planning
LF	Load factor
LNG	Liquefied Natural Gas
LPG	Liquid petroleum gas
LULUCF	Land use, land use change and forestry

M

MAED	Model for analysis of energy demand
MARKAL	Market Allocation model
MBMS	Multi-buyer multi-seller
MC	Marginal cost
MENA	Middle East and North African countries
Mt	Million tons (metric)

N

NBP	National Balancing Point (UK)
NEMS	National Energy Modelling system
NGL	Natural Gas Liquids
NOC	National oil companies
NOx	Nitrous oxides
NPV	Net present value

O

OEB	Overall Energy Balance
OECD	Organisation for Economic Co-operation and Development
OPEC	Organisation of the Petroleum Exporting Countries
OTC	Ozone Transport Commission

P

PBR	Performance-based regulation
PDD	Project design document
PES (PEC)	Primary energy supply (Primary energy consumption)
PM	Particulate matters
PPP	Purchasing power parity

R

R&D	Research and development
RCEP	Royal Commission on Environmental Protection (UK)
RE	Renewable energies (if not otherwise indicated)
RES	Reference Energy System
RIM	Ratepayer impact test
RO	Renewable obligation
ROC	Renewables Obligation certificate
RPI	Retail price index

S

SAM	Social Accounting Matrix
SD	Sustainable development
SHS	Solar Home systems
SIP call	State Implementation Plan call
SOE	State owned enterprise
SOx	Sulphur Oxides
SWI	Shannon Wiener Index
SWNI	Shannon Wiener Neumann Index

T

T&D	Transmission and distribution
TCF	Trillion cubic feet
TE	Traditional energies
TFC	Total final consumption
TFP	Total factor productivity
TFS	Total final supply
TPA	Third party access

U

UN	United Nations
UNDP	United Nations Development Programme
UNEP	United Nations Environment Programme
UNFCCC	United Nations Framework Convention on Climate Change
USD	United States Dollar

V

VIM	Vertically integrated model
VOC	Volatile organic compounds

W

WACC	Weighted average cost of capital
WASP IV	Wien Automatic System Planning Package IV
WB	World Bank
WEC	World Energy Council
WEM	World Energy Model
WEO	World Energy Outlook
WHO	World Health Organisation

Chapter 1
Introduction to Energy Economics

1.1 Introduction

Energy economics or more precisely the economics of energy is a branch of applied economics where economic principles and tools are applied to "ask the right questions" (Stevens 2000), and to analyse them logically and systematically to develop a well-informed understanding of the issues.

The energy sector is complex because of a number of factors:

- The constituent industries tend to be highly technical in nature, requiring some understanding of the underlying processes and techniques for a good grasp of the economic issues.
- Each industry of the sector has its own specific features which require special attention.
- Energy being an ingredient for any economic activity, its availability or lack of it affects the society and consequently, there are greater societal concerns and influences affecting the sector.
- The sector is influenced by interactions at different levels (international, regional, national and even local), most of which go beyond the subject of one discipline.

Consequently, analyses of energy problems have attracted inter-disciplinary interests and researchers from various fields have left their impressions on these studies. The influence of engineering, operations research and other decision-support systems in the field of energy economics has been profound.

Energy issues have been analysed from an economic perspective for more than a century now. But energy economics did not develop as a specialised branch until the first oil shock in the 1970s (Edwards 2003). The dramatic increase in oil prices in the 1973–1974 highlighted the importance of energy in economic development of countries. Since then, researchers, academics and even policymakers have taken a keen interest in energy studies and today energy economics has emerged as a recognised branch on its own.

Like any branch of economics, energy economics is concerned with the basic economic issue of allocating scarce resources in the economy. Thus the microeconomic concerns of energy supply and demand and the macro-economic concerns of investment, financing and economic linkages with the rest of the economy form an essential part of the subject. However, the issues facing the energy industry change, bringing new issues to the fore. For example, in the 1970s, the focus was on understanding the energy industry (especially the oil industry), energy substitution and to some extent on renewable energies. Moreover, there was some focus on integrated planning for energy systems with a major emphasis on developing countries.

The scope of the work expanded in the 1980s. Environmental concerns of energy use and economic development became a major concern and the environmental dimension dominated the policy debate. This brought a major shift in the focus of energy studies as well- the issue of local, regional and global environmental effects of energy use became an integral part of the analysis.

In the 1990s, liberalisation of energy markets and restructuring swept through the entire world although climate change and other global and local environmental issues also continued. These changes brought new issues and challenges to the limelight and by the end of the decade, it became evident that unless the fundamental design is not well thought through, reforms cannot succeed.

In recent years, the focus has shifted to high oil prices, energy scarcity and the debate over state intervention as opposed to market-led energy supply. This swing of the pendulum in the policy debate is attributed to the concerns about security of supply in a carbon-constrained world.

Accordingly, the objective of this book is to present in a single volume basic economic tools and concepts that can be used to understand and analyse the issues facing the energy sector. The aim is to provide an overall understanding of the energy sector and to equip readers with the analytical tools that can be used to understand demand, supply, investments, energy-economy interactions and related policy aspects.

1.2 Energy and Multidimensional Interactions

The multidimensional nature of the energy-related interactions is indicated in Fig. 1.1. At the global level, three influences can be easily identified (Bhattacharyya 2007):

(a) Energy trade—All transactions involving energy commodities (especially that of oil and to a lesser extent that of coal and gas) are due to the differences in the natural endowments of energy resources across countries and the gaps in domestic supply and demands; similarly flow of technologies, human resources, financial and other resources as well as pollutants generated from energy and other material use can also be considered at this level.

1.2 Energy and Multidimensional Interactions

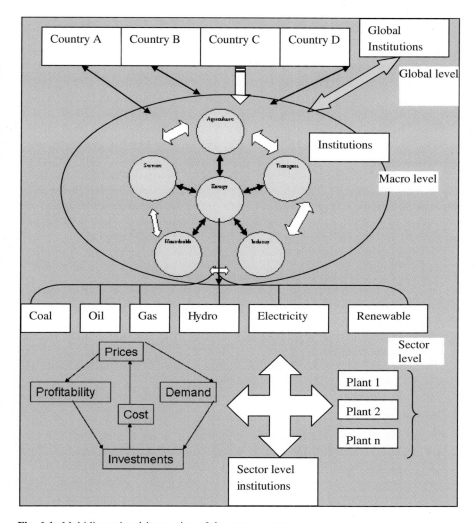

Fig. 1.1 Multidimensional interaction of the energy sector

(b) International institutional influences—Various influences through international institutions affect interactions among countries and govern transactions. These include the legal frameworks, treaties and conventions, international organisations such as the United Nations (UN), the World Bank and the International Monetary Fund (IMF), the judicial system and the like.
(c) Other interaction—Other interactions among countries (co-operation, competition and conflicts) involving their governments or other entities (such as the firms) also influence the energy sector.

These influences are neither mutually exclusive nor static in nature. Consequently, the relative importance of one or more of these influences on a particular country would vary and changes in the importance of one or the other over time could modify the relationships extensively.

The key role of the energy sector in the economic activities of any economy arises because of the mutual interdependence between economic activities and energy. For example, the energy sector uses inputs from various other sectors (industry, transport, households, etc.) and is also a key input for most of the sectors. These interrelations influence the demand for energy, possibilities of substitution within the energy and with other resources (capital, land, labour and material), supply of energy and other goods and services, investment decisions, and the macro-economic variables of a country (economic output, balance of payment situations, foreign trade, inflation, interest rate, etc.). Once again, the national level institutions (including the rules and organisations like government, judiciary, etc.) both influence and get influenced by these interactions.

Thus the macro-level influences arise broadly from:

(a) The level of economic activities and its evolution over time;
(b) Interdependence of energy and other economic activities as well as interactions among economic activities;
(c) The structure of each activity and its evolution over time;
(d) The technical composition and characteristics of the economic activities and its evolution over time;
(e) The institutional arrangement that provides the enabling environment for different activities to flourish and its evolution;
(f) Macro-management of the economy and its interaction with the institutional arrangement.

Finally, the energy sector itself is composed of different industries (or sub-sectors), each of which has different technical and economic characteristics. They are also interdependent to some extent and each industry attempts to achieve a balanced operation considering demand, investment, prices, supply and the institutional environment. The operating decisions are highly influenced by the objectives and goals of the operators and the operating constraints faced by them (including the resource related and socio-political constraints). The ownership pattern as well as institutional factors also influences the decisions.

Thus the sector faces both micro-level operating issues which are short-term in nature as well as those involving the medium and long-term future. Because of specific characteristics of the energy sector such as reliance on non-renewable energies, capital intensiveness of investments, discrete plant sizes, long gestation period, scale economies, tradability of certain goods leading to high revenue generation potential compared to other economic activities, and the boom-bust cycle phenomenon, the decisions need to be taken well in advance for the future and the present greatly shapes the future outcomes, although with a greater level of uncertainty. While the above outline of interaction is generic, the specifics vary depending on the circumstances (e.g. resource rich or resource poor country), economic conditions (developed or developing country), time dimension, and the like.

Various chapters of this book focus on the above aspects. The book is organized into six parts each covering a specific theme.

(1) Part 1 presents the topics related to energy demand analysis and forecasting. This part covers energy statistics, concepts about energy demand and presents simple methods for demand forecasting. It also covers the ideas related to demand-side management.
(2) Part 2 is devoted to the economics of energy supply. It starts with the concepts of economic evaluation of projects and uses this framework to understand the economics of fossil fuel, renewable energy and electricity supply.
(3) Part 3 is concerned with energy markets. An introductory chapter provides the basic ideas of markets and extends this to include the specific features of the energy sector. This is followed by an analysis of energy pricing, taxation and subsidies. Subsequent chapters present the specific aspects of oil, gas and coal markets. Finally, a chapter is devoted to an integrated analysis of energy systems.
(4) Part 4 deals with important issues and challenges facing the energy industries. Although the issues vary from one country to another, this section picks up a few common issues such as energy security, effects of high oil prices on the economy, energy investments and energy access, that are widely analysed and discussed in the current policy debate.
(5) Part 5 introduces the concepts of environmental economics as applied to the energy sector. It covers the mitigation options for pollution from stationary and mobile sources, and introduces the issues of climate change from an economic perspective. It also touches on the Clean Development Mechanism.
(6) Finally, Part 6 considers the regulatory and governance issues related to the energy sector. The regulatory options commonly used in the network industries and the approaches to reform and restructuring of the sector are presented in this part.

References

Bhattacharyya SC (2007) Energy sector management issues: an overview. Int J Energy Sect Manag 1(1):13–33
Edwards BK (2003) The economics of hydroelectric power. Edward Elgar, Cheltenham
Stevens P (2000) An introduction to energy economics. In: Stevens P (ed) The economics of energy, vol 1. Edward Elgar, Cheltenham

Part I
Energy Demand Analysis and Forecasting

Chapter 2
Energy Data and Energy Balance

2.1 Introduction

This chapter first defines some terms commonly used in any energy study. It then introduces the energy system and presents the energy accounting framework. The data issues related to the energy sector are considered next. Finally, a few ratios are considered to analyse the energy situation of a country.

2.2 Energy Basics

2.2.1 Energy Defined

Energy is commonly defined as the ability to do work or to produce heat. Normally heat could be derived by burning a fuel—i.e. a substance that contains internal energy which upon burning generates heat, or through other means—such as by capturing the sun's rays, or from the rocks below the earth's surface (IEA 2004). Similarly, the ability to do work may represent the capability (or potential) of doing work (known as potential energy as in stored water in a dam) or its manifestation in terms of conversion to motive power (known as kinetic energy as in the case of wind or tidal waves).

Thus energy manifests itself in many forms: heat, light, motive force, chemical transformation, etc. Energy can be captured and harnessed from very diverse sources that can be found in various physical states, and with varying degrees of ease or difficulty of capturing their potential energies. Initially the mankind relied on solar energy and the energy of flowing water and air. Then with the discovery of the fire-making process, the use of biomass began. The use of coal and subsequently oil and natural gas began quite recently—a few hundred years ago.

According to the physical sciences, two basic laws of thermodynamics govern energy flows. The first law of thermodynamics is a statement of material balance—a mass or energy can neither be created nor destroyed—it can only be transformed. This indicates the overall balance of energy at all times. The second law of thermodynamics on the other hand introduces the concept of quality of energy. It suggests that any conversion involves generation of low grade energy that cannot be used for useful work and this cannot be eliminated altogether. This imposes physical restriction on the use of energy.

2.2.2 Alternative Classifications of Energy

As energy can be obtained from various sources, it is customary to classify them under different categories, as discussed below.

2.2.2.1 Primary and Secondary Forms of Energy

The term primary energy is used to designate an energy source that is extracted from a stock of natural resources or captured from a flow of resources and that has not undergone any transformation or conversion other than separation and cleaning (IEA 2004). Examples include coal, crude oil, natural gas, solar power, nuclear power, etc.

Secondary energy on the other hand refers to any energy that is obtained from a primary energy source employing a transformation or conversion process. Thus oil products or electricity are secondary energies as these require refining or electric generators to produce them.

Both electricity and heat can be obtained as primary and secondary energies.

2.2.2.2 Renewable and Non-Renewable Forms of Energy

A non-renewable source of energy is one where the primary energy comes from a finite stock of resources. Drawing down one unit of the stock leaves lesser units for future consumption in this case. For example, coal or crude oil comes from a finite physical stock that was formed under the earth's crust in the geological past and hence these are non-renewable energies.

On the other hand, if any primary energy is obtained from a constantly available flow of energy, the energy is known as renewable energy. Solar energy, wind, and the like are renewable energies.

Some stocks could be renewed and used like a renewable energy if its consumption (or extraction) does not exceed a certain limit. For example, firewood comes from a stock that could be replenished naturally if the extraction is less than the natural growth of the forest. If however, the extraction is above the natural forest growth, the stock would deplete and the resource turns into a non-renewable one.

2.2.2.3 Commercial and Non-Commercial Energies

Commercial energies are those that are traded wholly or almost entirely in the market place and therefore would command a market price. Examples include coal, oil, gas and electricity.

On the other hand, non-commercial energies are those which do not pass through the market place and accordingly, do not have a market price. Common examples include energies collected by people for their own use (see Stevens (2000) for more details).

But when a non-commercial energy enters the market, by the above definition, the fuel becomes a commercial form of energy. The boundary could change over time and depending on the location. For example, earlier fuel-wood was just collected and not sold in the market. It was hence a non-commercial form of energy. Now in many urban (and even in rural) areas, fuel-wood is sold in the market and hence it has become a commercial energy. At other places, it is still collected and hence a non-commercial form of energy. This creates overlaps in coverage.

Another term which is commonly used is modern and traditional energies. Modern energies are those which are obtained from some extraction and/or transformation processes and require modern technologies to use them. On the other hand, traditional energies are those which are obtained using traditional simple methods and can be used without modern gadgets. Often modern fuels are commercial energies and traditional energies are non-commercial. But this definition does not prevent traditional energies to be commercial either. Thus if a traditional energy is sold in the market it can still remain traditional. Thus it reduces some overlap but the definition remains subjective as the practices and uses vary over time and across cultures and regions.

2.2.2.4 Conventional and Non-Conventional Energies

This classification is based on the technologies used to capture or harness energy sources. Conventional energies are those which are obtained through commonly used technologies. Non-conventional energies are those obtained using new and novel technologies or sources. Once again the definition is quite ambiguous as conventions are subject to change over time, allowing non-conventional forms of energies to become quite conventional at a different point in time.

Based on the above discussion, it is possible to group all forms of energy in two basic dimensions: renewability as one dimension and conventionality as the other. Table 2.1 provides such a classification.

2.3 Introduction to the Energy System

The energy system today is highly dependent on fossil fuels, with coal, oil and gas accounting for about 80% of world primary energy demand. A number of physical

Table 2.1 Energy classifications

Conventionality	Renewability	
	Renewable	Non-renewable
Commercial	Large scale hydro	Fossil fuels
	Geothermal	Other nuclear
	Nuclear	
Traditional/non-commercial	Animal residues	Unsustainable fuelwood
	Crop residues	
	Windmills and watermills	
	Fuelwood (sustainable)	
New and novel	Solar	Oil from oil sands
	Mini and micro hydro	Oil from coal or gas
	Tidal and wave	
	Ocean thermal	

Source Codoni et al. (1985) and Siddayao (1986)

and economic activities are involved to capture the energy and to deliver it in a usable form to the users. The chain of systems or activities required to ensure supply of energy is known as the energy supply system. The supply system is made up of the supply sector, the energy transforming sector and the energy consuming sector. The supply involves indigenous production, imports or exports of fuel and changes in stock levels (either stock pileup or stock draw down). Transformation converts different forms of primary energies to secondary energies for ease of use by consumers. Transformation processes normally involve a significant amount of losses. Transportation and transmission of energy also involve losses. The final users utilise various forms of energies to meet the needs of cooling, heating, lighting, motive power, etc.

The relative importance of the above segments varies from one country to another and even from one fuel to another depending, to a large extent, on the availability of resources in a particular country. For a resource-rich country, the supply segment is evidently well developed, while for a resource-poor country the transformation and final use segments tend to be more developed.

The activities vary by the type of energy. For non-renewable energies, exploration, development and production of fuel(s) constitute the first step. A variety of exploratory techniques are used to identify the location of the resource but drilling a hole only can confirm the existence of the stock. Upon confirmation of the economic viability and technical feasibility of extraction of the stock, the field is developed and production follows.

The fuel so produced often requires cleaning, beneficiation and processing to make it usable. Cleaning and beneficiation processes are used to remove impurities using simple cleaning processes. The fuel is then transported to the centres of conversion or use. Most forms of energies cannot be used as such and require processing (e.g. crude oil to petroleum products). Similarly, depending on consumers' demand, fuels also undergo conversion processes to convert them in

2.3 Introduction to the Energy System

preferred forms (e.g. to electricity). Conversion involves a significant amount of energy losses.

The processed and converted energy then needs to be transported to consumers. This also involves transmission losses. Before consumption, some storage may be required for some forms of energy, while for electricity no practical and economic storage solution exists.

Final consumers use energy for various purposes. Normally these are the end-users who cannot sell or transfer the energy to others. These consumers are grouped into different broad categories: industrial, transport, residential, commercial, and agricultural. Some energy is also used as feedstock in production processes or as non-energy purposes (e.g. tar is used in roads). Figure 2.1 captures the above chain of activities.

As energy is used for meeting certain needs, it is used in conjunction with appliances. The efficiency of the appliance affects the demand. The consumer is interested in the useful energy (i.e. the energy required to meet the need and not the final or primary energies). Reducing losses can reduce pressure on fuel demand.

An example will clarify this point. For example, a person wants to travel 10 kilometres in a car which uses one litre of fuel per 10 km. Transportation of fuel from refinery to the user involves a 5% loss. The refinery operates at an efficiency level of 95% and produces 30% gasoline from the crude oil it uses. The crude oil recovery rate from the national fields is 20% at present. How much crude oil is required to complete the distance?

As the car requires a litre of oil per 10 km, one litre of oil needs to be put in the car. To supply this amount of oil, the refinery needs to produce = 1.05 l of oil. The input requirement of the refinery is = 1.05/0.95 = 1.10 l. As 30% gasoline is produced from the crude, the crude requirement is = 1.10/0.3 = 3.68 l. To produce

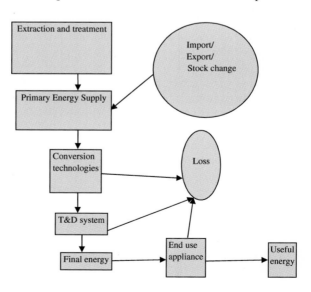

Fig. 2.1 Energy supply chain

3.68 l of crude, 18.4 l of crude oil has to be found (=3.68/0.2). This shows the importance of efficiencies at various stages in the energy chain.

2.4 Energy Information

Information is crucial for any decision-making: be it development planning decisions or business decisions or decisions by individual consumers. Reliable and quality information facilitates decision-making and improves the decision-making process. Any decision-making process requires analysis of the past and present status of the sector (or sub-sector or specific area of concern) and a vision about the future. This implies that a large amount of both historical and projected data would be required related to the specific components and subsystems of the energy sector. While the information requirement would vary by stakeholders, broadly the common requirement would include, inter alia,

(1) energy use by various economic activities;
(2) energy production, transformation and delivery to various users,
(3) technical and operating statistics of the plants and installations;
(4) financial and cost information, and
(5) macro-economic and other social information.

Although information requirements vary from one application to another and generalization of the requirement is difficult, it is possible to indicate some needs based on the important energy issues facing many countries. The information required may be categorized as follows (Codoni et al. 1985):

(a) Energy pricing: Despite the liberalization of energy markets, energy pricing continues to be a very sensitive and contentious issue because of social and political implications. Regulators and price-setting agencies require considerable information to make correct pricing decisions. This includes: consumption of fuels by various consumers, consumption pattern by income groups, rural–urban divide in consumption and supply, cost of supply to various consumers, impact of price revision on consumers, etc.
(b) Energy investment: Energy investment decisions have high visibility because of their size. Investment decisions require an understanding of the evolution of demand, pricing policies, business environment, viability of alternative options, and various types of impacts. Historical and forecast data are required for such exercises.
(c) Energy research and development (R&D): Decisions on R&D require information on resources of various kinds of fuels, cost of production and conversion, evolution of demand for various kinds of fuels, costs and benefits of investment in R&D activities, etc.
(d) System management: Decisions on energy system management would normally be taken by the operators themselves but quite often there would be

some regulatory or governmental supervision/involvement. A large volume of information is required for efficient and most economic system operation. The requirement is significantly high in the case of electricity where supply and demand balancing has to be ensured every moment. The information required includes supply and demand positions, system availability, technical constraints, etc.
(e) Contingency plan: Any system should remain prepared to deal with a number of contingencies. Complete or partial system failure, supply failure due to technical or other problems, erratic change in demand, etc. are some such contingencies. Preparation of a contingency plan would require information on geography of energy supply, distribution and consumption, technical features of the system, knowledge of social and economic impacts of energy disruption, etc.
(f) Long-term planning: This involves developing a view of the possible future evolution of energy demand and the possibilities of fulfilling that demand in various ways. This requires a proper understanding of current consumption activities and consumption pattern, possible changes in the activities in terms of efficiency and structure, possible supply alternatives, possible technological changes, etc.

Consequently, reliable and consistent information is a pre-requisite for any serious analysis.

2.5 Energy Accounting Framework

The energy accounting framework is one that enables a complete accounting of energy flows from original supply sources through conversion processes to end-use demands with all double-counting avoided. By accounting for all conversion losses this framework provides an exhaustive accounting for itemizing the sources and uses of energy. The energy flow considered in this framework is indicated in Fig. 2.2.

Normally the framework is applied to each individual fuel or energy type used in an economy and thus the energy account is essentially a matrix where

- Each type of fuel is considered along the columns. The columns are chosen based on the importance of energy commodities in the country under consideration. More diversified the energy system, more detailed accounting is required.
- Each row captures the flow of energy. The rows are organised in three main blocks to indicate the supply of energy, its transformation and final use (see Fig. 2.3).

It is quite common however to focus on the commercial energies given the ease of data collection and flow measurement. Information in the columns is also arranged in terms physical attributes (such as solid fuel, liquid fuel and gaseous

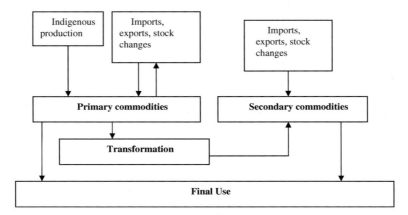

Fig. 2.2 Energy commodity flow. *Source* IEA (2004)

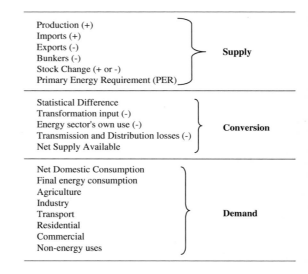

Fig. 2.3 Main flows considered in energy accounting. *Source* UN (1982), Codoni et al. (1985), UN (1991) and IEA (2004)

fuels), various forms of electricity, nuclear power and renewable energies. A brief description of the three major components of the energy accounting system is provided below.

2.5.1 Components of the Energy Account

An energy balance table has three main building blocks: the supply-side information, conversion details and the demand information. The supply-side

2.5 Energy Accounting Framework

information captures domestic supply of energy products through production, international trade, and stock change. Energy production provides the marketable quantities of energy domestically produced in a country. Marketable quantities exclude any part of the production that is not available for use or stock. Examples include wastes (gas flaring), re-injection as part of production process (gas re-injection), removal of impurities, etc.

The external trade information captures the transactions of energies taking place across the national boundary of a country, irrespective of whether customs clearance was taken or not. Imports are those quantities that enter the country for domestic use (this excludes transits). Exports are those quantities leaving the country for use by outsiders. As imports expand domestic supply, it is considered as a positive flow in the energy accounts whereas exports are considered as a negative flow.

Fuel used by ships for international voyages is considered as a special item and included as bunker. This is treated in a similar manner as international trade and any quantity delivered to ships, irrespective of their country of registration, undertaking international voyages is eligible for this treatment.

Stocks of fuels serve as cushions to cover fluctuations in supply and demand and are maintained by the suppliers, importers/exporters and the consumers. A stock rise represents a diminution in available supplies, and a stock fall represents an increase in supplies. For this reason, a minus sign is used to denote a rise and that a plus sign is used to denote a stock fall. The net position of domestic supply considering the above elements gives the primary energy supply of any energy.

The transformation section of the energy accounting captures the conversion of primary energies into secondary energies either through physical or chemical changes. Normally the inputs used in the transformation process are given a negative sign while the outputs are given a positive sign. If a single output comes from a number of energy sources, the clarity of input–output relation may be lost when the information is placed in a single row. In such cases, further details are presented as memo items or in additional rows. Commonly used transformation processes are oil refining, electricity generation, gas separation and conversion, coke production from coal, etc. However, as with supply information, transformation or conversion is also a country specific section of the energy account and would normally vary across countries.

The conversion section also captures information on energy used by the energy industries and transmission and distribution losses. Both these elements carry a negative sign as they represent reduction in energy flows for use by consumers. Energy sector own use is the energy used in the production process (say in refineries, power plants, coal mines, oil fields, etc.). Although this is essentially energy consumption and hence part of energy demand, sector's own use is treated separately to obtain a clear picture of the energy use in the rest of the economy. Transmission and distribution (T&D) losses are the wastes in the delivery system—such as pipelines, electric networks—that cannot be eliminated altogether.

The final section captures the energy flows available to final consumers. In terms of accounting balance, this is the residual amount available for domestic consumption from primary supplies after accounting for conversion. Generally, net supply is calculated from the supply side while the net demand is calculated from the demand side and these two figures should match, thus ensuring correctness of the accounting. However, it is quite rare that the two items are exactly same. The statistical difference term is used as the balancing item. Its sign would indicate whether the supply-side total is higher (thus requiring a deduction of some balancing amount) or lower (thus requiring some balancing amount) than the demand-side total.

The above representation is not the only possible arrangement and in practice, wide variations in the energy balance representation can be found. Generally, the treatment of stocks, import-exports, primary electricity generation, electricity consumption, non-energy use and conversion process details creates differences. Similarly, the coverage and emphasis may be different: for example, some cover only commercial energies while some others include newer and traditional energies.

2.5.2 Commodity Accounts and Overall Energy Balance

Two core accounts are set up to record energy transactions for each and all energy sources, from production or import to final consumption.

(a) Energy Commodity Accounts (ECA)—This shows all the flows in the appropriate original unit of measurement (tons, barrels, cubic meters, etc.). See Table 2.2 for an example. Normally, each energy producing, transforming and distribution industry has its own particular way of presenting statistics on its activities according to the purposes for which it needs data. This information forms the basis for any Energy Commodity Account but often the raw data requires reconciliation and harmonization to ensure correctness. The columns of an ECA cannot be directly compared or summed up because of differences in the units. Hence, such an account does not permit any overall appraisal of the energy system. The overall energy balance removes this problem.

(b) Overall Energy Balance (OEB)—This shows all the flows in terms of a common accounting unit (like Joule, kilocalories, Btu, etc.). See Table 2.3 for an example. The ECA is the starting point for an overall energy balance and using appropriate conversion factors, a suitably designed overall energy balance can be developed from it.

The OEB constructed on this basis can then be used for the analysis of changes in the level and mix of energy sources used for particular purposes before and after transformation. It can also be used for the study of changes in the use of pattern of different fuels, for the examination of the extent of or scope for substitution between fuels at different stages of the flow from primary supplies to final energy uses, and as a source for the generation of time series tables.

2.5 Energy Accounting Framework

Table 2.2 Abridged Energy Commodity Account of the U.K. for 2008

	Coal	Coal products (solid)	Coal gas	Primary Oil	Petro products	NG	Electricity
	(kt)	(kt)	(GWh)	(kt)	(kt)	(GWh)	(GWh)
Supply							
Production	17604	4661	15345	71665	80435	810284	385560
Other sources	449	0	0	0	3135	0	4089
Imports	43875	738	0	60074	23916	407054	12294
Exports	−599	−210	0	−48410	−28811	−122670	−1272
Marine bunkers	0	0	0	0	−2594	0	0
Stock change	−3395	206	0	232	14	−3087	0
Transfers	0	0	−3	−2928	−208	−68	0
Total supply	57935	5395	15342	80633	75887	1091513	400671
Statistical difference	−278	−3	−139	−91	−64	876	1053
Total demand	58212	5398	15481	80725	75951	1090637	399619
Transformation	55621	4363	7900	80725	1170	397246	0
Energy industry use	5	0	4759	0	4531	69196	30632
Losses	0	0	2332	0	0	13634	27425
Final consumption	2586	1036	490	0	70249	610561	341561
Industry	1872	728	490	0	5807	132501	113558
Transport	0	0	0	0	51924	0	8434
Other sectors	714	308	0	0	4035	468788	219570
Non-energy use	0	0	0		8483	9273	0

Source Digest of UK Energy Statistics, 2009

As countries often use their own assumptions and accounting conventions, international comparison of energy statistics can be difficult. However, organisations like the International Energy Agency (IEA), United Nations Statistics Division, Eurostat, Energy Information Administration (EIA) of US Department of Energy and the Asian Development Bank regularly publish standardized data sets the produces such information. In addition, the BP Statistical Review of World Energy is also a widely used dataset, mostly on the supply of commercial energies.

2.5.3 Units, Conversion Factors and Aggregation of Energy Flows

From Table 2.2, it is clear that various units of measurement are used in constructing the ECA. For example, coal is presented in thousand tonnes, gas and electricity in GWh, and so on. The conversion from one unit to another is a common feature while working with energy data. Therefore, it is important to have some familiarity with the conversion factors. Some commonly used factors for

Table 2.3 Abridged Overall Energy Balance for the U.K. for 2008 (unit: ktoe)

	Coal and products	Oil and products	Natural gas	Renewable and waste	Electricity and heat	Total
Supply						
Indigenous production	11362	78580	69672	4361	12965	176939
Imports	28918	91683	35000	948	1057	157606
Exports	−599	−84325	−10548	0	−109	−95581
Marine bunkers	0	−2733	0	0	0	−2733
Stock change	−1996	268	−265	0	0	−1993
Primary supply	37684	83473	93859	5309	13913	234238
Statistical difference	−171	−195	75	0	91	−200
Primary demand	37855	83668	93784	5309	13822	234438
	0	0	0	0	0	0
Transfers	−126	6	−6	0	0	−125
Transformation	−33882	−1530	−34157	−3537	21468	−51638
Electricity generation	−30769	−989	−32165	−3537	20187	−47274
Heat generation	−336	−60	−1992	0	1281	−1107
Petroleum refineries	0	−264	0	0	0	−264
Coke manufacture	−218	0	0	0	0	−218
Blast furnaces	−2570	−217	0	0	0	−2787
Patent fuel manufacture	12	0	0	0	0	12
Energy industry use	853	4777	5950	0	2354	13934
Losses	236	0	1172	0	2358	3766
Final consumption	2758	77367	52499	1772	30578	164974
Industry	1990	6360	11393	336	10537	30616
Transport	0	57268	0	821	725	58814
Other	768	4454	40308	615	19316	65461
Non energy use	0	9284	797	0	0	10081

Source Digest of UK Energy Statistics, 2009

conversion of mass and volume are presented in Table 2.4 (see IEA (2004) and IPCC (2006) for more detailed conversion tables).

Natural gas data is generally reported using both the metric system and the imperial system. In common industry and business transactions, both the systems are widely used and in some cases, the units used may be non-standard as well. Table 2.5 indicates some conversion factors specific to gas.

Similarly, heating values of fuels vary and it is important to have an understanding of the heat content of different types of fuels. Table 2.6 gives some

2.5 Energy Accounting Framework

Table 2.4 Some conversion factors

Unit	Values
Volume conversion	
1 US gallon	3.785 l
1 UK gallon	4.546 l
1 Barrel	158.9 l (or 42 US gallons)
1 cubic foot	0.0283 cubic metres
Cubic metre	1000 l
Mass conversion	
1 kilogram (kg)	2.2036 lb
1 Metric tonne	1,000 kg
1 long ton	1016 kg
1 short ton	907.2 kg
1 lb	453.6 grams
1 tonne of crude oil	7.33 barrels

Table 2.5 Some conversion factors for natural gas

Description	Conversion factor
1 cubic metre of Natural gas	35.3 cubic feet of natural gas
1 cubic feet of natural gas	0.028 cubic metre of natural gas
1 billion cubic metre of natural gas	0.9 Mtoe or 35.7 trillion Btu
1 billion cubic feet of natural gas	0.025 Mtoe or 1.01 trillion Btu
1 cubic feet of natural gas	1000 Btu
1 million tonne of LNG	1.36 billion cubic metres of gas or 48.0 billion cubic feet of gas
1 million tonnes of LNG	1.22 Mtoe or 48.6 trillion Btu

Source BP Statistical Review of World Energy 2009

Table 2.6 Gross Calorific values of different energies

Fuel	Gross calorific value	Carbon content
Anthracite coal	7000–7250 kcal/kg	778–782 kg/t
Coking coals	6600–7350 kcal/kg	674–771 kg/t
Other bituminous coals	5700–6400 kcal/kg	590–657 kg/t
Metallurgical coke	6600 kcal/kg	820 kg/t
Coke-oven gas	19 MJ/m^3	464 kg/t
Crude oil	10^7 kcal/t	
Petroleum products	$(1.05–1.24) \times 10^7$ kcal/t	
Natural gas	37.5–40.5 MJ/m^3	

Source UN (1987), IEA (2004) and IEA (2010)

representative values for commonly used fossil fuels. See Annex 2.1 for worked out examples of unit conversion.

Energy is measured on the basis of the heat which a fuel can make available. But at the time of combustion some amount of heat is absorbed in the evaporation of moisture present in the fuel, thereby reducing the amount of energy available for

Table 2.7 Scientific units and their relations

1 calorie	4.1868 J
1 Btu	252 cal
1 kWh	3.6 MJ = 859.845 kcal

practical use. Measurement on the basis of total energy availability is called gross calorific value; measurement on the basis of energy available for practical use is called net calorific value. The differences between the two bases are about 5% for solid fuels and about 10% for liquid and gaseous fuels.

To obtain a clear picture about energy supply and demand, a common unit is required. Different alternative approaches are possible. Similar to national accounts, national currency unit of the country concerned or an international currency such as US$ or euro can be used. This however would make the results highly sensitive to price changes and such an aggregation would obscure the energy demand and supply information by highlighting the monetary value.

Energy balances use a simple aggregation method where each energy source is converted to a common energy unit and aggregated by simple addition. Two types of units are commonly used:

(a) Precise (or scientific) units: Scientific units include calorie, joule, Btu and kWh. These indicate the heat or work measures of the energy. A calorie is equal to the amount of heat required to raise the temperature of one gram of water at 14.5°C by one degree Celsius. A joule is a measure of work done and is approximately one-forth of a calorie and one thousandth of a Btu. A British thermal unit is equal to the amount of heat required to raise the temperature of one pound of water at 60°F by one degree Fahrenheit. Its multiple of 10^5 is the therm. A kilowatt-hour is the work equivalent of 1000 J/s over a one-hour period. Thus one kilowatt-hour equals 3.6 million joules (see Table 2.7).

(b) Imprecise (or commercial) units: These units provide a sense of physical quantities of the energy. Ton oil equivalent is the most commonly used commercial unit but ton of coal equivalent is also used in some areas. Commercial units are imprecise because the commodities on which these are based are not uniform in energy content. For example, energy content of coal varies from one type to another and can even vary from one year to another. But such units are easily understood and hence most frequently used.

Conversion to scientific units is easy: it requires information on heat content (i.e. calorific value) of the energy. For example, the energy content of 20 Mt of coal having a calorific value 5 Gcal/t is 100 Petacalories (or equivalent to 418.68 PJ). Some common prefixes used in the metric system are indicated in Table 2.8.

Commercial units on the other hand require establishing equivalence between the chosen fuel and the rest. For example, IEA defines the ton of oil equivalent as one metric ton of crude oil having a net calorific value of 10 Gcal (=41.9 GJ). The energy content of all other energies has to be converted to oil equivalence using

2.5 Energy Accounting Framework

Table 2.8 Prefixes used in metric (or SI) system

Prefix	Abbreviation	Multiplier
Kilo	k	10^3
Mega	M	10^6
Giga	G	10^9
Tera	T	10^{12}
Peta	P	10^{15}
Exa	E	10^{18}

Table 2.9 Conversion from precise to imprecise units

Energy units	
1 Mtoe	10^7 Gcal
1 Mtoe	3.968×10^7 MBtu
1 GWh	860 Gcal
1 GWh	3412 MBtu
1 TJ	238.8 Gcal
1 TJ	947.8 MBtu
1 MBtu	0.252 Gcal
1 MBtu	2.52×10^{-8} Mtoe
1 Gcal	10^7 Mtoe
1 Gcal	3.968 MBtu

this rule. As a thumb rule, seven barrels of oil equal one ton. Some conversion factors from precise to imprecise units are given in Table 2.9.[1]

The ton of coal equivalent (tce) is the oldest of the commercial units and is mainly used in China. The coal equivalent is equated with one tone of coal with a calorific value of 7 million kilocalories. However, its popularity is declining due to declining importance of coal in many regions and disparities in the energy values of coal from one area to another.

The heating value-based simple aggregation scheme is easy to understand but has certain limitations. For example, it just focuses on the energy content and energy flow but ignores other attributes that influence choice of energy use. It assigns same weight to all forms of energies without taking into consideration their differences in quality, efficiency of use, substitution possibility, environmental effects, etc.

An alternative that tries to capture some of the above aspects is the useful energy concept. For example, the Indian statistics at certain times used the ton of coal replacement (tcr) measure. Coal replacement is the quantity of coal that generates the same useful energy as any other fuel would produce when used for a particular purpose. Thus, it attempted to combine the heating value of different energies and the efficiency of conversion to capture the effective output or useful work.

---

[1] A reliable unit converter can be found at http://www.iea.org/interenerstat_v2/converter.asp.

(a) Evidently, this approach requires additional information in the form of efficiency of conversion processes and appliances. As such information tends to be site specific for both resources and technologies, cross-country comparisons become difficult.
(b) As use efficiency changes over time, this approach makes inter-temporal comparisons difficult.

However, in some analysis, the concept of useful energy is still used.

Another alternative to the above aggregation method was suggested by Brendt (1978). This was to include the price information of energy in the aggregation scheme to account for the variations in the attributes of different energies. The inherent assumption here is that prices of different fuels capture the differences in qualities and other attributes relevant for making preferences by the consumers. This however would make the results highly sensitive to price changes. Such an aggregation would obscure the energy demand and supply information by highlighting the monetary value.

2.6 Accounting of Traditional Energies

So far, the focus has been on commercial energies but traditional energies (TE) play an important role in many countries, including most of the developing countries. The share of traditional energies varies considerably from one country to another, but its contribution ranges from one-third to one-half of total energy demand of many developing countries. The share can be even higher in rural areas where commercial/modern energies are less used.

A wide variety of TEs can be identified for use in domestic as well as agricultural activities. Some of these fuels have alternative uses and accordingly can have different impacts on the society. For example, animal wastes can be used as a fuel or as fertilizer or for both purposes. As a natural fertilizer, animal waste provides some minerals and rare elements that may not be available from chemical fertilizers. Thus there is an allocation problem of different resources in order to optimize the welfare.

TEs can be acquired through purchase or collection or a combination of purchase and collection. For this reason, traditional energies are also classified as non-commercial energies. There has, however, been a progressive tendency towards monetization of TEs, especially of fuel wood, even in rural areas. But the rate of monetisation is lower in rural areas compared to urban peripheries. When a traditional energy participates in trading, by definition it becomes a commercial energy, although it remains a traditional fuel. To avoid confusion, we are using the term traditional energies in this chapter.

Traditional energies already compete with other fuels in productive activities in domestic, commercial and industrial uses. Most of the households, especially in rural areas of developing countries, depend on these sources not only for their cooking needs but also for farm operations, water heating, etc. Although TEs predominate in rural areas, their use is not limited there only. Urban and industrial consumers also use a considerable amount of TEs.

Despite playing an important role in the economy of many developing countries, TEs do not receive adequate attention. In many cases, the energy balances do not include such energies. Wherever they are included, the coverage may be limited to traded TEs and not all. Similarly, the national accounts also do not cover activities involved in TE supply, mainly because of the valuation problem. Non-money activities often occupy a far greater share than the monetised part in rural energy of many developing countries. The problem is further complicated by the fact that energy is not for direct consumption but used to derive some end-uses, which can be satisfied by a number of substitutes. Evaluating the contribution in monetary terms when some are acquired through non-monetary activities remains problematic (Bhattacharyya 1995; Bhattacharyya and Timilsina 2009).

Yet, neglecting TEs and their contribution is not a solution as this creates a number of serious problems. First, neglecting a dominant source of energy from any analysis underestimates the energy demand and supply situation in a country. Second, it reduces the credibility of the analysis and introduces errors in analysis and policy prescriptions. For example, any policy on internalization of environmental costs of commercial fuels loses credibility in a developing economy when TEs are excluded from the analysis, since commercial and traditional fuels are complementary to one another. This not only underestimates the problem but also misinterprets it. Third, excluding the contribution of activities involved in TE supply from the national accounts underestimates the national product and the importance of these economic activities.

In what follows, the accounting of traditional energies and issues involved with this are presented briefly.

2.6.1 Features of TEs

There are some specific characteristics of traditional energy sources which require special attention. Some such features are (Denman 1998).

Most of the biomass resources are obtained as by-products of the overall activities relating to agricultural production, crop processing and livestock maintenance. These resources have multiple uses as fodder, fuel, fertilizers and construction materials. A detailed understanding of the agricultural processes and systems is required for any reliable information on the production, availability and use of such resources.

Traditional fuels are often collected by the users and this constitutes the principal mode of traditional energy supply. It is difficult to have any record of supplies of traditional fuels from different sources and due to the localised nature of this activity, and differences in the geographical, weather and other user conditions, it is also difficult to generalize based on small samples of ad-hoc surveys.

Non-standard units (such as bundles, bags, headloads, backloads, baskets, buckets, etc.) used to describe and measure traditional energies make it very difficult to get precise information about energy supply and use.

Typically, consumers do not keep records of consumption of these fuels because these are collected, stored and used by various family members over different seasons, primarily for domestic purposes.

The gross energy content of each category of fuel is different and it varies from one season to another due to moisture content. Hence, norms of energy content of these fuels cannot be used for obtaining estimates of energy consumption.

The efficiency of different end-use devices used for TE consumption varies and hence, the useful energy available from such fuels varies with end-use devices. Norms based on laboratory tests or ad-hoc surveys will not be of much use.

As mentioned before, most of these sources are not traded in the market.

2.6.2 Data availability, Data Collection and Reporting

At the national level, some countries produce good information. In Asia, Nepal, Thailand, Sri Lanka and Philippines have reliable time series data on traditional energy data. Most other countries have had several studies or surveys on TEs but do not have a consistent time series data.

At the international level, United Nations, Food and Agricultural Organisation, International Energy Agency and World Bank are active in data collection and reporting. With the renewed emphasis on energy planning in the mid-seventies, some attempts have been made to quantify the role of traditional energies in developing countries. But that effort was not sustained beyond a few surveys and quick estimates in most cases. In late 1990s, traditional energies came to focus once again in the debate over sustainable energy development. The workshop on Biomass energy organized by IEA in 1997 attempted to understand the role, level and sustainability of biomass use for energy purposes in non-OECD countries. IEA has since started to play an important role in collecting and reporting data on TEs. It has already started collecting data for all non-OECD countries.

At the national level, generally data on consumption of TEs is available from special purpose surveys. These surveys can be specifically for TEs or as part of overall energy survey. The scope and coverage of surveys to be conducted depends on the objectives of the survey in question. To assess the level and pattern of TE consumption, a large-scale extensive survey at the national/regional level would be required. On the other hand, rural level surveys would be required if the objective is to assess the possibility of improvements in the existing use-patterns and introduction of new technologies.

Normally the consumption is first estimated either on the basis of sample surveys or on the basis of other estimates. Production is then considered equal to consumption, ignoring transformation processes. Some energy balances also use production equivalence to reflect the amount of fossil fuel that would be required to supply the energy made available by traditional sources.

2.7 Special Treatments of Some Entries in the Energy Balance

Certain entries to the energy balance require special attention. The most important ones relate to electricity production and use, self-generation and traditional energies. These are discussed below.

2.7.1 Treatment of Primary Electricity Production

Production and use of electricity pose certain problems for energy balance tables. This is because for other fuels the total energy content is measured rather than the available energy, while for electricity generated from hydroelectric power, nuclear power or geothermal, the available energy is essentially measured. This leads to an inconsistency of approach.

In general two approaches are used to resolve this problem:

1. Consumption equivalence: Here the OEB records the direct heat equivalent of the electricity (i.e. converting 1 kWh to kcal or kJ using the calorific value of electricity, note that 1 kWh = 860 kcal). This is done on the premise that the energy could essentially be harnessed by transforming it into electricity and that electricity is the practically first usable form of the energy under consideration. This approach is known as consumption equivalence of electricity treatment for energy balances.
2. Production equivalence: The second method attempts to measure the equivalent or comparable fossil fuel requirement of primary electricity production. This is done on the premise of consistency in approach. This method estimates the amount of fossil fuel input that would be required to provide the same energy as produced by the primary electricity sources. This approach is known as fossil fuel input equivalent approach or simply production equivalence approach (or partial substitution approach). A two step procedure is followed to determine the input primary energy requirement:

    (a) the overall thermal efficiency of thermal power generation for the country concerned is estimated first;
    (b) this efficiency is applied to primary electricity generation to arrive at the input energy requirement.

For example, assume that a country produced 1 GWh of primary electricity in a year. If the OEB shows the physical energy, it records $1*860*10^6/10^{10}$ = 0.086 ktoe. However, if the production equivalence is used, assuming a thermal electricity generation efficiency of 30%, the input primary energy would be 0.086/0.3 = 0.287 ktoe.

Both the above approaches are used in practice. Normally countries with high hydro or nuclear energy share tend to use the production equivalence concept, while others tend to use consumption equivalence concept. According to IEA (2004), the production equivalence approach has been abandoned now.

2.7.2 Treatment of Electricity in Final Consumption

A similar issue arises regarding the treatment of electricity consumption in the energy balance. This arises because:

- Electricity, being a secondary form of energy, is a high grade energy compared to other forms of energy; and
- the appliance efficiency is often much higher than other types of energy-using appliances.

The issue is whether electricity should be treated like any other source of energy or differently.

Again two options are available to rectify the problem:

- Useful energy basis: If the OEB is expanded to include useful energy used by consumers, differences in appliance efficiencies can be taken into consideration. However, data availability may be a constraint in implementing this approach.
- Fossil fuel equivalence: The other alternative is to express all electricity delivered to consumers in terms of its fossil fuel input equivalent. This would follow an approach similar to the production equivalent approach discussed above. This is however rarely followed in practice.

The common practice is to reflect the electricity consumed in its direct heat equivalent without accounting for differences in appliance efficiency, although this may underestimate the contribution of electricity in the final consumption.

2.7.3 Self Generation

Self generation or auto-production means production of energy (electrical or otherwise) by the user itself essentially for its own consumption. However, in some cases the excess energy or some by-products may even be sold outside as well. Auto-production of electricity plays a significant role in many countries (e.g. coking plants in integrated steel industries, captive or stand-by generating capacities in many developing countries, etc.). Information on this auto-production is extremely important for a complete picture of energy transformation and use.

Information on auto-producers is difficult to collect, as there is no compulsory reporting of this activity in many countries. Fuels used for electricity generation or for producing other energy may be shown as final consumption in the sector (industry or others). This lack of information can badly distort analysis of energy statistics at the national level. It also makes comparison with other countries difficult as definitions would not tally between countries.

Some energy balances account for self-generation of electricity in the transformation part. This approach is consistent with the logic of the overall energy balance and represents correctly the energy consumption of a country. A separate row is added in the transformation section of the energy balance to report auto-production of electricity. Thus electricity production is split into public electricity and auto-producers of electricity.

2.8 Analysis of Energy Balance Information

Energy balances provide a great deal of information about the energy situation of a country. They are also a source of consistent information that could be used to analyse the supply and demand situations of a country and with appropriate care, can be used for international comparisons.

As the energy balance is organized in three sections (supply, transformation and use), it is possible to gain insight in these areas, depending on the need and purpose of the analysis. For example, the primary energy requirement indicates the total energy requirement of the country to meet final demand and transformation needs in the economy. The trend of primary energy requirement of a country shows how the internal aggregate demand has changed over time. Similarly, the transformation section of the energy balance provides information on energy conversion efficiency and how the technical efficiency of aggregate conversion has changed over the study period could be easily analysed from energy balance tables. Final consumption data can be used to analyse the evolution of final energy demand of the country by fuel type and by sector of use. Such analyses provide better understanding of the demand pattern of each sector and energy source.

In addition to any descriptive analysis using trends or growth rates, further insights can be obtained by analyzing various ratios. IAEA (2005) has compiled a large set of useful ratios that could be examined and analysed. A few of these ratios are discussed below[2]:

(a) Energy supply mix: As primary energy supply comes from various types of energies, it is important to know the contribution of each type and its evolution over time. The share of each energy source in primary consumption (i.e. the ratio coal, oil, gas or electricity supply in the total) characterises the energy

---
[2] See also Energy Efficiency Indicators Europe project website (http://www.odyssee-indicators.org/index.php).

supply mix of a country. This share shows the diversity of the supply mix (or lack of it) in a country. It is normally considered that a diversified energy mix is better and preferable compared to a highly concentrated mix.

For example, from Table 2.3, the share of coal in the British primary energy supply in 2008 was 16.1%, while that of oil and gas was 35.6% and 40.0% respectively. Thus the share of fossil fuels in the primary energy supply in that year w therefore 91.7%, showing the overwhelming dominance of such energies.

(b) Self-reliance in supply: As the supply can come from local production or imports, independence of a country in terms of supply is considered an important characteristic of the supply system. The rate of energy independence (or self-reliance) is the ratio of indigenous production to total primary energy requirement. For importers, self-reliance would be less than 100% while for exporters, the value would be more than 100%. This analysis can be done at a more disaggregated level by considering the self-reliance in respect of each type of energy.

Again using the British example, Table 2.3 indicates that in 2008, about 30% of coal supply came from local sources, while 94% and 74% respectively of oil and gas came from domestic sources. So, the country had an overall self-sufficiency of over 75% in that year.

(c) Share of renewable energies in supply: Where the energy balance covers the renewable energies, this could be examined to see the role of alternative energies in the supply mix.

(d) Efficiency of electricity generation: The Overall efficiency of power generation can be determined from the ratio of electricity output to energy input for electricity generation. Where input and output values are available by energy type, efficiency can be determined by fuel type as well. This indicator can reflect how the electricity conversion is evolving in the country and whether there is any improvement in this important area.

Using the British example again, in 2008, the electricity system efficiency comes to 40%.

(e) Power generation mix: The power generation mix of a country can be obtained from the share of electricity production by type of fuel. The higher the concentration of power generation technology, the more vulnerable a country could be in terms of supply risk.

For example, in the British case, the electricity generation mix for 2008 was as follows: 38% came from natural gas, 36% from coal, 22.5% from nuclear and the rest from renewable sources including hydropower.[3]

(f) Refining efficiency: This is determined from the ratio of output of refineries to refinery throughput. This indicator could be easily compared internationally to

---

[3] This is based on more detailed information about electricity available in the energy statistics, DUKES 2009.

2.8 Analysis of Energy Balance Information

see how the refineries are performing in a country.

In the British case, the refinery efficiency for 2008 was 99.7%.

(g) Overall energy transformation efficiency: This is determined as the ratio of final energy consumption to primary energy requirement. This shows how much of energy is lost in the conversion process. Lower the loss, more efficient the system is. In the case of UK, the overall transformation efficiency was 70%, which represents a high level of performance.

(h) Per capita consumption of primary energy and final energy: These two indicators are frequently used in cross country comparisons. The ratio of primary (or Final) energy consumption to population in a country gives the per capita consumption. Generally per capita consumption of energy is higher in developed countries than in developing countries and this index is often used as a rough measure of prosperity. Similarly, per capita electricity (or other fuel consumption) could be used to see the level of electricity (or fuel) use in a country.

In 2008, the UK population was 61.38 million. Accordingly, energy consumption per person was 3.82 toe.

(i) Energy intensity: This indicator is used to analyse the importance of energy to economic growth. Energy intensity is the ratio of energy consumption to output of economic activities. When energy intensity is determined on a national basis using GDP, it is termed as GDP intensity. GDP intensity can be defined in a number of ways: using primary energy consumption or final energy consumption, using national GDP value or GDP expressed in an international currency or in purchasing power parity. Accordingly, the intensity would vary and one has to be careful in using intensity values for cross-country comparisons. This is considered in more detail in Chap. 3.

In 2008 the GDP of UK was £1332.7 billion (at 2005 prices). This leads to an energy intensity of 175.8 toe per million pounds of output.

The OEB constructed on this basis can then be used for the analysis of changes in the level and mix of energy sources used for particular purposes before and after transformation. It can also be used for the study of changes in the use of pattern of different fuels, for the examination of the extent of or scope for substitution between fuels at different stages of the flow from primary supplies to final energy uses, and as a source for the generation of time series tables.

2.9 Alternative Presentation of Energy Accounting Information

2.9.1 Energy Flow Diagrams

Flow diagrams present the energy balance information in a pictorial form. There are various diagrammatic ways of presenting the energy balance information and

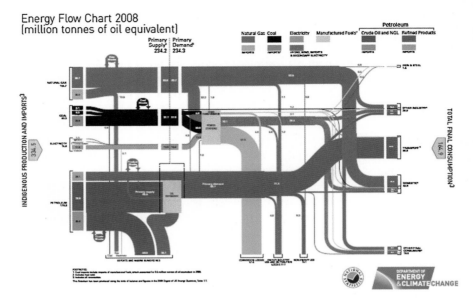

Fig. 2.4 Example of an energy flow diagram. *Source* DECC website at http://www.decc.gov.uk/en/content/cms/statistics/publications/flow/flow.aspx

the general practice is to vary the width of the flow to reflect the importance of different energies. Wider bands would represent larger flow and inversely, narrow bands represent smaller flow. Different shades or colours are used to reflect flow of different fuels. A simple energy flow chart provides the basic information of availability and demand in an aggregated manner. Normally, different fuels would be shown separately but inter-relationships and substitution of fuels is not shown. A simple energy flow chart is shown in Fig. 2.4.

More complex charts describe the flow interactions of various fuels at the end-use level and indicate alternative uses of fuels. These charts would provide a clearer picture of energy flow in the economy and the relative importance of various fuels.

Flow diagrams are valuable presentation aids that are useful for training sessions and briefing of high-level officials and general public. They do not serve much analytical purpose and cannot be prepared unless the energy balances are ready.

2.9.2 Reference Energy Systems (RES)

The Reference Energy System is a more formal analytical tool than flow diagrams. RES was developed in the US in 1971 for energy analysis. RES employs a network form to represent the activities and relationships of an energy system. The network follows the energy commodity flow path from its origin through transformation to end-uses.

2.9 Alternative Presentation of Energy Accounting Information

The RES is organised as follows: for each form of energy, a separate line is used. Different forms of energies are presented vertically while the processes and technologies employed for any energy commodity are specified along horizontal axis. RES would cover all phases of energy flow: resource extraction, refining or treatment, transport, conversion, distribution, and utilization in an end-use device for each energy type.

A network has two essential elements: a node and a flow represented by an arrow. A node represents the start or completion of a process. For example, the start and end of refining will be shown by two nodes and a line joining the nodes will indicate the flow of energy. The size of the flow is indicated by a numerical value and where appropriate, conversion efficiency is also indicated in the brackets. Processes that are occurring at the same stage of progress of an energy commodity would be shown vertically. Diagonal arrows denote transportation of an energy commodity from (or to) outside system or to a conversion process. Solid line arrows indicate that the fuel undergoes the process or activity; dotted arrows are used when the process or activity is not applicable to the fuel. Energy flows from left to right in a RES. The left side indicates all extraction and production of energies. The right side indicates the demands. Figure 2.5 presents an example of a RES.

Each path indicates a possible route of energy flow from a given energy source to a demand. Multiple flows for any end-use or activity indicated by alternative paths and branches would reflect substitution possibilities of various resources and technologies. For example, Fig. 2.5 allows determination of the amount of oil products used for agricultural purposes. Both traditional and modern energy sources can be accommodated in this framework.

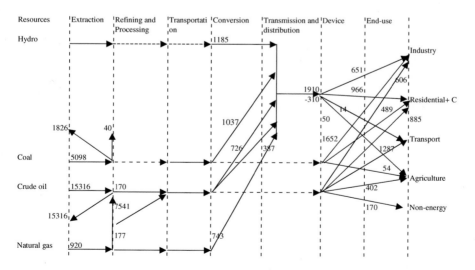

Fig. 2.5 A simple RES diagram

A RES contains all the information available in an overall energy balance. It also provides information about conversion efficiencies of each conversion and end-use devices. RES can be used for analytical studies of energy systems using optimization and other techniques. RES can cover sectoral energy balances as well and detailed representation of end-uses and appliances is possible in this framework. It allows one to visualize the entire energy system of a country and analyse it comprehensively. RES is widely used in system-wide modelling tools and packages.

RES can be drawn using historical data and also using forecasts. For forecasting and future system analysis applications, RES would incorporate future technological options and possible changes in energy types.

The main drawback of RES is that the pictorial presentation becomes unmanageable as the energy system becomes complex and inter-relationships increase. It becomes difficult to incorporate all flows.

2.9.3 Common Energy Data Issues

A number of conceptual, technical problems and data-related issues are confronted while dealing with energy data (Codoni et al. 1985; Siddayao 1986; IEA 1998; Ailawadi and Bhattacharyya 2002).

Data availability: Often multiple agencies collect and publish data. Collection and reporting involves some time lag and delayed publication of information is quite common. Delays reduce usefulness of the information and its value. Data on energy use is often sketchy and inadequate. Even in cases where a network is used for supply, reliable information on consumer category-wise usage is not available. Manual systems for recording and storing information coupled with managerial incompetence are responsible for such poor state of affairs.

Data quality: There is doubt about the quality of information whenever data is available. This is because in absence correct sales and consumption information, estimates are used and their basis is often questionable. Besides, consistency problems also arise in data and arithmetic errors, internal inconsistency, logical errors, etc. are not uncommon. For example, in the case of natural gas, production may be reported on gross (i.e. including gas vented, flared and re-injected) or net basis.

Similarly, it is almost impossible to obtain export and import statistics that match. For example, exporters' records of destination of the gas do not correspond to the origin of the gas according to the importer. Trade discrepancies may also arise from use of different conversion factors for different origin of imports. In LNG trade, both methane and natural gas liquids (NGL) are involved but at the receiving point NGL may be separated. So production and export will cover more than import and consumption in the country of destination.

For coal, due to differences in the basic characteristics (such as calorific value, ash content, content of impurities, suitability for coking, etc.), a wide variety of classification is used and the systems followed by different countries and international organisations are not necessarily identical. This leads to compatibility problem for data.

Boundary problem: This is generally encountered while using data from a number of sources, especially from different countries. Countries use different conventions about energy classifications and consumer categorization. The boundary problem arises due to: (1) exclusion of or inclusion of traditional fuels; (2) different terminologies used for the same product; (3) different user sectors identified for different data (e.g. electricity end use sectors may be different from that for petroleum products); (4) accounting for differences in energy efficiencies, efficiencies of energy delivering equipment, etc.

Common measurement unit: Aggregating energy sources of different characteristics is a difficulty faced in energy data. The problem is how to aggregate energy forms of different qualities in a way that will allow appropriate cross-country comparisons. In order to present the variety of units on a comparable basis, a common denominator for all fuels is required. Traditionally, the common denominator is their energy or heat content, expressed in Joules, Btu or kWh. Units like tons of coal or oil, or barrels equivalent are derivatives of the heat content.

Conversion factors: This is related to common measurement unit. Once a choice is made about the common denominator, the next question comes is how precise does the conversion factor need to be and how much will the overall picture change if one factor is used rather than the other. The quality of certain products such as coal varies significantly from one country to another and also from one extraction site to another. This necessitates a specific factor for each country and often for each time period as the domination of different extraction sites vary from year to year. For other products, the variation may not be significant and a common factor may be used.

To resolve these issues, a number of initiatives have been taken. For oil statistics, the Joint Oil Data Initiative (JODI) has created a platform for interaction of various stakeholders. Similarly, the UN Statistical Commission and UN Statistics Division are working on the challenges facing the energy statistics. The UN organisations are working towards revising the older manuals and recommendations for international energy statistics. However, this is a more recent development and the consultations and preparatory works have just completed in 2009.

2.10 Conclusion

This chapter introduced the concepts related to the energy systems and presented the energy accounting principles in simple terms. It has also covered the energy conversion issues and treatment of some special elements of the energy data, including that of traditional energies. The issues related to data are also presented and a few indicators are discussed to describe and analyse the information. This chapter lays the data foundation for the rest of the book.

Annex 2.1: Worked Out Examples

Example 1

Table 2.10 provides information on indigenous production of energy of a country in 2009. Present the information in ktoe and PJ.

Table 2.10 Primary energy production in 2009

Fuel	Quantity	Calorific value
Coal	72 Mt	6200 kcal/kg
Crude oil	495 Million barrels	8000 kcal/l
Natural gas	2860 Billion cft	900 btu/cft
Hydro general	11600 GWh	860 kcal/kWh
Geothermal	2900 GWh	860 kcal/kWh

Answer

Table 2.11 Answers to the unit conversion problem

Fuel	Quantity	Calorific value	Energy content		
			Pcal	PJ	ktoe
Coal	72 Mt	6200 kcal/kg	446.4	1868.9875	44640
Crude oil	495 Million barrels	8000 kcal/litre	629.64	2636.1768	62964
Natural gas	2860 Billion cft	900 btu/cft	648.648	2715.7594	64864.8
Hydro general	11600 GWh	860 kcal/kWh	9.976	41.7675	997.6
Geothermal	2900 GWh	860 kcal/kWh	2.494	10.4418	249.4

Example 2

The refinery input and output are given in Table 2.12. Present the information in a common unit (ktoe).

Table 2.12 Refinery statistics example

Refinery	Quantity	Calorific value
Refining Input		
Crude oil ('000 bbls)	−345868	8000 kcal/l
Natural gas (MNCFT)	−13219	900 btu/cft
Refining output ('000 bbls)		kcal/bbl
Gasoline	73642	1339000
ATF (Avturbo)	6432	1378000
Kerosene	58490	1437000
Diesel	99781	1501000
Fuel oil	24444	1576000
LPG (ktons)	546	12.96 ktoe/ktons
OPP/Non energy	61735	1272000

Annex 2.1: Worked Out Examples

Answer

Table 2.13 Presents the answer to the above example

Refinery	Quantity	Calorific value	Energy	
Refining input			Pcal	ktoe
Crude oil ('000 bbls)	−345868	8000 kcal/l	−439.9441	−43994.4096
Natural gas (MNCFT)	−13219	900 btu/cft	−2.9981	−299.8069
Refining output ('000 bbls)		kcal/bbl		
Gasoline	73642	1339000	98.6066	9860.6638
ATF (Avturbo)	6432	1378000	8.8633	886.3296
Kerosene	58490	1437000	84.0501	8405.0130
Diesel	99781	1501000	149.7713	14977.1281
Fuel oil	24444	1576000	38.5237	3852.3744
LPG (ktons)	546	12.96 ktoe/ktons	70.7616	7076.16
OPP/Non energy	61735	1272000	78.5269	7852.692
Total output			529.1036	52910.3609
Total input			−442.9422	−44294.2165
Refinery gain			86.1614	8616.1444

Example 3: Energy Balance Preparation

A small island country does not have any natural resources for energy production. It depends mostly on imported crude oil for its energy needs. Some natural gas is also imported and is used for power generation only. The details are given below for 2008.

Crude oil imported: 52 Mtoe; Gas imported: 1.2 Mtoe
Import of Petroleum products: 21 Mtoe; Export of petroleum products: 46.4 Mtoe

The following are uses of petroleum products (Mtoe) in the country:

Road transport 1;
International transport 16.7, and
Industry 5.

The details of electricity production and use are given in Table 2.14.

Table 2.14 Data about electricity system

Electricity generation	Electricity consumption (Mtoe)
Production 22 TWh	Residential and commercial 0.9
Efficiency 40%	Transport 0.02
Losses and own-use 0.17 Mtoe	Industry 0.8

Based on the given information, prepare the overall energy balance of the country for the year 2000. Show only those rows and columns, which are relevant for this case.

Answer:
Table 2.15 provides the results

Table 2.15 Energy balance for the island country (Mtoe)

	Crude oil	Petro prod	Gas	Electricity	Total
Imports	52	21	1.2		74.2
Exports		−46.4			−46.4
Bunkers		−16.7			−16.7
PES	52	−42.1	1.2	0	11.1
Refining	−52	51.63			−0.37
Electricity		−3.53	−1.2	1.892	−2.838
Own use & T&D				−0.17	−0.17
Stat diff				−0.002	−0.002
TFS	0	6	0	1.72	7.72
TFC		6		1.72	7.72
Ind		5		0.8	5.8
Trans		1		0.02	1.02
Rescom				0.9	0.9

References

Ailawadi VS, Bhattacharyya SC (2002) Regulating the power sector in a regime of incomplete information: lessons from the Indian experience. Int J Regul Gov 2(2):1–26

Bhattacharyya SC (1995) Internalising externalities of energy use through price mechanism: a developing country perspective. Energy Environ 6(3):211–221

Bhattacharyya SC, Timilsina GR (2009) Energy demand models for policy formulation: a comparative study of energy demand models. World Bank Policy Research Working Paper No. WPS4866, Mar 2009

Brendt ER (1978) Aggregate energy, efficiency and productivity measurement. Annu Rev Energy 3:225–273

Codoni R, Park HC, Ramani KV (eds) (1985) Integrated energy planning: a manual. Asian and Pacific Development Centre, Kuala Lumpur

Denman J (1998) IEA biomass energy data: system, methodology and initial results. In: Biomass energy: data, analysis and trends, Conference Proceedings, International Energy Agency, Paris, 23–24 Mar, 1998

Edwards BK (2003) The Economics of hydroelectric power. Edward Elgar, Cheltenham

IAEA (2005) Energy indicators for sustainable development, Austria (see http://www.iea.org/textbase/nppdf/free/2005/Energy_Indicators_Web.pdf. Also see IAEA website)

IEA (1998) Biomass energy: data, analysis and trends, conference proceedings, International Energy Agency, Paris, 23–24 March, 1998

IEA (2004) Energy statistics manual, International Energy Agency, Paris (see http://www.iea.org/textbase/nppdf/free/2005/statistics_manual.pdf, Accessed on 26 July 2006

IEA (2010) Key world energy statistics. International Energy Agency, Paris

References

IPCC (2006) 2006 IPCC Guidelines for national greenhouse gas inventories, vol 2, energy, National Greenhouse Gas Inventories Programme, IGES, Japan
Siddayao CM (1986) Energy demand and economic growth: measurement and conceptual issues in policy analysis. Westview Press, Boulder
Stevens PJ (2000) An introduction to energy economics. In: Stevens P (ed) The economics of energy, vol 1. Edward Elgar, Cheltenham. Also reproduced in two parts in the J Energy Lit vol 6(2) Dec 2000 and vol 7(1) June 2001
UN (1982) Concepts and methods in energy statistics, with special reference to energy accounts and balances: a technical report, Series F No. 29, Department of International Economic and Social Affairs, UN, New York (see http://unstats.un.org/unsd/publication/SeriesF/SeriesF_29E.pdf)
UN (1987) Energy statistics: definitions, units of measure and conversion factors, Series F No. 44, Department of International Economic and Social Affairs, UN, New York (see http://unstats.un.org/unsd/publication/SeriesF/SeriesF_44E.pdf)
UN (1991) Energy statistics: a manual for developing countries, UN, New York (see http://unstats.un.org/unsd/publication/SeriesF/SeriesF_56E.pdf)

Further Reading

Karbuz S (2004) Conversion factors and oil statistics. Energy Policy 32(1):41–45
Natural Resources Forum—Indicators of sustainable energy development, Nov 2005 issue
Sinton J (2001) Accuracy and reliability of China's energy statistics. China Econ Rev 12(4): 373–383

Some Data Sources

International Energy Agency (http://www.iea.org/Textbase/stats/index.asp)
EU: Eurostat
Department for Business, Enterprise and Regulatory Reform, UK (http://www.dti.gov.uk/energy/statistics/source/index.html)
BP Statistical Review of World Energy (http://www.bp.com/productlanding.do?categoryId=6848&contentId=7033471)
US Department of Energy (http://www.energy.gov/)

Chapter 3
Understanding and Analysing Energy Demand

3.1 Introduction

The term "energy demand" can mean different things to different users. Normally it refers to any kind of energy used to satisfy individual energy needs for cooking, heating, travelling, etc., in which case, energy products are used as fuel and therefore generate demand for energy purposes. Energy products are also used as raw materials (i.e. for non-energy purposes) in petrochemical industries or elsewhere and the demand for energy here is to exploit certain chemical properties rather than its heat content.

Similarly, the focus may be quite different for different users: a scientist may focus on equipment or process level energy demand (i.e. energy used in a chemical reaction) while planners and policy-makers would view the aggregate demand from a regional or national point of view. Energy demand can correspond to the amount of energy required in a country (i.e. primary energy demand) or to the amount supplied to the consumers (i.e. final energy demand). Often the context would clarify the meaning of the term but to avoid confusion, it is better to define the term clearly whenever used.

A distinction is sometimes made between energy consumption and energy demand. Energy demand describes a relationship between price (or income or some such economic variable) and quantity of energy either for an energy carrier (e.g. electricity) or for final use (such as cooking). It exists before the purchasing decision is made (i.e. it is an ex ante concept—once a good is purchased, consumption starts). Demand indicates what quantities will be purchased at a given price and how price changes will affect the quantities sought. It can include an unsatisfied portion but the demand that would exist in absence of any supply restrictions is not observable. Consumption on the other hand takes place once the decision is made to purchase and consume (i.e. it is an ex post concept). It refers to the manifestation of satisfied demand and can be measured. However, demand and consumption are used interchangeably in this chapter despite their subtle differences.

Energy demand is a derived demand as energy is consumed through equipment. Energy is not consumed for the sake of consuming it but for an ulterior purpose (e.g. for mobility, for producing goods and services, or for obtaining a certain level of comforts, etc.). Need is specific with respect to location, technology and user. The derived nature of demand influences energy demand in a number of ways (discussed below), which in turn has influenced the demand analysis by creating two distinct traditions—one following the neoclassical economic tradition while the other focusing on the engineering principles coupled with economic information (Worrel et al. 2004).

This chapter intends to provide a basic understanding of various concepts related to energy demand and show how energy demand could be analysed using simple tools covering both the traditions indicated above.

3.2 Evolution of Demand Analysis

Prior to the first oil shock, the energy sector had a supply-oriented focus where the objective was to meet a given exogenous energy demand by expanding the supply. Since early 1970s, when energy caught the attention of policymakers because of sudden price increases, the research on energy has grown significantly in size. From a level of limited understanding of the nature of demand and demand response due to presence of external shocks (Pindyck 1979) and energy system interactions, there has been a significant build-up of knowledge. Energy models were however not developed for the same purpose—some were concerned with better energy supply system design given a level of demand forecast, better understanding of the present and future demand–supply interactions, energy and environment interactions, energy-economy interactions and energy system planning. Others had focused on energy demand analysis and forecasting.

In the three decades that followed since the first oil shock, the energy sector has experienced a wide range of influences that have greatly influenced energy analysis and modelling activities (Worrel et al. 2004; Laitner et al. 2003):

Firstly, the rise in concerns about global warming which required a very long term understanding of the implications of energy use. This has led to the developments in very long term analysis covering 50–100 years.

Secondly, due to the changes in the market operations with the arrival of competitive market segments in various energy industries, especially in the case of electricity, the focus has shifted to short-term analysis, covering hours or days, essentially for operational purposes.

Thirdly, there are growing concerns about future security of fuel supplies and large capacity expansion needs globally. This is evident from the European decision to create an Energy Market Observatory and the UK decision indicated in the White Paper on Energy in 2007 to establish its own energy data observatory (DTI 2007). The twin concerns of the day, namely that of security of energy supply and environmental concerns of energy use, are contributing to a paradigm shift (Helm, 2005), which in turn is fuelling a closer look at the energy infrastructure development both in developed and developing countries either for replacing age-old, sweated assets or for meeting new demand.

We can add a fourth influence as well—that of vast improvements in computing and communications facilities. The emergence of low cost computing and internet facilities has dramatically changed the data processing and analytical capabilities.

Energy projects tend to be capital intensive and often require long lead time. For example, a thermal power station may need 3–4 years to build, a nuclear or hydro power station requires typically 7–10 years, if not more, and a refinery project can easily take 2–4 years. Given the long gestation period of energy investments and diversity of technologies as well as economic conditions of countries and consequent constraints on the analytical choices, medium to long-term analysis is essential for energy system-related decisions.

Moreover, mobilizing resources for energy projects is not always easy. Hence, correct timing of supply capacity additions is important, for which correct demand projection is a pre-requisite. Lumpiness of investment implies that for such projects huge sums of capital are tied up in advance and no return or output is obtained until the project is completed. Consequently, the decision-makers have to form a view about the future well in advance and plan for new projects and actions. The decision-making depends to a large extent on demand forecasting and misjudgements can lead to costly gaps or equally costly over capacities.

In this respect, developing countries have certain distinct features. Bhatia (1987) indicated a number of difficulties experienced in analyzing the energy demand of developing countries as follows:

(a) Data on traditional energies used widely in rural areas may be lacking and may have to be estimated.
(b) Many poor consumers lacking purchasing power may not enter the commercial energy ladder but over time a shift to commercial energies takes place. This needs to be captured.
(c) Supply shortage in many developing countries implies that consumption may not represent the actual demand due to the existence of unfulfilled demand.
(d) The availability and consumption of commercial energy may be greatly influenced by a few large consumers.
(e) Response to price changes is more difficult to assess due to "difficulties of obtaining complimentary non-energy inputs, the absolute shortages of certain fuels, imperfect product and capital markets."

Moreover, the demand for commercial energies tends to grow faster here compared to the developed countries and generally, the demand for liquid fuels grows faster than other fuels, suggesting a shift from solid to liquid fuels. As an adequate supply of energy is vital for the smooth running of a country, and because of long lead times and capital intensive nature of energy projects, developing countries need to analyze the past trends to forecast the likely paths of energy demand growth.

At the other end of the spectrum, there is need for forecasts for day-to-day operation and management of energy systems. How much electricity needs to be generated next hour or tomorrow? This information forms the basis for unit commitment exercise (i.e. to find out which plants should be used to produce the required electricity so that the operating cost is minimised). Similarly, forecasts for six months to one year are required for business planning purposes, for regulatory approvals, to assess the prospects of the business in the coming year, etc. Thus, short term forecasting is also important in addition to medium to long-term analysis and forecasting.

More recently, in the late 1980s, the emphasis was increasingly on sustainable development. Some of the problems associated with energy use, such as the possibility of global warming, have long-term implications and any strategy to deal with them has to be seen in a long-term context. Very long-term demand forecasts for more than 50 years have also become necessary.

3.3 Overview of Energy Demand Decisions

Energy is not consumed for the sake of it but is used for satisfying some need and is done through use of appliances. Any commercial energy requires monetary exchanges and the decision to switch to commercial energies can be considered as a three-stage decision-making process [see Hartman (1979), Stevens (2000) and Bhattacharyya (2006)].

- First, the household has to decide whether to switch or not (i.e. switching decision).
- Second, it decides about the types of appliances to be used (i.e. appliance selection decision).
- In the third stage, consumption decision is made by deciding the usage pattern of each appliance (i.e. consumption decision). All these stages influence energy demand. This is shown in Fig. 3.1.

As Hartman (1979) indicates, any demand analysis and forecasting should consider this three stage decision-making process and capture related policy variables so that interventions, if required, could be properly designed. There are two decision outcomes at the first stage: to purchase an energy consuming appliance or not to purchase. If appliance is not purchased, demand for that particular use does not arise for that consumer. The switching decision is largely determined by monetary factors: the amount and regularity of money income, alternative uses of money and willingness to spend part of the income to consume commercial energies as opposed to allocating the money to other competing needs.

Once a buying decision is made, two important parameters are to be decided next. If alternative fuel choices are available, which fuel would be used and what type of appliance for this fuel? Once a decision is made to buy an appliance and the appliance is purchased, the only variable leaves in the hand of the user is its

3.3 Overview of Energy Demand Decisions

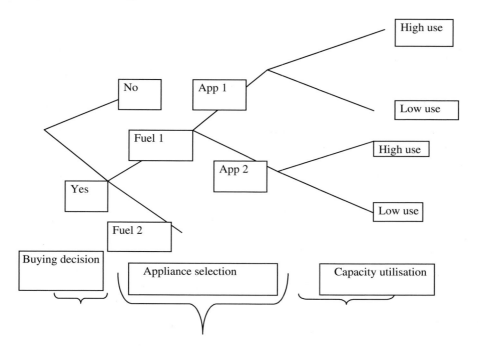

Fig. 3.1 Three-stage decision-making. *Source* Author

utilisation. The level of utilisation varies from consumer to consumer and consumers can adjust utilisation in response to changes in external factors. Box 3.1 provides the implications of each stage of decision making on the demand analysis.

Box 3.1: Implication of the Three-Stage Decision-Making on Energy Demand

For energy demand, information related to appliance holding pattern is important for two reasons:

(1) to understand consumption behaviour: If there is lack of interest in a particular use, it may be that there are important barriers which need to be looked into. These barriers include: cost, financing options, user friendliness, etc.
(2) to understand growth potential: If a particular segment of market is saturated, demand growth from new consumers would be less and vice versa.

Appliance stock and its growth potential are important determinants of demand. For example, in a developing country there is only one car per hundred thousand people. If the government provides cheap gasoline to promote energy access, would it work? The cheap gasoline would go to those having cars and would not benefit the rest. The barriers to owning car need to be looked into first to promote motorized transport.

The second stage has a deciding influence on demand. Often equipment has a long life time (5–10 years) and is costly. Once an appliance is purchased, it will be in operation for sometime. This introduces strong path dependence in energy demand (meaning that the choice of appliance forecloses certain options and influences the demand path). Strong path dependence affects fuel switching possibility and responsiveness of the consumers to external changes. Fuel switching option would be limited by the appliance choice decision and involves capital expenditure, at times of considerable amounts. Limited responsiveness: The rigidity or strong path dependence leaves limited options to consumers in the event of sudden changes in prices or supply conditions in the short run. They have to depend on their existing stock of appliances in any case. The full reaction to external changes is not instantaneous. It is spread over a number of periods because of the rigidity of the system. This process is called lagged reaction (i.e. the reaction lags behind the action) and only over a number of period, the accumulated effect gives the full reaction.

The short term response arises from this factor and its scope is not very broad. Therefore, short-term response is quite limited. This can have a social dimension as low capacity utilisation may lead to deprivation of essential energy services.

The three-stage decision process therefore influences: access to energy services, market growth potential in a particular service or use, path dependence, responsiveness in the short run, reaction response, and consumer's usage behaviour. The above discussion also suggests that technology matters: because energy demand is dependent on technical efficiency, substitution possibility depends on technical options available.

3.4 Economic Foundations of Energy Demand[1]

From the point of view of economics, the principle for estimating and analyzing the demand for energy is not different from that for any other commodity. There

---

[1] This section relies on Bohi (1981), Chapter 2, Estimating the demand for energy: Issues and Methodologies. Similar treatments are also provided in Hartman (1979), Munasinghe and Meier (1993)

3.4 Economic Foundations of Energy Demand

are characteristics of energy demand, institutional features of energy markets, and problems of measurement that require particular attention in analyzing energy markets. But the microeconomic foundation of energy demand is same as for other commodities.

Demand for energy can arise for different reasons. Households consume energy to satisfy certain needs and they do so by allocating their income among various competing needs so as to obtain the greatest degree of satisfaction from total expenditure. Industries and commercial users demand energy as an input of production and their objective is to minimize the total cost of production. Therefore the motivation is not same for the households and the productive users of energy and any analysis of energy demand should treat these categories separately.

From basic microeconomic theory, the demand for a good is represented through a demand function which establishes the relation between various amounts of the good consumed and the determinants of those amounts. The main determinants of demand are: price of the good, prices of related goods (including appliances), prices of other goods, disposable income of the consumer, preferences and tastes, etc. To facilitate the analysis, a convenient assumption (known as ceteris paribus) is made which holds other determinants constant (or unchanged) and the relation between price and the quantity of good consumed is considered. This simple functional form can be written as follows:

$q = f(p)$, where q is the quantity demanded and p is the price of the good. The familiar demand curve is the depiction of the above function.

3.4.1 Consumer Demand for Energy: Utility Maximization Problem

The microeconomic basis for consumer energy demand relies on consumers' utility maximization principles. Such an analysis assumes that

- Consumers know their preference sets and ordering of the preferences.
- Preference ordering can be represented by some utility function and
- The consumer is rational in that she will always choose a most preferred bundle from the set of feasible alternatives.

Following consumer theory, it is considered that an incremental increase in consumption of a good, keeping consumption of other goods constant, increases the satisfaction level but this marginal utility (or increment) decreases as the quantity of consumption increases. Moreover, maximum utility achievable given the prices and income requires marginal rate of substitution to be equal to the economic rate of substitution. This in turn requires that the marginal utility per dollar paid for each good be the same. If the marginal utility per dollar is greater for good A than for good B, then transferring a dollar of expenditure from B to A will increase the total utility for the same expenditure. It follows that reduction in the relative price of good A will tend to increase the demand for good A and vice versa.

Fig. 3.2 Budget constraint

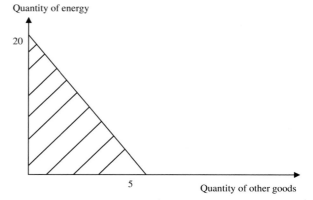

We shall use this basic idea in a graphical example to explain how the consumer demand curve for energy could be developed. The mathematical development is provided in Annex 3.1 for interested readers.[2]

Assume that an individual has 100 dollars to allocate between energy E and other goods X. One unit of energy costs 5 dollars while one unit of other goods costs 20 dollars. Accordingly, the individual can buy 20 units of energy or 5 units of other goods or a combination of these goods as shown by the shaded area of Fig. 3.2.

$$\text{In equation form this is written as } 100 = 5E + 20X \qquad (3.1)$$

$$\text{Consider a utility function } U = X^{0.5}E^{0.5} \qquad (3.2)$$

The combinations of X and E for various levels of utility (e.g. $U = 2, 3, 4$ and 5) can be easily determined for this function (see Fig. 3.3). These curves are called indifference curves. The optimal demand for energy and other commodities could be determined for the given individual from the budget line and the indifference curves (see Fig. 3.3).

The budget line is tangent to the indifference curve ($U = 5$) and the optimal combinations of energy and other goods can be found from this (which turns out to be 10 units of energy and 2.5 units of other goods). Hence, when the energy price is 5 per unit, given the budget constraint, the individual consumes 10 units of energy. This forms one pair of data set for his/her demand curve.

Now consider that the price of energy changes to 10 per unit while the price for other goods remains unchanged. Naturally, the consumer now will be able to consume only 10 units of energy or 5 units of other goods or some combinations of energy and other goods (as shown in Fig. 3.4). Following the method indicated above, the new optimal combination is found and in this particular case, the individual would consume 5 units of energy and 2.5 units of other goods (i.e. just

---

[2] See also Chapter 2 of Bohi (1981), Munasinghe and Meier (1993) and Medlock III (2009).

3.4 Economic Foundations of Energy Demand

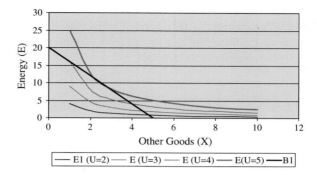

Fig. 3.3 Utility maximisation

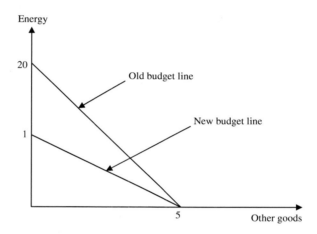

Fig. 3.4 Effect of changes in energy price on the budget line

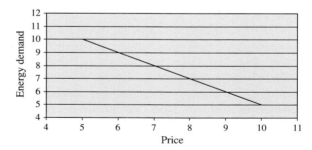

Fig. 3.5 Energy demand curve of an individual

50% reduction of energy demand). This gives another pair of points on the demand curve.

The individual's energy demand schedule can now be drawn using these points (see Fig. 3.5). As you have noticed, in the entire process, we have only changed energy prices while keeping other variables unchanged (i.e. assumed that ceteris paribus condition holds). In Fig. 3.5, the demand curve is downward sloped as is expected.

The market demand function for a particular good is the sum of each individual's demand for that good. The market demand curve for the good is constructed from the demand function by varying the price of the good while holding all other determinants constant.

3.4.2 Cost Minimization Problem of the Producer

In the case of producers, the theory of the producers is used to determine the demand for factors of production. In the production process, it is normally possible to replace one input by the other and the producer would try to find the combination of inputs that would minimize the cost of production. Once again, we use a graphical approach for the general description, while a more mathematical presentation is given in Annex 3.2.

Consider that a producer uses capital and energy to produce her output which follows the production function given in Eq. 3.3.

$$Q = 10 K^{0.5} E^{0.5} \tag{3.3}$$

The isoquant map for this production function can be graphed by setting Q at different levels (say 50 or 100) and then finding the combinations of K and E that would produce the given level of outputs (see Fig. 3.6).

Assume that the price of capital and energy per unit is $1 each. If K units of capital and E units of energy are used in the production process, the total cost will be $K + E$. The cost lines are shown as constraints in Fig. 3.6. As can be seen from the figure, the optimal choice would be at the point where the cost line is tangent to the isoquant. For a given level of output, the demand for input energy can then be determined.

While the above theoretical concepts provide some understanding of energy demand, these theoretical ideas are based on quite restrictive assumptions. While the econometric modelling tradition explicitly follows the economic principles for energy demand analysis and forecasting purposes, this is not the only economic philosophy followed in energy demand modelling. Although price, rationality and optimising behaviour within the neoclassical tradition greatly influence the econometric tradition, others do not always believe in the crucial role of these

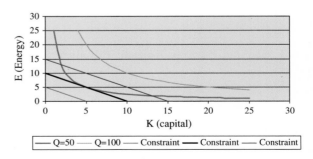

Fig. 3.6 Optimal input selection for the firm

factors. Accordingly, other behavioural assumptions (such as "satisficing" approach in the sense of Herbert Simon or evolutionary approach for technological change) and beliefs are used in some approaches,[3] especially in the "bottom-up" approach or "engineering-economic" approach.

3.5 Alternative Approaches for Energy Demand Analysis

Analysis of the historical evolution of energy demand and its interpretation is an essential part of energy demand analysis. Such an analysis allows identification of the underlying factors affecting energy demand. Various analytical methods are used to analyze energy demand. Three approaches are presented below: simple descriptive analysis, factor (or decomposition) analysis, and econometric analysis.

3.5.1 Descriptive Analysis[4]

Here we present three simple but commonly used indicators that are used to describe the change in demand or its relationship with an economic variable. These are growth rates, demand elasticities and energy intensities.

Any demand analysis starts with a general description of the overall energy demand trends in the past. It enables qualitative characterization of the pattern of energy demand evolution and identification of periods of marked changes in the demand pattern (such as ruptures, inflexions, etc.). This preliminary step could set the scope and the priorities of the analysis (see Fig. 3.7). Such a historical analysis is first based on a graphical presentation of the evolution of demand through time. Two types of graphs are generally used:

- energy demand in absolute value (Mtoe, PJ, etc.) and time;
- energy demand in index and time.

The graph in absolute value provides an indication of the trend while that in index allows comparison with respect to the base year. Index also allows comparison of trends of different fuels and energy groups.

3.5.1.1 Growth Rates

Annual growth rate is another indicator commonly used to describe the trend. This can be on an annual basis or an average over a period. Table 3.1 presents the

---

[3] See Wilson and Dowlatabadi (2007).
[4] This section is based on UN (1991). See also IEA (1997).

Fig. 3.7 Evolution of global primary energy demand. *Data source* BP Statistical Review of World Energy 2009

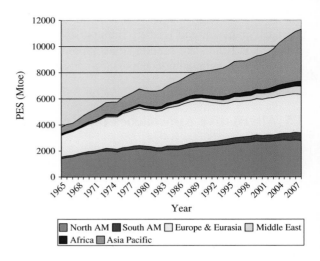

Table 3.1 Mathematical relationships for simple indicators of trend

Indicator	Formula	Parameter description
Year-on-year growth rate	$a = (E_{t+1} - E_t)/E_t$	Where a = annual growth in demand, E_{t+1} = energy consumption in year $t+1$ and E_t = energy demand in year t
Annual average growth rate over a period	$E_{T1} = E_{T0}(1+a_g)^{(T1-T0)}$ $a_g = \left(\frac{E_{T1}}{E_{T0}}\right)^{1/(T1-T0)} - 1$	Where E_{T1} = energy demand in period $T1$ and E_{T0} = energy demand in period $T0$, a_g = annual growth rate
Demand elasticities	$e_t = \frac{(\Delta EC_t/EC_t)}{(\Delta I_t/I_t)}$	Where t is a period given EC is energy consumption I is the driving variable of energy consumption such as GDP, value-added, price, income etc. Δ is the change in the variable
Energy intensity (for a single energy)	$EI_t = \frac{E_t}{I_t}$	EI_t = energy intensity for year t, E_t = energy consumption in year t and I_t = value of the driving variable (say GDP or value added)
Energy intensity in case of aggregated fuels	$EI_t = \frac{\sum_{i=1}^{n} E_{it}}{I_t}$	Where E_{it} = energy consumption of ith type of fuel in year t

formula commonly used for this purpose. The year-on-year growth rates are calculated year after year so as to get a historical series. The average growth rate over a period on the other hand provides a picture for the entire period. Although an arithmetic average of the annual year-on-year growth rates can be calculated, this is not done generally. Instead, a geometric average is calculated for the period. Annual growth rates can also be calculated at any level of disaggregation. This is an easily understood indicator capturing the speed of change in demand.

3.5 Alternative Approaches for Energy Demand Analysis 53

Example According to BP Statistical Review of World Energy, the world primary energy consumption was 9,262.6 Mtoe in 2000. The demand increased to 11,104.4 Mtoe in 2007 and 11,294.9 Mtoe in 2008. Calculate the growth rate of demand between 2007 and 2008. Also calculate the annual average growth rate between 2000 and 2008.

Answer: The primary energy demand increased from 11,104.4 Mtoe in 2007 to 11,294.9 Mtoe in 2008. This amounts to a growth of $= (11{,}294.9 - 11{,}104.4)/11{,}104.4 = 0.017$ or 1.7%.

The annual average growth rate between 2000 and 2008 is $= (11{,}294.9/9{,}262.6)^{\wedge}(1/8) - 1 = 0.0251$ or 2.51%.

3.5.1.2 Demand Elasticities

Elasticities measure how much (in percent) the demand would change if the determining variable changes by 1%. In any economic analysis, three major variables are considered for elasticities: output or economic activity (GDP), price and income. Accordingly, three elasticities can be determined. The general formulation is given in Table 3.1. There are two basic ways of measuring elasticities: using annual growth rates of energy consumption and the driving variable, or using econometric relationships estimated from time series data. The first provides a point estimate while the second provides an average over a period, and accordingly, the two will not give the exactly same result.

Output or GDP elasticities of energy demand indicate the rate of change of energy demand for every 1% change in economic output (GDP or value added). Normally the GDP growth is positively related to energy demand but the value of elasticity varies depending on the stage of development of an economy. It is normally believed that the developed countries tend to have an inelastic demand with respect to income (i.e. the elasticity less than 1) while developing countries have an elastic energy demand with respect to income.

Example The primary energy consumption in China increased from 1,970 Mtoe in 2004 to 2,225 Mtoe in 2005. The GDP increased from 14,197 Billion Yuan in 2004 to 15,603 Billion Yuan in 2005 at constant 2,000 prices. What was the GDP elasticity of energy demand in China?

$$\% \text{ change in energy demand} = (2{,}225 - 1{,}970)/1{,}970 = 12.9\%$$

$$\% \text{ change in GDP} = (15{,}603 - 14{,}197)/14{,}197 = 9.9\%$$

$$\text{GDP elasticity} = 12.9/9.9 = 1.31$$

Price elasticities indicate how much demand changes for every percent change in the energy price. Price elasticities are negative numbers, indicating that an increase in price results in a decrease in energy demand. As this elasticity aims to find out the responsiveness of consumers to price changes, the price to be used for elasticity purposes should reflect as closely as possible what consumers really pay (retail price or wholesale price as the case may be). A distinction is normally made between short-term and long-term price elasticities. The short-term price elasticity captures the instantaneous reaction to price changes. In the short run consumers do not have the possibility to change their capital stock and can only change their consumption behaviour and hence only a partial reaction is normally felt. The long-term elasticity would capture the effect of adjustments over a longer period. On the other hand, over the long run, consumers have the possibility of adjusting their capital stock as well as their consumption behaviour. This results in a better reflection of the reaction to price change.

3.5.1.3 Energy Intensities

Energy intensities (also called energy output ratios) measure the energy requirement per unit of a driving economic variable (e.g. GDP, value added, etc.). Energy consumption may refer to a particular energy or to various energy aggregates and is expressed as a ratio of energy demand per unit of economic output (see Table 3.1 for the formula). For an economic driving variable, normally the constant dollar values are used for better comparability across a time scale. Table 3.2 explains the choice of the economic driving variable (I) in different cases. In the productive sectors (industry, agriculture and commercial), the value-added of these sectors should be used to calculate their energy intensity. As for the case of non-productive energy consuming sectors such as household sector, the GDP of the whole country or the private consumption of the households should be used as driving economic variable. As the energy consumption of the transport sector includes the consumption of all vehicles, it is then irrelevant to use only the value-added of the transport companies to calculate the transport sector's energy intensity. Instead, GDP as the economic indicator is more appropriate for calculating transport energy intensity.

Although energy intensity (or energy GDP ratio) is widely used as a measure of relative performance of economies, the ratio is subject to various conceptual and measurement problems. The ratio is highly sensitive to the bases chosen for either of its components and any problem that may distort the size of either the numerator

Table 3.2 Selection of driving economic variable by sector

Sector	Driving economic variable
Whole country	GDP
Industry	Value-added of industry sector
Agriculture	Value-added of agriculture sector
Commercial	Value-added of commercial sector
Transport	GDP
Households	GDP or private consumption

3.5 Alternative Approaches for Energy Demand Analysis

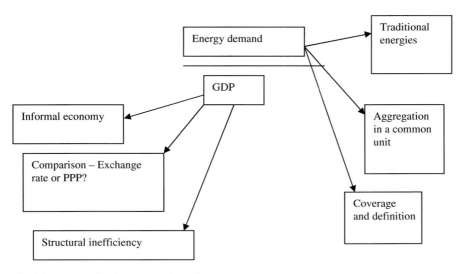

Fig. 3.8 Issues related to energy intensity

(energy consumption) or the denominator (GDP) distorts the picture presented by the ratio (see Fig. 3.8).[5]

The Gross Domestic Product measures the total output of a country's economy. This aggregate statistic represents all goods produced and services rendered within the political boundaries of a country. The GDP can be measured in three standard ways:

(1) by industrial origin, summing up value added by all industries (i.e. gross output minus input);
(2) by summing up the remuneration accruing to all income-producing sectors of the economy; and
(3) by summing up final expenditures to different sectors, that is, presenting aggregated final demand.

The problems related to GDP as a measure of output are:

The measure may be understated by the existence of an underground or informal economy, whose transactions may not be captured by national statistics. This is particularly true of developing countries where many transactions do not get reported in market statistics as they do not enter the market system.

Expenditure on various items may not represent efficient behaviour. In fact, inefficiency would try to increase expenditure and therefore increase GDP when

---

[5] This discussion is based on Chapter 3, Energy Demand and Economic Growth, Measurement and Conceptual Issues in Policy Analysis, by C. M. Siddayao, West View Press, 1986.

expenditure is used to measure GDP. The GDP statistics may obscure the structural inefficiencies of an economy.

For international comparisons, conversion of the GDP to a common unit is required. The use of foreign exchange rates is the obvious approach. Commonly, exchange rate is used to convert local currency GDP to US$. This faces two problems:

Currency values fluctuate but fluctuations in the exchange rate of a particular currency may not necessarily be related to real changes occurring in the domestic economy.

Exchange rates reflect only the values of internationally traded goods and services and not the entire economic price structure of the reference country.

Depending on the case, the GDP will be understated or overstated in the foreign currency and would lead to distorted intensity.

Studies suggested that purchasing power of low income countries was systematically greater than that suggested by their exchange rates when compared to the purchasing power/exchange rate relationships of high-income countries. For these reasons, various international organizations (World Bank, for example) use another measurement of GDP calculated by converting the national currencies into US dollars with "Purchasing Power Parities (PPP)". The PPP Values are based on a comparison of the purchasing power of a typical "basket" of goods and services, characteristic of each country's consumption pattern.

Problems related to measurement of energy consumption also affect energy intensity estimation. Common issues related to energy measurement are:

- Use of traditional energies in developing countries, data for which is often not accurate and not included in analysis; Exclusion of traditional energies can understate energy consumption and accordingly, energy intensity significantly.
- Aggregation of energies to a common unit can be a problem in itself. Simple summation of heat content of energies does not capture the factors that influence the choice of energy forms. Moreover, such aggregation reflects total energy content rather than available energy. As end-use efficiencies of appliances are different for different forms of energies, such an aggregation is biased towards inefficient technologies.
- Aggregation of hydropower, nuclear power, solar and other renewable energies also poses another problem. The amount of energy is measured only at the output end and not for inputs. But other fossil fuels are measured at the input and output ends. In order to measure hydropower, nuclear power and other such renewable energies on a comparable basis, an assumption has to be made about the amount of fossil fuel input that would be required to provide the same energy. Energy accounts however are not always presented using the production equivalence approach.
- The definition and coverage of energy forms and energy consuming sectors are not same and even within a country can vary from time to time. For comparability, comparable definitions and coverage are required.

Different end-use efficiencies of appliances complicate the problem further. Countries with different fuel mix and appliance use pattern cannot be appropriately compared using toe (oil equivalent values) as the measure of energy consumption. For example, one country relies on coal and traditional energies (e.g. India) to meet its energy demand, while another country is more dependent on natural gas for its needs. If the energy intensity of these two countries is compared using the standard energy intensity approach, the differences in the efficiencies of fuel utilisation will not be captured. A remedy for this problem is to introduce the concept of "oil replacement value", which expresses various fuels in terms of the quantity of oil products that would provide the same amount of useful energy (i.e. same energy service). This approach attempts to measure the effective output of useful work at the downstream end of the energy consumption process and considers the efficiency of the energy utilization equipment. As an example, if the efficiency of the end-use equipment using traditional fuels is 25% of that of oil-using equipment, then 4 toe of traditional fuels would be required to produce the same useful energy provided by1 toe (i.e. 1 ton of oil replacement). Similarly, if the relative efficiency of coal-using equipment is 40% of that of oil-using equipment, 2.5 toe of coal would be equal to 1 tor.

While the replacement value takes care of the differences in energy forms, its principal shortcomings are that it is site-specific (e.g. depends on the type of energy and appliance used) for the resource in question and is time-specific for both resource and technology (i.e. the factors used can change over time, making inter-temporal comparisons difficult). For international comparisons, this approach requires additional information (e.g. appliance efficiency), which may not be readily available.

3.6 Factor (or Decomposition) Analysis

The simple indicators discussed earlier capture the nature of the change in energy demand or use but do not explain the underlying cause. However, for a better understanding of energy use and future energy requirements, it is important to understand the causal factors. A large volume of literature has developed on devising methods and frameworks for explaining the demand. A particular method, known as decomposition method, has been widely used (see Ang and Zhang 2000 for a survey of application of this method).[6] Traditionally, these methods try to identify changes in energy demand arising from a number of factors, the commonly used ones are: changes in economic activity (the activity effect), changes in technological efficiency of energy use at the sector level (the intensity effect) and changes in the economic structure (the structural effect). The

---

[6] Also see ODYSSEE project for energy efficiency indicators in Europe (http://www.odyssee-indicators.org/). IEA (1997) also presents a large study for IEA Member countries.

decomposition models attempt to determine the contribution of each of these effects to changes in energy consumption.

Activity effect captures the influence of the changes in the economic activity of the country, assuming other factors do not change. Economic activities, captured by the output generated in an economy, do not remain constant between two periods. For example, if the total output of an economy increases, energy demand would increase, depending on the GDP elasticity of energy demand.

Structural change within the economy refers to shifts in the shares of economic activities at the sector level. For example, many developed counties have moved from energy intensive industries to service-related activities (e.g. U.K.). This change in the economic structure would reduce energy consumption.

The intensity effect captures the role of changing intensities within the sectors. Technical energy efficiencies are the major determinants of energy intensities and changes in the processes and product mixes affect the energy intensities of particular industries. For example, dry process and wet process of cement manufacturing have different energy requirements and accordingly, a shift from wet process to dry process would result in a change in energy intensity of the industry. Changes in the fuel mix also results in a change in the energy intensity due to different levels of efficiencies involved in conversion. For example, a shift away from coal to electricity would result in a reduction in energy intensity.

3.6.1 Analysis of Change in Total Energy Demand

The framework for analyzing the change in energy consumption between two time periods is based on the simple relation

$$E = Q \cdot \mathrm{EI} = Q \cdot \sum_i \left(\frac{E_i Q_i}{Q_i Q}\right) = Q \sum_i \mathrm{EI}_i S_i \qquad (3.4)$$

where,

EI_i = energy intensity in sector i (i.e. ratio of energy consumption in the sector to the driving economic activity of the sector), and
S_i = structure of sector i (i.e. share of the activity of sector i relative to the overall activity of the economy),
Q = overall economic activity with Q_i as the activity of sector i,
E = energy consumption and E_i is the energy consumption in sector i

As mentioned earlier, the contribution of one factor to the overall change is analysed by looking at how the factor under consideration has changed over time, while keeping the other factors constant. For this, let us consider two time periods,

3.6 Factor (or Decomposition) Analysis

base year 0 and end year t. Energy consumption in year t and base year 0 can then be written as

$$E^t = Q^t \sum_i \text{EI}_i^t S_i^t \quad \text{and} \quad E^0 = Q^0 \sum_i \text{EI}_i^0 S_i^0 \tag{3.5}$$

The contribution of activity changes to the overall change in energy consumption is given by:

$$Q_{\text{effect}} = (Q^t - Q^0) \sum \text{EI}_i^0 S_i^0 \tag{3.6}$$

This is called the activity effect.

The contribution of changes in intensity (or the intensity effect) is given by:

$$I_{\text{effect}} = (Q^0) \sum (\text{EI}_i^t - \text{EI}_i^0) S_i^0 \tag{3.7}$$

Finally, the contribution of changes in the sector level structure (or the structural effect) on the overall change in energy consumption is given by:

$$S_{\text{effect}} = (Q^0) \sum (S_i^t - S_i^0) \text{EI}_i^0 \tag{3.8}$$

The total change in energy consumption (or demand) is then

$$\Delta E = Q_{\text{effect}} + I_{\text{effect}} + S_{\text{effect}} \tag{3.9}$$

The decomposition leaves some residual, which is equal to the difference between the change of energy consumption actually recorded and the sum of the three components estimated above. The residual can be quite significant and the error could be great. To remove this problem, a number of sophistications are now available that distribute the residual to the components but these perfect or complete decomposition methods lose their intuitive appeal. Also note that in the above formulation, we have left the unchanged variables at their initial or base year values. This follows Laspeyres index method. The final year values could also be used (which follows Paasche index method). See Ang and Zhang (2000), Ang (2004) and Sun (1998).

Example The total primary energy consumption in the world has increased from 10,029 Mtoe in 2001 to 11,700 Mtoe in 2007. The regional distribution of the demand and other relevant information is given in Table 3.3. Analyse the change in energy demand using the decomposition method.

Answer: For simplicity, we are using the Laspeyres method of decomposition. First we calculate the energy intensity and GDP shares of each region. This is presented in Table 3.4.

Following the method outlined above, each effect is estimated. For the activity effect, we consider the change in the level of activity but keep energy intensity and GDP shares at the 2001 level. In the case of OECD region, this then leads to the following:

Table 3.3 Relevant data for the decomposition example

	2001		2007	
	GDP (Billion US$2,000)	TPES (Mtoe)	GDP (Billion US$2,000)	TPES (Mtoe)
OECD	30,271	5,333	30,110	5,497
ME	638	390	891	552
FSU	572	935	620	1,019
Europe-Non-OECD	151	99	174	106
China	1,392	1,156	2,623	1,970
Other Asia	1,917	1,152	2,308	1,377
LA	1,743	450	1,938	550
Africa	664	514	830	629
World	37,348	10,029	39,494	11,700

Note: *TPES* total primary energy supply, *GDP* gross domestic product, *OECD* Organisation for Economic Co-operation and Development, *ME* Middle East, *FSU* Former Soviet Union, *LA* Latin America
The GDP data for 2001 was converted from constant 1995 values to constant 2000 value using GDP deflator. *Data Source* IEA (2003) and IEA (2009)

Table 3.4 Energy intensities and GDP shares for the decomposition analysis

	GDP shares (%)		Intensities (toe/1,000 US$2,000)	
	2001	2007	2001	2007
OECD	0.8105	0.7624	0.1762	0.1826
ME	0.0171	0.0226	0.6109	0.6195
FSU	0.0153	0.0157	1.6341	1.6435
Europe-Non-OECD	0.0040	0.0044	0.6560	0.6092
China	0.0373	0.0664	0.8305	0.7510
Other Asia	0.0513	0.0584	0.6008	0.5966
LA	0.0467	0.0491	0.2582	0.2838
Africa	0.0178	0.0210	0.7736	0.7578
World	1.0000	1.0000	0.2685	0.2962

$$(39,494 - 37,348) \times (0.8105) \times (0.1762)$$
$$[\text{Billion US\$2,000} \times \text{toe}/(1,000\text{US\$2,000}) = \text{Mtoe}] = 306.4705\,\text{Mtoe}.$$

The structural effect in the OECD region is then obtained as follows:

$$(37,348) \times (0.7624 - 0.8105) \times (0.1762) = -316.5325\,\text{Mtoe}$$

The energy intensity effect is obtained as follows:

$$(37,348) \times (0.8105) \times (0.1826 - 0.1762) = 193.7315\,\text{Mtoe}.$$

The results for the rest of the regions are presented in Table 3.5.

3.6 Factor (or Decomposition) Analysis

Table 3.5 Results of the decomposition analysis

Regions	Activity (Mtoe)	Structural (Mtoe)	Intensity (Mtoe)	Total change Explained	Residue Actual	
OECD	306.4705	−316.5325	193.7315	183.6695	164.0000	−19.6695
ME	22.4326	125.5696	5.2369	153.2391	162.0000	8.7609
FSU	53.6701	24.4196	5.0857	83.1754	84.0000	0.8246
Europe-Non-OECD	5.6277	9.7941	−6.9318	8.4900	7.0000	−1.4900
China	66.4780	902.6097	−110.7499	858.3378	814.0000	−44.3378
Other Asia	66.1530	159.3411	−8.2386	217.2555	225.0000	7.7445
LA	25.8763	23.1438	44.6503	93.6704	100.0000	6.3296
Africa	29.5697	92.5155	−10.8361	111.2490	115.0000	3.7510
World	576.2779	1,020.8609	111.9480	1,709.0868	1,671.0000	−38.0868

It can be seen that the structural effect played an important role in the change of global energy demand between 2001 and 2007. The decline in the share of the developed world in the global economic output and the southward movement of activities led to an increase of over 1,000 Mtoe of energy demand. Most of this change however came from China, reflecting the Chinese influence in global energy demand. The intensity effect was the least important driver of energy demand change in the above example.

Also note that the residue in the analysis. The explained change differs from the actual change in energy demand but in this example, the residue is fairly small.

3.6.2 Analysis of Changes in Energy Intensity

Although decomposition of the changes in energy demand is a useful tool to gain understanding of the underlying factors of demand evolution, the academic literature has placed more emphasis on analysing the energy intensity and its decomposition. This identifies two factors—energy intensity and the structural change and finds out the contribution of each in the overall intensity change. There are two commonly used approaches: additive form decomposition and multiplicative form decomposition. The additive form is similar to the method indicated above for changes in energy demand and is not presented here.[7]

The multiplicative form of decomposition can be used at the aggregated level and at a disaggregated level, where the complexity of analysis increases in the latter case. Both the approaches require some understanding of mathematics and can be omitted by those who are not familiar with them. In what follows, the

---

[7] See Ang and Zhang (2000) for a detailed review. See also UN (1991).

Divisia index method is presented while the formulation for the disaggregated analysis is presented in Chap. 4.

This analysis is based on the Divisia index method (see UN 1991). The Divisia decomposition of energy intensity starts with the basic equation of energy intensity (Eq. 3.10).

$$\mathrm{EI} = \frac{E}{Q} = \sum_i \left(\frac{E_i Q_i}{Q_i Q}\right) = \sum_i \mathrm{EI}_i S_i \qquad (3.10)$$

where

E_i = energy consumption in sector i
Q_i = economic activity variable for sector i
E = energy consumption in all sectors
Q = economic activity of all sectors
S_i = (Q_i/Q) = share of sector i in economic value of all sectors
EI_i = Energy intensity of sector i (E_i/Q_i)

Differentiating Eq. 3.4 with respect to time t yields

$$\frac{d(\mathrm{EI})}{dt} = \sum_i \frac{d(\mathrm{EI}_i)}{dt} S_i + \sum_i \frac{d(S_i)}{dt} \mathrm{EI}_i \qquad (3.11)$$

Dividing all the terms by EI and noting that $\mathrm{EI} = \sum_i \mathrm{EI}_i S_i$, we can rewrite Eq. 3.11 as follows:

$$\frac{1}{\mathrm{EI}} \frac{d\mathrm{EI}}{dt} = \sum \frac{\mathrm{EI}_i}{\sum \mathrm{EI}_i S_i} \frac{dS_i}{dt} + \sum \frac{S_i}{\sum \mathrm{EI}_i S_i} \frac{d\mathrm{EI}_i}{dt} \qquad (3.12)$$

Equation 3.12 can be rearranged as

$$\frac{1}{\mathrm{EI}} \frac{d\mathrm{EI}}{dt} = \sum \frac{\mathrm{EI}_i S_i}{\sum \mathrm{EI}_i S_i} \frac{dS_i}{S_i dt} + \sum \frac{\mathrm{EI}_i S_i}{\sum \mathrm{EI}_i S_i} \frac{d\mathrm{EI}_i}{\mathrm{EI}_i dt} \qquad (3.13)$$

Let $w_i = \frac{\mathrm{EI}_i S_i}{\sum \mathrm{EI}_i S_i}$ = weight of energy intensity of sector i in the overall energy intensity.[8] Equation 3.7 then simplifies to

$$\frac{d\mathrm{EI}}{\mathrm{EI}dt} = \sum w_i \frac{dS_i}{S_i dt} + \sum w_i \frac{d\mathrm{EI}_i}{\mathrm{EI}_i dt} \qquad (3.14)$$

---

[8] This can also be expressed as the ratio of share of energy of a sector i and the total energy used in the economy (i.e. E_i/E). This can be seen from the following relationship $\frac{\mathrm{EI}_i S_i}{\sum_i \mathrm{EI}_i S_i} = \frac{\frac{E_i Q_i}{Q_i Q}}{\frac{E}{Q}} = \frac{E_i}{E}$.

3.6 Factor (or Decomposition) Analysis

Equation 3.8 can be rewritten as

$$\frac{d}{dt}\ln(\text{EI}) = \sum w_i \frac{d}{dt}\ln(S_i) + \sum w_i \frac{d}{dt}\ln(\text{EI}_i) \qquad (3.15)$$

Discrete integration of Eq. 3.15 leads to a relation given in Eq. 3.16.

$$\ln\left(\frac{\text{EI}^T}{\text{EI}^{T0}}\right) = \sum w_i \ln\left(\frac{S_i^T}{S_i^{T0}}\right) + \sum w_i \ln\left(\frac{\text{EI}_i^T}{\text{EI}_i^{T0}}\right) \qquad (3.16)$$

Equation 3.10 can be rewritten in multiplicative form as follows:

$$\frac{\text{EI}^T}{\text{EI}^{T0}} = e^{\sum w_i \ln\left(\frac{S_i^T}{S_i^{T0}}\right) + \sum w_i \ln\left(\frac{\text{EI}_i^T}{\text{EI}_i^{T0}}\right)}$$
$$= e^{\sum w_i \ln\left(\frac{S_i^T}{S_i^{T0}}\right)} \cdot e^{\sum w_i \ln\left(\frac{\text{EI}_i^T}{\text{EI}_i^{T0}}\right)} = D_{st}D_{\text{Int}} \qquad (3.17)$$

Since only discrete data are available in empirical studies, the weight function is often approximated by the arithmetic mean of the weights for year 0 and year T. This leaves a small residue but the residue is normally small compared to other approaches presented earlier. Following index number convention, the intensity of the initial year is set at 100. The final year intensity is then determined in comparison to the base year.

To avoid the residue problem, Ang (1997) proposed the use of the logarithmic mean scheme of weights $w_{i,0}$ and $w_{i,T}$ to yield.

$$L(w_{i,0}, w_{i,T}) = (w_{i,T} - w_{i,0})/\ln(w_{i,T}/w_{i,0}) \qquad (3.18)$$

However, the sum of this weight function, when taken over all sectors, is not unity. This sum is always slightly less than unity, which can be seen from property of the weight function mentioned above, i.e., $(xy)^{1/2} < L(x,y) < (x+y)2$. To fulfill the basic property of weight functions, that the sum is 1, Eq. (3.18) may be normalized. The normalized weight function can be written as

$$w_i^* = L(w_{i,0}, w_{i,T}) / \sum_k L(w_{k,0}, w_{k,T}) \qquad (3.19)$$

where the summation in the denominator on the right-hand side is taken over all sectors. Normalization is also common in other statistical applications when problems similar to the above arise. For instance, in the decomposition of a time series with seasonal variation, the estimated seasonal factors are normalized as the sum for all the seasons in a year is unity.

Example Use the data in Table 3.3 and analyse the change in global primary energy intensity between 2001 and 2007.

Answer: The basic data required for the analysis is presented in Table 3.6.

Table 3.6 essentially presents the structural shares, energy intensities and the share of regional energy demand within the world for two years 2001 and 2007.

Table 3.6 Transformed data for energy intensity decomposition analysis

	GDP shares (%)		Intensities (toe/1,000 US$2,000)		Energy shares	
	2001	2007	2001	2007	2001	2007
OECD	0.8105	0.7624	0.1762	0.1826	0.5318	0.4698
ME	0.0171	0.0226	0.6113	0.6195	0.0389	0.0472
FSU	0.0153	0.0157	1.6346	1.6435	0.0932	0.0871
Europe-Non-OECD	0.004	0.0044	0.6556	0.6092	0.0099	0.0091
China	0.0373	0.0664	0.8305	0.751	0.1153	0.1684
Other Asia	0.0513	0.0584	0.6009	0.5966	0.1149	0.1177
LA	0.0467	0.0491	0.2582	0.2838	0.0449	0.047
Africa	0.0178	0.021	0.7741	0.7578	0.0513	0.0538
World	1	1	0.2685	0.2962	1	1

Table 3.7 Weights and normalised weights for log-mean Divisia analysis

Regions	Log mean weight	Normalised weight (W^*)	$W^*\ln(S^t/S^0)$	$W^*\ln(EI)^t/(EI)^0$
OECD	0.50016	0.50134	−0.03067	0.01789
ME	0.04292	0.04302	0.012	0.00057
FSU	0.09012	0.09033	0.00233	0.00049
Europe-Non-OECD	0.00949	0.00951	0.00091	−0.0007
China	0.14018	0.14051	0.08103	−0.01414
Other Asia	0.11629	0.11657	0.01511	−0.00084
LA	0.04594	0.04605	0.00231	0.00435
Africa	0.05254	0.05266	0.00871	−0.00112
World	0.99764	0.99999	0.09173	0.0065

Note that we have used the ratio of regional energy share as weights in the Divisia analysis, as explained in footnote 8. The results of the weights calculation and their normalisation is presented are presented in Table 3.7.

Using these weights and the ratios of energy intensities and structural shares, the final decomposition into the structural factor and intensity factor is arrived at. This is compared with the actual change in the intensities. Note that there is no residue left for the factor being decomposed (i.e. the global energy intensity). The results are presented in Table 3.8.

3.7 Analysis Using Physical Indicators

In this approach, indicators are related to physical outputs and the analysis has focused on the analysis of unit consumption. The unit energy consumption measures the energy requirement per unit of a techno-economic driving variable: energy per ton of product, per car, per household, etc. It is calculated by simply diving the

3.7 Analysis using Physical Indicators

Table 3.8 Results of the decomposition analysis

Regions	D_s	D_{int}	D_{tot}	D_{act}	D_{rsd}
OECD	0.9698	1.01805	0.9873	1.03632	1.04965
ME	1.01207	1.00057	1.01265	1.01341	1.00075
FSU	1.00233	1.00049	1.00282	1.00544	1.00261
Europe-Non-OECD	1.00091	0.9993	1.00021	0.92923	0.92903
China	1.0844	0.98596	1.06918	0.90427	0.84576
Other Asia	1.01522	0.99916	1.01437	0.99284	0.97878
LA	1.00231	1.00436	1.00668	1.09915	1.09186
Africa	1.00875	0.99888	1.00762	0.97894	0.97154
World	1.0961	1.0065	1.1032	1.1032	1

annual energy consumption (E_t) by the value of the driving techno-economic variable (Q_t) measured in physical terms (i.e. tons of steel, vehicle-km, etc.).

$$UE_t = E_t/Q_t \tag{3.20}$$

The energy consumption refers to all types of energy involved in the same homogenous group (or sector). The concept of unit energy consumption is used in techno-economic analysis of demand and forecasting.

3.8 Energy Demand Analysis Using the Econometric Approach

The econometric approach relies on the economic foundation of energy demand discussed earlier to analyse energy demand and the effects of price and policy changes. The level of analysis can vary from a single equation system to simultaneous equation systems and the method has evolved over the past four decades to take advantage of the developments in the econometric analysis.

As indicated earlier in this chapter, any analysis of energy demand should consider three decisions made by the user—equipment buying decision, fuel and equipment choice decision and the capacity utilization decision. The econometric method tries to capture the above ideas of derived demand for energy using a variety of modeling techniques, leading to a widely varying level of effectiveness. Two commonly used modeling approaches are known as the reduced form models and structural models. These two forms of models are discussed below.

The starting point in this modeling approach is the identity that links energy consumption with the stock of capital equipment and its rate of utilisation indicated below (referred to as the Fisher and Kaysen (1962) model).

$$Q_i \equiv \sum_{k=1}^{M} R_{ki} A_{ki} \tag{3.21}$$

The equation indicates that total consumption of fuel i is the sum of fuel consumption of each of k types of appliances, where the fuel consumption by an appliance type is obtained as the product of the stock of such appliance (A) and the utilization rate (R). For example, electricity consumed by a household can originate from a number of white goods—refrigerators, cookers, blenders, entertainment equipment, etc. The stock of such appliances multiplied by the kWh of electricity used by each good gives the electricity consumption by appliance type. The total electricity consumption then is sum of electricity consumed by all the appliances used by the household.

A structural model considers the derived nature of energy demand explicitly by specifying separate demand functions for the appliance stock and the utilization rate. A_i and R_i may be expressed as follows:

$$A_i = f_1(P_i, P_j, P_a, Y, X)$$
$$R_i = f_2(P_i, Y, Z) \qquad (3.22)$$

where P_i is the price of fuel i, P_j is the price of competitive fuel j, P_a is the price of appliance, Y is income or output (in case of intermediate consumers), X and Z are other relevant variables. In the above equation, the appliance stock is hypothesized to depend on the fuel price, substitute fuel price, income and a vector of other variables. The rate of utilization is considered to depend on the own-price of fuel, income and other variables.[9]

A structural model will simultaneously estimate the above functions to determine the fuel demand. However, this requires information on equipment stocks and such information is not always available. In such cases a reduced-form fuel consumption model can be used. These models are most commonly used for energy demand analysis. Energy demand is estimated by combining the effects of inter-fuel substitution, stock adjustment of appliances and the rate of utilization of devices.

Substituting Eq. 3.22 in 3.21 leads to

$$Q_i = k(P_i, P_j, P_a, Y, X, Z) \qquad (3.23)$$

The above model is known as the reduced form static model as it assumes an instantaneous adjustment process in the capital stock. This in other words means that if the fuel demand increases, the capital stock increases magically to the matching level (Hartman 1979). Therefore such a model does not capture the distinction between the short and long run adjustment.

In the dynamic models, the instantaneous adjustment process is relaxed. In the partial stock adjustment models, it is assumed that the stock of appliances cannot adjustment rapidly due to time lags in the process of retirement and new capacity addition.

---

[9] Substitute fuel price is not relevant here as the appliance has a specific fuel use capacity. This ignores the dual-fuel capability of the appliance.

3.8 Energy Demand Analysis Using the Econometric Approach

A distinction is made between actual demand and the desired demand. It is assumed that the desired consumption Q_t^* is dependent on current price and other variables and specified as follows:

$$Q_t^* = a' + b'P_t + c'Z_t + e'_t \tag{3.24}$$

In any given period, the actual consumption may not adjust completely to obtain the desired level due to lack of knowledge, technical constraints and other factors. An adjustment process is assumed relating desired and actual consumption:

$$Q_t - Q_{t-1} = g(Q_t^* - Q_{t-1}), \quad 0 < g < 1$$

That is, as a result of a price change, the consumer will move partly from his initial consumption to the desired consumption. The closer g is to unity, the faster the adjustment process. Substituting for Q_t^* and rearranging leads to

$$Q_t = a'g + b'gP_t + c'gZ_t + (1-g)Q_{t-1} + ge'_t \tag{3.25}$$

The above equation can be rewritten as

$$Q_t = a + bP_t + cZ_t + dQ_{t-1} + e_t \tag{3.26}$$

This equation only contains observable variable Q and its lagged term.[10] The coefficient of the lagged term indicates the speed of the adjustment process. This can be used to determine the short and long-term response coefficients.

Probably the most widely used single equation specification takes the following form:

$$\log E_t = a + b\log(P_E) + c\log(Y_t) + u_t \tag{3.27}$$

where E (the per capita real energy consumption) is determined by the relative price of energy (P_E), per capita real income or output (Y) and the disturbance term. Advantages of this particular specification are its straightforward allowance for both price and real activity influences. This type of specification has been applied to total final energy demand, individual sector energy demand (industry, transport, residential, etc.).

In Eq. 3.27, the estimated coefficients b and c are price and output elasticities of energy demand, which measure percentage changes in industrial energy demand for a percentage change in energy price and economic output. The equation can be used to analyze the effect of changes in energy prices on energy demand. The magnitude of the effect of price changes would depend on price elasticities of energy demand. If the price elasticity is greater than −1, the demand is called elastic and when it is less than −1, it is called inelastic. For an elastic demand, the

---

[10] The same general result is also obtained using the adaptive price expectation specification. This is presented in Annex 3.3.

effect of price change would lead to more than proportionate reduction in demand. Energy conservation can be promoted for such energies through higher prices.

The economic foundation of the above specification can be traced in the tradition of Cobb–Douglas function, where the demand is considered to be a function of price and income. The log-linear form of the specification provides direct estimation of price and income elasticities and is better suited to energy demand than a simple linear specification.

Such a basic model can be extended to include more independent variables, to carry out disaggregated analysis at sub-sectoral level and to analyze different types of energy carriers. However, it does not consider the impact of changes in prices on economic growth, inflation and other macro-economic variables. It also does not analyze the inter-fuel substitution issue. Moreover, Ryan and Ploure (2009) argue that this functional form is not generally consistent with the optimizing behaviour and assumes that the capital stock adjusts instantaneously.

A simple extension of the basic model is the inclusion of a lagged variable. This is used very frequently in macro or sectoral demand analysis and can be written as follows:

$$\log E_t = \log a + b \log Q_t + c \log P_t + d \log E_{t-1} \tag{3.28}$$

where E_{t-1} is one time-lagged E_t, all other variables being the same as in Eq. 3.27.

Equation 3.28 assumes that the total energy demand at period t is not only a function of the real price of energy and the level of output of the same period, it also depends on the level of energy demand of the previous period. The lagged model often explains the variation in energy demand better than the basic equation. This may be due to the fact that the level of activity and the level of energy consumption at any period in time are highly correlated to and influenced by those of the previous period. However, this is generally considered to be an ad hoc specification. In addition, the specification assumes a constant elasticity of demand and the specification may not be consistent with the demand theory. If the income elasticity is greater than one, the demand can be grossly overestimated as income increases. This can yield inconsistent results.

From Eq. 3.28, the short run and long run elasticities are found as follows:
Short run income elasticity $= b$, long run income elasticity $= b/(1 - d)$
Short run price elasticity $= c$, long run price elasticity $= c/(1 - d)$

Example Table 3.9 presents the annual per person gasoline consumption, per person GDP and gasoline prices for Iran between 1980 and 2005. Using a simple specification, analyse the gasoline demand econometrically.[11]

Answer: We use the demand model shown below.

$$\log Q_{ijt} = a_{ijt} + b_1 \log P_{ijt} + b_2 \log Y_{jt} + b_3 \log Q_{ijt-1} + e_t \tag{3.29}$$

---

[11] This is based on Bhattacharyya and Blake (2009).

3.8 Energy Demand Analysis Using the Econometric Approach

Table 3.9 Data for gasoline demand analysis in Iran

Year	Per person gasoline consumption bbl/1,000 pop	Real gasoline price Rial/bbl	Per person GDP 000 Rial (1998 prices)
1980	2.15	106,591.12	4,110.73
1981	1.87	207,247.67	4,163.56
1982	1.84	174,657.88	4,580.45
1983	2.32	145,861.07	4,857.69
1984	2.47	129,559.95	4,824.86
1985	2.60	124,126.70	4,831.59
1986	2.36	100,335.38	4,236.52
1987	2.40	78,562.76	4,044.26
1988	2.35	60,948.52	3,391.94
1989	2.48	51,913.43	3,515.60
1990	2.62	47,625.81	4,103.93
1991	2.77	67,707.43	4,509.91
1992	2.74	54,429.42	4,264.76
1993	3.03	44,414.28	4,236.09
1994	3.18	32,851.09	4,160.47
1995	3.08	40,390.54	4,123.66
1996	3.26	43,990.43	4,528.88
1997	3.49	46,074.86	4,803.59
1998	3.82	48,955.51	4,862.37
1999	3.92	67,246.33	4,903.20
2000	4.19	78,695.10	5,062.65
2001	4.45	86,396.98	5,153.91
2002	4.85	76,426.75	5,477.74
2003	5.32	68,941.26	5,787.77
2004	5.64	59,772.15	5,992.18
2005	6.14	67,056.42	6,181.10

Data source Gasoline consumption from IEA Energy Statistics of Non-OECD Countries database, Gasoline prices from OPEC Annual Statistical Bulletin (various issues), IMF World Economics Outlook database for population, deflator and GDP data

where

Q_{ijt} = per capita consumption of petroleum product i in country j in period t
P_{ijt} = the real price of petroleum product i in country j in period t
Y_{jt} = per capital real gross domestic product in country j in period t
e_t = the error term, normally distributed with zero mean and constant variance

b_1, b_2 and b_3 are parameters.

The data given in Table 3.9 is first transformed into its logarithmic values and using ordinary least square method, the parameters are estimated. The results are given in Table 3.10

The estimated equation is

$$\ln Q_t = -3.793 - 0.124\ln(P_t) + 0.652\ln N(Y_t) + 0.748\ln(Q_{t-1}) \quad (3.30)$$

Table 3.10 Results of gasoline demand analysis (t values within brackets)

Short run		Long run		Lagged demand coefficient b_3	Adjusted R-square
Price elasticity b_1	Income elasticity b_2	Price elasticity $b_1/(1-b_3)$	Income elasticity $b_2/(1-b_3)$		
−0.124	0.652	−0.494	2.589	0.748	0.975
(−3.595)	(4.403)			(9.868)	

Although single equation, reduced form models have been widely used, researchers have often tried to adopt more sophisticated formulations as well. A commonly used refinement is the use of distributed lag structure on price and income variables. As the adjustment process takes time in the energy sector, the response is expected to be distributed over a number of periods. A series of lagged explanatory variables are then added in the model to capture this effect.[12] However, as the number of lag periods increases, the degrees of freedom of the model reduce, leading to imprecise estimates of the lagged coefficients. This often limits the number of lag period in a model.

Another common variant of the econometric models is the use of fuel share model, especially for the transport sector where a number of fuels are used and the total demand of the components has to add up to one. These models use a two-step approach in which the total demand is first analysed followed by the analysis of shares of different fuels. This approach was useful in taking the inter-fuel substitution into account. However, Mehra and Bharadwaj (2000) indicate that these models assume that the prices and quantities are independent of each other and a change in the fuel price will not affect the total demand. The staged estimation process also assumes that the fuel shares do not affect the total demand.

Another tradition of the econometric modeling is the use of time series models for demand analysis. These models do not use any independent variables as explanatory variables but rely on the past behaviour of the dependent variable to find out how the demand can be explained. This is a sophisticated extrapolation method when these models are used for forecasting. A wide range of specifications and formulations can be found in the literature, including ARMA (Auto regressive Moving Average), ARIMA (Integrated Auto Regressive Moving Average), Box–Jenkins method, etc.[13] The time series models are more commonly used in short-term demand analysis and where long data series can be found (e.g. electricity demand, gas demand, etc.).

In the 1990s, the econometric approach was swept by a wave of "cointegration" analysis. Traditionally, the econometric analysis did not explicitly consider the issue of "stationarity" of the economic data. The assumption was that the data came from processes with constant means and variances. However, demand, prices

---

[12] See any econometrics textbook, e.g. Pindyck and Rubinfeld (1998) for further discussions on this subject.

[13] See any econometrics textbook, e.g. Pindyck and Rubinfeld (1998) for further discussions on this subject.

and income data used in the energy studies often show strong "trends that are changing stochastically over time" (Bentzen and Engsted 1993) and are therefore considered to be integrated. In such a case, the ordinary least square regression analysis will yield "spurious" results, implying that despite having no meaningful relationships among the variables, the R^2 can be high. The spurious result occurs due to the presence of trends and therefore leads to inaccurate results (Chan and Lee 1997).

To address the above problem, the first difference of the variables are used instead of the original variables but Bentzen and Engsted (1993) argue that the above adjustment only removes the long-run variation of the data, leaving the analysis to capture only the short-run effects. The cointegration method developed by Engle and Granger (1987) offered a solution to the problem. Their idea was that although the individual variables are non-stationary, their linear combinations may be stationary. In such cases, the variables are cointegrated. The appropriate analytical method in such a case changes to the following: first, the stationary and cointegration properties of the variables have to be examined. If the variables are non-stationary but cointegrated, the long-run effect can be estimated using cointegration analysis. Then using an error correction model (ECM) short-run effects and the speed of adjustment can be estimated (Bentzen and Engsted 1993).[14]

The above approach has been extensively used in the energy literature, leading to a proliferation of research outputs. However, this wave has also meant that the demand analysis following the econometric tradition has returned back to a single equation variety where the issues of inter-fuel substitution, technical change and changes in the supply have not found any place. Although academically appealing, such outputs have limited practical, policy implications because the results often did not show major differences compared to the simple analyses carried out earlier. Moreover, the statistical properties of the cointegration techniques have been questioned (see Harvey 1997) and it has been argued that the OLS produces super consistent results even in the case where the variables are non-stationary but cointegrated.

3.9 Conclusion

This chapter has provided a basic overview of energy demand and its analysis. The derived nature of energy demand influences energy use and its demand. We have seen how economic theory can be used to understand energy demand. We have learnt that simple indicators could be used to analyse the evolution of demand. Similarly, a simple framework has been introduced to explain the underlying factors that contribute to changes in demand. The chapter has also introduced the basic econometric formulation for energy demand analysis.

While the chapter focused on the aggregate demand, the approaches and frameworks discussed could be easily extended at the sector level. In fact, there are

---

[14] Again, this book does not enter into this.

two distinct traditions in the energy field: one is known as top-down approach where the focus remains on the aggregate level of analysis and the other is known as bottom-up approach where the overall demand is aggregated from the sector and sub-sector level analysis. The next chapter deals with this aspect in some detail.

Annex 3.1: Consumer Demand for Energy—The Constrained Optimization Problem

Consider that the utility function of a consumer can be written as

$$\text{Utility } u = U(X_1, X_2, X_3, \ldots, X_n) \tag{3.31}$$

The consumer has the budget constraint

$$I = p_1 X_1 + p_2 X_2 + \cdots + p_n X_n \tag{3.32}$$

For maximization of the utility subject to the budget constraint, set the lagrange

$$L = U(X_1, X_2, X_3, \ldots, X_n) - \lambda(I - (p_1 X_1 + p_2 X_2 + \cdots + p_n X_n)) \tag{3.33}$$

Setting partial derivatives of L with respect to $X_1, X_2, X_3, \ldots X_n$ and λ equal to zero, $n + 1$ equations are obtained representing the necessary conditions for an interior maximum.

$$\begin{aligned} \delta L/\delta X_1 &= \delta U/\delta X_1 - \lambda p_1 = 0; \\ \delta L/\delta X_2 &= \delta U/\delta X_2 - \lambda p_2 = 0; \\ &\vdots \\ \delta L/\delta X_n &= \delta U/\delta X_n - \lambda p_n = 0 \\ \delta L/\delta \lambda &= I - p_1 X_1 + p_2 X_2 + \cdots + p_n X_n = 0 \end{aligned} \tag{3.34}$$

From above,

$$(\delta U/\delta X_1)/(\delta U/\delta X_2) = p_1/p_2 \text{ or MRS} = p_1/p_2 \tag{3.35}$$

$$\lambda = (\delta U/\delta X_1)/p_1 = (\delta U/\delta X_2)/p_2 = \cdots = (\delta U/\delta X_n)/p_n \tag{3.36}$$

Solving the necessary conditions yields demand functions in prices and income.

$$\begin{aligned} X_1^* &= d_1(p_1, p_2, p_3, \ldots p_n, I) \\ X_2^* &= d_2(p_1, p_2, p_3, \ldots p_n, I) \\ &\vdots \\ X_n^* &= d_n(p_1, p_2, p_3, \ldots p_n, I) \end{aligned} \tag{3.37}$$

Annex 3.1: Consumer Demand for Energy—The Constrained Optimization Problem

An individual demand curve shows the relationship between the price of a good and the quantity of that good purchased, assuming that all other determinants of demand are held constant.

Annex 3.2: Cost Minimization Problem of Producers

Consider a firm with single output, which is produced with two inputs X_1 and X_2. The cost of production is given by

$$TC = c_1 X_1 + c_2 X_2 \tag{3.38}$$

This is subject to

$$St \; q_0 = f(X_1, X_2) \tag{3.39}$$

Write the Lagrangian expression as follows:

$$L = c_1 X_1 + c_2 X_2 + \lambda(q_0 - f(X_1, X_2)) \tag{3.40}$$

The first order conditions for a constrained minimum are:

$$\delta L / \delta X_1 = c_1 - \lambda \, \delta f / \delta X_1 = 0$$
$$\delta L / \delta X_2 = c_2 - \lambda \delta f / \delta X_2 = 0$$

From above,

$$c_1 / c_2 = (\delta f / \delta X_1)/(\delta f / \delta X_2) = \text{RTS}\,(X_1 \text{ for } X_2) \tag{3.41}$$

In order to minimize the cost of any given level of input, the firm should produce at that point for which the rate of technical substitution is equal to the ratio of the inputs' rental prices.

The solution of the conditions leads to factor demand functions.

Annex 3.3: Adaptive Price Expectation Model

Consider that Q_t is related to price expectation and not the actual price level in time t.

$$Q_t = a^* + b^* P_t^* + e_t^* \tag{3.42}$$

where P^* represents expected level of prices, not actual prices

A second relationship defines the expected level of P^*. It is assumed that in each time period, the expectation changes based on an adjustment process between the current observed value of P and the previous expected value of P^*. The relationship is

$$P_t^* - P_{t-1}^* = c(P_t - P_{t-1}^*) \tag{3.43}$$

$$\text{or } P_t^* = cP_t + (1-c)P_{t-1}^* \tag{3.44}$$

This implies that the expected price is a weighted average of present price and the previous expected level of price.

For econometric estimation, the above equation is rearranged as follows:

$$\begin{aligned}(1-c)P_{t-1}^* &= c(1-c)P_{t-1} + (1-c)^2 P_{t-2}^* \\ (1-c)^2 P_{t-2}^* &= c(1-c)^2 P_{t-2} + (1-c)^3 P_{t-3}^*\end{aligned} \tag{3.45}$$

Substituting and combing we obtain

$$P_t^* = c[P_t + (1-c)P_{t-1} + c(1-c)^2 P_{t-2} + \cdots] = c\sum (1-c)^s P_{t-s} \tag{3.46}$$

Substituting P_t^* in Q_t, we get

$$Q_t = a^* + b^*c \sum (1-c)^s P_{t-s} + e_t^* \tag{3.47}$$

$$Q_t = a^* + b^*c \sum_{s=0}^{\infty} (1-c)^s P_{t-s} + e_t^* \tag{3.48}$$

Letting $a = a^*$, $b = b^*c$, $w = (1-c)$, and $e_t = e_t^*$
$Q_t = a + b \sum (w)^s P_{t-s} + e_t$, the original geometric lag model
For econometric estimation, model is rewritten as:

$$Q_{t-1} = a^* + b^*c \sum (1-c)^s P_{t-s-1} + e_{t-1}^* \tag{3.49}$$

We calculate $Q_t - (1-c)Q_{t-1}$ to obtain

$$Q_t - (1-c)Q_{t-1} = a^*c + b^*cP_t + e_t^* - (1-c)e_{t-1}^* \tag{3.50}$$

$$Q_t = a*c + b*c\,P_t + (1-c)Q_{t-1} + u_t* \tag{3.51}$$

References

Ang BW (1997) Decomposition of aggregate energy intensity of industry with application to China, Korea and Taiwan. Energy Environ 8(1):1–11

Ang BW, Zhang FQ (2000) A survey of index decomposition analysis in energy and environmental studies. Energy 25:1149–1176

Ang BW (2004) Decomposition analysis for policymaking in energy: which is the preferred method? Energy Policy 32:1131–1139

References

Bentzen J, Engsted T (1993) Short- and long-run elasticities in energy demand: a cointegration approach. Energy Econ 15(1):9–16

Bhatia R (1987) Energy demand analysis in developing countries: a review. Energy J 8:1–33

Bhattacharyya SC, Blake A (2009) Domestic demand for petroleum products in MENA countries. Energy Policy 37(4):1552–1560

Bhattacharyya SC (2006) Renewable energies and the poor: Niche or Nexus. Energy Policy 34(6):659–663

Bohi D (1981) Analyzing demand behaviour: a study of energy elasticities. John Hopkins University Press, Baltimore

Chan HN, Lee SK (1997) Modelling and forecasting the demand for coal in China. Energy Econ 19:271–287

DTI (2007) Meeting the energy challenge: a white paper on energy. Department of Trade and Industry, London

Engle RE, Granger CW (1987) Cointegration and error-correction: representation, estimation and testing. Econometrica 55:251–276

Fisher FM, Kaysen C (1962) A study in econometrics: the demand for electricity in the United States. North-Holland, Amsterdam

Hartman RS (1979) Frontiers in energy demand modelling. Annu Rev Energy 4:433–466

Harvey AC (1997) Trends, cycles and autoregressions. Economic J 107:192–201

Helm D (2005) The assessment: The new energy paradigm. Oxford. Rev. Econ. Pol 21(1), pp. 1–18.

IEA (1997) Indicators of energy use and efficiency: understanding the link between energy and human activity, International Energy Agency, Paris (see http://www.iea.org/textbase/nppdf/free/1990/indicators1997.pdf)

IEA (1997) The link between energy and human activity. International Energy Agency, Paris (see http://www.iea.org/textbase/nppdf/free/1990/link97.pdf)

IEA (2003) Key world energy statistics 2003. International Energy Agency, Paris

IEA (2009) Key world energy statistics 2009. International Energy Agency, Paris

Laitner JA, DeCanio SJ, Coomey JG, Sanstand AH (2003) Room for improvement: increasing the value of energy modeling for policy analysis. Utilities Policy 11:87–94

Medlock III K (2009) Chapter 5: Energy demand theory. In: Evans J, Hunt LC (eds.) International handbook on the economics of energy. Edward Elgar, Cheltenham

Mehra M, Bharadwaj A (2000) Demand forecasting for electricity. The Energy Resources Institute, New Delhi (http://www.regulationbodyofknowledge.org/documents/044.pdf)

Munasinghe M, Meier P (1993) Energy policy analysis and modeling. Cambridge University Press, Cambridge

Pindyck RS (1979) The structure of world energy demand. The MIT Press, Cambridge

Pindyck RS, Rubinfeld DL (1998) Econometric models and economic forecasts. McGraw-Hill, Boston

Ryan DL, Ploure A (2009) Chapter 6: Empirical modeling of energy demand. In: Evans J, Hunt LC (eds) International handbook on the economics of energy. Edward Elgar, Cheltenham

Siddayao CM (1986) Energy demand and economic growth, measurement and conceptual issues in policy analysis. West View Press, London

Stevens P (2000) An introduction to energy economics. In: Stevens P (ed.) The economics of energy, vol. 1. Edward Elgar, Cheltenham. Also reproduced in two parts in the Journal of Energy Literature, vol. VI(2), December 2000, vol. VII(1) June 2001

Sun JW (1998) Changes in energy consumption and energy intensity: a complete decomposition model. Energy Econ 20(1):85–100

UN (1991) Sectoral energy demand studies: application of the end-use approach to Asian Countries. Energy Resources and Development Series, No. 33, United Nations

Ussanarassamee A, Bhattacharyya SC (2005) Changes in energy demand in Thai industry between 1981 and 2000. Energy 30(10):1845–1857

Wilson C, Dowlatabadi H (2007) Models of decision making and residential energy use. Annu Rev Environ Res 32:169–203

Worrel E, Ramesohl S, Boyd G (2004) Advances in energy forecasting models based on engineering economics. Annu Rev Environ Res 29:345–381

Further Reading

Bauer M, Mar E (2005) Transport and energy demand in the developing world. Energy Environ 16(5):825–844 (access to PDF file through DU Library)
Cooper JCB (2003) Price elasticity of demand for crude oil: estimates for 23 countries. OPEC Rev 27(1), 1–8 (see OPEC Review publisher website through DU Library for access to PDF file)
Dahl C (1994) A survey of oil product demand elasticities for developing countries. OPEC Rev Spring:47–85
Hannesson R (2002) Energy use and GDP growth. 1950–1997 OPEC Rev 26(3):215–233
Moroney JR (ed) (1997) Advances in the economics of energy and resources. Energy demand and supply, vol 10. Jai Press, London
Price L, de la Rue du Can S, Sinton J, Worrell E, Nan Z, Sathaye J, Levine M (2006) Sectoral trends in global energy use and greenhouse gas emissions. Lawrence Berkeley National Laboratory, CA (see http://ies.lbl.gov/iespubs/56144.pdf)
Watkins GC (1992) The economic analysis of energy demand: perspectives of a practitioner. In: Hawdon D (ed) Energy demand: evidence and expectations. Academic Press, London

Chapter 4
Energy Demand Analysis at a Disaggregated Level

4.1 Introduction

The previous chapter has presented an aggregated analysis of energy demand. A macro level analysis is useful to understand the general and global issues and policy directions. However, such analyses fail to capture the specific features and characteristics of each sector or sub-sectors. A sector-level analysis provides further detailed information and understanding useful for formulating specific energy policies related to, among others, energy conservation, fuel substitution and technology promotion.

Energy demand is not homogeneous across different uses and therefore, the demand determinants and the behavioural factors influencing them vary from one sector to another. Similarly, different energy carriers can be consumed interchangeably for certain uses and processes but not for others. The technological characteristics vary widely from one use to another across sectors. Macro demand analysis cannot treat these issues properly; this justifies disaggregated demand analysis independently of or in addition to a macro demand analysis.

This chapter deals with such a disaggregated or detailed analysis of energy demand at the sector or fuel level. It introduces the sectoral accounts for this purpose and presents the commonly used approaches for a detailed analysis.

4.2 Disaggregation of Demand

The final energy demand in the energy balance is generally split into a number of sectors: industry, transport, residential, commercial, agriculture and non-energy uses. However, very often because of data limitations, residential and commercial sectors are grouped together. In some tables, agriculture is combined with the residential-commercial sectors to give a big "other" sector.

Within each sector, further disaggregation is done to enhance homogeneity in consumption and demand behaviour. Usually the industrial sector is split into three major sub-sectors: mining, manufacturing and construction. Manufacturing is then broken down into various categories of industries to understand the demand pattern of energy intensive and non-intensive industries. The proportion of final energy consumed by the sector varies from one country to another depending on the degree of industrialization and stage of economic development of the country. Therefore, the detail will vary by country but the usual disaggregation of manufacturing industry is shown in Fig. 4.1.

Note that the energy consumption of the energy sector should not be included in the final energy consumption of industry since these energy sector own uses are accounted for in the transformation of primary energy into final energy. The electricity generated by industrial sector itself, known as captive power or self-generation, should also be included in the transformation sector and the consumption of this energy should be included in electricity consumption of industry.

For the transport sector, the final consumption of energy arises for different modes of transport, namely domestic air, water, rail and road transport. Within each mode, the need generally arises for two reasons—to transport passengers or goods. Each mode also uses different types of technologies for the transport activity and different fuels. Thus the energy demand is usually split according to the main transport modes and types of vehicles or according to the nature of transport activity. The usual disaggregation of the transport sector is shown in Fig. 4.2.

The consumption of all road vehicles is by definition allocated to transport, including agricultural and industrial trucks as well as commercial and government vehicles. The consumption of energy for international air and water transport should be excluded since this consumption is considered as exports.

For the residential and commercial sectors, the main difference arises from the urban versus rural setting of the consumer. The nature of the demand and often the

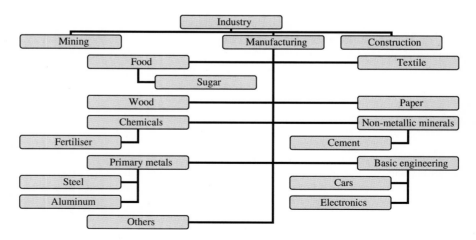

Fig. 4.1 Usual disaggregation of the industrial sector

4.2 Disaggregation of Demand

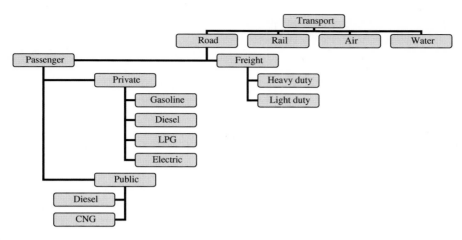

Fig. 4.2 Disaggregation of the transport sector

appliance holding and consumption pattern will vary substantially as a result. This is however often not captured in most analyses. Similarly, the demand pattern varies by income group or size of the commercial activity.

Clearly, the level of disaggregation is decided based on the availability of data required for the analysis, importance of the activities of a sector and subsector, the purpose of the study. Detailed sector level information requires additional sources of information. For the industrial sector, major sources like the IEA provides at the 2-digit level of industrial classification but for other sectors, data has to be gather from surveys and other published statistics. A separate set of accounts called sectoral energy accounts are created for this, which we consider next.

4.3 Sectoral Energy Accounting

As mentioned above, an overall energy balance provides a summary that is helpful in understanding the broad picture but it lacks information for carrying out a detailed investigation. An overall picture of the energy balance masks the finer details of consumption and demand patterns. Sectoral energy accounts attempt to remedy this problem by extending the final energy demand part of the energy balance. These accounts provide disaggregated information by sub-sector level, by process type and/or by end-uses.

Due to heterogeneity of energy consuming sub-sectors or end-uses among countries, there cannot exist only one single standardized framework for these accounts. The breakdown among energy products should be specific to each sector and should only take into account the most important energy products.

For example, to facilitate a detailed analysis of the transport sector, information is required by mode of transport, fuel used in each mode and by category of vehicles.

The matrix tabulating the consumption of various fuels by mode and vehicle category will be referred to as the sectoral energy account for the transport sector. A simple matrix is shown in Table 4.1, which contains imaginary data for the purpose of illustration.

The data necessary for these accounts comes from transport companies (rail, air, and bus) and other transport agencies. The data problem is generally less for rail and air transport. However, data for the road transport is a major problem due to its decentralized nature of demand, for which energy consumption statistics are not maintained. This has to be estimated through surveys and other means.

The sectoral energy account can be prepared for any sector of interest. For the industrial and residential sectors, specific surveys such as household energy survey or industrial energy consumption survey provide useful information. However, in most cases data from a number of sources will be required to get a complete picture. Wenzel et al. (1997) provide an example of a data sourcebook for the US residential sector. Part of the data will rely on recorded information (statistics from the historical database), part has to be extrapolated from surveys, and part has to be estimated on the basis of additional information on equipment and the pattern of use. The main problem with statistics comes from the possible overlaps or gaps in the definition of sub-sectors or end-uses. Extrapolation from surveys can give quite reliable information if the samples are well designed. Where estimates are used the reliability can be an issue.

Table 4.1 Example of a sectoral energy account for the transport sector (Mtoe)

Mode of transport	LPG	Gasoline	Jet fuel	Diesel	Total
Private passenger transport	45	1,335		245	1,625
Motorcycles		550			550
Cars	45	785		245	1,075
Public passenger transport	120	40	40	710	910
Tricycles	30	5			35
Taxis	90	5		10	105
Small buses		20		60	80
Buses				100	100
Inter city buses		10		490	500
Rail				50	50
Air			40		40
Freight transport	10	210		2,690	2,910
Light trucks	10	210		740	960
Medium trucks				720	720
Heavy trucks				1,100	1,100
Rail				30	30
Water				100	100
Total	175	1,590	40	3,650	5,455

A distinction is often made between the energy delivered to the consumer and the energy usefully consumed by her. The useful energy is the energy output of the appliances used by the final consumer to satisfy her requirement. But there is no direct measure of useful energy and hence, the amount of useful energy is derived by applying additional factors to the final energy consumption. These factors reflect the average or estimated efficiency values for conversion by the final appliances. This presupposes a thorough knowledge of appliance stock and their efficiencies.

4.4 Analysis at the Sectoral Level

The analytical approaches considered in Chap. 3 are also applicable at the disaggregated sector level. However, to take care of the specific features of each sector, the emphasis of the analytical tools changes, which in turn leads to adjustments or modifications or new developments. In what follows, a brief account of such modifications or specific features is presented without repeating the basic details presented earlier in Chap. 3.

4.4.1 Industrial Energy Demand Analysis

Industrial energy demand is a major constituent of final energy demand of many countries. According to IEA (2008), the global industrial energy use in 2005 was 116 EJ, which represented an increase of 21% compared to the level in 1990. Generally, the heavy industries (chemical and petrochemicals, iron and steel, paper and pulp, non-metallic minerals and non-ferrous metals) account for two-thirds of the industrial energy demand (see Table 4.2) while light industries consume the rest (IEA 2007). Three major consumers the US, China and Western Europe account for about one half of the global industrial energy demand (IEA 2007). China has accounted for about 80% of the increase in industrial production for the past 25 years and is now the largest industrial energy consumer in the world.

Energy is consumed by industry as an intermediate product to produce other commodities. Although some energy consumption does not bear direct relationship with production, these indirect uses are in most cases insignificant or negligible.

4.4.1.1 Decomposition of Energy Demand and Intensities

The method discussed earlier can be applied here directly. Both the overall industrial energy demand and industrial energy intensities can be decomposed using Laspeyres, Paasche or Divisia methods. The examples below explain the methodology for this type of work. Similarly, the physical indicators can also be used.

Table 4.2 Final energy use in industry, 2004 (EJ)

Industry	China	USA	Western Europe	Japan	World
Chemical and petrochemical	4.53	7.65	5.45	2.35	33.62
Iron and steel	7.11	1.46	2.74	1.89	21.44
Non-metallic minerals	4.53	1.07	1.70	0.31	10.61
Paper and pulp	0.66	2.24	1.52	0.37	6.45
Non-ferrous metals	0.93	0.52	0.57	0.08	4.21
Food and tobacco	0.77	1.24	1.30	0.18	5.98
Machinery	1.11	0.85	0.82	0.36	4.25
Textile and leather	0.91	0.26	0.40	–	2.17
Wood and products	0.13	0.48	0.23	–	1.36
Transport equipment	0.34	0.40	0.37	–	1.28
Mining	0.35	0.09	0.13	0.03	1.81
Construction	0.39	0.08	0.35	0.15	1.41
Others	0.72	1.09	1.62	0.95	18.65
Total	22.48	17.43	17.20	6.66	113.25

Source IEA (2007), p. 40

Table 4.3 Data for the factor analysis example

	1981			2000		
	En cons (ktoe)	Share of VA (%)	En intensity (ktoe/million baht)	En cons (ktoe)	Share of VA (%)	En intensity (ktoe/million baht)
Mining	71	0.0274	0.0102	85	0.0137	0.0060
Construction	125	0.1749	0.0028	149	0.0729	0.0020
Manufacturing	4,293	0.7977	0.0211	16,208	0.9134	0.0170
Food and beverage	2,171	0.2343	0.0363	4,865	0.1457	0.0320
Textiles	391	0.1939	0.0079	1,139	0.1772	0.0062
Wood and furniture	51	0.0456	0.0044	124	0.0128	0.0093
Paper	129	0.0287	0.0176	701	0.0296	0.0227
Chemical	240	0.0450	0.0209	2,124	0.0863	0.0236
Non-metallic	968	0.0420	0.0901	3,936	0.0454	0.0831
Basic metal	118	0.0209	0.0221	820	0.0131	0.0600
Fabricated metal	82	0.0203	0.0158	948	0.0296	0.0307
Others	143	0.1669	0.0034	1,551	0.3738	0.0040

Example The value addition of Thai industries grew from 255.45 billion baht (in 1988 constant prices) in 1981 to 1,043.185 billion in 2000. Table 4.3 provides the relevant details about energy demand, intensities and changes in the value additions by industry.

Using the above information, analyse the influence of activity, structure and intensity on energy demand in the Thai industrial sector between 1981 and 2000.[1]

---

[1] This is a simplified version based on Ussanarassamee and Bhattacharyya (2005). Please see the original work for a more detailed exposition.

4.4 Analysis at the Sectoral Level

Table 4.4 Results of decomposition (ktoe)

1981–2000	Activity effect	Structural effect	Intensity effect	Total	Actual change	Residue
Mining	218.94	−35.55	−29.32	154.08	14	−140.08
Construction	385.46	−72.92	−37.43	275.12	24	−251.12
Manufacturing	13,238.39	622.95	−827.01	13,034.33	11,915	−1,119.33
Food and bev	6,694.75	−821.29	−254.77	5,618.68	2,694	−2,924.68
Textile	1,205.73	−33.76	−85.73	1,086.24	748	−338.24
Wood	157.27	−36.64	56.81	177.45	73	−104.45
Paper	397.80	3.96	37.55	439.31	572	132.69
Chem	740.09	219.98	31.38	991.45	1,884	892.55
Non-metal	2,985.04	77.68	−75.77	2,986.95	2,968	−18.95
Basic metal	363.88	−44.15	202.85	522.58	702	179.42
Fabricated	252.86	37.65	77.10	367.61	866	498.39
Others	440.97	177.22	26.61	644.80	1,408	763.20

Answer Using the Laspeyres decomposition method, changes in energy demand between 1981 and 2000 are determined following the method indicated Chap. 3. Results are presented in Table 4.4.

The process followed is as follows:

For activity effect, changes in the level of activity between 1981 and 2000 are considered, keeping the intensity and share of the sector in value addition unchanged in the initial year values. For example, for mining, this amounts to = (1,043.185 − 255.45) * 1,000 * 0.0274 * 0.0102 = 218.94 ktoe. Note that the GDP is given in billions of baht and this required a conversion factor of 1,000 to bring it to millions. This implies that if the activity would have changed alone, the energy demand for the mining activities would have increased by about 219 ktoe.

For the structural effect, the structural change (i.e. the change in the share of value addition of the sector) within the period is considered while keeping the other two factors unchanged. In the case of mining this amounts to = 255.45 * 1,000 * (0.0137 − 0.0274) * 0.0102 = −35.55 ktoe

This suggests that the share of the mining activity in the industrial output has reduced and if this only had changed, the energy demand would have reduced by 35 ktoe.

Finally, for the intensity effect, we look at the changes in energy intensity within the period and keep the other factors at their initial values. For the mining industry, this can be written as = 255.45 * 1,000 * 0.0274 * (0.0060 −0.0102) = −29.32 ktoe.

This suggests that the intensity in the mining industry has reduced and this reduced intensity would have reduced the energy demand by 29 ktoe between 1981 and 2000 if other things did not change.

Example Continuing with the previous example of Thai industry, energy intensity changed from 17.57 toe per million baht (1988 prices) in 1981 to 15.76

Table 4.5 Intermediate outputs of the analysis

1981–2000	Share of energy		Calculations for decomposition analysis			
	1981	2000	Log mean weight	Normalised weight (W^*)	Structural factor $W^*\ln(S_t/S_0)$	Intensity factor $W^*\ln(EI_t/EI_0)$
Mining	0.0158	0.0052	0.0095	0.0098	−0.0068	−0.0052
Construction	0.0278	0.0091	0.0167	0.0172	−0.0151	−0.0061
Food and bev	0.4836	0.2959	0.3821	0.3929	−0.1867	−0.0490
Textile	0.0871	0.0693	0.0778	0.0800	−0.0072	−0.0198
Wood	0.0114	0.0075	0.0093	0.0096	−0.0121	0.0072
Paper	0.0287	0.0426	0.0352	0.0362	0.0011	0.0093
Chem	0.0535	0.1292	0.0858	0.0883	0.0574	0.0108
Non-metal	0.2156	0.2394	0.2273	0.2337	0.0180	−0.0191
Basic metal	0.0263	0.0499	0.0368	0.0379	−0.0177	0.0379
Fabricated	0.0183	0.0577	0.0343	0.0352	0.0133	0.0234
Others	0.0319	0.0943	0.0575	0.0592	0.0477	0.0101

Table 4.6 Results of the analysis

1981–2000	D_{str}	D_{EI}	D_{tot}
Mining	0.9932	0.9948	0.9881
Construction	0.9850	0.9939	0.9790
Food and bev	0.8297	0.9521	0.7900
Textile	0.9928	0.9804	0.9733
Wood	0.9879	1.0072	0.9950
Paper	1.0011	1.0093	1.0104
Chem	1.0591	1.0109	1.0706
Non-metal	1.0182	0.9811	0.9990
Basic metal	0.9824	1.0386	1.0203
Fabricated	1.0134	1.0236	1.0373
Others	1.0489	1.0101	1.0595
Total	0.8975	0.9994	0.8969

toe/million baht in 2000. Determine the influence of changes in industry-level energy intensity and industry structure between 1981 and 2000 on the overall industrial energy intensity change.

Answer The share of energy consumption by industry is first calculated for 1981 and 2000. This is used to determine the weights to be used in the log-mean divisia index. The weight is normalised to achieve a sum of 1. The normalised weight is then used to calculate the structural and intensity factor (see Table 4.5).

The final results are obtained by converting the structural and intensity factors to their exponential values. The multiplication of these decomposed factors yields the total effect. It can be observed that the total aggregate effect is fully explained by this method and there is no residue in this case. Table 4.6 gives the final results.

In addition, it is also possible to use a hierarchical approach where some activities are disaggregated while others are not. For example, in the industry sector, mining and construction can be analysed at an aggregated level while the

4.4 Analysis at the Sectoral Level

manufacturing industry could be analysed at a more detailed level. The extension of the methodology is presented below and an example clarifies the finer points.

The level of disaggregation affects the results of decomposition. The more disaggregated the group, the more relevant and reliable is the measurement. But the limit for disaggregation is given by data availability. In such a case, a hierarchical measurement of the effects is done.

The structural effect identified with aggregated analysis can be further split into sub-structural effect and sub-intensity effect. The total structural effect is obtained by the multiplication of macro structural effect and sub-structural effect. The real intensity effect is obtained by the multiplication of macro sectoral intensity effect (for sectors without sub-sectors) and sub-sectoral intensity effect (for sectors with sub-sectors).

In such a case, the extension of the basic energy intensity equation takes the following form:

$$\text{EI} = \sum_i \sum_j \frac{e_{ij}}{Q_{ij}} \frac{Q_{ij}}{Q_i} \frac{Q_i}{Q} = \sum_i S_i \sum (\text{SEI}_{ij} \text{SS}_{ij}) \tag{4.1}$$

where e_{ij} = energy consumption in subsector j of sector i; Q_{ij} = activity of subsector j in sector i; SEI_{ij} = subsectoral energy intensity in subsector j of sector i; SS_{ij} = subsectoral share of subsector j in sector i. Other variables have same meaning as before.

Differentiating Eq. 4.1 yields

$$\frac{1}{\text{EI}} \frac{d\text{EI}}{dt} = \sum_i \frac{W_i}{S_i} \frac{dS_i}{dt} + \sum_i W_i \left[\sum_j \frac{w_{ij}}{s_{ij}} \frac{ds_{ij}}{dt} + \sum_j \frac{w_{ij}}{e_{ij}} \frac{d\text{EI}_{ij}}{dt} \right] \tag{4.2}$$

where S_i = sectoral share at the overall level; s_{ij} = subsectoral share of sub-sector j in sector I; W_i = weight at the sectoral level; w_{ij} = weight at the subsectoral level; EI_{ij} = energy intensity of sub-sector j in sector i.

Equation 4.2 can be rewritten as

$$\frac{d\text{EI}}{\text{EI}} = \sum_i \frac{W_i}{S_i} dS_i + \sum_i \sum_j \left[W_i \frac{w_{ij}}{s_{ij}} ds_{ij} + \sum_i \sum_j W_i \frac{w_{ij}}{e_{ij}} de_{ij} \right] \tag{4.3}$$

Integration of Eq. 4.3 between 2 years in discrete form results in the following equation:

$$\ln \left(\frac{\text{EI}^T}{\text{EI}^0} \right) = \sum_i W_i \ln \left(\frac{S_i^T}{S_i^0} \right) + \sum_i \sum_j W_i w_{ij} \ln \left(\frac{s_{ij}^T}{s_{ij}^0} \right) + \sum_i \sum_j W_i w_{ij} \ln \left(\frac{e_{ij}^T}{e_{ij}^0} \right) \tag{4.4}$$

The first term measures the structural effect at the upper level (i.e. sectoral level), the second term measures the intra-sectoral structural effect and the third term measures the intensity effect (which is also called the real intensity effect).

Table 4.7 Energy and activity details at sectoral level

	T_0 VA	T_0 EC	T_1 VA	T_1 EC	T_2 VA	T_2 EC
Agriculture	20	25	30	10	15	20
Commercial	30	40	50	20	25	30
Industry	60	80	100	50	65	80
Total	110	145	180	80	105	130

Table 4.8 Sub-sectoral details for industry

	T_0 VA	T_0 EC	T_1 VA	T_1 EC	T_2 VA	T_2 EC
Sub-sector 1	30	40	50	10	15	18
Sub-sector 2	20	25	30	20	25	30
Sub-sector 3	10	15	20	20	25	32
Total	60	80	100	50	65	80

Example Table 4.7 presents value added (million dollars) and energy consumption (Mtoe) for various sectors for 3 years (T_0, T_1 and T_2). Table 4.8 provides the break up of industrial energy consumption and value additions. Required is decomposition of energy intensity between T_0 and T_2.

Value addition is in million dollars in 1990 prices and energy consumption in Mtoe.

Answer For the decomposition, first sectoral shares and energy intensities have to be calculated. Table 4.9 provides the results. Using the information in Table 4.10, the necessary calculations are performed for decomposition. Results are given in Table 4.11.

4.4.1.2 Econometric Approach

As indicated in the previous chapter, the econometric analysis has undergone a significant evolution over the past three decades. Although single equation reduced form analysis played its role, the industrial sector witnessed a tremendous effort in terms of application of the flexible functional form called the translog model. In the 1970s this was pioneered by Brendt and Wood (1975) and there was a proliferation of application of this model in both developed and developing world until mid-1980s. The details of the translog function are presented in Box 4.1 for the interested readers. The preference for this functional form derived from the theoretical underpinning of the function, the flexibility of avoiding pre-specification of any particular relationships, and the imposition of minimum restrictions on the parameters. However, the disadvantages of this function include: (a) local approximation of the demand that may not be plausible globally, (b) loss of degrees of freedom, and (c) complicated estimation techniques (Wirl and Szirucsek 1990). Further, many of them relied on pre-1970 data, thereby missed the opportunity to consider the sudden price changes in the 1970s. In addition, the static translog model does not describe the adjustment process to the long-term.

4.4 Analysis at the Sectoral Level

Table 4.9 Shares, intensities and weights for decomposition calculations

	T_0		T_1		T_2		Average weights								
	Shares	EI	EI	W	W		Shares	EI	W	Shares	EI	W	T_1-T_0	T_2-T_0	T_2-T_1
Agri	0.18	1.25	0.17	0.17	0.33	0.13	0.14	1.33	0.15	0.15	0.14	0.16
Com	0.27	1.33	0.28	0.28	0.40	0.25	0.24	1.20	0.23	0.26	0.24	0.25
Ind-1	0.50	1.33	0.50	0.50	0.20	0.20	0.23	1.20	0.23	0.35	0.21	0.36
Ind-2	0.33	1.25	0.31	0.30	0.67	0.40	0.38	1.20	0.38	0.36	0.39	0.34
Ind-3	0.17	1.50	0.19	0.20	1.00	0.40	0.38	1.28	0.40	0.29	0.40	0.29
Ind	0.55	1.33	0.55	0.56	0.50	0.63	0.62	1.23	0.62	0.59	0.62	0.58

Table 4.10 Results of energy intensity decomposition

	T_1-T_0 Struc eff	T_2-T_1 Intrs st eff	T_2-T_0 RE eff	Struc eff	Intra st eff	RE eff	Struc eff	Intrs st eff	RE eff
Agri	−0.0129		−0.1966	−0.0215		0.1933	−0.0393		0.0105
Com	0.0048		−0.3166	−0.0371		0.2641	−0.0344		−0.0267
Ind-1		0.0000	−0.3907		−0.1019	0.2361		−0.1636	−0.0223
Ind-2		−0.0221	−0.1318		0.0597	0.1413		0.0287	−0.0082
Ind-3		0.0315	−0.0701		0.1622	0.0612		0.1433	−0.0272
Ind	0.0108			0.0671			0.0739		
Total	0.0027	0.0094	−1.1056	0.0086	0.1200	0.8960	0.0001	0.0085	−0.0738
Exponential	1.0027	1.0095	0.3310	1.0086	1.1275	2.4498	1.0001	1.0085	0.9288
Total effect	0.3350			2.7860			0.9369		

Table 4.11 Example of calculations for equivalent cars for gasoline vehicles

Vehicle type	No. of vehicles	Specific cons (km/l)	Mileage (km)	Yearly cons (Ml)	Unit consumption (kl/vehicle)	Equivalent car
Cars	12,000	14	25,000	21.43	1.79	1.00
Jeeps	1,000	9	20,000	2.22	2.22	1.24
Motorcycles	100,000	40	10,000	25.00	0.25	0.14

Box 4.1: Translog Cost Function

The translog cost function is considered to be the second order approximation of an arbitrary cost function. It is written in general form as follows:

$$\ln C = \alpha_0 + \sum_i \alpha_i \ln P_i + 0.5 \sum_i \sum_j \gamma_{ij} \ln P_i \ln P_j$$
$$+ \alpha_Q \ln Q + 0.5 \gamma_{QQ} (\ln Q)^2 + \sum_i \gamma_{Qi} \ln Q \ln P_i \qquad (4.5)$$

where C = total cost, Q is output, P_i are factor prices, i and j = factor inputs. This cost function must satisfy certain properties:

- Homogeneous of degree 1 in prices.
- Satisfy conditions corresponding to a well-behaved production function.
- Cost function is homothetic (separable function of output and factor prices) and homogeneous.

Accordingly, the following parameter restrictions have to be imposed:

$$\sum \alpha_i = 1$$

$$\gamma_{ij} = \gamma_{ij}, i \neq j$$

$$\sum_i \gamma_{ij} = \sum_j \gamma_{ij} = 0$$

$$\sum_i \gamma_{Qi} = 0$$

$$\gamma_{Qi} = 0 \quad \text{and}$$

$$\gamma_{QQ} = 0 \qquad (4.6)$$

The derived demand functions can be obtained from Shepherd's lemma

$$X_i = \delta C / \delta P_i \qquad (4.7)$$

Although these functions are non-linear in the unknown parameters, the factor cost shares ($M_i = P_i X_i / C$) are linear in parameters.

$$M_i = \alpha_i + \sum_j \gamma_{ij} (\ln P_j) \text{ for } i = \text{factor inputs}, j = \text{factor inputs}, i \neq j \qquad (4.8)$$

These share equations are estimated to obtain the parameters. Only $n-1$ such equations need to be estimated as the shares must add to 1.

The own price elasticity of factor demand is obtained as follows:

$$E_{ii} = \partial \ln X_i / \partial \ln P_i \qquad (4.9)$$

$$X_i = \frac{C}{P_i} \left(\alpha_i + \sum_j \gamma_{ij} (\ln P_j) \right) \qquad (4.10)$$

$$\ln X_i = \ln C - \ln P_i + \ln \left(\alpha_i + \sum \gamma_{ij} \ln P_j \right)$$
$$= \ln C - \ln P_i + \ln M_i \qquad (4.11)$$

$$\partial \ln X_i / \partial \ln P_i = \frac{\partial \ln C}{\partial \ln P_i} - 1 + \frac{\gamma_{ii}}{M_i}$$
$$E_{ii} = M_i + \frac{\gamma_{ii}}{M_i} - 1 \qquad (4.12)$$

$$E_{ii} = (M_i^2 - M_i + \gamma_{ii}) / M_i \qquad (4.13)$$

The cross-price elasticity can be derived similarly as

$$E_{ij} = (\gamma_{ij} + M_i M_j) / M_i \qquad (4.14)$$

Allen partial elasticity of substitution is given by:

$$\sigma_{ij} = (\gamma_{ij} + M_i M_j) / M_i M_j \qquad (4.15)$$

Source Pindyck (1979).

Parallel to the developments in the translog approach, the use of multinomial logit models became popular in the energy studies. The logit model is not derived from the utility maximisation theory but derives its appeal from its interesting properties (Pindyck 1979; Urga and Walters 2003):

- It is relatively easy to estimate.
- It ensures that the outcomes are non-negative and add to one.

- As the share of a component becomes small, it requires increasingly large changes to make it smaller.
- Flexible for incorporating a dynamic structure.

> **Box 4.2: Logit Model Description**
>
> The logit model for fuel share, S_i, can be written as
>
> $$\frac{Q_i}{Q_T} = S_i = \frac{\exp(f_i)}{\sum_{j=1}^{n} \exp(f_j)} \quad (4.16)$$
>
> where Q_i is the quantity of fuel i, $Q_T = \sum Q_i$, and f is the function representing consumers preference choices.
>
> The share equation for any two fuels can be written as
>
> $$\log\left(\frac{Q_i}{Q_j}\right) = \log\left(\frac{S_i}{S_j}\right) = f_i - f_j \quad (4.17)$$
>
> As the sum of the shares adds up to one, only $(n-1)$ equations need to be estimated simultaneously.
>
> For estimation purposes, a specific functional form has to be chosen. This is often done arbitrarily and we use a linear specification of relative fuel pries, income and temperature as given below.
>
> $$f_i = a_i + b_i \tilde{P}_i + c_i Y + d_i T \quad (4.18)$$
>
> where $\tilde{P}$ is (P_i/P_E)—ratio of price of fuel i to the aggregate fuel price P_E, Y is income, T is the temperature.
>
> Substitution of (3) in (2) yields the equations to be estimated:
>
> $$\log\left(\frac{S_i}{S_n}\right) = (a_i - a_n) + b_i \tilde{P}_i - b_n \tilde{P}_n + (c_i - c_n)Y + (d_i - d_n)T \quad (4.19)$$
>
> where $i = 1, 2, 3, \ldots, (n-1)$
>
> A dynamic version of the equation can be easily written by including the lagged shares in the functional form
>
> $$f_i = a_i + b_i \tilde{P}_i + c_i Y + d_i S_{i,t-1} \quad (4.20)$$
>
> The equations for dynamic estimation in that case turns out as
>
> $$\log\left(\frac{S_i}{S_n}\right) = (a_i - a_n) + b_i \tilde{P}_i - b_n \tilde{P}_n + (c_i - c_n)Y + d_i S_{i,t-1} - d_n S_{n,t-1} \quad (4.21)$$
>
> where $i = 1, 2, 3, \ldots, (n-1)$
>
> *Source* Pindyck (1979).

4.4 Analysis at the Sectoral Level

In the case of industrial energy use this has been used to analyse fuel shares or market shares of fuels. Some details about the logit model are presented in Box 4.2.

In recent times, the analysis has returned to the use of single equation variety where co-integration and error correction models are being used. However, it appears that there is slow down in the econometric analysis of industrial energy demand and most of the studies are at an aggregated level thereby ignoring the specificites of each sub-sector.

4.4.1.3 Techno-Economic Approach

This method is often applied in energy demand analysis either by itself or in conjunction with econometric methods. There are several advantages to this approach as compared with the econometric approach. First, it can easily incorporate engineering and technical characteristics of energy consumption into modeling, which cannot be done by the econometric method. Second, it can also be undertaken even when no reliable time series data on sectoral economic and energy demand are available for a sufficiently long period of time.

For example, IEA (2007, 2008) and provide detailed studies of this type for the industry sector. Take the example of iron and steel industry, which is the second largest energy consumer of industrial energy globally. Iron and steel industry uses a number of technologies (see Box 4.3) for steel making and a number of processes to reach the final products. The complexity of the industry varies depending on the activities undertaken and the effectiveness of energy use varies by technology, inputs used and the plant size and configuration. The energy use per tonne of steel varies as a result of differences in these technical aspects and IEA (2008) indicates that if the best available technology is applied

---

Box 4.3: Steel-Making Technologies

In the steel industry, a number of technologies are used: Blast furnace (bf)—basic oxygen furnace (bof) process, direct reduction (DR) technologies, smelting reduction technologies and electric arc furnaces.

The most common steel making technology is the Bf-Bof Route. Coke is used in Blast Furnace (BF) both as a reducing agent and as a source of thermal energy. It involves reduction of ore to liquid metal in the blast furnace and refining in convertor to form steel.

Direct reduction processes are of significance and form an alternative to the conventional Blast furnace. Solid sponge iron (direct reduced iron or DRI) is produced by removing oxygen from the ore. The technique of Direct reduction varies according to the type of reducing agents employed and the metallurgical vessel in which they are reduced. The reducing agents used normally are: Gas reduction and Solid reducing agents. Gas reduction is more commonly used.

Smelting reduction usually produces hot metal from ore in two steps. Ores are partly reduced in the first step and then final reduction and melting takes place in the second stage. The commonly used smelting processes are: coal reduction process (corex), direct iron smelting reduction (Dios), etc.

The electric arc furnace, as the name suggests, is a furnace in which heat is generated with the aid of electric arc produced by graphite electrodes. The main components of the electric arc furnace are the furnace shell with tapping device and work opening, the removable roof with the electrodes and a tilting device. This process uses electricity as the major source of energy.

In terms of energy requirement, steel making processes are generally energy intensive but the older processes using blast furnace technology are less efficient. Coking coal is the main source of energy in these processes. In direct reduction process, natural gas is used as the source of energy and reducing agent. Electric arc furnaces are highly intensive in electricity use. According to IEA (2007), 62% of steel is produced from pig iron, about 5% using direct reduction approaches and the rest from scrap.

in all steel producing countries, 4.5 EJ can be saved, which in turn could reduce 340 Mt of CO_2.

Similarly, there are specific technological features for other energy intensive industries (see Box 4.4). These specific features are generally taken into consideration where relevant in the techno-economic analyses.

The general approach followed in the end-use method is to identify the homogenous energy demands and their determinants. Depending on the nature of the industry, these end-use demands could be analysed for industrial processes and for buildings. Normally, the process energy demand would be significant for energy intensive industries while the building-related energy demand could be important for labour intensive industries. This approach is shown schematically in Fig. 4.3.

These models work on snap-shots at a specific time to calibrate the model and perform "what-if" or scenario-based exercises for policy analysis. Although they do not require time-series information in general, these models are highly data intensive and the availability of such information and their reasonableness can affect the overall performance of the model.

Box 4.4: Other Energy Intensive Industries

In cement industry, two processes commonly used are dry process and wet process and each has different energy requirement. The dry process requires less energy per tonne compared to the older wet process. A shift towards dry process would result in a reduction in energy intensity. Similarly, an increase in the size of cement producing furnaces would reduce energy requirement.

A distinction can also be made between the energy used in the processes (at the level of production which is affected by rate of utilization of capacity), and the energy used in the buildings (for lighting, heating or air conditioning, etc.). To understand the possibility of energy substitution, the energy used may be broken down into thermal uses and mechanical uses. The thermal use can then be further broken down by temperature ranges: high temperature (such as in furnaces), medium temperature (as in boiler) and low temperature (in processing food or textile mills, etc.).

In the nonferrous metal industry, aluminum production is an energy intensive activity. An increase in aluminum production capacity or production in a period can easily affect industrial energy demand.

In the chemicals industry, some activities are highly energy intensive. For example, production of chlorine and caustic soda is around three times energy intensive and ammonia-based fertilizers are nearly four times as energy intensive as chemicals production as a whole. Natural gas can be used as feedstock in fertilizer production and depending on the size of the plant, the efficiency varies.

4.4.2 Energy Demand Analysis in the Transport Sector

The final energy consumption in the transport sector includes the energy used for domestic air and water transport, rail and road transport. The consumption of all road vehicles is by definition allocated to transport, including those used in agriculture, industry, commercial sectors and government offices. It however excludes the consumption of energy for international air and water transport.

Different approaches are used to analyse energy demand in the transport sector. Many studies, mostly using econometric approach, just focus on aggregate fuel demand in the transport sector—focusing mainly on gasoline and diesel demand. A simple model of this type is indicated first. Then another style of models where the derived nature of the demand is recognized is presented. Finally, a more detailed end-use approach is outlined.

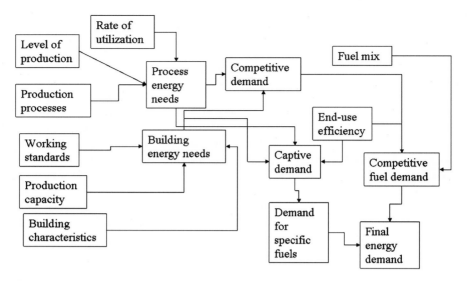

Fig. 4.3 Energy demand analysis using the techno-economic approach. *Source* UN (1991) and Bhattacharyya and Timilsina (2009)

4.4.2.1 A Simple Transport Fuel Demand Model[2]

Consider that two substitutable fuels—diesel and gasoline—are used for transport purposes. The market share approach is used to estimate the demand. The model has two components: first, the total fuel demand for transport is estimated; then, the demand for individual fuels is estimated using their market share.

The total demand for diesel and gasoline is considered to be a function of weighted average price of fuels in real terms, real per capita GDP and the total consumption of both fuels in the previous year. The equation in log-linear form can be written as

$$\ln TC = a_0 + a_1 \ln P + a_2 \ln GDP + a_3 \ln TC - 1, \qquad (4.22)$$

where $P = (DC/TC) \cdot DP + (GC/TC) \cdot GP$, TC = total consumption of diesel and gasoline, P is the average price, GDP is the real per capita GDP, DC is the diesel consumption, GC is the gasoline consumption, DP is the price of diesel and GP is the price of gasoline.

The market share of a fuel is assumed to be a function of its real price, the price of the substitute fuel, the per capita GDP and the share of the fuel in the previous year. The equation for gasoline can be written as follows:

---

[2] This is based on Miklius et al. (1986).

4.4 Analysis at the Sectoral Level

$$\ln(GC/TC) = b_0 + b_1 \ln DP + b_2 \ln GP + b_3 \ln GDP + b_4 \ln(GC/TC) - 1 \tag{4.23}$$

As there are two fuels in this case, the total share has to be 100. The diesel share is thus obtained

$$DC/TC = 100 - \exp[\ln(GC/TC)] \tag{4.24}$$

By estimating these equations the fuel demand can be analysed.

4.4.2.2 Energy Demand Analysis Through Vehicle Ownership Modeling

Energy demand in this type of analysis is based on the following relationship (Dargay and Gately 1997; Medlock and Soligo 2002; Bouachera and Mazraati 2007):

$$F = C \times U/SC \tag{4.25}$$

where F is fuel consumption (million litres), C is the number of cars (million), U is the annual car usage per car (km/year/car), and SC is the specific fuel consumption (km/litre).

Various approaches have been used to determine fuel demand using the above approach. Early studies such as those by Adams et al. (1974) and Pindyck (1979) have attempted to formulate transport fuel demand taking these variables into account. The fuel demand is obtained as a product of the above three variables, each of which is estimated using a function of other explanatory variables. Accordingly, the demand is not obtained from the utility or cost functions or from the perspective of any optimisation process (Pindyck 1979, p. 61). Some other studies have used simplified assumptions about car usage and specific fuel consumption while the car ownership is estimated using an econometric model (Dargay and Gately 1997).

It is generally believed that the car ownership in the long-run follows a S-shaped curve when plotted against per person income. Various functional forms have been used to analyse car ownership. These can be grouped under different categories (Bouachera and Mazraati 2007; Han 2001): time trend models which can be captured through the following equation

$$C_t = \frac{S}{\left[1 + \frac{k}{m}\left(1 + \frac{at}{n}\right)^{-n}\right]^m} \tag{4.26}$$

where C is car ownership level at any time t, S is the saturation level, k, m and n are parameters that determine the shape of the curve, a is a constant. Depending on the values of m and n, the function takes different forms.

Any S-shaped curve implies that at low income levels the car ownership is low but as the income increases the car ownership grows fast until it reaches a saturation level. A number of functional forms can be used—logistic, Gompertz and quasi-logistic models.

The following equation describes the logistic function

$$C_t = \frac{S}{1 + a \cdot e^{b \cdot \text{GDP}}} \quad (4.27)$$

where C is car ownership (say vehicles per 1,000 people), GDP is per capita income (in real terms), S is the saturation level (vehicles per 1,000 people), a and b are parameters (negative) defining the shape or curvature of the function.

The Gompertz model can be written as follows:

$$C_t = S \cdot e^{a \cdot e^{b \cdot \text{GDP}_t}} \quad (4.28)$$

The long-run elasticity of vehicle ownership with respect to GDP (or income elasticity of vehicle ownership) is not constant and for the Gompertz model is given by

$$e_t = a \cdot b \cdot \text{GDP}_t e^{b \cdot \text{GDP}_t} \quad (4.29)$$

The above elasticity is always positive because a and b are negative. The elasticity is zero for a zero GDP and increases with an increase in GDP. It reaches its maximum and then declines as the saturation is reached. Figure 4.4 indicates the shape of ownership function for various values of S, a and b. The growth,

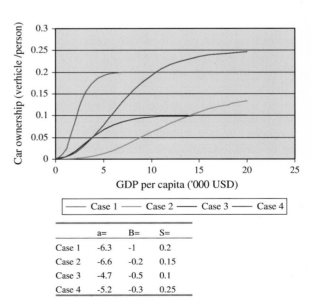

Fig. 4.4 Gompertz model shapes

	a=	B=	S=
Case 1	-6.3	-1	0.2
Case 2	-6.6	-0.2	0.15
Case 3	-4.7	-0.5	0.1
Case 4	-5.2	-0.3	0.25

4.4 Analysis at the Sectoral Level

saturation level and the time required to reach the saturation level vary depending on these parameters.

The quasi-logistic model is written as follows:

$$C_t = \frac{S}{1 + a \cdot \text{GDP}^b} \qquad (4.30)$$

The definitions given earlier for the variables and parameters hold in all these cases.

The above functional forms can be transformed into their linear equivalents as follows:

$$\text{Logistic}: \ln\left(\frac{(S-C)}{C}\right) = \ln(a) + b \cdot \text{GDP} \qquad (4.31)$$

$$\text{Gompertz}: \ln\left(\ln\left(\frac{S}{C}\right)\right) = \ln(-a) + b \cdot \text{GDP} \qquad (4.32)$$

$$\text{Quasi-logistic}: \ln\left(\frac{(S-C)}{C}\right) = \ln(a) + b \cdot \ln(\text{GDP}) \qquad (4.33)$$

The estimation of the parameters using country specific data or regional data provides a basis for estimating the car ownership.

4.4.2.3 End-Use Analysis of Transport Energy Demand

In the transport sector, energy is mainly used for passenger transport and freight transport. In less developed countries, the frequency of passenger trips and volume of shipment of freight are low. Moreover, traditional methods such as human and animal-powered transport systems co-exist in these countries alongside modern systems. The energy demand for passenger and freight transportation tends to increase rapidly, often at a rate higher than the growth rate of GDP, due to economic growth. This also leads to growth in ownership of cars and personalized transportation modes. The increase in demand for vehicles in turn causes higher demand for oil.

It is conventional practice for the transport sector's energy demand analysis to divide the sector into passenger and freight transports. The determinants of energy demand and the units of measurement of outputs are different in these two types of transport activities. On a macro or national level, energy consumption for passenger transport depends on the number of passengers traveling, the frequency and average length of trips, the distribution of trips among various modes of transport (i.e. air, sea, rail, road) and the technical characteristics of the carriers and their conditions of use. Figure 4.5 presents these determinants in a schematic form.

The development of transport modes and the modal distribution of a country are greatly affected by energy as well as general economic policy. The energy

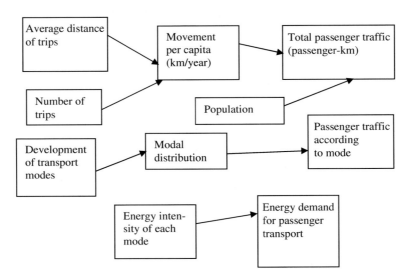

Fig. 4.5 Determinants of passenger transport energy demand

consumption per passenger-km varies greatly by mode of transformation. The energy consumption per unit of driving (i.e. l/km) is in principle a function of the power of the engine and of engine efficiency. The weight of the vehicle, traffic, speed, and driving style are further important factors affecting the energy intensity of the modes. If all these remain constant over time, the determinant of the energy intensity of each mode reduces to the fuel consumption efficiency.

Energy demand for freight transport depends on the volume of commodities, average distance of shipping, the modal structure of freight transport, and the economic and technical characteristics of each transport mode. The relationship among these variables is shown in Fig. 4.6.

Based on the above, the disaggregation of transport demand is shown in Fig. 4.7.

The definition of unit consumption in the transport sector varies by mode of transport and depending on the purpose of use of the transport service. For passenger transport, unit consumption is expressed as fuel consumption per passenger km. For freight transport, fuel consumption per ton-km is the usual measure.

The usual way to calculate the average unit consumption is to divide the consumption by the number of vehicles (or freight transported, ton-km). For road transport, this type of evaluation puts all types of vehicles on the same level: motorcycles, cars, trucks and buses are added together where one motorcycle is considered as same as one truck. If the number of motorcycles increases more rapidly, the average unit consumption will decrease rapidly. Any such structural change will affect the unit consumption.

To avoid this problem, an indicator called unit consumption per equivalent vehicle is used. The conversion of the actual stock of vehicles to a stock of

4.4 Analysis at the Sectoral Level

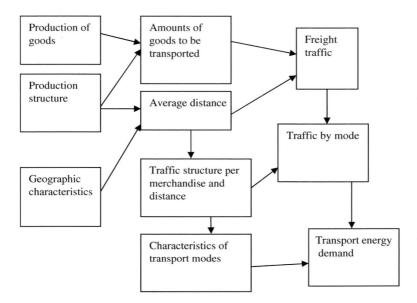

Fig. 4.6 Determinants of energy demand for freight transport

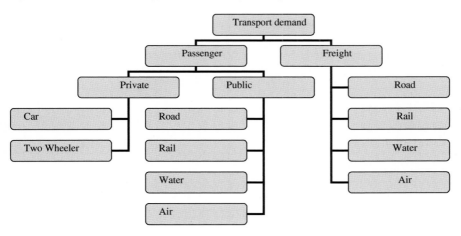

Fig. 4.7 Level of transport demand analysis

equivalent vehicles is based on a coefficient reflecting the difference in the average yearly consumption between each type of vehicle and the reference vehicle. For example, if a motorcycle consumes on average 0.2 toe/year and a car 1 toe/year, a car is equivalent to 5 motorcycles or one motorcycle is equivalent to 0.2 car.

Assume for example that the specific consumption (km/l) and mileage of different types of gasoline vehicles are obtained from a survey as given in Table 4.11. The yearly unit consumption (i.e. kl/vehicle) shown in the last column varies from 0.25 kl for motorcycles to 2.22 for jeeps. If the car is considered as the base

Table 4.12 Equivalent trucks for diesel vehicles

Vehicle type	No. of vehicles	Specific cons (km/l)	Mileage (km)	Yearly cons (Ml)	Unit consumption (kl/vehicle)	Equivalent car
Trucks	20,000	3.5	100,000	571.4286	28.57142857	1
Buses	2,500	3.5	25,000	17.85714	7.142857143	0.25
Minibuses	1,000	5	40,000	8	8	0.28

vehicle, motorcycle is equivalent to 0.14 car (=0.25/1.79). Similarly, equivalent vehicles can be estimated for diesel vehicles as shown in Table 4.12.

Using the coefficients of equivalent vehicles, it is possible to calculate time series for the stock of all vehicles in terms of equivalent vehicles. The overall unit consumption can then be estimated in terms of equivalent vehicles. Equivalent unit consumption assumes that the coefficients remain constant over time, which implicitly assumes that the mileage and the specific consumption remain constant (or changes in same proportion for all vehicles).

4.4.2.4 Decomposition of Energy Consumption Variation in the Transport Sector

In the case of road transport, the energy consumption depends on three variables: the composition of the fleet (Q_i), distance traveled (mileage, D_i) and the specific consumption of the vehicle (SP_i). This can be expressed as

$$TED_i = Q_i * D_i * SP_i \tag{4.34}$$

Total demand of the road transport is the sum of energy demand of different types of vehicles.

$$TED = \sum_i (Q_i D_i SP_i) \tag{4.35}$$

Multiplying the right hand side by (Q/Q, where Q is the total fleet size) and noting that $S_i = Q_i/Q$ is the share of ith type of vehicle in total fleet size, the above equation can be rewritten as:

$$TED = Q \sum_i (Q_i D_i SP_i)/Q = Q \sum_i (S_i D_i SP)_i \tag{4.36}$$

Decomposition of Eq. 4.36 would give four effects:

- a fleet effect, showing the influence of road fleet variation on the energy consumption;
- a structural effect, reflecting the impact of variations in the fleet composition on the demand;
- a mileage effect, reflecting the impact of changes in the distance traveled;

4.4 Analysis at the Sectoral Level

- a technical efficiency effect, reflecting changes in specific consumption of the vehicles.

In cases where disaggregated information is not available about mileage and specific fuel consumption but unit consumption (U_i) is available (or can be estimated), an alternative formulation of the road transport energy demand could be

$$\text{TED} = Q \sum_i (Q_i U_i)/Q = Q \sum_i (S_i U_i) \tag{4.37}$$

Using the above equation, the variation in energy demand can be explained by three factors:

- a fleet effect, showing the influence of road fleet variation on the energy consumption;
- a structural effect, reflecting the impact of variations in the fleet composition on the demand; and
- a unit consumption effect, reflecting the impact of changes in the yearly consumption per vehicle.

The fleet effect and the structural effect can be considered as the influence of changes in the economy and the transport system on the energy demand. The other effects are due to changes in energy policy and technical changes.

For rail or air or water transport, the energy demand is related to the traffic volume (pass-km or ton-km) and the unit consumption (litre/pass-km or litre/ton-km). Accordingly, the change in energy consumption is decomposed into two components: one reflecting the influence of traffic variation and the other indicating unit consumption effect.

4.4.3 Energy of Energy Demand in the Residential and Commercial Sectors

4.4.3.1 Purpose of Energy Use

Energy is used in residential and commercial sectors for four main purposes:

1. for maintaining a certain inside temperature;
2. for heating water;
3. for cooking, and
4. for electrical appliances including lighting.

Energy used by the residential sector is in most cases a final demand, while energy used by the commercial sector is an intermediate product used in output production. Various socio-economic and technological factors such as level of income, population size, average size of homes and buildings, customs and habits of customers, and efficiency of end-use device affect the energy consumption of these sectors. Climatic conditions also affect energy demand by these sectors.

The main difficulty often faced by the residential and commercial sectors in analyzing energy demand is the availability of data, especially of end-use breakdowns of energy consumption. Moreover, traditional fuels play a vital role in many countries to meet the energy demand of residential and commercial sectors but data is often not available in a systematic and regular manner. In addition, as the end-use efficiency of traditional fuel use is comparatively low, the final energy consumption including traditional energies may hide certain changes taking place within the sectoral energy consumption pattern. To avoid this problem, energy consumption in terms of replacement values (coal or oil) may provide a better indication.

The importance of different end-uses for energy varies significantly from one country to another due to different levels of economic development, climatic conditions, policies and other factors. Even within a country, similar variations can be found among different climatic zones and economic regions. Consumption behaviour also depends on income level of households.

In many developing countries cooking constitutes the main source of final energy demand in the residential sector. This is mainly due to prevalence of traditional energies for cooking purposes. Energy needs for cooking are directly related to individual households' eating habits and these needs for a family may not change significantly with level of income. But it is normally observed that as the standard of living improves, households switch from traditional energies to coal or kerosene, LPG and electricity. Such a fuel switch is accompanied by changes in end-use devices, which are more efficient. This can lead to a reduction in the useful energy demand for cooking.

Space heating and cooling energy requirements can also be important in many countries. Energy used for this purpose is induced by the need for inside temperature maintenance, which is a function of the insulation and ventilation characteristics of homes and buildings, the volume to be heated, and the free heat generated by the occupants and the sun, as well as the heat losses of the electrical appliances. This need is translated into a demand for energy products, which is determined by the efficiency of the heating equipment. The actual demand is eventually determined by need, subject to the level of income and to energy prices. The determinants of energy demand for heating and air conditioning can be classified into two categories, operating in opposite directions with respect to energy conservation. For example, improvements of the insulation and ventilation characteristics of a home and building and improvements of end-use devices tend to reduce the energy requirements per unit space over time. On the other hand, as the level of income improves, average dwelling space per home tends to increase and so does the demand for more comfortable inside temperatures, both of which result in increases in energy demand per home and building.

Hot water is needed in the residential and commercial sectors for bathing, washing (laundry and dish-washing). Energy demand for these needs is not very significant initially, but it is a historical trend that hot water needs increase rapidly up to a certain stage of development as standards of living improve and more and more homes are equipped with their own bathing facilities, washing machines and dishwashers. After that stage, energy demand for hot water needs levels off as ownership of equipment and machines is saturated. Actual energy demand induced by this need is determined by the efficiency of hot-water producing end-use devices and subject to income and energy prices.

The last category of energy needs in the residential and commercial sectors is for electrical appliances and lighting. The energy demand of a family for these needs depends on the ownership of these appliances (refrigerators, television sets, other audio-visual systems, irons, cleaners, electric bulbs, etc.), average utilization hours per period of time, and technical characteristics and efficiency of those appliances. The total energy demand of a country for these needs is determined by the number of families and by family ownership. The electricity demand of a family for these needs increases very fast up to the stage when the ownership of these appliances reaches saturation. Significant improvements have been made in manufacturing electrical appliances, which have contributed to reduction of energy consumption per appliance. However, in a stage of rapid economic development (as in developing countries), increases in the ownership of the appliances exceed the efficiency improvement of the appliances and thus the net effect is an increase in energy demand for these needs.

4.4.3.2 Econometric Analysis of Residential/Commercial Energy Demand

Energy demand in the residential and commercial sectors can be analyzed at an aggregate or a disaggregate level. Due to lack of data, aggregate analysis is undertaken more often than micro-level analysis. The aggregate approach is relatively simple and less expensive than the disaggregate approach.

The econometric approach explained previously can be used equally for energy demand analysis in the residential and commercial sectors by appropriately changing the variables. For example, weather conditions often influence energy demand for space heating or cooling and heating degree days (HDD) or cooling degree days (CDD) are included as additional variables.[3] The average energy consumption per household (which is E divided by number of households) can be analyzed as function of per capita income (GDP divided by the number of households), prices and the changes in fuel efficiency.

$$\mathrm{Ehh} = f(\mathrm{Zhh}, P, T) \tag{4.38}$$

---

[3] HDD is used in relation to any analysis of the heating requirement whereas CDD is used to that for cooling. Both of them are calculated with respect to a base temperature. For example, in the UK, the commonly used base temperature is 15.5°C. HDD indicates how many days within a period had temperatures below the base level whereas CDD indicates for how many days the temperature was above the base level, thereby requiring cooling.

where Ehh is the average energy consumption per household, Zhh is the GDP per household, P is the energy price and T is weather variable.

The functional relationship can be specified in various functional forms, but the following specification is commonly used in analysis.

$$\log E = \log a + b \log Y + c \log Z + d \log P + e \log T + e \qquad (4.39)$$

where the variables have the same meaning as before.

Similarly, the dynamic adjustment process can be captured by including a lagged demand variable in the model as discussed earlier.

More sophisticated studies, especially in the developed country context, have used expenditure functions analogous to the translog cost function. The almost ideal demand system (AIDS) or its linear approximation has been used in such studies.[4] In recent times, studies have relied on co-integration and error correction models but as indicated earlier, the focus of such studies have reduced to estimation price and income elasticities of demand and accordingly, reduced the variety and policy-relevance of academic studies.

4.4.3.3 End-Use Method of Residential Energy Demand Analysis

Total energy demand in the disaggregated approach estimated by summing up end-use energy demands for space heating, air conditioning, water heating, cooking and use of electrical appliances including lighting. Total energy consumption for space heating and air-conditioning of a country for a given year is determined by the average energy consumption per household and per building for those purposes, and the total number of households and buildings for that year. Similarly, energy demand for cooking is related to unit demand per household and number of households. The lighting requirement can be expressed as a function of household area, lighting requirement per unit area and the number of households.

As the demand pattern in households vary with income level and geographical location (rural/urban), better results are obtained by disaggregating the demand by income level and rural/urban areas. The total demand in that case would be sum of demands by all categories and locations.

For commercial sector, the drivers of energy demand are similar to that of residential demand. But instead of number of households as the activity parameter, value addition of the commercial sector can be taken as the measure of activity. For techno-economic analysis, the physical output is taken as the driving variable.

---

[4] See Ryan and Plourde (2009) for more details on this.

4.4.3.4 Decomposition of Energy Demand in the Residential and Commercial Sectors

Changes in energy demand in the residential sector are often related to the population change, and changes in demand per capita. Measuring activity is difficult since there are many different energy-using activities that take place in homes but no single measure. For that reason, population is used as an indicator of residential activity. Energy consumption in this case is considered as follows:

$$E = \text{POP} * S * \text{EI}; \qquad (4.40)$$

where E = total energy demand; POP = population; S = structural parameter indicating per capita ownership of energy-using appliance or dwelling area per person; EI = energy intensity expressed in terms of energy use per unit of an application.

As there are different end-uses (e.g. space heating, water heating, lighting, electric appliances, etc.) and different appliances or applications within end-uses, the total energy demand is obtained by summing all applications in an end-use and then adding demand in all end-uses.

$$E = \text{POP} \sum_i \sum_j S_{ij} \text{EI}_{ij} \qquad (4.41)$$

Following the decomposition approaches indicated earlier, the above relationship can be analysed to produce the activity effect, structural effect and the intensity effect.

4.5 Conclusion

This chapter has extended the analysis presented in the previous chapter by looking at the sector-level analysis. As the demand drivers vary, aggregated analyses cannot capture the essential aspects of sector level demand. More detailed analyses at the sector level can provide better insights. This chapter presented methodologies for such studies. However, the data requirement increases in such cases and data remains a problem in many countries.

References

Adams FG, Graham H, Griffin JM (1974) Demand elasticities for gasoline: another view, discussion paper 279. Department of Economics, University of Pennsylvania

Bhattacharyya SC, Timilsina GR (2009) Energy demand models for policy formulation: a comparative study of energy demand models. World Bank Policy Research Working Paper WPS4866, March 2009

Bouachera T, Mazraati M (2007) Fuel demand and car ownership modeling in India. OPEC Rev 31(1):27–51

Brendt ER, Wood DO (1975) Technology, prices and the derived demand for energy. Rev Econ Stat 57(3):259–268

Dargay J, Gately D (1997) Vehicle ownership to 2015: implications for energy use and emissions. Energy Policy 25(14–15):1121–1127

Han B (2001) Analyzing car ownership and route choices using discrete choice model, Department of Infrastructure and Planning, Royal Institute of Technology, Stockholm, TRITA-IP-FR01-95

IEA (2007) Tracking industrial energy efficiency and CO_2 emissions. IEA, Paris (see http://www.iea.org/textbase/nppdf/free/2007/tracking_emissions.pdf)

IEA (2008) Worldwide trends in energy use and efficiency: key insights from IEA indicators analysis. International Energy Agency, Paris (see http://www.iea.org/Papers/2008/Indicators_2008.pdf)

Medlock K, Soligo R (2002) Car ownership and economic development with forecasts to the year 2015. J Transp Econ Policy 36(2):163–188

Miklius W, Leung P, Siddayao CM (1986) Analysing demand for petroleum-based fuels in the transport sectors of developing countries. In: Miyata M, Matsui K (eds) Energy decisions for the future: challenge and opportunities, vol II, proceedings of 8th annual international conference of IAEE, Tokyo, Japan

Pindyck RS (1979) The structure of world energy demand. The MIT Press, Cambridge

Ryan DL, Ploure A (2009) Empirical modeling of energy demand, Chap. 6. In: Evans J, Hunt LC (eds) International handbook on the economics of energy. Edward Elgar, Cheltenham

UN (1991) Sectoral energy demand studies: application of the end-use approach to Asian countries, report 33. Energy Resources Development Series, United Nations, New York

Urga G, Walters C (2003) Dynamic translog and linear logit models: a factor demand analysis of interfuel substitution in US industrial energy demand. Energy Econ 25:1–21

Ussanarassamee A, Bhattacharyya SC (2005) Changes in energy demand in Thai industry between 1981 and 2000, Energy 30: 1845–57

Wenzel TP, Koomey JG, Rosenquist GJ, Sanchez M, Hanford JW (1997) Energy data sourcebook for the US residential sector, LBL-40297. Lawrence Berkeley National Laboratory, University of California, Berkeley

Wirl F, Szirucsek E (1990) Energy modelling - a survey of related topics, OPEC Review, Autumn, pp. 361–78

Chapter 5
Energy Demand Forecasting

5.1 Introduction

In the academic literature we find a large number of approaches for forecasting energy demand. Some of them are relatively simple, easy to use and less sophisticated approaches, while others employ more advanced methodologies. Some approaches are static in nature while others consider the dynamic adjustment process. Similarly, some approaches use a probabilistic framework while others are deterministic in nature (Lipinsky 1990). Here a brief overview of some methodologies is presented retaining a simple classification—simple and sophisticated (or advanced) techniques and their advantages and disadvantages are discussed.

5.1.1 Simple Approaches

Under this heading, simple indicators discussed previously are presented as a forecasting tool. We also discuss the trend analysis and consider how direct surveys can be used for forecasting purposes.

5.1.1.1 Forecasting Using Simple Indicators

The simple approaches are easy to use indicators that can provide a quick understanding. Such techniques are relatively less common in academic literature although practitioners rely on them in many cases. Four such simple indicators commonly used for forecasting are: growth rates, elasticities (especially income elasticity), specific or unit consumption and energy intensity (See Box 5.1 for details). In addition, trend analysis that finds the growth trend by fitting a time trend line is also commonly used. All of these approaches rely on a single indicator and the forecast is informed by the assumed changes in the indicator during the forecast period. Clearly these methods lack explanatory power and being based on extrapolation or arbitrary assumption, their attractiveness for any long-term work is rather low.

Box 5.1: Simple Approaches for Energy Demand Forecasting

Growth-rate based method

Let g be the growth rate in demand and D_0 is the demand in year 0, then D_t can be obtained by

$$D_t = D_0(1+g)^t \tag{5.1}$$

Elasticity-based demand forecasting

Elasticity is generally defined as follows:

$$e_t = \frac{(\Delta EC_t/EC_t)}{(\Delta I_t/I_t)} \tag{5.2}$$

where t is a period given; EC is energy consumption; I is the driving variable of energy consumption such as GDP, value-added, price, income etc.; Δ is the change in the variable.

In forecasting, output elasticity or income elasticity is commonly used. The change in energy demand can be estimated by assuming the percentage change in the output and the output elasticity. Normally, the elasticity is estimated from past data or gathered using judgement. The output change is taken from economic forecasts or planning documents.

Specific consumption method

Energy demand is given by the product of economic activity and unit consumption (or specific consumption) for the activity.

This can be written as

$$E = A \times U \tag{5.3}$$

where A is level of activity (in physical terms); U is the energy requirement per unit of activity. These two factors are independently forecast and the product of the two gives the demand.

Ratio or intensity method

Energy intensity is defined as follows:

$$EI = E/Q \tag{5.4}$$

where EI = energy intensity, E = energy demand, and Q = output.

This can be rearranged to forecast energy demand $E = EI \cdot Q$.

Using the estimates for Q for the future and assumptions about future energy intensity, the future energy demand can be estimated.

5.1 Introduction

Clearly, simple methods can be applied for both commercial and traditional energies and can be used both in urban and rural areas. They could be used to include the effects of informal activities and unsatisfied demand. However, they neither explain the demand drivers, nor consider technologies specifically. They only rely on the value judgements of the modeller, wherein lies the problem. Further, these methods do not rely on any theoretical foundation and accordingly, they are ad-hoc approaches.

Example A study used the growth rate method for forecasting energy demand between 1980 and 2008. It used three macro-economic growth scenarios given in Table 5.1. Using the GDP elasticity of industrial demand given in Table 5.2, it determined the energy demand growth rates for the industrial sector for the period between 1980 and 2008 (see Table 5.2).

5.1.1.2 Trend Analysis

The trend analysis extrapolates the past growth trends and is normally done by fitting some form of time trend to past behaviour. The analysis:

(a) Assumes that there will be little change in the growth pattern or in the determinants of demand such as incomes, prices, consumer tastes, etc.
(b) Finds the best trend line that fits the data. This is usually estimated by a least square fit of past consumption data or by some similar statistical methodology.

Table 5.1 Macro economic growth scenarios

Scenario	Annual growth rates				
	1980	1981	1982–1986	1987–1991	Post 1992
High growth	−1.0	5.5	8.0	7.0	6.0
Medium growth	−1.2	5.5	6.5	5.0	4.0
Low growth	−2.0	5.0	5.0	4.0	4.0

Source Codoni et al. (1985)

Table 5.2 Industrial demand growth forecast

Period	GDP elasticity	Energy demand growth rates		
		High	Medium	Low
1980	1.254	−1.3	−1.5	−2.5
1981	1.254	6.9	6.9	6.3
1982–1983	1.254	10.0	8.2	6.3
1984–1986	1.108	8.9	7.2	5.5
1987–1988	1.108	7.8	5.5	4.4
1989–1991	1.043	7.3	5.2	4.2
1992–1993	1.043	6.3	4.2	4.2
1994–1998	1.022	6.1	4.1	4.1
1999–2008	1.000	6.0	4.0	4.0

Source Codoni et al. (1985)

(c) The fitted trend is then used to forecast the future. Frequently, ad hoc adjustments are made to account for substantial changes in expected future demands due to specific reasons.

Depending on the availability of data, the analysis can be

- Performed at the national level for a given energy source or they may be broken down by region, by consuming sector or by both.
- Used on its own or in combination with another method. For example, if energy demand is estimated using per capita consumption (i.e. unit consumption approach), the trend of population growth and per capita consumption can be estimated using trend analysis. The results can then be used in the unit consumption approach to get the final results.

This is the most commonly used approach for forecasting and appears to be more useful at relatively aggregate levels. The simplicity of use is its main advantage. It can be applied at aggregate and disaggregate levels and can be based on whatever data is available. Its disadvantages include:

- future demand cannot be expected to depend on the past trend;
- it takes insufficient account of structural changes;
- it does not explain what determines demand as it does not explicitly include variables on price, income, etc.; and
- it is not suitable for policy analysis work (Codoni et al. 1985).

Example Using the global primary energy consumption data available from the BP Statistical Review of World Energy 2010, forecast the global primary energy demand using the trend method.

The original data and the fitted linear trend lines are shown in Fig. 5.1. The linear relationship is also shown in the figure.

$$Y = 155.21X - 300876 \qquad (5.5)$$

where X is year.

Using the forecast year as X in the above relation, the forecast can be obtained. The fitted trend has achieved a good level of fit as is evident from the R^2 information. The demand forecast for the period up to 2010 using this trend is given in Table 5.3.

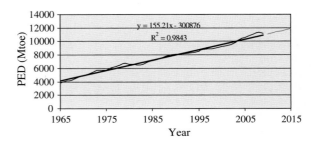

Fig. 5.1 Trend analysis of global PEC

Table 5.3 Forecast of global PEC using trend line

Year	Forecast (Mtoe)
2010	11,096.1
2011	11,251.3
2012	11,406.5
2013	11,561.7
2014	11,716.9
2015	11,872.2

5.1.1.3 Direct Surveys[1]

Direct surveys are generally used to generate primary information essentially for the short term but surveys can also be used as a direct and reliable tool for demand analysis and forecasting.

Such surveys ask major energy users to reveal their present consumption and future consumption plans. Through this, surveys try to account for changes in the energy-consuming sectors themselves that would affect the demand. In addition, by analyzing the investment plans and programmes, changes in the supply and demand are captured. But as surveys are costly undertakings, their use is generally limited to major energy consumers such as medium to large size industrial plants, mines and smelters, large transportation companies, utility companies, important governmental users, etc. Often, these major consumers account for a large percentage of total energy consumption in many countries.

In many developed and developing countries, it is quite common to use industrial surveys to generate information on activity levels, energy use by fuel type, efficiency of energy use, cost of energy in the production process and future plans for energy use. For example, the Energy Information Administration of the US government conducts the manufacturing energy consumption survey every four years (see http://www.eia.doe.gov/emeu/mecs/contents.html for details). Often, this is the source of information for developing disaggregated sector-level picture of energy use. A comparison of survey results at different points in time could provide a better picture about demand patterns and changes. The effects of new economic development programs, such as industrial settlement, mining and hydrocarbon developments, etc. could also be assessed using surveys but care must be taken to assess the realism of these specific projected development programs.

While industrial surveys are more common, household energy surveys are less common because of cost and time implications. In general, such surveys are less frequent (once in 10 years or so). In many cases, special energy-related questions are added in the national population census, which could bring out certain special information. However, such less frequent surveys are less useful for forecasting purposes.

---

[1] See Munasinghe and Meier (1993) for further details.

The major problems with surveys as a tool for demand forecasting are as follows:

(a) This is a time consuming process and consequently involves high cost.
(b) It requires skilled staff to undertake the survey and analyse the results;
(c) As the survey depends on information provided by the respondents, the quality of responses influences the results. It is quite possible that respondents are unwilling to divulge information and can provide inaccurate information deliberately. Many may not know the correct answers to the questions as well.
(d) Future energy use plans may be vague, or too optimistic/pessimistic. Relying on this information for forecasting could result in over/under capacities.

5.1.2 Advanced or Sophisticated Techniques

Sophisticated demand forecasting techniques rely on more advanced methodologies. Such techniques can be classified using alternative criteria: for example, a common method of classification is the top-down and bottom-up models. Top-down models tend to focus on an aggregated level of analysis while the bottom-up models identify the homogeneous activities or end-uses for which demand is forecast. Another classification relies on the modelling philosophy:

- econometric models are grounded in the economic theories and try to validate the economic rules empirically;
- engineering–economy models (or end-use models on the other hand attempt to establish accounting coherence using detailed engineering representation of the energy system; and
- input–output models which rely on forward and backward linkages in any economy to determine the demand for energy.
- combined or hybrid models that attempt to reduce the methodological divergence between the econometric and engineering models by combining the features of the two traditions.[2]

Some other models are also indicated in the literature—system dynamics models, scenario approaches,[3] decomposition models,[4] process models,[5] and artificial neural networks.[6] However, for the sake of simplicity we shall not cover all alternative options here.

---

[2] See Reister (1990) for an example. We also discuss The POLES model can also be considered a hybrid model.

[3] See Ghanadan and Koomey (2005) for an example.

[4] See Sun (2001). The results of the study show a significant divergence with actual EU15 demand.

[5] See Munasinghe and Meier (1993) and Labys and Asano (1990).

[6] Al-Saba and El-Amin (1999) for an application.

5.1.3 Econometric Approach to Energy Demand Forecasting

The econometric tradition of demand forecasting extends the demand analysis considered in Chaps. 3 and 4. The relationship determined for the demand can then be used for forecasting simply by changing the independent variables and determining their effect on the dependent variable.

The main step involved at the forecasting level is to decide a systematic way of forecasting the independent variables. This is however very poorly described in the literature. One could choose from a number of alternative options:

- forecast them using judgements, which could be based on a literature review or a survey of expert views or otherwise;
- use simple indicators (such as growth rates) to generate a set of data for the future; the growth rates can be based on historical levels or expected levels as suggested by other agencies or experts;
- use a trend analysis of the independent variables to extrapolate the future values;
- use a combination of above or any other plausible method.

Generally, alternative scenarios are used to analyse a range of plausible outcomes. The choice of alternative scenarios needs care—although a base or reference case and two alternative cases (high/low) are frequently used.

As indicated in previous Chaps. 3 and 4, a number of econometric forms (such as the reduced form and structural models) can be used for demand analysis. Whatever the option chosen, the forecasting follows the above steps.

One measure of accuracy of forecasting is given by the root mean square (RMS) error. This measures the deviation of the forecast from its actual value and can be written as

RMS forecast error

$$\text{RMS Error} = \sqrt{\frac{1}{T}\sum_{t=1}^{T}(Y_t^s - Y_t^a)^2} \qquad (5.6)$$

where Y_t^s = forecast of Y_t; Y_t^a = actual value; and T = number of periods.

The range of error here can be between zero and infinity. As this squares up the errors, larger errors influence the result more than the smaller errors. Theil's inequality coefficient reorganizes RMS error to fall within a range of zero and one. This is also widely used in the economic literature. This is written as follows:

$$U = \frac{\sqrt{\frac{1}{T}\sum_{t=1}^{T}(Y_t^s - Y_t^a)^2}}{\sqrt{\frac{1}{T}\sum_{t=1}^{T}(Y_t^s)^2} + \sqrt{\frac{1}{T}\sum_{t=1}^{T}(Y_t^a)^2}} \qquad (5.7)$$

The econometric approach has certain advantages: (1) this method can be used for both short run and long run projections and policy analyses; (2) this is a flexible method which can be applied at the country (economy) level or at disaggregated sector level, (3) this is perhaps the only method that can capture the effect of price on energy demand and inter-fuel substitution; (4) it can identify important determinants of demand; and (5) this can be used for energy–economy interactions (Codoni et al. 1985).

The main difficulties associated with the econometric method are that:

(a) correct use of this method requires experienced econometricians, soundly trained in both economic and econometric theory. Such people are relatively scarce in developing countries;
(b) statistical analysis of energy demand requires consistent data of sufficient quality. This may not be available in many cases;
(c) the method essentially relies on past demand behaviour to determine future demand. However, as the economies undergo structural changes, the above assumption often becomes difficult to satisfy. Consequently, the extrapolation from the past may lead to poor forecasts.
(d) econometric methods are not ideally suited for capturing in detail technological change. This is specially the case for new technologies and for commodities not already in existence.
(e) the basic theoretical assumptions behind the demand functions may not hold true in many cases due to government intervention, reliability and availability of supply and similar constraints (Codoni et al. 1985).

Example Using the specification shown in Chap. 3, forecast the gasoline demand for the next 3 years. Use appropriate assumption as required.

Answer: The estimated equation for gasoline demand was

$$\ln(Q) = -3.793 - 0.1244\ \ln(P) + 0.6522\ \ln(GDPc) + 0.7481\ \ln(Q_{t-1}) \quad (5.8)$$

For forecasting, we use the actual price and GDP per person information for 2006–2008 obtained from IEA Energy Statistics and IMF World Economic Outlook. The data required for the estimation is shown in Table 5.4.

These inputs were converted to their log forms and used in the estimated equation to produce the forecasts. The results are then compared with the actual consumption data reported by IEA to see the accuracy of forecasting. These are presented in Table 5.5.

Clearly, the result for the first year of forecasting is very accurate—understates the demand by one quarter of a percent. But the error increases significantly in the second year and the demand is overstated by 27%. As GDP per person increases,

Table 5.4 Data for independent variables

	Local prices (Rial/bbl)	CPI inflation (2000 base)	Real price (Rial/bbl)	GDP per capita (1000 Rial/person, 1998 base)
2006	159000.00	212.199	74929.665	6356.491
2007	174900.00	249.315	70152.217	6740.194
2008	174900.00	300.881	58129.294	6778.79

5.1 Introduction

Table 5.5 Forecasting results

ln(Qt)	Qt (bbl/1000p)	Pop (M)	Gasoline 1000 bbl/day	Actual	% difference
1.879804	6.55222	70.473	461.7546	462.93	−0.2539
1.974867	7.205661	71.662	516.3721	405.35	27.3892
2.073095	7.949385	72.871	579.2797	NA	NA

the demand increases exponentially following the constant elasticity function and the accuracy of forecasting decreases. The deviation would reduce if the demand function is re-estimated updating the data but this makes the analysis cumbersome.

5.1.4 End-Use Method of Forecasting

Contrary to the econometric approach which focuses on the aggregate level of activities, the basic idea behind the end-use approach of energy demand is to disaggregate the demand into homogeneous modules and sectors and to link the demand of each module to technical and economic indicators. The basic element of analysis is the end-use and demand is estimated working backwards. Hence the approach is called end-use approach and is also known as a bottom-up approach. For instance, to estimate the demand for gasoline, the focus would be on the final use of gasoline (i.e. transport) in cars and motorcycles. The analysis would consider the number and types of cars, average unit energy consumption of each type, average traveling habit, etc. to arrive at an estimate of the demand.

The end-use analysis puts emphasis on

- the role of technology (e.g. fuel economy of vehicles, unit consumption of electrical appliances or of industrial processes),
- behaviour of consumers (mileage of vehicles), and
- the economic environment (load factors, ratio value added/physical output) in the demand analysis.

The role of prices in the demand is traditionally not considered in these models, as the focus is not on the transitory phase. From this perspective, this family of models is more suitable for medium to long term demand analysis and forecasting.

The end-use models normally have the following features:

(a) They contain a detailed representation of energy end-uses: being a disaggregated method of analysis, this approach breaks down the demand in small components and includes a technical picture of energy use at each level.

- For example, the demand would be broken down into a number of sectors: industry, transport, residential and commercial.
- For each sector, further disaggregation would be made. For instance, in case of residential demand, a distinction between urban and rural consumption

would be made. Within each zone, the demand for cooking, heating and lighting could be considered.
- For each type of use, different types of possible fuels would be allowed to be used.

(b) A few key driver variables would be used at each level: for example, per capita consumption could be used for the residential sector; value addition could be used for the industrial sector. These variables are forecast exogenously using scenarios or judgements.
(c) These models use a few policy variables, which could be changed to see the effect on the overall demand. Examples of such variables could include energy intensity, unit consumption, fuel mix, etc.
(d) The past data and information is used to establish a base or reference case. The data could be used to calibrate the model as well. The analysis is done in terms of a number of scenarios which are compared with the reference case. The results show how different policies could influence the future demand.
(e) The analysis is normally done providing snapshots of the future and does not provide a path to reach various ends.

The first generation models specified most of the variables as exogenous and were used to perform a set of multiplications and additions. They were nothing more than energy accounting models Recent developments of end-use models have attempted to overcome these limitations either by introducing econometric relationships to account for energy savings and substitutions or by relying on models simulating consumer behaviour.

5.1.5 Input–Output Model

The input–output table has long been used for economic analysis. It provides a consistent framework of analysis and can capture the contribution of related activities through inter-industry linkages in the economy. The following example will explain the basic approach while more mathematically oriented readers could refer to Annex 5.1 for a formal treatment.

Consider a small three sector economy with agriculture, industry and energy as the main activities. The interdependence of one sector on the other is indicated through purchases from other sectors (see Table 5.6). In this table, a column entry indicates the purchase of one sector from the others. For example, the agriculture sector uses $10 from agriculture, $10 worth of inputs from energy, and another $10 of inputs from industry to produce an output worth of $80. In the process, the agriculture adds a value of $50. Similarly, along the row, the uses of the output of a sector are indicated. For example, the output of energy sector is used in agriculture, energy, industry and to meet final demand.

The direct requirements (or technical coefficients) of each sector are obtained by dividing each inter-sectoral purchase by the sector total output. Table 5.7 gives

5.1 Introduction

Table 5.6 Transactions table

	Agriculture	Energy	Industry	Final demand	Total output
Agriculture	10	10	30	30	80
Energy	10	20	20	50	100
Industry	10	30	30	50	120
Value added	50	60	40	150	
Total output	80	100	120		

Table 5.7 Direct requirement matrix

	Agriculture	Energy	Industry
Agriculture	0.125	0.1	0.25
Energy	0.125	0.2	0.166667
Industry	0.125	0.3	0.25

the direct requirement matrix corresponding to Table 5.1. This basically indicates the combination of inputs required to produce $1 output of the sector. For example, the energy sector requires $0.1 of agricultural input, $0.2 of energy input and $0.3 of industrial input.

Table 5.7 provides an easy way of linking the outputs of the sector and the purchases from the sectors. For example, $80 of agricultural output is used by different industries and the final users (final demand). As agriculture, energy and industry use $0.125, $0.1 and $0.25 of agricultural output respectively for each $ of output from these sectors, and there is final demand of $30, the overall output demand can be expressed as follows:

$$0.125 * 80 + 0.1 * 100 + 0.25 * 120 + 30 = 80 \tag{5.9}$$

Similar expressions can be written for other sectors as well.

$$0.125 * 80 + 0.2 * 100 + 0.166667 * 120 + 50 = 100 \tag{5.10}$$

$$0.125 * 80 + 0.3 * 100 + 0.25 * 120 + 50 = 120 \tag{5.11}$$

In the above expressions, the output of the sectors appears on both sides and it is possible to rearrange the expressions to put these terms on one side of the equation. This results in the following set of expressions:

$$(1 - 0.125) * 80 - 0.1 * 100 - 0.25 * 120 = 30 \tag{5.12}$$

$$-0.125 * 80 + (1 - 0.2) * 100 - 0.166667 * 120 = 50 \tag{5.13}$$

$$-0.125 * 80 - 0.3 * 100 + (1 - 0.25) * 120 = 50 \tag{5.14}$$

Note that we now have all the outputs in one side and the final demand on the other side of the expression. Also note that only the coefficients in the diagonal of the set has become positive now and the rest are negative after transformation.

In mathematical terms, this is represented in matrix notation as $(I - A)X = F$, where I is the identity matrix, A is the direct coefficient matrix, X is the output vector and F is the final demand vector.

The output vector is then obtained by inversing the $(I - A)$ matrix and multiplying with final demand vector. This basic process of input-out operation was first suggested by Leontief and the inverse matrix is known as Leontief inverse matrix. Table 5.8 presents the Leontief inverse matrix for the above example.

Note that all the entries in the above table are given in monetary terms. A simple way to use the above method for energy purposes is to consider the average energy content of energy products per $ of output. Let us assume that energy is sold with an average energy content of 10 cal/$. In our simple example, the agricultural sector consumes $10 of energy or 100 cal and produces $80 in output. Thus, per unit of output, the energy consumption is $100/80 = 1.25$ cal/$. Similarly, the energy sector uses $20 worth of energy (or 200 cal of energy) to produce $100 of output. Thus energy use per unit of output is 2 cal/$. This energy consumption information can be presented in a tabular form as shown in Table 5.9. Note that this is a table where the elements are zero except for the diagonal (i.e. a diagonal matrix).

Exercise: Find the energy consumption in industry per $ of output.

The information generated so far could be used to find out the energy implications of $1 output of various activities. Table 5.10 presents the results for $1 output of the industry sector. It can be seen that the energy consumption including the indirect effects is much higher than the direct consumption.

Thus the input–output method is able to capture the direct energy demand as well as indirect energy demand through inter-industry transactions. This feature makes this method an interesting analytical tool.

We have used a simplified method to include energy in this example. In practice, mixed type input–output tables are used where energy can be used in physical units alongside monetary values for other activities. For a practical large-scale input–output application, refer to www.eiolca.net.

Although this method has its appeal, it has certain limitations as well. The method is data intensive and requires a disaggregated sector-level break down of information to generate the technical coefficients. This could become messy as the number of sectors increases. In addition, such tables are not prepared every now

Table 5.8 Leontief inverse matrix

1.26	0.34	0.50
0.26	1.44	0.41
0.32	0.63	1.58

Table 5.9 Energy consumption matrix

	Agriculture	Energy	Industry
Agriculture	1.25	0	0
Energy	0	2	0
Industry	0	0	1.6666

5.1 Introduction

Table 5.10 Energy consumption and sector outputs for $1 purchase of industry output

	Output ($)	Energy cons (cal)
Agriculture	0.50	0.62
Energy	0.41	0.81
Industry	1.58	2.63
Total	2.48	4.07

and again, which means that changes in the economic structure are not easily captured. This method requires specialized knowledge in matrix algebra and large data manipulation skills.

5.1.6 Scenario Approach[7]

The scenario approach has been widely used in climate change and energy efficiency policy making (Ghanadan and Koomey 2005). The scenario approach has its origin in the strategic management where it has been used since 1960. In the energy and climate change area, the use of scenarios by the Intergovernmental Panel of Climate Change (IPCC) has played an important role in the policy debate. Similarly, studies by the World Energy Council (WEC), Inter-laboratory Working Group of US and similar studies in Australia have brought the approach to limelight. See for example Jefferson (2000), Brown et al. (2001), Saddler et al. (2004, 2007) and Shell Studies[8] (Shell 2008), among others. Scenarios are an integral part of the end-use approach as well and accordingly, they are not new to energy analysis.

"A scenario is a story that describes a possible future" (Shell 2003). In simple terms, scenarios refer to a "set of illustrative pathways" that indicate how "the future may unfold" (Ghanadan and Koomey 2005). Evidently, they do not try to capture all possible eventualities but try to indicate how things could evolve. It is a particularly suitable approach in a changing and uncertain world (Leydon et al. 1996).

> Scenarios give the analyst the opportunity to highlight different combinations of various influences, so that alternative future contexts can be sketched out, and the energy implications examined (Leydon et al. 1996, p. 5).

"Scenarios are based on intuition, but crafted as analytical structures...They do not provide a consensus view of the future, nor are they predictions" (Shell 2003). Clearly, "scenarios are distinct from forecasts in that they explore a range of possible outcomes resulting from uncertainty; in contrast, forecasts aim to identify the most likely pathway and estimate uncertainties" (Ghanadan and Koomey 2005).

---

[7] This section and the section on hybrid method are based on Bhattacharyya and Timilsina (2009).

[8] see http://www.shell.com/home/content/aboutshell/our_strategy/shell_global_scenarios/previous_scenarios/previous_scenarios_30102006.html for details). In its latest scenario study, Shell has introduced the new catch phrase "There are no ideal answers" (TAN!A) in 2008.

Jefferson (2000) presents a brief history of the WEC efforts in understanding the future energy demand and describes the scenarios used in a number of studies. In 1978 study, the Council called for actions to ensure a sustainable future. Until 1989 the Council used two scenarios—high growth and middle course. Since 1993, an ecologically driven scenario was added which were further refined subsequently in 1998 to develop six scenarios—three high growth, one middle course and two ecologically driven scenarios. The fully integrated scenarios present a range of possible rational outcomes and forecast energy and environmental indicators up to 2100.

Similarly, Shell was active in using scenario technique for strategic management and planning. It is the ex-Shell planners who have brought this method to the wider public (Ghanadan and Koomey 2005). Shell produced its first energy scenario studies in the 1992 and produced the catch phrase "There is no alternative" (TINA). This was followed by a number of studies in 1995, 1998, 2002 and 2005.

The strength of the scenario approach is its ability to capture structural changes explicitly by considering sudden or abrupt changes in the development paths. The actual level of disaggregation and inclusion of traditional energies and informal sector activities depend on model implementation. Theoretically it is possible to include these aspects but how much is actually done in reality cannot be generalised. Moreover, the development of plausible scenarios that could capture structural changes, emergence of new economic activities or disappearance of activities is not an easy task.

5.1.7 Artificial Neural Networks[9]

An artificial Neural Network (ANN) is an information paradigm that is inspired by biological nervous systems. The system structure is composed of a large number of highly interconnected processing elements (neurons) working together to solve specific problems. These units are connected by communication channels referred to as "connections" which carry numeric data between nodes. Each unit operates only on its local data and on the inputs they receive via the connections. The processing ability of the network as a whole is stored in the inter-unit connection strengths, or weights. These weights are obtained by a process of adaptation to a set of training patterns.

Feed-forward Neural Networks allow signals to flow through the network in one direction only: from input to output. Data enter the neural network through the input units on the left. The output values of the units are modulated by the connection weights. Most widely used learning method in Feed-forward Networks is the back-propagation of error. Back-propagation is a form of supervised learning in which the network's connections are adjusted to minimize the error between the

---

[9] This is based on Bhattacharyya and Thanh (2004).

5.1 Introduction

actual and the correct output. In back-propagation, the input data is repeatedly presented to the neural network. The output of the neural network is compared to the desired output and an error is computed. This error is then fed back (back-propagated) to the neural network and used to adjust the weights such that the error decreases with each iteration until achieves to acceptable value (gradient descent) and the neural model gets closer and closer to producing the desired output. This process is known as "training."

A multi-layered neural network consists of one or more layer of hidden units between the input and output layer. The number of hidden layers and the number of nodes per layer are not fixed; each layer may have different number of nodes depending on the application. A trial-and-error method is commonly used to determine how many layers and how many nodes per layers for the application. Figure 5.2 is a graphical representation of a multi-layered network, which is the most commonly used in forecasting. Each neuron can have multiple inputs $(x_1, x_2, \ldots, x_{nj})$, while there can be only one output (Y_i). The output of neuron i is connected to the input of neuron j through the interconnection weight W_{ij}.

There are several typical output functions used in multi layer networks such as Sigmoid function and Tanh(x) function. A sigmoid function takes values from 0 to 1 and is S-shaped, monotonically increasing.

ANNs have been successfully used for forecasting economic problems, electric load forecasting, water resource and hydrologic time series modelling (Gerson and David 1995; Ho et al. 1992; Park and Damborg 1991). This method is more commonly used for short-term demand forecasting (see Metaxiotis et al. 2003, for a review).

5.1.8 Hybrid Approach

This, as the term indicates, approach relies on a combination of two or more methods discussed above with the objective of exploring the future in a better way.

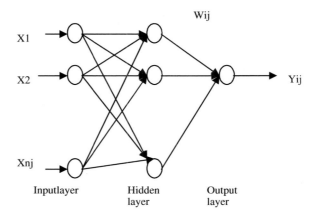

Fig. 5.2 Neural network structure for forecasting

The hybrid methods have emerged to overcome the specific limitations of individual approaches. These models have become very widespread now and it is really difficult to classify any particular model into a specific category. For example, econometric models now adopt disaggregated representation of the economy and have internalised the idea of detailed representation of the energy–economy activities. Similarly, engineering–economy models use econometric relationships at the disaggregated levels thereby taking advantages of the econometric estimation method. The end-use approach heavily relies on the scenario building approach to enrich itself.

There is a growing interest in the hybrid energy models in recent times with the objective of reconciling the differences between the top-down and bottom-up approaches. This is evident from a set of recent studies[10]:

- To reconcile the "efficiency gap," models with top-down structure are using bottom-up information to estimate parameters. See for example Koopmans and te Velde (2001) for such an exercise.
- To capture the technological details of bottom-up models and micro- and macro-economic details of econometric models, the hybrid option is being adopted. NEMS falls in this category. NEMS is the model used by the U.S. Department of Energy for its Annual Energy Outlook. NEMS uses the details found in engineering-economic models but retains the behavioural analysis found in top-down models, making it a hybrid model. Other examples include the CIMS model (see Bataille et al. 2006).
- To enhance the capability of price-induced policies in a bottom-up model, price information is explicitly included in the bottom-up structure. The POLES model is such an example, which is widely used by the European Union for its long-term energy policy analysis.

This approach has now been extended beyond demand analysis and forecasting to include energy–economy interactions and even more recent concerns such as renewable energy penetration and technology choice.

5.2 Review of Some Common Energy Demand Analysis Models[11]

This section provides a brief review of four of models have appeared in the literature. This is not intended to be exhaustive and more complete description is available in the reference manual of each model.

---

[10] See for example Special Issue of *Energy Journal* (November 2006) on this theme.
[11] This section is based on Bhattacharyya and Timilsina (2009).

5.2.1 MAED Model

This is a widely used bottom-up model for forecasting medium to long-term energy demand. MAED (acronym for Model for Analysis of Energy Demand) falls in the MEDEE family of model developed by B. Chateau and B. Lapillonne (IAEA 2006; Lapillonne 1978) but has been modified now to run on PCs and using EXCEL.

The earlier versions of the model were built around a pre-defined set of economic activities and end-uses. Manufacturing industry was broken into four sub-sectors while the transport sector considered passenger and freight transports separately. Various types of households could also be considered but they were aggregated at the national or regional level. An aggregated representation was used for other sectors.

Given the diversity of needs of the users from across different countries around the world, the more recent version has been developed to provide a more flexible structure where the user can add more sub-sectors, transport modes and fuel types, and household types.

The model follows the end-use demand forecasting steps typical for an engineering–economy model. It relies on the systematic development of consistent scenarios for the demand forecasts where the socio-economic and technological factors are explicitly taken into consideration. Through scenarios, the model specifically captures structural changes and evolution in the end-use demand markets. For competing forms of energies, the demand is first calculated in useful energy form and the final demand is derived taking market penetration and end-use efficiency into consideration. The model does not use pricing and elasticity information for the inter-fuel substitution as is common in the econometric tradition. This is a deliberate decision of the model developers as the long-term price evolution is uncertain, the elasticity estimates vary widely and because energy policies of the governments tend to influence demand significantly.

The energy demand is aggregated into four sectors: industry, transport, households and service. The industrial demand includes agriculture, mining, manufacturing and construction activities (or sub-sector). The demand is essentially determined by relating the activity level of an economic activity to the energy intensity. However, the demand is determined separately for non-substitutable energy forms (electricity, motor fuels, etc.) and substitutable forms (thermal energies). The need for feedstock or other specific needs can also be considered.

The demand is first determined at the disaggregated level and then added up using a consistent accounting framework to arrive at the overall final demand. The model focuses only on the final demand and does not cover the energy used in the energy conversion sector. The general framework of analysis of the MAED model is presented in Fig. 5.3. The detailed list of principal equations used in the model is provided in IAEA (2006).

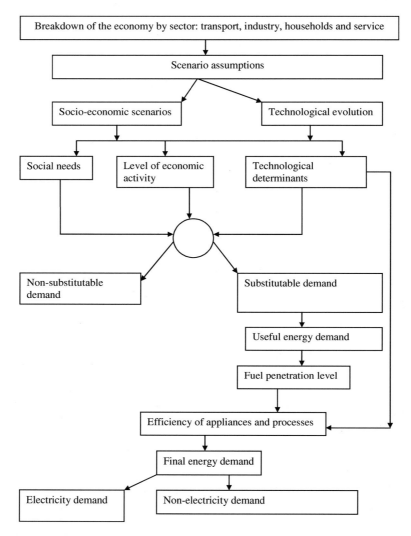

Fig. 5.3 MAED framework of analysis. *Source* IAEA (2006)

5.2.2 LEAP Model

The Long-range Energy Alternatives Planning (LEAP) is a flexible modelling environment that allows building specific applications suited to particular problems at various geographical levels (cities, state, country, region or global). As an integrated energy planning model LEAP covers both the demand and supply sides of the energy system. However, we briefly outline the demand forecasting features of the LEAP model here.

5.2 Review of Some Common Energy Demand Analysis Models

The model follows the accounting framework approach to generate a consistent view of energy demand (and supply) based on the physical description of the energy system. It also relies on the scenario approach to develop a consistent storyline of the possible paths of energy system evolution. Thus for the demand forecasting, the model does not optimise or simulate the market shares but analyses the implications of possible alternative market shares on the demand.

The demand analysis, following the end-use approach, is carried out as follows (Heaps 2002):

- The analysis is carried out at a disaggregated level, where the level of disaggregation can be decided by the users.
- The disaggregated structure of energy consumption is organised as a "hierarchical tree," where the total or overall activity is presented at the top level and the lowest level reflects the fuels and devices used. An example of such a tree will be: sectors, sub-sectors, end-uses and fuels/devices.
- The socio-economic drivers of energy demand are identified. The distribution of these activities at the disaggregated level following the "hierarchical tree" is also developed.
- Generally, the product of activity and the energy intensity (i.e. demand per unit of the activity) determines the demand at the disaggregated level. However, the model allows alternative options:
  - At the end-use level, useful energy can be considered to forecast the demand.
  - Stock analysis allows the possibility of capturing the evolution of the stock of appliances/devices or capital equipment and the device energy intensity.
  - For the transport sector, the fuel efficiency of the vehicle stock and distance travelled can be used to determine the demand.

The model can be run independently on a stand alone mode and can be used for specific sector analysis or for analysing the energy system of a given geographic region. The model has been widely used and it is reported that 85 countries have chosen the model for their UNFCCC reporting requirements.

5.2.3 Demand Module in NEMS (National Energy Modeling System)

The National Energy Modeling System (NEMS) was designed and primarily used by the US Department of Energy for preparing the Annual Energy Outlook. The demand analysis component is divided into four modules (residential, commercial, industrial and transport) and each module captures the diversity at the regional level to a great extent (see Table 5.11).

The residential demand module forecasts energy demand using a structural model based on housing stock and the appliance stock. The demand is driven by four drivers: economic and demographic factors, structural effects, technology, and

Table 5.11 Demand representation in NEMS

Energy activity	Categories	Regions
Residential demand	16 end-use services Three housing types 34 end-use technologies	Nine Census divisions
Commercial demand	Ten end-use services 11 building types Ten distributed generation technologies 64 end-use technologies	Nine Census divisions
Industrial demand	Seven energy-intensive industries Eight non-energy-intensive industries Cogeneration	Four Census regions, shared to nine Census divisions
Transportation demand	Six car sizes Six light truck sizes Sixty-three conventional fuel-saving technologies For light-duty vehicles Gasoline, diesel, and thirteen alternative-fuel Vehicle technologies for light-duty vehicles Twenty vintages for light-duty vehicles Narrow and wide-body aircraft Six advanced aircraft technologies Medium and heavy freight trucks 37 advanced freight truck technologies	Nine Census divisions

Source EIA (2003)

market effects. The housing stock and appliance stock information from the Residential Energy Consumption Survey is used to capture the diversity of stock holding and usage patterns across the country. It projects the demand for various end-uses by fuel type.

The commercial sector demand module projects energy demand in the commercial sector by taking into account building and non-building demand. It also captures the appliance stock and technological advancements and their effects on energy demand for three major fuels, namely electricity, natural gas and distillate oil. For the remaining minor fuels, the demand is projected using a simple econometric method. The demand by fuels for various end-uses is projected by the module.

The industrial demand module projects energy demand in the industrial sector using a hybrid approach: it uses the technological representation found in the end-use method and incorporates the behavioural aspects of a top-down approach. The demand is analysed at a disaggregated level—with a greater focus on energy intensive industries which are analysed at the three-digit level of industrial classification. Within each industry, three elements of demand are considered—building, boiler and process/assembly activities. The demand for each element is estimated separately using a combination of approaches ranging from simple growth rates to more involved methods.

The transport demand module projects the fuel demand in the transport sector by mode and includes alternative energy demand. A disaggregated approach is used in demand forecasting where personal car usage, light truck, freight transport, air transport and miscellaneous transport are considered separately. A nested multinomial logit model is used to predict the vehicle sales by technology. The vehicle miles per capita is estimated based on fuel costs of driving, disposable income per capita and an adjustment for men to women driving ratio. The model captures the regional variation in transport demand as well.

EIA (2007) presented a retrospective review of the projections contained in the Annual Energy Outlooks between 1982 and 2007. The review shows that the overall energy demand was quite close to the actual demand but the difference was somewhat high for natural gas demand and energy price forecasts. The main driving variable, GDP, was less accurately forecast, which influenced other outcomes directly.

Although NEMS is a detailed model, its use has remained confined to government agencies and a limited number of research laboratories because of the model's reliance on costly proprietary software packages and complex model design.

5.2.4 Demand Modelling in WEM (World Energy Model)

The International Energy Agency uses the WEM for making long-term forecasts of energy demand and supply. The demand part of the model follows a hybrid approach where the econometric tradition is combined with the and end-use methodology. The model follows an energy balance or accounting approach in its overall demand forecasting and covers the final demand and the energy demand for the transformation sector separately. The final demand is broken down into a number of sectorsGú—industry, transport, residential and services. Further disaggregation is used in each sectorGú—for example, the industrial sector considers the main sub-sectors, the transport sector is analysed by mode and type of fuel, while the residential and services sectors consider fuels by end-uses.

In line with end-use models, the WEM uses activity variables (GDP or per capita GDP) and structural variables to take care of specific features of demand. However, unlike end-use models, WEM also uses price variables for energy end-uses by linking them to international energy prices and energy taxes. GDP, population, technological changes and international price variables are exogenous to the model and the model often uses scenarios to take care of a range of possibilities.

The demand equations have been econometrically estimated but adjustments are also made to take care of changes in the structure, policy or technologies. The model produces annual forecasts and the demand module can be run independently or in combination with other modules.

The model has undergone significant modifications in recent times and many new features and greater level of disaggregation have been added.

5.3 Conclusion

Forecasting is a venture into the unknown and therefore it is not a precise science. Consequently, the forecasts, more specifically the long-term energy demand forecasts, are rarely very accurate. Craig et al. (2002) reviewed some past long-term energy demand forecasts for the USA and shows how many forecasts were inaccurate. Often the forecasts overestimated the demand by 100% due to failure to anticipate structural changes, technological changes and "break-points." Similarly Bentzen and Linderoth (2001) reported that although OECD countries have recorded some improvements in energy forecasting, especially at the aggregated level, "the forecasting failure increases with the length of the forecasting horizon."

Koomey (2002) also indicated that forecasters need to avoid a number of big mistakes. These are:

(a) Using historical data for the determination of parameters and reliance on outdated assumptions. Both econometric tradition and engineering–economy tradition of demand forecasting suffers from this problem as "the fundamental relationships upon which they depend are in flux." "People and institutions can adapt to new realities" and "historically determined relationships... can become invalid."
(b) Conducting an analysis with an incomplete technology portfolio due to "data limitations, ideological precommitments, or lack of familiarity with these technologies." Consequently, the future scenario that deviates from the business-as-usual case appears to be costlier than that would happen in reality.

Koomey (2002) also suggested that the modellers should ask

(a) Whether they have made provisions for policy or other external influence induced changes to the historical relationships or not.
(b) Whether the model relies on a single forecast or a set of forecasts.
(c) Whether the modelling tool is driving or supporting the process of developing a coherent scenario and credence to the analysis or not.
(d) Whether the assumptions are "recorded in a form that can be evaluated, reproduced and used by others" or not.
(e) Whether the model has used robust strategies in the face of uncertain and imperfect forecasts.

5.3 Conclusion

Therefore it is important to understand the purpose of forecasting. As Craig et al. (2002) emphasise, "A good forecast can illuminate the consequences of action or inaction and thus lead to changes in behaviour. Although these changes may invalidate a specific numerical prediction, they emphasize, rather than detract from, the forecast's importance. One may judge a forecast successful if it (a) helps energy planners, (b) influences the perceptions of the public or the energy policy community, (c) captures the current understanding of underlying physical and economic principles, or (d) highlights key emerging social or economic trends."

Accordingly, for any forecasting exercise it is important to keep the following guidelines suggested by Craig et al. (2002) in mind:

(a) Instead of burying analytical assumptions in "black box," it is important to document them in "a form that can be evaluated, reproduced and used by others."
(b) Clearly define the audience of the forecasting exercise and the decisions they will make using the forecasts.
(c) Instead of focusing on complex programming or esoteric mathematics, it is important to focus on data and careful scenario creation. They support the view of Armstrong (2001) "that simple models can sometimes yield results as accurate as more complicated techniques."
(d) "Discontinuities are inherently difficult or impossible to predict, but they remain important to consider, particularly when they might lead to large, irreversible or catastrophic impacts."
(e) "Assuming that human behaviour is immutable will inevitably lead to errors in forecasting, no matter which kind of modelling exercise you undertake."
(f) "In the face of inevitably imperfect forecasts, the most important way to create robust conclusions is to create many well-considered scenarios. No credible analysis should rely on just one or two forecasts." "Quantitative analysis can lend coherence and credence to scenario exercises by elaborating on consequences of future events, but modelling tools should support that process and not drive it."
(g) Use combined approaches as "these techniques seem to be able to take advantage of the best characteristics of all techniques which comprise the combination. Combining different approaches allows biases in one technique to offset biases in other techniques."
(h) Identify and adopt strategies that are robust in the face of the inevitably imperfect and uncertain forecasts.
(i) Effective communication is essential to achieve greater influence in policy debates.
(j) "We need to be humble in the face of our modest abilities to foresee the future."

Annex 5.1: Mathematical Representation of Demand Forecasting Using the Input–Output Model

The value of output relations in a set of inter-industry accounts can be defined as:

$$X_i = \sum_{j=1}^{n} X_{ij} + \sum_{k=1}^{r} F_{ik}; \quad i = 1, 2, \ldots, n \tag{5.15}$$

where X_i is the value of total output of industry i, X_{ij} is the value of intermediate goods' output of industry i sold to industry j, and F_{ik} is the value of final goods' output of industry i sold to final demand category k (net of competitive import sales).

The final demand arises from a number of sources, which is shown in Eq. 5.16:

$$\sum_{k=1}^{r} F_{ik} = C_i + \Delta V_i + I_i + G_i + E_i - M_{Fi} \tag{5.16}$$

where C_i is the value of private consumer demand for industry i final output, V_i is the value of inventory investment demand for industry i final output, I_i is the value of private fixed investment demand for industry i final output, G_i is the value of government demand for industry i final output, E_i is the value of export demand for industry i final output and M_{Fi} is the value of imports of industry i final output (and often referred to as competitive imports).

It is assumed that intermediate input requirements are a constant proportion of total output, which is expressed as:

$$a_{ij} = \frac{x_{ij}}{X_j} \tag{5.17}$$

where a_{ij} is the fixed input–output coefficient or technical coefficient of production.

Equations 5.15–5.17 can be written more concisely in matrix form as

$$X = AX + F \tag{5.18}$$

where F = vector of final demand; A = matrix of inter-industry coefficients; and X = vector of gross outputs.

The well-known solution for gross output of each sector is given by

$$X = (I - A)^{-1} F \tag{5.19}$$

where I is the identity matrix, and $(I - A)^{-1}$ is the Leontief inverse matrix.

Thus, given the input–output coefficient matrix A, and given various final demand scenarios for F, it is straightforward to calculate from Eq. 5.18, the corresponding new values required for total output X, and intermediate outputs x_{ij} of each industry.

Annex 5.1: Mathematical Representation of Demand Forecasting

For energy analysis, the basic input output model is extended to include energy services. It is considered that the input–output coefficient matrix can be decomposed and expanded to account for energy supply industries (e.g. crude oil, traditional fuels, etc.), energy services or product equations (e.g. agriculture, iron and steel, water transportation).

Equation 5.19 is modified to a more general system as shown in Eq. 5.20

$$A_{ss}X_s + A_{sp}X_p + F_s = X_s$$

$$A_{ps}X_s + A_{pi}X_i + F_p = X_p$$

$$A_{is}X_s + A_{ii}X_i + F_i = X_i \quad (5.20)$$

where X_s = output vector for energy supply; X_p = output vector for energy products; X_i = output vector for non-energy sectors; F_s = final demand for energy supply; F_p = final demand for energy products; F_i = final demand for non-energy sectors. A_{ss} = I/O coefficients describing sales of the output of one energy/supply conversion sector to another energy conversion sector; A_{sp} = I/O coefficients describing how distributed energy products are converted to end-use forms; A_{si} = 0 implying that energy supplies are not used by non-energy producing sectors. Energy is distributed to the non-energy producing sectors via energy product sectors; A_{ps} = I/O coefficients describing how energy products—final energy forms— are used by the energy supplying industries; A_{pp} = 0 implying that energy products are not used to produce energy products; A_{pi} = I/O coefficients describing how energy products—final energy forms are used by non-energy producing sectors; A_{is} = I/O coefficients describing the uses of non-energy materials and services by the energy industry; A_{ip} = 0 implying that energy product sectors equipment require no material or service inputs. This is because they are pseudo sectors and not real producing sectors; A_{ii} = I/O coefficients describing how non-energy products are used in the non-energy producing sectors.

If we rewrite Eq. 5.19 in the summary form,
$X^E = A^E X^E + F^E$, where the superscript E indicates energy input–output matrices, the equivalent equation of Eq. 5.19 is

$$X^E = (I - A^E)^{-1} F^E \quad (5.21)$$

We could then calculate the various alternative final energy demand scenarios, the corresponding new total output requirements for non-energy industry and energy supply industry and energy services output and their respective intermediate outputs. Some perspective on inter-fuel substitution could also be gained, if one were satisfied that prices would not significantly influence the substitution process and if one were satisfied that the assumption of constant input output relationships would be true in practice.[12]

---

[12] *Source*: Based on Chap. 7, Macro-Demand Analysis, of Codoni et al. (1985). See also Miller and Blair (1985).

References

Al-Saba T, El-Amin I (1999) Artificial neural networks as applied to long-term demand forecasting. Artif Intell Eng 13:189–197

Armstrong JS (ed) (2001) Principles of forecasting: a handbook for researchers and practitioners. Kluwer Academic, Norwell

Bataille C, Jaccard M, Nyboer J, Rivers N (2006) Towards general-equilibrium in a technology-rich model with empirically estimated behaviour parameters. Energy J (Special Issue) 93–112

Bauer M, Mar E (2005) Transport and energy demand in the developing world. Energy Environ 16(5):825–844 (access to PDF file through DU library)

Bentzen J, Linderoth H (2001) Has the accuracy of energy projections in OECD countries improved since the 1970s? OPEC Rev 25:105–116

Bhatia R (1987) Energy demand analysis in developing countries: a review. Energy J 8:1–32

Bhattacharyya SC, Thanh LT (2004) Short term electric load forecasting using an artificial neural network: case of Northern Vietnam. Int J Energy Res 28(5):463–472

Bhattacharyya SC, Timilsina GR (2009) Energy demand models for policy formulation: a comparative study of energy demand models. World Bank Policy Research Working Paper WPS4866, March 2009

Brown MA, Levine MD, Short W, Koomey JG (2001) Scenarios for a clean energy future. Energy Policy 29:1179–1196

Codoni R, Park HC, Ramani KV (1985) Integrated energy planning—a manual. Asia Pacific Development Centre, Kuala Lumpur

Craig PP, Gadgil A, Koomey JG (2002) What can history teach us? A retrospective examination of long-term energy forecasts for the United States. Annu Rev Energy Environ 27:83–118

EIA (2003) The national energy modeling system: an overview 2003, Energy Information Administration, Department of Energy, report no. DOE/EIA-0581, Washington, DC. http://www.eia.doe.gov/oiaf/aeo/overview/overview.html

EIA (2007) Annual energy outlook retrospective review: evaluation of projections in past editions (1982–2006), Energy Information Administration, Department of Energy, DOE/EIA-0640 (2006), Washington, DC. http://www.eia.doe.gov/oiaf/analysispaper/retrospective/

Fletcher K, Marshall M (1995) Forecasting regional industrial energy demand: the ENUSIM end-use model. Reg Stud 29(8):801–811 (access to PDF file through DU Library)

Gerson L, David J (1995) Backpropagation in time series forecasting. J Forecast 14:381–393

Ghanadan R, Koomey JG (2005) Using energy scenarios to explore alternative energy pathways in California. Energy Policy 33:1117–1142

Hainoun A, Seif-Eldin MK, Almoustafa S (2006) Analysis of the Syrian long-term energy and electricity demand projection using the end-use methodology. Energy Policy 34(14): 1958–1970

Heaps C (2002) Integrated energy–environment modelling and LEAP, SEI. http://www.energycommunity.org/default.asp?action=42

Ho KL, Hsu YY, Yang CC (1992) Short-term load forecasting using a multilayer neural network with an adaptive learning algorithm. IEEE Trans Power Syst 7(1):141–149

IAEA (2006) Model for analysis of energy demand (MAED-2), manual 18. International Atomic Energy Agency, Vienna

Jefferson M (2000) Long-term energy scenarios: the approach of the World Energy Council. Int J Global Energy Issues 13(1–3):277–284

Koomey JG (2002) From my perspective: avoiding "the Big Mistake" in forecasting technology adoption. Technol Forecast Soc Change 69:511–518

Koopmans CC, te Velde DW (2001) Bridging the energy efficiency gap: using bottom-up information in a top-down energy model. Energy Econ 23:57–75

Labys WC, Asano H (1990) Process models, special issue of energy. Energy 15(3&4):237–248

Lapillonne B (1978) MEDEE-2: a model for long-term energy demand evaluation, RR-78-17. International Institute for Applied Systems Analysis (IIASA), Laxenburg. http://www.iiasa.ac.at/Publications/Documents/RR-78-017.pdf

Leydon K, Decker M, Waterlaw J (1996) European energy to 2020: a scenario approach, Directorate General For Energy (DG XVII), European Commission, Brussels. http://ec.europa.eu/energy/library/e2020fd.pdf

Lipinsky A (1990) Introduction, section 2, demand forecasting methodologies, special issue of energy. Energy 15(3&4):207–211

Metaxiotis K, Kagiannas A, Askounis D, Psarras J (2003) Artificial intelligence in short-term load forecasting: a state-of-the art survey for the researcher. Energy Convers Manag 44(9):1525–1534

Miller R, Blair P (1985) Input output analysis: foundations and extensions. Prentice-Hall, Englewood Cliffs

Munasinghe M, Meier P (1993) Energy policy analysis and modeling. Cambridge University Press, London

Park DC, Damborg MJ (1991) Electric load forecasting using an artificial neural network. IEEE Trans Power Syst 6(2):442–447

Reister DB (1990) The hybrid approach to demand modeling, special issue of energy. Energy 15(3&4):249–260

Saddler H, Diesendorf M, Denniss R (2004) A clean energy future for Australia, Clean Energy Group, Sydney, WWF Australia. http://www.bioenergyaustralia.org/reports/Clean_Energy_Future_Report.pdf

Saddler H, Diesendorf M, Denniss R (2007) Clean energy scenarios for Australia. Energy Policy 35:1245–1256

Shell (2003) Scenarios: an explorer's guide. Shell International, Rijswijk. http://www-static.shell.com/static/aboutshell/downloads/our_strategy/shell_global_scenarios/scenario_explorersguide.pdf

Shell (2008) Shell energy scenarios to 2050. Shell International BV, Rijswijk. http://www-static.shell.com/static/aboutshell/downloads/our_strategy/shell_global_scenarios/SES%20booklet%2025%20of%20July%202008.pdf

Sun JW (2001) Energy demand in the fifteen European Union countries by 2010: a forecasting model based on the decomposition approach. Energy 26(6):549–560

Sureerattanan S (2000) Backpropagation networks for forecasting. Ph.D. Thesis, AIT, Bangkok

Swisher JN, Jannuzzi GM, Redlinger RY (1997) Tools and methods for integrated resource planning: improving energy efficiency and protecting the environment, UCCEE, Riso. http://www.uneprisoe.org/IRPManual/IRPmanual.pdf. Accessed 15 Jan 2005

Further reading

Chakravorty U et al (2000) Domestic demand for petroleum in OPEC countries. OPEC Rev 24:23–52

Dahl C (1994) A survey of oil product demand elasticities for developing countries. OPEC Rev 18:47–85

Moroney JR (ed) (1997) Advances in the economics of energy and resources. Energy demand and supply, vol 10. Jai Press, London

Sterner T (ed) (1992) International energy economics. Chapman & Hall, London

Sun JW (1998) Changes in energy consumption and energy intensity: a complete decomposition model. Energy Econ 20(1):85–100

Chapter 6
Energy Demand Management

6.1 Introduction

So far, we have considered the evolution of energy demand, demand analysis and demand forecasting methods. In this chapter, our focus is on energy demand management. However, before we talk about demand management, we need to get an idea of the future energy demand. Various organisations produce such forecasts using different models. While there are some differences in the final outcomes, often there is a generic consensus in the pattern of demand growth. In this section, some information from the World Energy Outlook 2008 (WEO, 2008) is presented.

According the WEO (2008), the global energy demand is expected to reach about 17,000 Mtoe in 2030 from about 11,500 Mtoe in 2006 (see Fig. 6.1). This represents an annual average growth of about 1.6% for the entire period.

Oil is projected to remain the dominant fuel with a 30% share of the demand, followed by coal with a 29% share. Natural gas will occupy the third place with a 22% share. The renewable energies will see an increase in their share but are unlikely to play a major role.

The regional demand shares are expected to change as well with developing countries demanding more energy (see Fig. 6.2). The share of the non-OECD economies in global energy demand is expected to cross that of the OECD by 2010 and by 2030, non-OECD demand will account for 63% of the global energy demand. Asian developing countries as a regional block would represent the second most important energy demand centre in the world with a share of 38% of the global demand.

As fossil fuels continue to dominate the future energy scene, the supply-related concerns (due to the uneven distribution of resource endowments, infrastructure related constraints and possible depletion of some resources) coupled with environmental problems that these fuels bring have prompted to look at the demand-side management. This is what we consider in this chapter.

Fig. 6.1 Global energy demand forecast up to 2030. Source: WEO (2008)

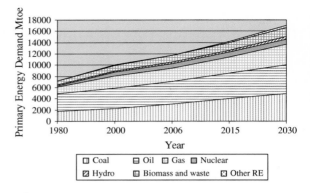

Fig. 6.2 Regional distribution of future energy demand. Source: WEO (2008)

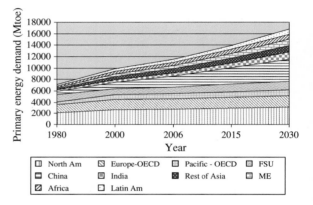

6.2 Energy Demand Management

We first define the concept, and then explain the need for managing demand, present the historical development in the area and discuss the alternative types of ways of managing demand in turn.

6.2.1 Definition

As discussed previously in Chap. 2, the energy system consists of both supply-side and demand-side activities. In the early days when energy prices were cheap, the focus on the energy sector was on the supply-side. This meant that for any given demand, the objective was to arrange for adequate supply so that the demand is satisfied. The demand-side was considered as given and there was a presumption that the supply-side is easily influenced and managed (perhaps due to less number of actors involved) than the demand-side.

However, with rising prices in the 1970s, researchers, governments and the utilities started to look at the entire gamut of the problem and it became apparent that ignoring the demand-side of the equation may not be an efficient way of managing the energy problem. The electric utilities in the USA were the first to experiment with this idea and the concept started to gain importance in other energy industries as well. Now the concept is used in the gas industry, transport sector, water industry and elsewhere.

Demand-side management (DSM)[1] of energy is the "systematic utility and government activities designed to change the amount and/or timing of customer's use" (CRA 2005) of energy for the overall benefit of the society. This is a generic term that is used to encompass various categories of activities such as (CSPM 2001; CRA 2005):

- Load management: Load management aims at reducing or changing the size or timing of the demand.
- Energy conservation: Energy conservation aims at reducing the demand, essentially through technical efficiency improvements.
- Fuel substitution: Fuel substitution aims at replacing one fuel by another and thereby modifies the demand.
- Load building: Load building implies developing load for strategic purposes which could help manage the system better. Although this appears to be in contradiction to the demand reduction objectives followed in the DSM, it could be relevant in certain circumstances.

This chapter focuses on load management and energy conservation as these represent the widely-used actions. Given the rich literature in this subject, this chapter would provide an overall understanding of the topic and draw examples mostly from the electricity industry. However, where appropriate other references for more information will also be provided.

6.2.2 Evolution of DSM

Over the past 30 years, DSM has evolved considerably. High oil prices in the 1970s provided justification for efforts directed towards reducing demand. The initial programmes were essentially aimed at energy conservation and load management, although the emphasis was on providing information on energy saving options and better understanding of energy demand through energy audits. This period also saw efforts towards fuel substitution, so that demand for imported oil is reduced by moving towards locally available fuels.

---

[1] Useful additional information is available from IEA DSM programme website (http://www.ieadsm.org/Home.aspx).

The 1980s saw a more systematic use of the DSM in the electricity sector through the least-cost capacity expansion and integrated resource planning programmes.[2] The least-cost capacity expansion programme attempted to identify the cheapest options for the utility considering the supply-side options while the IRP combined both supply and demand-side options. Systematic use of these options also led to the concern about revenue loss and regulatory treatment of the costs.

In the 1990s, as environmental concerns emerged, DSM received further support because of perceived benefits of these programmes. Yet, this was also the period of energy sector reform and DSM investments started to decline as the competitive markets started to emerge. The objective of price reduction through competition was in direct conflict with the demand reduction objective of DSM.

More recently, as energy prices have once again risen, the focus on better utilisation of energy has resurfaced (see for example EC green paper EC 2005). DSM activities received another lease of life as a result. Both energy efficiency and price responsive programmes are now being promoted as new breed of options.

6.2.3 Justification for DSM

The demand-side management can be justified for a number of reasons:

- The appeal for DSM arises from the fact that 1 MWh of energy saved is more than 1 MWh of electricity produced. This is because of the system losses. Any supply of energy involves transmission and distribution losses and consequently, any reduction in demand places lower pressure on system expansion.
- Similarly, because of generally low technical efficiency of the conversion process, the pressure on resource requirement reduces through demand reduction. This then brings additional benefits by reducing the pressure on additional infrastructure and by reducing the accompanied environmental damage.
- Demand management also improves the utilisation of the available infrastructure by distributing the demand over time and can reduce congestion or improve reliability of supply.
- As many countries depend on imports to meet their needs, a reduction in demand also reduces the import dependence, which in turn reduces the vulnerability to price fluctuations and thereby improves the supply security.
- Integration of the demand-side response in the market operation leads to better resource utilisation and therefore improves market operation.

---

[2] For details on the Integrated resource Planning see Swisher et al. (1997).

6.3 Load Management

Load management is one of the demand-side options that try to alter the load shapes so that the demand during the peak period is reduced (thereby reduce the demand for investment in new peaking capacities) and the facilities are better utilized at other times (which reduces the costs of production). A number of commonly used load management options are shown in Table 6.1.

Table 6.1 Load management options

(a) Peak clipping—this aims demand reduction during the on-peak hours. Often this is done either by restricting use of appliances during peak hours or by encouraging the consumers to change their demand behaviour by providing appropriate price signals. In the context of transport this can be done by disallowing certain vehicles to enter the road network or imposing congestion charges during peak hours.

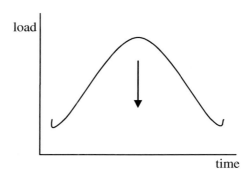

(b) Valley filling—the aim here is to promote use of energy during off-peak periods so that the level of average utilization of the facilities improves. This can be achieved by encouraging consumers to undertake for example charging and filling activities during off-peak periods when the utilities tend to use less capacity to meet demand. For example, large-scale charging of batteries for electric cars at night can improve the load shape greatly.

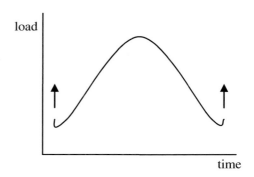

(c) Load shifting—this aims at moving loads from on-peak to off-peak periods without changing the pattern of energy usage. For example, consumers could store thermal heat during off-peak hours and use the heat to maintain the desired room temperate throughout the day. Similarly, certain activities in the households such as dishwashing and clothes washing can be operated at night to avoid peak loading.

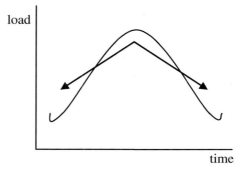

Table 6.1 (continued)

(d) Electrification (or enhancing energy access)—the objective here is to increase demand strategically by providing access to new areas or consumers.	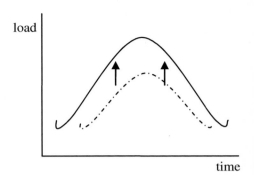
(e) Energy conservation—aims at reducing energy demand through efficient use of energy in efficient appliances or by changing lifestyles. Switching to efficient appliances can reduce the demand and change the load shape. In the case of transport, this amounts to better mileage per litre of fuel.	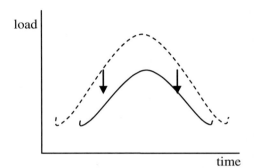
(f) Flexible load shape—this makes the load shape responsive to reliability conditions, meaning that loads could be modified as a function of system reliability.	

Source: CRA (2005) and Swisher et al. (1997)

Normally a utility employs a number of alternative ways to manage end-use demand. But it is customary to consider them in two broad categories: direct load control and indirect load control.

6.3.1 Direct Load Control Method

The utility reduces the demand by directly disconnecting, reconnecting or modifying the operation of the end-use device.

- The major energy consuming devices such as air conditioners, water heaters, space heating devices, etc. are directly controlled by the utility by sending appropriate signal for their operation.

6.3 Load Management

- The utility normally encourages customers to participate in its load management activities by providing incentives.
- An extreme form of direct control is the load shedding where the utility cuts supply to an area for a certain period to manage the demand. This is more common in developing countries which face chronic power shortages.

In the transport sector, restricting vehicle movement during peak hours would constitute a load management exercise in the road transport. For example, many cities do not allow lorries during the peak period to avoid congestion. Similarly, allowing only even (or odd) numbered vehicles to ply on a day could be another example.

In addition to direct curtailment or control of load by the utility, price-based mechanisms are also used to control loads. For example, EdF uses tariffs for interruptible loads where the consumer accepts that the utility would be able to disconnect a part of the load during peak periods of the year by providing the consumer a short notice. Similar price-based load curtailment schemes are also being used by other utilities as well. A more recent development in this direction is the use of demand bidding or buyback programmes in competitive markets. This offers incentives to consumers to reduce demand at certain times for payment of certain monetary incentives (see CRA 2005 for more details). Normally large consumers participate in such programmes, who often tend to have some sort of self-generating capacities. However, the demand bidding has not yet proved to be very popular because of the difficulty in evaluating the benefits for the customers.

The economic cost–benefit analysis for a direct control revolves around the following (Stoll 1989):

(a) Demand reduction leads to savings in fuel cost and there is saving in production, transmission and distribution related capacity costs.
(b) But the utility suffers revenue loss due to loss of demand and reduction in energy sales. It also incurs costs for managing the direct control systems and often has to provide incentives to customers to participate in the direct control options.

If the benefits exceed the costs, the option becomes suitable for the utility.

6.3.2 Indirect Load Control

The indirect load management on the other hand provides price signals to consumers to induce changes in energy demand patterns. This relies on the logic that consumers should be made aware of the varying costs of supply of the utility through appropriate price signals. Consumers should be allowed to decide the most economic level of consumption for them instead of forcing consumers to change their behaviour. This approach therefore requires designing appropriate energy charges such as time-of-day tariffs.

Indirect control of load is achieved through appropriately designed pricing systems. But the experience has remained uneven in this regard. In the American system, the electricity rates were based on traditional regulation system and the end-use tariffs were not marginal cost based. The presumption was that changing the price is more difficult and that any DSM option has to look for other incentives. However, experience from other countries as well as simple economic logic suggests that the price could be the most important driver for energy demand. For example, EdF in France relies on the time of use rates for electricity and this price has generated high degree of awareness amongst consumers.

Although TOU rates are discussed in the textbooks, which essentially follow the marginal cost-based pricing principle, they are relatively less used in practice. Yet, recent experience indicates that these rates are effective in managing the demand. [See CRA (2005) for a list of a number of alternative tariff options.]

6.4 Energy Efficiency Improvements and Energy Conservation

Energy conservation can be defined as (Munasinghe and Schramm 1983) "the deliberate reduction in the use of energy below some level that would prevail otherwise". This deliberate reduction often involves a trade-off involving comfort and other factor inputs (capital or labour) and in the extreme case, conservation may lead to deprivation, especially for the vulnerable section of the population.

In this section, we focus on the energy efficiency improvements as a means of energy conservation. The objective of energy efficiency improvements is to reduce energy demand through better use of energy in energy consuming devices. It is commonly believed that the efficiency of end-use appliances is not often high, which results in losses, higher demand for inputs and consequently environmental damages. By improving efficiency of energy utilisation, energy demand could be managed along with environmental benefits.

6.4.1 What is Energy Efficiency?[3]

Although the term is frequently used, different users tend to attach different meanings to the term depending on the focus of analysis. A number of definitions can be found in the literature having the following basic origin:

$$\text{Energy efficiency} = \frac{\text{Useful output of a process}}{\text{Energy input into a process}} \quad (6.1)$$

---

[3] This section is based on Paterson (1996). See also Herring (2006).

6.4 Energy Efficiency Improvements and Energy Conservation

The above definition is very similar to the one used in thermodynamics, the science of energy and energy processes. In thermodynamics, efficiency is defined as the ratio of heat content of the output to that of inputs. For example, if the efficiency of an incandescent lamp is 6% in the thermodynamic sense, only 6% of the input energy is converted to light while the rest 94% is lost to the environment.

However, the above thermodynamic definition does not distinguish between low and high quality of energy. For example, electricity as high quality energy could produce different levels of productive output than the low quality solar energy. This creates the problem of comparing the non-comparables.

To avoid this problem, another concept of efficiency is used which compares the efficiency of an actual process with that of an ideal process of doing the same work. For example, if the actual efficiency of power generation is 30% and the theoretically achievable ideal efficiency is 72%, the power generation process can be considered as $30/72 = 42\%$ efficient.

However, there are problems with this concept as well:

- the ideal may not be easy to determine and define;
- the real-life processes may not follow the standard systems considered in the thermodynamics;
- perfect reversibility assumption used in the ideal system does not really hold in practice.

Consequently, the thermodynamic efficiencies find limited use outside engineering design.

For energy analysis, emphasis is laid on energy services provided per unit of input (Gillingham et al. 2009). Energy analysts have adjusted the above energy efficiency definition by modifying the numerator to capture the physical outputs (such as tonnes of a product or ton-kilometres of freight transport, which implies one tonne of a good transported over a kilometre). This measure allows objective measurement of the output and is amenable to time series analysis. At times, the inverse of the above ratio is used: energy input per unit of output measures the efficiency. For example, the efficiency of a car is often expressed as litres of fuel used per 100 km travelled. Lower the input requirement, more efficient is the energy use. Due to heterogeneity of outputs or activities, sector or activity specific indicators are required to measure energy efficiency in physical terms.

(a) In the residential and commercial sectors, the commonly used indicator is energy input per square metre, although this assumes that energy requirement is directly proportional to the area of the building. This assumption may not be correct, as cooking, water heating and such needs may not have a direct relation to the area or volume of the building.

(b) The commonly used physical indicators in the transport sector are energy input per passenger-kilometres for passenger transport and energy input per tonne kilometres for freight transport. Although these indicators are often used, they ignore the importance of time dimension in the sense that the objective of minimisation of transport time is not captured.

(c) In the industrial sector, energy input per tonne of output is the common indicator, although in some cases, energy input per volume of output is also used.

Energy intensity is also used as a measure of energy efficiency. Here, the output is measured in money terms and as in the case of physical indicators, the ratio is reversed (energy input per $ of output is used) here as well (hence lower the energy intensity, energy efficient the system is). Although this is widely used, there are various issues related to this indicator which were discussed in Chap. 3.

A European project maintains information on energy efficiency indicators for 15 countries in the ODYSSEE database.[4] Three types of indicators are used in this project: economic indicator such as energy intensity which relates energy consumption per unit of economic activity, techno-economic indicators that relate energy consumption to an activity measured in physical terms (such as unit or specific consumption) and energy savings. The environmental benefits of efficient energy use are also captured through some environmental indicators.

6.4.2 Opportunities for Energy Saving

A number of end-use areas can be easily identified for better utilisation of energy resources:

(a) Lighting is an area which has caught attention for quite sometime because of low efficiency of lighting appliances.

  a. In general, the standard incandescent lamps have a low technical efficiency (around 10–15%) whereas energy efficient lamps could reach significantly higher levels of technical performance.
  b. In addition, users often leave lights on even when they are not required (for example street lights during day time, outside lights throughout night, and lights in unused rooms, etc.).
  c. Inefficient technology and unnecessary use are often promoted by poor pricing policies for the energy carrier, high initial cost barrier, poor building designs where daylights are incorrectly used, and lack of awareness.

(b) Space heating and cooling is another area with saving potential. Energy saving in heating or cooling can be achieved through better insulation and prevention of leakages. As the buildings tend to have a very long life, the older buildings tend to be less efficient compared to the relatively newer ones. Often the regulations allow derogations for older constructions, allowing them to continue with inefficient use of energy. Building codes and their effective implementation and building design play an important role here.

---

[4] See http://www.odyssee-indicators.org/

(c) Appliance efficiency and usage: Inefficient appliances and un-mindful use provide scope for energy saving at a relatively low cost. Improving efficiency of domestic appliances (refrigerators, freezers, washing machines, vacuum cleaners, cookers, water heaters and dishwashers) and their appropriate use could save electric energy. Given that electricity is often generated inefficiently and that there are transmission and distribution losses to supply electricity, any saving in electricity at the end-user level, leads to savings in capacity usage and fuel use, thereby providing environmental benefits as well.

(d) Transportation is another area where losses remained high. As transportation is required for moving personnel or goods over distances, energy can be saved by:

   a. switching to less energy intensive modes of travel,
   b. improving the efficiency of transportation and
   c. by changing travel behaviour (Munasinghe and Schramm 1983).

The efficiency of a car is normally around 15% and this level remained so for a relatively long period. Better engines, better roads and adequate traffic management are essential to reduce fuel consumption in transport. Similarly, energy intensity of transport reduces with better capacity utilisation of the fleet and use of mass transit systems. Even behavioural changes as well as expansion of electronic communication can reduce the need for undertaking a travel: for example, bill payments could be done over telephone instead of visiting an office, thereby eliminating the need for travel. By encouraging people to share vehicles with colleagues (to improve capacity utilisation and reduce the number of vehicle on road), to live close to work place and encouraging (or allowing) to work from home could save energy.

However, energy saving requires better awareness as well as proper pricing of complimentary and substitute goods as the demand for transport use tends to be inelastic for a wide band of fuel price, making fuel pricing a less effective policy to induce change. For example, higher parking charges and reduction in free parking entitlements tend to reduce short distance travel. Similarly, adequate and flexible public transport system allows substitution of personal car usage for commuting. Moreover, the long life of a vehicle implies that it takes a long time to upgrade the vehicle stock and achieve an efficient capital stock. Therefore, changes take a longer time to show results.

(e) Electricity generation: Although this is a supply-side activity, we discuss this here given its importance as a primary energy user.

   a. Electricity being a high grade, versatile energy, its demand is increasing in the final use relatively fast. However, the efficiency of electricity generation processes is relatively low (often less than 30%).
   b. In addition, a significant amount of electricity is lost in transmission and distribution, causing further losses.
   c. Improving the efficiency of electricity generation process by using modern technologies (like combined cycle gas turbine, supercritical pressure

boilers, etc.) and reducing losses in transmission and distribution can save significant amounts of fossil fuels.

At the sector level, each sector shows certain scope for improvements. For example,

(a) Industries could save energy through process changes (e.g. moving from wet process of cement making to dry process), utilising waste heat, efficiently using heat and power through cogeneration, and using efficient lights and motors. As the cost of production and competitiveness of energy-intensive industries depend on the energy input costs, efficient energy use becomes a necessity for survival of such industries, unless such industries are protected through from international trade. The economic justification for saving tends to be pronounced for such industries. For the rest, the signal may be less clear, requiring incentives and awareness generation for any improvement.
(b) Commercial sector uses electricity in large quantities and could save energy through better housekeeping, avoiding leakages, using efficient lights and through better building designs.
(c) Transport: In addition to passenger transport, major energy saving is possible in freight transport by reducing poor capacity utilisation, through improvements in road and traffic conditions, better signs and signals, and taking advantages of economies of scale.
(d) Agriculture does not figure as a major energy consumer in many countries but efficient use of energy can be a concern for major agricultural economies. Generally, water pumping for irrigation purposes consumes a major share of energy in agriculture. This has the potential of improving through use of efficient motors and pumps, reducing wasteful use of water and improving the irrigation technology.

6.4.3 Economics of Energy Efficiency Improvements

The basic idea of energy efficiency improvements can be captured using the production function concept. Assume that capital and energy are two inputs into a production process. The cost minimising level of energy use is obtained at the point of tangency of the isoquant and the total cost line. At this point, the rate of technical substitution is equal to the ratio of the inputs' rental prices. As shown in Fig. 6.3, a change in relative prices from P0 to P1 results in a substitution process where capital is substituted for energy by moving along the same isoquant. Here, lesser energy input is used per unit of output but results in a higher use of the other factor of production, capital.

It is also possible that the production possibility moves to a different isoquant consequent to a technological change. This is shown in Fig. 6.4 where a technological change results in a shift from I0 to I1. It results in a lower level of energy

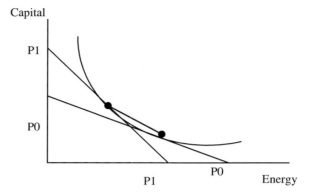

Fig. 6.3 Energy efficiency improving substitution. Source: Gillingham et al. (2009)

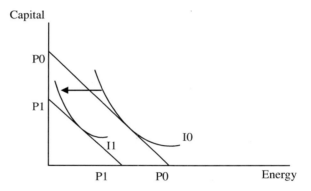

Fig. 6.4 Energy saving technological change. Source: Gillingham et al. (2009)

input for any combination of capital used, and therefore produces greater energy efficiency.

The basic economic principle used for evaluating energy efficiency investments is the cost–benefit analysis. In such an analysis, the total system cost of any energy saving activity is compared with the total benefits and as long as the benefits and greater than the costs, the conservation programme is economically justified. The benefits and costs are not just the present day costs or benefits; the expected lifetime values have to be considered for a meaningful analysis.

In mathematical terms, this can be written as

$$\sum_{t=1}^{n} b_t \left(\frac{1}{1+r}\right)^t > \sum_{t=1}^{n} c_t \left(\frac{1}{1+r}\right)^t \tag{6.2}$$

where b_t and c_t are annual savings and costs respectively for year t, where t varies from 1 to n, and r is the discount rate.

When different alternative options are considered, if they are assumed to generate same or similar benefits, the above decision criterion can be modified to minimisation of total system costs. The least costly option is then considered as the appropriate solution.

It is clear from the above equation that the choice of discount rate would influence the decision. A low discount rate attaches higher value to the future cost-savings through lower discounting, which would make many programmes viable. As the discount rate is essentially a reflection of the cost of capital faced by the users, this has significant policy implications. "Energy users who confront high opportunity costs of capital (as for example, in many developing countries) will find capital-cost-intensive energy conservation measures less attractive than users who have access to low-cost sources of capital" (Munasinghe and Schramm 1983).

6.5 Analysing Cost Effectiveness of DSM Options

The basic economic tool used in determining the effectiveness of DSM options is the cost–benefit analysis. However, as the cost-effectiveness often depends on the eyes of the beholder (Swisher et al. 1997), the analysis is done from different perspectives. The California Public Utilities Commission has been active in formulating standards for use by the state utilities in this regard since late 1980s and has developed a set of tests discussed below to consider various perspectives (CSPM 2001). These tests are briefly discussed below (a summary is given in Table 6.2).

Table 6.2 Summary of tests for DSM effectiveness

Test	Measures	Costs	Benefits	Selection criteria
Participant	NPV Discounted pay-back Benefit–cost	Costs directly incurred by the participant	Reduction in bills, incentive receipts	Acceptable when cost to the participant is lower than the benefits
Rate impact measure	Lifecycle revenue impact per unit of energy NPV Annual revenue impact	Costs to utility for administering the programme and lost revenue for lower sales	Avoided costs for supply and capacity expansion	Acceptable when rates do not rise after DSM
Total resource cost	NPV Benefit–cost ratio	Cost incurred by the utility and the participant in administering the programme	Avoided costs of supply and capacity expansion	Acceptable when the net benefits are higher than the costs
Utility cost	NPV Benefit–cost ratio	Cost incurred by the utility	Avoided costs of supply and capacity expansion	Acceptable when costs to the utility are lower than the benefits

Source: Based on CSPM (2001) and Swisher et al. (1997)

6.5 Analysing Cost Effectiveness of DSM Options

6.5.1 Participant Test

This measures the costs and benefits of participating in the DSM programme. The test can be performed either for an average consumer or for all participants as a whole. By participating in the programme, a consumer would

- See a reduction in energy consumption and hence a reduction in her utility bill by participating in the programme.
- She may also receive other incentives from the utility (or government) and perhaps tax breaks. All these would have to be considered over the life of the investment.
- On the other hand, the costs involve purchase of new equipment (or appliance), operating and maintaining the equipment and any other costs related to removal at the end of the life.

This cost and benefit information is used in different indicators like net present value (NPV), discounted pay-back period or cost–benefit ratio.

6.5.2 Ratepayer Impact Measure (RIM)

This test measures how the customer rates will be affected as a result of the programme. If the revenues to the utility increase compared to the costs, the rates are expected to fall and vice versa.

- Normally through DSM the utility saves in terms of supply capacity expansion for generation, transmission and distribution facilities. The avoided expenses (or cost) reduce the revenue requirement of the utility for providing the regulated service.
- On the other hand, the utility incurs additional costs
  - in administering the DSM programme
  - for providing incentives to the customers and
  - for ensuring increased supply during periods when the demand increases (due to load shifts)
- The utility also suffers revenue reduction due to loss of demand. These costs have to be included in the overall impact assessment.

The impact can be estimated over the life time of the programme. Other measures include NPV, annual changes in revenue, impact in the first year and the benefit–cost ratio.

In order for a DSM programme to be cost-effective under the RIM test, the utility rate must not increase after the introduction of DSM. As Swisher et al. (1997) indicate, the ratepayers will not see the rates going up:

- As long as the cost of saving one kilowatt-hour of electricity is less than the benefits earned (i.e. the difference between the marginal cost and the average cost).

- When the marginal cost is higher than the average cost, this relationship will hold. When the marginal cost is lower than the average cost, the rates will rise.

6.5.3 Total Resource Cost Test

This measures the net cost of the DSM programme combining the net costs incurred by the participant and the utility. Net costs are obtained as the difference between the benefits and the costs for programme.

- The benefits include the costs avoided by the utility for supply capacity expansion and for providing the supply.
- The costs include costs of equipment, operation and maintenance, administrative costs and any removal cost at the end of life of the equipment.

As this provides an overall picture of costs and benefits of the programme, this measure is used widely. The indicators used for this test are NPV and cost–benefit ratios.

For example, if the utility spends $0.02/kWh in rebates for promoting efficient lamps and if the consumer invests $0.03/kWh for switching to efficient lamps, the total cost for the programme is $0.05/kWh. If the overall benefit is more than $0.05/kWh then the programme passes the total resource cost test (Swisher et al. 1997).

6.5.4 Programme Administrator Cost or Utility Cost Test

In this test, the net costs incurred by the programme administrator (often the utility) are taken into consideration and ignores the costs borne by the participants. Thus this test looks at the costs from a narrower angle. If the utility saves money by implementing a DSM programme, its revenue requirement will reduce and will justify the utility participation in the programme. NPV and benefit–cost ratios are the indicators commonly used in this test.

As indicated earlier, it is often difficult for any DSM programme to pass all the tests. For example, a programme may pass the total resource cost test but fail the ratepayer impact measure test. Similarly, it may pass the utility test but fail the total resource cost test. In such a case, a final judgement becomes important.

Example To reduce electricity demand a utility intends to introduce compact fluorescent lamp (CFL) costing $15 each for a standard 18 W lamp (with effective output of 75 W). The utility proposes to offer a $5 rebate on each lamp and expects that one million consumers would take advantage of this scheme. A CFL lasts for 10,000 h. The utility avoids the cost of electricity purchase at 4 cents per kWh but

6.5 Analysing Cost Effectiveness of DSM Options

Table 6.3 Comparison of test results

Test	Costs	Benefits	Net position
Participant test	$10 M	$14.28 M	$4.28 M gain
Utility cost test	$6 M	$22.8 M	$16.28 M gain
Rate impact test			$14.28 M transfer to consumers
Total resource cost test	$16 M	$22.8 M	$6.28 M

incurs an administration charge of $1 per customer. If the utility intends to pass 85% of the net savings to the consumers, determine how the proposal fares from the perspective of different stakeholders?

Answer: Cost to the utility = $(5 + 1) * 1 = \$6$ million.
Cost avoided by the utility = $[(75 - 18) * 10,000/1,000] * 0.04 * 1 = \$22.8$ million.
Net benefit to the utility = $(22.8 - 6) = \$16.8$ million.
Benefits passed to participants = $\$16.8 * 0.85 = \$14.28$ million.
Cost to participants = $(10 * 1) = 10$ million.
Table 6.3 provides the overall positions for different stakeholders.
This example leads to a win–win situation for all.

6.6 Energy Efficiency Debate

There are a number of debatable areas in the area of energy efficiency including the following: (a) market barriers and intervention, (b) energy efficiency versus economic efficiency, and (c) rebound effect (and back-fire)[5] and declining benefits of energy efficiency. We discuss these issues briefly below.

6.6.1 Market Barriers and Intervention Debate

As with many economic policies, energy efficiency faces the debate whether government intervention is required or not. The proponents of interventionist policies believe that there are important market barriers which justify government policies promoting energy efficiency. Others believe that markets should take care of the problem if they are properly allowed to operate. There is a huge body of literature on this issue (Gillingham et al. 2009 for a recent review), and only a brief summary is presented below.

The discussions on energy barriers often use the term efficiency gap. This is the difference between the cost-effective level of investment in energy efficiency

---

[5] These terms are defined later in the chapter.

Fig. 6.5 A stylised conserved energy supply curve

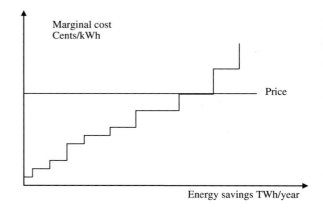

based on engineering-economic analysis and the level of investment actually made (Golove and Eto 1997).[6] The gap is usually presented graphically in a supply curve of conserved energy which is a schedule of potential conservation of energy ranked by the marginal cost of supply (see Fig. 6.5). The curve indicates the amount of energy that could be saved at different levels of costs and the price line shows which efforts are cost effective at the prevailing price.[7] Normally, these curves are based on technical potential analysis that assumes an instantaneous switchover to a new technology. As such a switchover is not practically feasible and also because different users use different discount rates and face different barriers and market imperfections, the actual level of investment in energy saving is lower than that perhaps would be ideal. Accordingly, the theoretical potential is quite different from that actually feasible. Grubb et al. (1993) attribute this gap to a number of factors as shown in Fig. 6.6. From that diagram it is clear that only a part of the potential is really realisable even after putting in place correct policies.

6.6.2 What are the Market Barriers to Energy Efficiency?

Generally six types of barriers are identified in the literature (Golove and Eto 1997; Brown 2001): (1) misplaced incentives, (2) lack of access to financing, (3) flaws in market structure, (4) inappropriate pricing and regulation, (5) gold plating, and (6) lack of information or misinformation.

When the energy user and the investor in efficient energy are not the same, the incentives targeted to efficiency may not reach its target and would result in

---

[6] See also Jaffe and Stavins (1994).

[7] See Stoft (1995) for further details on the conserved energy supply curve. Also see Grubb et al. (1993).

6.6 Energy Efficiency Debate

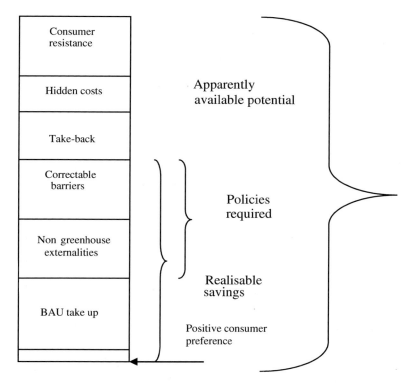

Fig. 6.6 Realisable versus potential energy savings. Source: Grubb et al. (1993)

misplaced incentives. For example, in rented properties, the landlord makes the decision for most energy intensive appliances but as she does not benefit from energy conservation, the landlord does not have any incentive to invest in costly energy efficient appliances. This behaviour can also be found in the commercial sector activities.

Financing barrier applies to various types of energy users but may be more relevant for poor (or low-income) group. Given limited financial resources available to meet unlimited demands, each consumer would make ranking of their investment choices. Investment in energy efficiency often does not rank high up in the agenda for many consumers. Moreover, a certain section of the population cannot afford investment in energy efficient technologies due to low creditworthiness and inability to access financial markets.

Existence of market power and non-competitive markets also could affect penetration of energy efficient technologies. Appliance market can see such imperfections in a number of ways: domination of the local market by a single company (for example, lighting market may be dominated by a major supplier who may not want to promote efficient technology), protectionist regulation (for example in many developing countries cooking appliances are manufactured by small scale industries, thereby preventing technological updating), trade barriers (e.g. imposition of quotas and tariffs on imports to protect local producers), etc.

Pricing issues are often related to the industry structure as well. If the utility charges inappropriate prices that do not provide consumers with adequate signals, it is unlikely that consumers would recognise the benefits of energy efficiency and conservation. Subsidised supply prevailing in many countries does not make investments in energy efficiency attractive. In addition, non-internalisation of externalities distort prices of energy and leads to non-optimal use of resources.

The gold plating argument suggests that energy efficient options do not come alone but are often bundled with other features. The choice is between an efficient technology with other bundled features or to settle for an inefficient technology. Another example of this type of features is the standby operation of appliances. A large number of appliances now come with sleep mode or standby mode although a significant amount of energy is consumed in this way as these appliances are always on. An IEA study indicates that standby power consumption account for between 3 and 13 per cent of residential energy demand in the OECD countries (IEA 2001).

Finally, a number of information-related hindrances to energy efficiency investments has been identified (Golove and Eto 1997): (1) the lack of information, (2) the cost of information, (3) the accuracy of information and (4) the ability to use the information. In standard economic discourse, it is assumed that information is available freely, is costless and the users have the ability to treat them as appropriate. However, these assumptions do not hold in practice as information gathering and supply is a costly process. Similarly, the users have a limited capacity to use the information and often the required level of information is not available. As a consequence, the information constraint affects the decision-making of a common user of energy. Information also has the features of a public good whose consumption by one does not diminish its availability to others and often can be made available at a very low marginal cost.

An example will make this point clear. For example, a consumer wants to buy a refrigerator for domestic use. She can buy a commonly used one or an energy efficient one at a premium. While the cost difference is apparent to the consumer, the savings in operating cost is not so apparent. It requires an analysis involving discounted stream of benefits which depend on the price of electricity at various point of time in the future, annual saving in electricity and the discount rate. Evidently, the consumer finds it difficult to perform such a calculation mentally or otherwise. Moreover, any consumer would find it difficult to find information on electricity prices, electricity consumption by the appliance (which depends on the usage pattern of the user) and the appropriate discount rate. Thus, the decision to buy an energy efficient appliance may not find favour because of initial cost difference, although the decision could be different if the user could perform the involved calculation convincingly and effortlessly.

6.6.3 Government Intervention and Its Nature

Existence of market failures normally implies that some corrective actions are required, giving ground for government intervention. In the case of energy efficiency, information related problem is an area where government intervention can be legitimate as private parties may not be interested in providing such a public good. Also, in the presence of externalities when prices do not reflect the correct cost to users, the consumption decision will be incorrect and governments can intervene to correct such externalities through pricing or regulation.

Swisher et al. (1997) suggest a number of government initiatives: (1) information and labelling, (2) standards and regulation, and (3) financial and fiscal mechanisms. See WEC (2001) for a review of policies in various parts of the world.

Information and labelling programmes aim at disseminating information on energy technologies, efficiency measures and incentives, as well as generating awareness about efficient use of energy. These could include educational and training programmes, performance labelling for specific products, awareness raising campaigns through leaflets, advertisements and seminar conferences (Swisher et al. 1997). The costs and the efficacy of the information programmes remain debatable as the impact is often short-lived unless they are coupled with initiatives. For example, educational programmes may have a better impact if decision-makers are trained to appreciate the value of energy efficiency through their curriculum. Clear and understandable labelling can make consumers aware about the choice. In the EU, labelling is used for a large number of electrical appliances to indicate their level of energy efficiency. Despite a significant effort in this direction, the effectiveness of the labelling remains unclear.

Standards and regulation tend to influence behaviour by stipulating certain minimum levels of performance or acceptable level of behaviour. Accordingly, there are two types of standards: prescriptive standards and performance standards. The standards can be used in a large number of areas to induce better energy efficiency. Examples include standards for appliances, buildings, vehicle and transport systems and lighting. The advantage of standards is that it forces all concerned to move to the desired direction. Although reliance on regulation and standards is quite common in both developing and developed world, mere having them does not ensure better energy efficiency or better performance. Monitoring and enforcement are essential for the success of regulations and standards but lack of adequate regulatory enforcement arrangements makes their implementation difficult in many countries. Moreover, the cost of regulatory compliance increases as the standards become stricter.

Financial and fiscal incentives form another type of government intervention in the energy efficiency market. A number of financial incentives are provided in many countries: grants and subsidies, tax relief, favourable depreciation rates and loans are some such options (see Price et al. 2005 for further details and an international comparison of such incentives). Most of these incentives often apply to the industrial sector or large users. Direct incentives for retail consumers are not widespread, although taxes, pricing and fiscal measures are used to promote one technology in detriment of the other. For example, different registration charges for cars and tax differences for fuels are used to influence consumer behaviour (see WEC 2001 for further details). In some places, banks and lenders are encouraged to promote energy efficient housing through specially designed mortgages. However, fiscal incentives when used without careful consideration can lead to undesirable behaviour or distortion.

6.6.4 Energy Efficiency Versus Economic Efficiency Debate

The objective of energy efficiency came under criticism from some economists who questioned the implicit assumption that equates energy efficiency to economic efficiency. The proponents of energy efficiency tend to suggest that since energy efficiency reduces wasteful use of energy, it is a reasonable objective and by promoting such policies, economic efficiency will be improved. However, this argument was questioned on the ground that energy efficiency is an engineering concept which considers only a single factor while economic efficiency aims at choosing the most appropriate combination of factor inputs to generate an output. Thus minimising use of a single input may not be a desirable objective for economic development because (Sutherland 1994):

(a) Normally for economic development, higher inputs of labour, capital and other factors are targeted. The objective of reducing energy inputs would be contradictory to this standard policy objective.
(b) In doing so the capital investment in one input increases without taking into account the substitution possibilities of other factor inputs, which results in an economic distortion.
(c) Selecting a technology based on minimisation of total costs is a rationale choice rather than just minimising the energy cost.

In this respect, four possibilities could be considered in the energy efficiency–economic efficiency debate:

(a) Those investments in energy efficiency which improve economic efficiency are the desirable ones.
(b) There may be investments which improve energy efficiency but reduces economic efficiency.
(c) There could be cases where economic efficiency improves with higher energy use (i.e. reduced energy efficiency) but a focus on energy efficiency would not allow such investments.
(d) Finally, there may be a case where neither energy efficiency nor economic efficiency improves and such investments would be rejected (see Table 6.4) for a summary.

6.6 Energy Efficiency Debate

Table 6.4 Energy efficiency versus economic efficiency

Options	Increases energy efficiency	Decreases energy efficiency
Increases economic efficiency	Energy efficiency scenario	Energy assisted growth
Decreases economic efficiency	Not promoted	Rejected as undesirable

Source: Golove and Eto (1997)

This suggests that the focus on energy efficiency investments could be misleading in cases when they are not promoting economic efficiency (i.e. when economic efficiency decreases or investments are not made despite economic efficiency).

Therefore, energy efficiency policies which are consistent with the economic efficiency should be considered to achieve better utilisation of resources.

Saunders (2009) presents a theoretical exposition of the problem following a simple framework presented by Hogan and Manne (1977). Assume that an economy produces output Y using energy E and other inputs R and can be expressed as

$$Y = f(E, R) \tag{6.3}$$

Where the production function is assumed to have usual properties commonly assumed in the literature.

Assuming R as fixed, the effect of changes in energy input on the economic output can be captured graphically (see Fig. 6.7). At any point on the curve the economic output corresponding to the energy input is indicated. The tangent to the curve at this point indicates the marginal productivity of energy and in a competitive market, this must be equal to the energy price (Saunders 2009). Any movement to the left of the curve indicates an increase in the slope of the tangent (from B to A), implying an increase in price, and therefore a reduction in energy use, and a consequent reduction in output. But reduction in economic output is not a one-for-one reduction or proportional to the energy input. This therefore represents an energy efficiency gain.

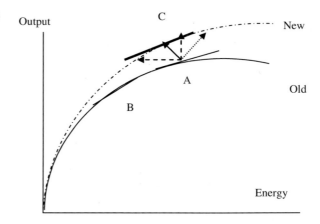

Fig. 6.7 Economic output as a function of energy input. Source: Based on Saunders (2009)

The production possibility changes as efficient technologies are introduced. Accordingly, more output can be produced with the same amount of energy. This is shown by the curve in broken lines. At point C the slope of the tangent to the new curve equals that at point A, implying equal energy prices at both the positions. Although energy input has fallen, the economic output has increased. Therefore, while the energy intensity has decreased with new technology, the output effect has offset part of the benefits and the net effect is the cumulative effect of the two.

6.6.5 Rebound Effect

Take-back or rebound effect implies that a part of the energy saved manifests itself in higher consumption. This is what is shown in Fig. 6.7 using the arrows. If the energy efficiency gain leads to an equivalent percentage change in energy input, then there is no rebound or take back. The figure indicates a number of alternatives: between zero and 100%. If the rebound is more than 100%, it is called as back-fire. The right-most arrow indicates such a situation. These are all theoretical possibilities but the extent of their occurrence in real-life is more difficult to ascertain.

For example, efficient lighting could result in higher consumption of electricity through other appliances. The same is true for transportation energy use. Three types of rebound effects have been identified in the literature (Herring 2006):

(a) Direct rebound effect—is the increase in energy consumption due to fall in energy prices as a result of better efficiency. The effect is similar to that of energy price reductions.
(b) Indirect rebound effect—which results from the saving in expenses from the reduction in energy-related costs, allowing the consumer to spend more on other goods and services, including energy.
(c) Tertiary or general-equilibrium effect—is the result of adjustments in supply and demand involving all producers and consumers in all sectors.

It is clear that the efficacy of the energy efficiency programme would ultimately depend on the magnitude of the rebound effect. If the size is relatively small (say 10%), then the energy efficiency programmes could still be viable. But if the rebound effect is 80–90%, then clearly the viability of the projects would be questioned.

The empirical verification of the rebound effect is not easy and most studies often focus only on the direct effects, which are found to be relatively small (less than 20%). But ignoring the other two components leads to underestimation of the take-back effect. A recent review by Sorrel (2009) confirms the difficulties in ascertaining the rebound effect but suggests that it is significant. For consumer energy services in OECD countries, the direct effect is less than 30%. Sorrel (2009) reports that the economy-wide rebound effect has been estimated by a few studies as 26% or more and some have indicated back-fire. However, the results come from a small sample of studies and the methodologies often influence the results.

6.6.6 Use of Market-Based Incentives for Energy Efficiency

Finally, we focus briefly on market-based tools for energy efficiency. Instead of relying on command and control tools to promote energy efficiency, market-based mechanisms could also be used. One such tool is the use of white certificates. This relies on the Cap and Trade principle where the participants are given an obligation for energy saving which they can achieve either by investing in energy saving technologies, by paying penalties for non-compliance or by buying white certificates from others who have saved more than their obligation and earned certificates for sale or for future use (called banking). This allows the participants to decide their strategy and does not force any option on them. This normally is a less costly option. This is being used in certain European countries (such as Italy, France and the UK) as a market-based mechanism to promote energy efficiency. A study by Bertoldi and Rezessy (2006) indicates that although the regulatory and transaction cost element can be high in this case, it offers greater potential benefits in terms of certainty of outcomes and economic efficiency.

6.7 Conclusion

This chapter has introduced the concepts of demand-side management for energy and considered two major options—load management and energy efficiency. The tools and tests for economic analysis are also presented and the debate around energy efficiency is discussed. The divergence between the technical view and economic view about energy efficiency arises from the differences in the perspectives used by the two sides and through a careful application of DSM options, both economic and environmental benefits are possible at a lower cost.

References

Bertoldi P, Rezessy S (2006) Tradable certificates for energy savings (White Certificates): theory and practice. EUR 22196 EN, Institute for Environment and Sustainability, Directorate General, Joint Research Centre, European Commission (see http://www.jrc.cec.eu.int)
Brown M (2001) Market failures and barriers as a basis for clean energy policies. Energy Policy 29:1197–1207
CRA (2005) Primer on demand-side management, with an emphasis on price-responsive programs. Charles River Associates, California (see http://siteresources.worldbank.org/INTENERGY/Resources/PrimeronDemand-SideManagement.pdf)

CSPM (2001) Economic analysis of demand-side management programs and projects. California Standard Practice Manual, California, USA (see http://www.energy.ca.gov/greenbuilding/documents/background/07-J_CPUC_STANDARD_PRACTICE_MANUAL.PDF)

EC (2005) Doing more with less: green paper on energy efficiency. European Commission, Brussels (see http://ec.europa.eu/energy/efficiency/doc/2005_06_green_paper_book_en.pdf)

Gillingham K, Newell R, Palmer K (2009) Energy efficiency economics and policy. RFF-DP-13, Resources for the Future, Washington

Golove WH, Eto JH (1997) Market barriers to energy efficiency: a critical reappraisal of the rationale for public policies to promote energy efficiency. Lawrence Berkeley National Laboratory, California (see http://eetd.lbl.gov/EA/EMP/reports/38059.pdf)

Grubb M, Edmonds J, Brink P, Morrison M (1993) The costs of limiting fossil fuel CO_2 emissions: a survey and analysis. Annual Review of Energy and the Environment pp 397–478 (see http://sedac.ciesin.columbia.edu/mva/iamcc.tg/articles/GE1993/GE1993.html#mod)

Herring H (2006) Energy efficiency—a critical review. Energy 31(1):10–20

Hogan W, Manne A (1977) Energy–Economy interaction: the fable of the elephant and the rabbit? in Energy and the Economy, EMF Report 1 of the Energy Modeling Forum, Stanford University

IEA (2001) Things that go blip in the night: standby power and how to limit it. International Energy Agency, Paris (see http://www.iea.org/textbase/nppdf/free/2000/blipinthenight01.pdf)

Jaffe AB, Stavins RN (1994) The energy-efficiency gap: What does it mean? Energy Policy 22(10):804–810 (see http://ksghome.harvard.edu/~rstavins/Papers/The%20Energy%20Efficiency%20Gap.pdf#search=%22sutherland%20market%20barriers%20energy%20efficiency%22)

Munasinghe M, Schramm G (1983) Chapter 6: Energy conservation and efficiency. In: Energy economics, demand management and conservation policy. Von Nostrand Reinhold Company, New York

Paterson MG (1996) What is energy efficiency? Concepts, indicators and methodological issues. Energy Policy 24(5):377–390

Price L, Galitsky C, Sinton J, Worrell E, Graus W (2005) Tax and fiscal policies for promotion of industrial energy efficiency: a survey of international experience. LNBL 58128, California (see http://repositories.cdlib.org/cgi/viewcontent.cgi?article=3747&context=lbnl)

Saunders H (2009) Chapter 8: Theoretical foundations of the rebound effect. In: Evans J, Hunt L (eds) International handbook on the economics of energy. Edward Elgar, Cheltenham

Sorrel S (2009) Chapter 9: The rebound effect: definition and estimation. In: Evans J, Hunt L (eds) International handbook on the economics of energy. Edward Elgar, Cheltenham

Stoft S (1995) The economics of conserved energy supply curves. University of California Energy Institute, California (see http://www.ucei.berkeley.edu/PDF/pwp028.pdf#search=%22conserved%20energy%20supply%20curve%22)

Stoll HG (1989) Least-cost electric utility planning. Wiley, New York

Sutherland RJ (1994) Energy efficiency or the efficient use of energy resources. Energy Sources 16(2):257–268

Swisher JN, Jannuzzi GM, Redlinger RY (1997) Tools and methods for integrated resource planning: improving energy efficiency and protecting the environment. UCCEE, Riso (see http://www.uneprisoe.org/IRPManual/IRPmanual.pdf)

WEC (2001) Energy efficiency indicators and policies: a report by the WEC, London (see http://www.worldenergy.org/wec-geis/publications/reports/eepi/download/download.asp)

WEO (2008) World Energy Outlook, International Energy Agency, Paris

Part II
Economics of Energy Supply

Chapter 7
Economic Analysis of Energy Investments

7.1 Introduction

The economic problem of allocating limited resources to various needs often requires decision-making about appropriate investments. Energy sector being a major demander of investment funds, choices have to be made among competing investment opportunities. This assumes greater importance for energy supply investments given their large sizes and capital intensiveness. The basic analytical framework is the cost-benefit analysis—where costs and benefits over the lifetime of the project are evaluated and investments with positive net benefits are considered to be acceptable investments (Squire and van der Tak 1984). While the financial analysis uses a similar framework and is more widely used for private investments, the economic analysis is important for investments in the public sector or for those with public or near-public good characteristics.[1] This chapter provides an introduction to the economic analysis of energy investments and highlight various important aspects related to such an analysis.

7.1.1 Main Characteristics of Energy Projects

Energy projects, especially those related to commercial energies, share a number of important features:

(a) *Capital intensiveness*: Energy projects tend to be capital intensive as the initial investment requirement is often high. According to IEA (2004), electricity industry is two to three times capital intensive compared to the manufacturing industry. Similarly, fossil fuel extraction is also relatively capital intensive (see

---

[1] A public good is characterised by its jointness of supply and non-exclusive nature of supply. See Chap. 23 for further details.

Table 7.1 for some data on capital costs and gestation periods of electricity generating technologies).

(b) *Asset specificity*: The assets in the energy industry tend to have a high degree of specificity, implying they are less re-deployable in nature. This means that they do not have alternative uses other than their use in the energy sector. This specificity makes assets vulnerable to risks.

(c) *Long-life of assets*: Most energy investments live long; for example, a conventional power plan can easily operate for 25 years, a hydro power plant can live 50 years, even a diesel plant can operate for more than 10 years. As the life increases, the uncertainty about the future costs and benefits increases.

(d) *Long gestation period*: Energy projects take longer to build; for example a nuclear power plant can easily take 8–10 years to construct, a dam can also take such a long period, even relatively faster gas plants can take 2 years to build. Any changes in the business environment during the construction could jeopardize the investment. Similarly, it requires investment decisions to be made well in advance, so that the assets are brought into operation at the time of need. This requires that any investment decision has to be based on projected market conditions at a relatively long future date, making the decision-making vulnerable.

Table 7.1 Capital costs and other relevant information for electricity generating technologies

Technology	Size (MW)	Lead time (years)	Costs (2008 $/kW)
Scrubbed coal new	600	4	2,078
Integrated coal-gasification comb cycle (IGCC)	550	4	2,401
IGCC with carbon sequestration	380	4	3,427
Conv gas/oil comb cycle	250	3	937
Adv gas/oil comb cycle (CC)	400	3	897
Adv CC with carbon sequestration	400	3	1,720
Conv comb turbine	160	2	653
Adv comb turbine	230	2	617
Fuel cells	10	3	4,744
Adv nuclear	1,350	6	3,308
Distributed generation—base	2	3	1,334
Distributed generation—peak	1	2	1,601
Biomass	80	4	3,414
Geothermal	50	4	1,666
MSW—landfill gas	30	3	2,430
Conventional hydropower	500	4	2,084
Wind	50	3	1,837
Wind offshore	100	4	3,492
Solar thermal	100	3	4,798
Photovoltaic	5	2	5,879

Source Table 8.2, Assumptions to the Annual Energy Outlook 2010, http://www.eia.doe.gov/oiaf/aeo/assumption/electricity.html

(e) *Big size*: Often energy projects tend to be big to take advantage of scale economies (i.e. big is beautiful). As a result, the capital outlay increases even where the capital cost per unit is low.

Because of these features and the importance of energy sector investments in the economy, an economic analysis of energy sector investments is essential.

7.2 Basics of the Economic Analysis of Projects

The economic analysis of projects aims at ranking projects so that economically effective projects can be identified and selected for better allocation of resources. Given the non-competitive environment prevailing in the energy sector in many countries and because the costs and benefits of energy sector projects may go beyond the project boundary, a systematic method of evaluation is important. According to World Bank (1996), an economic analysis can help answer the following issues:

- whether the project should be undertaken by the public sector or the private sector;
- the fiscal impacts of the project;
- the efficiency and equity of cost recovery; and
- the environmental impacts of a project.

Thus an economic analysis is aimed at analysing the welfare impacts of a project.

An economic analysis essentially involves three elements (Lovei 1992; ADB 1997; World Bank 1996):

(a) Identification and estimation of costs related to an investment,
(b) Identification and estimation of the benefits to be obtained from the investment and
(c) Comparing the costs with benefits to determine the appropriateness of the investment. If the benefits exceed the costs, investment in the project is acceptable, otherwise it is not.

The first two elements form the core of the analysis and normally require detailed investigations for identification, quantification and valuation of costs and benefits (ADB 1997).

Project-related costs and benefits are identified by considering and comparing two cases or situations: with-project case and without-project case. Implementation of any project will reduce the supply of inputs and increase the supply of outputs (Squire and van der Tak 1984). The difference in the availability of inputs and outputs in two situations forms the basis for identification of costs and benefits of a project in economic terms.

Similarly, a distinction is made between non-incremental and incremental outputs: Incremental outputs are obtained by expanding the supply to meet the demand whereas non-incremental outputs replace the existing production. Similarly, incremental and non-incremental inputs are also distinguished: incremental inputs come from an increase in the supply of inputs while the non-incremental inputs compete with the existing input supply and not through an increase in the input supply. As these aspects affect the overall economic viability, these important distinctions need to be kept in mind while identifying the costs and benefits.

Identification and valuation of costs and benefits could start from the financial statement of the project entity but two types of adjustments are normally made for economic analysis: (a) certain costs and benefits are either excluded or included and (b) the valuation of certain items would change from financial to economic values.

7.2.1 Identification of Costs

Cost means different things to different people (see Box 7.1 for some cost definitions). In an economic analysis, care has to be taken about the certain costs. Normally those costs which impose additional cost burdens are considered in the economic analysis. Some of these elements are as follows (see ADB 1997; World Bank 1996):

(a) *Sunk costs*: If a project uses already existing facilities, for which investments have already been made (i.e. sunk), the economic analysis would exclude these costs as they do not represent any additional costs for the project. These existing facilities would exist even without the project and hence they impose no extra burden to the project.
(b) *Contingencies*: The part of the contingency that represent additional claims on resources for the project would be included in the economic analysis. As the economic costs are measured in constant price terms (as opposed to nominal terms in financial calculations), the price-related contingencies are not included in the economic analysis.
(c) *Working capital*: The same logic for contingencies applies here. For economic analysis, the costs that represent real claim on national economic resources would qualify for inclusion. Any transfer payments would have to be excluded (see below).
(d) *Transfer payments*: These are payments which "transfer command over resources from one party to another without reducing or increasing the amount of resources available as a whole" (ADB 1997). Examples include taxes, duties and subsidies which in most circumstances would be considered as transfer payments. As they do not put any addition claim on the resources, they are not considered in economic analysis. However, taxes (duties and subsidies)

would be included in the price if the demand for inputs is non-incremental, or if the output is incremental or if the government tries to internalize externalities through the tax.

(e) *Depreciation*: The economic analysis uses the initial cost of an asset less the residual value (discounted). This fully reflects the cost of using an asset and does not pay any attention to the funding of the resource and its repayment. Accordingly, depreciation is not considered in the economic analysis.

(f) *Depletion premium*: For non-renewable resources, the economic analysis includes the depletion rent to reflect the economic cost to the society of using such resources. The opportunity cost of the resource includes the cost of substitutes at the time of exhaustion.

(g) *External costs*: As energy projects often generate externalities the cost of which are not borne by the users but the society as a whole, the economic analysis includes such external costs to arrive at the full economic costs of using the outputs. An economic analysis is incomplete without taking this into account.

Box 7.1: Cost Concepts

Cost has different meanings in different settings.

Historical costs: Costs recorded in the books of accounts can provide some idea about the cost under similar circumstances but this should not be relied on unquestioningly. The financial analysis uses this cost.

Future or replacement costs: The cost expected in the future to replace a given asset can be significantly different from historical costs. This is the cost to be paid to rebuild the asset at a future date.

Opportunity costs: Is the value of the foregone benefits; the cost of using a resource is the benefit lost for not having it available for an alternative use. For example, if crude oil is used in the domestic market, the same quantity is not available for exports. Thus an opportunity to earn in foreign currency is foregone. This is a measure of the maximum benefit that can be obtained from an alternative use. This is the appropriate measure of cost of a resource for economic analysis of alternatives but opportunity costs can be hard to estimate.

Marginal cost: Is the change in cost for a small change in the output. In mathematical terms, $MC = \partial C/\partial q$, where MC = marginal cost, C = total cost, and q = output. If total cost is composed of a fixed part F and a variable part v such that $C = F + v \cdot q$, the marginal cost is equal to v.

Some of these elements (such as external costs) are often difficult to value in monetary terms. However, attempts should be made to include such costs to the extent possible.

7.2.2 Identification of Benefits

The output of a project which is sold in the market generally constitutes the main benefit of a project. When the output is incremental, the project does not affect the market price and is considered as a price taker. This is the case when the project output is relatively small compared to the market size and the product is tradable. On the other hand, for non-tradable goods, the supply tends to be non-incremental and the project output could influence the market price.

In some cases, projects lead to directly productive and indirectly productive outputs. For example, a dam can provide recreational facilities in addition to electricity generation, irrigation water or drinking water supply. These additional benefits may be difficult to value in monetary terms but should be considered.

A project benefit may also include any changes in consumer surplus. This captures the difference between what consumers are willing to pay and what they actually pay for a good or service. However, care has to be taken to include only that part which benefits the society as a whole. For example, if a hydro power plant reduces electricity price and increases the demand for electricity, it generates consumer surplus. But the price reduction produces revenue loss to utility and therefore has an off-setting effect. The net surplus should be considered in such cases (World Bank 1996).

7.2.3 Valuation of Costs and Benefits

Once the costs and benefits are identified, they are valued using appropriate economic prices that represent the value to the national economy. Consequently, the valuation could be different from that used in the financial analysis. As an economic analysis intends to determine the costs and benefits to the society (or national economy) rather than to the suppliers or buyers, the true economic price has to be used in the analysis.

7.2.3.1 Valuation of Project Inputs and Outputs

One of the adjustments made to the market prices in the economic analysis is the use of shadow prices. The shadow price is the price that would exist if the market operated perfectly and allocated resources efficiently.

Distortions exist in all economies due to lack of competition or inadequate competition, government intervention (to tax or provide subsidies, to protect suppliers through duties or quotas, to control prices, to control foreign exchanges or wages, etc.) or the presence of externalities. As the energy sector is often characterised by these market failures or government interventions, prices in the market do not reflect the true value of energy project inputs and outputs.

7.2 Basics of the Economic Analysis of Projects

In order to understand the valuation, it is important to look at the effect of a project on the economy. As the project starts producing, it shifts the supply curve rightwards (see Fig. 7.1). This has two effects:

(a) it reduces the price of the good in the market and
(b) increases demand.

The output of the project displaces some existing supplier and substitutes their output (shown by Qwo-Qe) and meets the incremental demand (Qw-Qwo). Thus, the output of the project can be considered in two parts: non-incremental (the first component which displaces existing supply) and incremental (the second component that is used to meet the incremental demand).

The valuation of the non-incremental output (i.e. the output which substitutes alternative supply) is based on the adjusted supply price for the alternative supply because in absence of the project consumers would have paid that price. Adjustments to the market price have to be made to take care of any taxes and subsidies as well as for market imperfections or government controls. On the other hand, the economic price for the incremental output is the adjusted demand price for the output adjusted for taxes, subsidies and other market imperfections or government interventions.

Similarly, the project inputs also bring changes to the economy. The demand for inputs moves outwards as a result of the new project. This causes prices to increase for the inputs and the demand for inputs increases (see Fig. 7.2). Once again, the effect has two elements: incremental demand (Qw-Qwo) where the project introduces additional demand for inputs and non-incremental demand (which moves existing supply to the project).

The valuation of non-incremental inputs would be based on the adjusted demand price of the alternative supply that is substituted, because the project increases the willingness to pay for the inputs compared to the without project situation. For the incremental inputs, the adjusted supply price would apply.

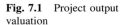

Fig. 7.1 Project output valuation

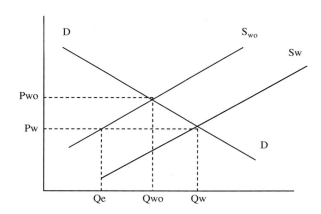

Fig. 7.2 Project inputs pricing

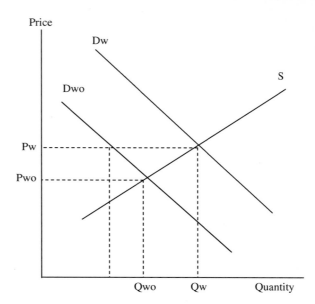

7.2.3.2 Border Prices for Traded Goods

Given that some energy products are traded internationally (e.g. oil) while others are essentially non-traded (e.g. electricity), the effect of the international trade on prices would have to be considered. The possibility of participation in the international trade allows the project output to be exported, project inputs to be imported or imported (exported) goods can be substituted. Accordingly, four situations can be considered for traded goods:

(a) Exportable project output—in this case, if the size of the project is small compared to the world market, the demand faced by the project output is infinitely elastic. Consequently, the output is incremental and the appropriate price for valuation is the free on board price at the port of delivery, as the project is unable to influence the prices in the global market.
(b) Import-substitute output—when a project output replaces imported goods, the opportunity cost is the foreign exchange saved and the appropriate price is the CIF price of the goods.
(c) Imported input—the supply faced by an imported input is infinitely elastic and as a consequence an additional demand does not influence the world price as long as the project input demand is small compared to the world supply. The relevant price is the CIF price.
(d) Exportable input—in absence of the project, the input would be exported and accordingly, the opportunity cost (i.e. the benefits foregone) for the input in using it in the project is the FOB price of the input.

Adjustments are made to take account of processing, transportation and handling charges.

7.2 Basics of the Economic Analysis of Projects

Example Calculate the economic and financial prices at the project site for the following two cases. Note LM means local money, CIF is cost insurance freight (Tables 7.2, 7.3).

Table 7.2 Case 1: Exportable output: crude oil

Description	Value
FOB price	20 $/barrel
Export tax	5%
Handling charge	20 LM/bbl
Transport cost from field to port	60 LM/bbl
Official exchange rate	30 LM/$
Shadow exchange rate	36 LM/$
Economic value of handling and transport	90% of the value

Table 7.3 Case 2: Importable input: coal

Description	Value
CIF price	28 $/t
Import tax	30% of CIF
Handling charge	60 LM/t
Transport cost from port to project site	200 LM/t
Official exchange rate	30 LM/$
Shadow exchange rate	36 LM/$
Economic value of handling and transport	90% of the value

Answer (Table 7.4)

Table 7.4 Case 1: Exportable output: crude oil

Description	Value	Economic (LM)	Financial (LM)
FOB price	20 $/barrel	$= 20 \times 36 = 720$	$= 20 \times 30 = 600$
Less export tax	5%		$-5\% \times 600 = -30$
Less handling charge	20 LM/bbl	$= -20 \times 0.9 = -18$	$= -20$
Less transport cost from field to port	60 LM/bbl	$= -60 \times 0.9 = -54$	$= -60$
Price at the project site		648	510

The economic value is more to the society than to the private owner (Table 7.5).

Table 7.5 Case 2: Importable input: coal

Description	Value	Economic	Financial
CIF price	28 $/t	$= 28 \times 36 = 1{,}008$	$= 28 \times 30 = 840$
Add: import tax	30% of CIF		$= 0.3 \times 840 = 252$
Add: handling charge	60 LM/t	$= 60 \times 0.9 = 54$	$= 60$
Transport cost from port to project site	200 LM/t	$= 200 \times 0.9 = 180$	200
Cost to the project		$= 1{,}242$	1,352

The value of the import has less economic value than that to the project owner.

7.2.3.3 Economic Prices of Non-Traded Goods

Non-traded goods are produced and consumed locally. This may arise due to absence of international market, trade restrictions and the nature of the industry.

(a) If non-traded goods used as project inputs increase the supply for the inputs (i.e. incremental inputs), the appropriate price is the marginal cost of supply. When excess supply capacity exists, the marginal cost is the variable cost of supply but the capacity is constrained, the marginal cost would include both variable and capacity costs.
(b) If the non-traded inputs are non-incremental in nature, the willingness to pay principle has to be applied.

For non-traded outputs the pricing principle changes as follows:

(a) For incremental outputs (i.e. which increase the supply to meet additional demand) the value to the consumers is used as the price (i.e. the demand price.) with proper adjustment for taxes and subsidies.
(b) Non-incremental outputs on the other hand should be valued at the cost of supply of the displaced output in absence of the project. The relevant pricing principle for inputs and outputs is summarized in Table 7.6.

7.2.3.4 Economic Price of Labour

Labour being an important component of any project, its economic valuation is important because of the distortions in the labour market. In a situation of full

Table 7.6 Summary of economic prices for project inputs and outputs

Category	Project impact	Basis of economic price	Basis of valuation
Output			
Tradable	Incremental	Demand price	FOB
	Non-incremental	Supply price	CIF
Non-tradable	Incremental	Demand price	Market price + net consumption tax
	Non-incremental	Supply price	Market price − production tax − operating surplus
Input			
Tradable	Incremental	Supply price	CIF
	Non-incremental	Demand price	FOB
Non-tradable	Incremental	Supply price	Market price − production tax − operating surplus
	Non-incremental	Demand price	Market price + consumer tax

Source ADB (1997)

employment and perfect competition, the cost of labour is market determined. However, in reality this is hardly the case. In general, it is often found that some labour is scarce (mostly the skilled labour) and it is easy for them to find alternative jobs. In such a case the opportunity cost of the labour has to be considered which would be the price at which the labour is willing to work, adjusted for any distortion in the market due to government control. For other types of labour, there is generally oversupply and this situation could exist due to government wage control (or intervention) in the labour market (see Fig. 7.3) in the organised sector or due to other factors.

The supply and demand of labour would clear at the wage rate W1 but the wage control in place requires the wage to be W2. This high wage attracts more labour to the market, thereby increasing the supply to q3 but the demand for labour falls to q2. Consequently, the involuntary unemployment is created to the tune of q2–q3. If the project employs unemployed labour from the controlled segment, the economic price should be the market clearing wage. But given that it may be difficult to determine the market clearing wage, alternative information (such as remuneration from alternative activities undertaken by the labour) could be used. This could come from employment in the informal sector, subsistence activities or seasonal works. The shadow wage rate for labour is the estimation of the economic wage rate of labour used in the project.

7.2.3.5 Economic Price of Land

Any project would use land and its economic valuation is essential to reflect the correct economic value of this resource. The appropriate price for land is the opportunity cost—the best alternative use foregone to develop the project (i.e. by comparing with without-project situation). For rural areas, this would imply the cost of agricultural output foregone valued at the economic price of the output. In the urban areas where the project may displace industrial and commercial activities, or residential housing or other amenities, the loss in economic activities or the willingness to pay for the amenities could give the economic value of the land.

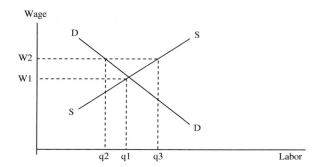

Fig. 7.3 Involuntary unemployment of labour

Fig. 7.4 Shadow price for exchange rate

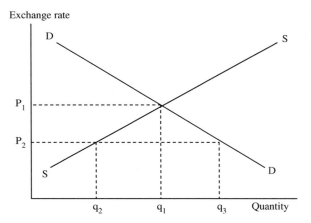

7.2.3.6 Economic Price of Foreign Exchange

The valuation of project inputs and outputs can be done in the local currency or in a foreign currency. Either way, the exchange rate plays a role as in the case of local currency valuation, all imported and exported goods have to be converted to the local currency using the foreign exchange rate.

In a competitive market, the foreign exchange rate is decided by the interaction of supply and demand for currencies. However, often the market is not competitive and governments intervene either by fixing a rate or by rationing the supply. Figure 7.4 indicates that if the rate is fixed at P2 or if the supply is restricted to q2 artificially, the true cost to the economy is P1, which would be the shadow exchange rate. The official exchange rate should not be used in such cases.

ADB (1997) suggests that the shadow exchange rate is the "weighted average imports and exports in domestic prices to the border price equivalent of the same goods". As it does not include income and capital flows but just the goods and service trading, the shadow rate would be different from the free market exchange rate as well.

7.3 Economic Versus Financial Investment Analysis

It should be clear from the discussion so far that the economic analysis of projects differs significantly from the financial analysis. Although both the analyses aim at appraising the profitability of an investment, the concept of profit is not same in the two disciplines. The financial analysis is concerned with the profitability of the owners of the project. An economic analysis considers the national economic interest. In this respect, the economic analysis has a much broader scope than the financial analysis.

The other difference lies in the valuation of costs and benefits.

7.3 Economic Versus Financial Investment Analysis

Table 7.7 Economic versus financial analysis

Criteria	Economic appraisal	Financial appraisal
Cost elements	Costs to the economy including external costs	Only costs relevant for the project that involve money outgo are considered
Benefits	Benefits to the national economy are relevant	Only benefits to the owners are relevant
Valuation	Costs and benefits are valued at the willingness to pay or willingness to accept compensation reflecting the opportunity cost of the resource	Valuation at market price is relevant
Coverage	Broad	Narrow
Viability	Financial viability is necessary for economic viability but not a sufficient condition	Considers financial profitability

(a) The financial analysis focuses on the monetary flows of incomes and expenditures and their timing,
(b) The economic analysis is based on the willingness to pay and the willingness to accept compensation rather than prices actually paid. Table 7.7 provides the differences in a summary form.

7.4 Indicators of Cost-Benefit Comparison

A number of methods are commonly used for comparing costs and benefits of a project which can be broadly grouped into two categories: methods ignoring time value of money and those employing time value of money. These methods are discussed briefly.

7.4.1 Methods Without Time Value

These are simple, easy-to-use methods that are widely used and understood. If carefully used, they can provide reasonable results. Two common indicators of this type are pay-back period and average rate of return on investment.

(a) Payback period provides the time required to amortise the initial investment through internal cash flow generation of a project. As a shorter payback period ensures quick recovery of the investor's funds, such a project if preferred. This is frequently used as an initial guide for screening projects. But the method does not pay attention to the distribution of cash flow and ignores any cash flow beyond the payback period. Accordingly, this method is biased towards front-loaded cash flows and does not consider the cash flow over the project life.

Example An energy saving lamp which consumes 14 W costs £2 while an incandescent lamp of equivalent luminosity (75 W) costs £0.5. Assuming 10 h of lighting per day and a cost of 10 pence per kWh of electricity consumption, determine the payback period.

Answer The difference in initial cost is £1.5. The saving in daily electricity consumption is $(75-14) \times 10 = 610$ Watt-hours or 0.61 kWh. The saving in monetary terms is 6.1 pence. The payback period to recover the investment is $= 1.5/(6.1/100) = 24.6$ days (say 25 days).

(b) The average simple rate of return on investment is the net profit per year as a ratio of initial investment. Thus if a project makes an annual profit of $1,200 on an investment of $10,000, the rate of return is 12%. Once again, this is a simple method that provides commonly understood information but this generally focuses on the commercial profitability of the project and hence may be less useful for a project that provides predominantly social benefits.

7.4.2 Methods Employing Time Value

As project cash flows occur at different points in time and because individuals prefer present to future (i.e. £1 now is valued more favourably than £1 tomorrow), ignoring the time value of money is not appropriate for project evaluation. Consequently, a set of indicators using time value of money are also available. These include indicators involving net present value (or worth) and internal rate of return. These are briefly presented below. Often costs or benefit streams are not occurring at the same point in time and hence any project analyst requires an understanding of equivalence of various amounts using discount factor/interest rates. Annex 7.1 provides some commonly used formulae in such instances.

7.4.2.1 Net Present Value Based Indicators

Net present value involves conversion of all benefit and cost streams occurring at different points in time to their present value equivalents and aggregating them to get the overall worth of the benefits and costs of the projects. This provides a comprehensive measure of net benefits (or costs) and is widely used for project ranking and decision-making. Mathematically,

$$\text{NPV} = \sum_{t=1}^{N} \frac{(R_t - C_t)}{(1+i)^t} - I_0 \qquad (7.1)$$

where R_t, revenue in year t; C_t, costs in year t; i, discount rate; I_0, initial investment.

7.5 Uncertainty and Risk in Projects

indicator in terms of percent change from the initial value and a switching value that causes the decision to change. Sensitivity analysis should focus on alternative assumptions that have unfavourable effects on the project. Either a single parameter or a combination of parameters could be considered simultaneously.

Although commonly used, sensitivity analysis has a number of shortcomings. These include:

(a) Even if a long list of variables is tested, the analysis may fail to identify the variable that by itself significantly affects the results.
(b) Correlations among variables may not be captured in one-variable at-a-time sensitivity analysis.
(c) It does not take the probability of occurrence of events into consideration.
(d) It is a subjective technique.

In addition to sensitivity analysis, a systematic risk analysis is particularly useful for large and important projects. Such an analysis can be qualitative or quantitative in nature. A qualitative risk analysis provides insights to the risks faced by projects and can be used at the early stages of project analysis to identify sources of project risks. These techniques are useful for project designing processes as well.

Risk Matrix is the most useful qualitative risk analysis technique. Risk matrix typically is a two dimensional matrix where one axis categorizes the risk in terms of qualitative probability (low, medium and high) of occurrence and the other axis identifies the seriousness of impacts of these risks in qualitative terms (low, medium, high).

For example, for a project the following risks are identified

Risk 1: Delay in implementation
Risk 2: Lack of political support
Risk 3: Lack of inter-connection facilities
Risk 4: Volatility of input prices
Risk 5: Control of output prices

Table 7.8 indicates the distribution of the risk in terms of probability of impacts.

Such a matrix is useful for allocating risks to different project participants. Appropriate systems of rewards and penalties can be designed for risk control.

A quantitative risk analysis quantifies the risk by assigning a probability distribution to the parameters and determines the expected outcome. Such an analysis can be a simple extension of the qualitative analysis. This requires specifying the frequency (or probability) of occurrence of an event identified as low, medium and

Table 7.8 Risk matrix: impact—probability analysis

Probability/impacts	Low	Medium	High
Low	Risk 4	Risk 3	
Medium	Risk 1	Risk 2	Risk 5
High			

high. Such three point estimates can be made from past experience, expert's opinion, etc. The expected value can then be calculated by weighted sum of the outcomes.

The expected value approach provides a single estimate of the present worth and does not indicate the degree of uncertainty or the range of the values to be expected. For practical purposes, the distribution of the present worth of net cash flows is required. Monte Carlo simulation is an approach to estimate the entire worth distribution by considering the possible combinations of parameters in proportion to their probability of occurring. This can provide a better understanding of the project value for long-life assets, although it requires a higher level of skill compared to an ordinary static project analysis.

7.6 Conclusion

This chapter has introduced the concept of economic analysis of projects and has explained the differences between the financial analysis and the economic analysis. It has elaborated the concepts related to cost and benefit valuation and presented the commonly used indicators to present the results. The chapter has finally outlined the need for incorporating risk and uncertainty in the project analysis.

7.7 Example of a Project Evaluation Exercise

7.7.1 Problem Statement

A project for a diesel power station has the following characteristics:

Cost of the plant per kW = $400 (CIF); Capacity = 50 MW.

The plant has to be imported and installation charges would be equivalent to $200 per kW. 75% of the installation would be done by a local contractor and the rest would be spent on foreign contractors. The local contractor would engage unskilled workers for 35% of the work while the rest of the installation work by the local contractor would be done by semi-skilled workers.

The plant is expected to consume 1.7 barrels of diesel per MWh for the first 5 years and then the consumption would increase to 1.8 barrels/MWh for the next 5 years. Imported diesel costs $50 (CIF) per barrel while the domestic refineries sell diesel at local money (LM) 2,400 per barrel. Transportation and handling would require $10 per barrel from the port to the project site. It is expected that the CIF price would increase 2.5% every year and the domestic price would increase at the rate of inflation.

The plant would require 5 skilled workers and 10 unskilled workers. The skilled workers command a salary of LM 10,000 per month. Salary of the employees

7.7 Example of a Project Evaluation Exercise

grows at an average rate of 5% every year. The average rate of tax is 10% on salary. Repair and maintenance cost is expected to be 3% of the plant cost per year. Other costs would be 5% of the initial investment (including installation) per year.

The official exchange rate is LM 40 to a dollar. However, the parallel market rate is LM 44 per dollar. The unskilled workers are paid LM 3,000 per month while the semi-skilled workers get LM 7,000 per month. The shadow price of the unskilled and semi-skilled labour wage is estimated to be 75% of their respective market values.

The plant is expected to operate 4,000 h per year. The output would be sold to the electric utility at a rate of LM 2.5 per kWh. The contract provides for an annual increase of 3% in the sale price.

Determine whether the investment is economically viable and financially sound. Use a nominal discount rate of 10%. Assume annual inflation rate of 4% per year. Assume any other information required.

7.7.2 Answer

A project for a diesel power station has the following characteristics:

Cost of the plant per kW = $400 (CIF); Capacity = 50 MW.

Plant cost = $400 \times 50 \times 1,000/10^6 = \$20$ Million. Economic cost of the plant = $44 \times 20 = 880$ million. Financial cost of the plant = $20 \times 40 = 800$ million.

Financial installation charges = $200 \times 50 \times 1,000/10^6 = 10$ million $\$ = 10 \times 40 = 400$ million LM.

Economic Installation charges

Foreign contractor gets = $0.35 \times 200 \times 50 \times 1,000/10^6 \times 44 = 154$ million LM
Local contractor cost = $0.65 \times 200 \times 50 \times 1,000/10^6 \times 44 \times 0.75 = 214.5$ million LM
(Local contractor uses semi-skilled and unskilled workers whose shadow rate is 75% of the financial cost.). Total installation charges = 368.5 million LM.

The plant is expected to consume 1.7 barrels of diesel per MWh for the first 5 years and then the consumption would increase to 1.8 barrels/MWh for the next 5 years. The plant is expected to run 4,000 h. Total generation would be = $4,000 \times 50 = 200,000$ MWh. The fuel consumption would be = $1.7 \times 200,000$ barrels (or 0.34 million barrels during the first 5 years and 0.36 million barrels during the last 5 years of the project life.

The financial cost of fuel in the base year is $50 + \$10 = \$60$ per barrel at the exchange rate of 40 per dollar, which is equal to 2,400 LM per barrel. As the cost is same with locally available diesel, local market price is considered in the analysis. The local price would increase at a rate of 4% per year.

Table 7.9 Economic and financial fuel costs and prices (million LM)

Year	Fuel price		Cost for fuel (Million LM)	
	Eco	Financial	Eco	Financial
1	2640.00	2400.00	897.60	816.00
2	2601.92	2496.00	884.65	848.64
3	2564.40	2595.84	871.89	882.59
4	2527.41	2699.67	859.32	917.89
5	2490.96	2807.66	846.92	954.60
6	2455.03	2919.97	883.81	1051.19
7	2419.62	3036.77	871.06	1093.24
8	2384.72	3158.24	858.50	1136.97
9	2350.33	3284.57	846.12	1182.44
10	2316.43	3415.95	833.91	1229.74

The economic cost of fuel is based on the import price (CIF), as diesel is a traded commodity. Cost per barrel is $= 60 \times 44 = 2{,}640$ LM/barrel. The current price increase in border price of diesel is 2.5%. Diesel price in constant terms is calculated by discounting the current price for inflation.

Table 7.9 presents the fuel prices and cost of fuel for 10 years.

Note that economic cost is expressed in constant prices and financial cost in current prices.

7.7.2.1 Salary and Other Operating Costs

The plant would require 5 skilled workers and 10 unskilled workers. The skilled workers command a salary of LM 10,000 per month. Salary of the employees grows at an average rate of 5% every year. The average rate of tax is 10% on salary.

The annual financial cost of salary to skilled workers is $= 5 \times 10{,}000 \times 12 = 600{,}000$ for the first year.

The annual financial cost of other workers is $= 10 \times 3{,}000 \times 12 = 360{,}000$ in the first year.

Total salary cost in the first year $= 960{,}000$ for the first year.

Tax on income is excluded for economic cost calculations. Economic cost of skilled workers $= 10{,}000 \times 12 \times 5 \times (1 - 0.1) = 540{,}000$ in the first year.

Unskilled labours are assumed to earn below taxing zone and no tax is considered on their incomes. But their wage has a shadow price which is 75% of the financial cost. The cost due to salary of unskilled labour for the first year is $= 3{,}000 \times 12 \times 10 \times 0.75 = 270{,}000$ LM.

Total economic cost of salary for the first year is $= 810{,}000$.

Repair and maintenance cost is expected to be 3% of the plant cost per year. Other costs would be 5% of the initial investment (including installation) per year. Therefore, other costs are estimated at 8% of the total investment costs. The

7.7 Example of a Project Evaluation Exercise

financial costs would be equal to $0.08 \times 1{,}200 = 96$ million per year in current money terms. The economic cost for the first year is $= 0.08 \times (880 + 368.5) = 99.88$ million. The constant price value for each year is obtained by removing the effect of inflation. Table 7.10 presents the salary and O&M costs for economic and financial cases.

The plant is expected to operate 4,000 h per year. The output would be sold to the electric utility at a rate of LM 2.5 per kWh. The contract provides for an annual increase of 3% in the sale price. The economic as well as financial value of the output is $(50 \times 1{,}000 \times 4{,}000 \times 2.5/10^6) = 500$ million LM for the first year. Assuming that the real increase in price is 3%, the economic and financial value of the revenues are calculated and presented in Table 7.11.

Table 7.12 presents the net economic and financial benefits from the project for various years.

The discount rate is 10% for financial calculations and 5.77% for the economic calculations. The NPV is clearly negative in both the cases (-4142.4 million LM in the economic case and -2543.6 million LM for the financial case).

Table 7.10 Economic and financial salary and O&M costs (million LM)

Year	Salary		O&M costs	
	Eco	Financial	Eco	Financial
1	0.81	0.96	99.88	96
2	0.82	1.01	96.03846	96
3	0.83	1.06	92.34467	96
4	0.83	1.11	88.79296	96
5	0.84	1.17	85.37784	96
6	0.85	1.23	82.09408	96
7	0.86	1.29	78.93661	96
8	0.87	1.35	75.90059	96
9	0.87	1.42	72.98134	96
10	0.88	1.49	70.17436	96

Table 7.11 Output value (million LM)

Year	Economic	Financial
1	500.00	500.00
2	515.00	535.60
3	530.45	573.73
4	546.36	614.58
5	562.75	658.34
6	579.64	705.22
7	597.03	755.43
8	614.94	809.22
9	633.39	866.83
10	652.39	928.55

Table 7.12 Net economic and financial benefits from the project (million, LM)

Year	Economic	Financial
0	−1248.5	−1,200
1	−498.29	−219.04
2	−466.51	−216.03
3	−434.61	−211.79
4	−402.58	−206.19
5	−370.39	−199.09
6	−387.12	−248.75
7	−353.83	−240.52
8	−320.33	−230.40
9	−286.59	−218.19
10	−252.58	−203.70

Annex 7.1: Some Commonly Used Interest Formulae

Single Compound Amount Formula

If P dollars are deposited now in an account earning $i\%$ per period for N periods, then

$$F = P(1+i)^N \tag{7.6}$$

This expression is used to move any amount forward in time.

Example A firm borrows $1,000 for 5 years. How much must it repay in a lump sum at the end of the fifth year? Assume interest rate is 5%.

With $P = 1{,}000$, $i = 0.05$, $N = 5$; $F = \$1276.28$

Single Present-Worth Formula

The present worth P of a sum F which would be available N periods in the future is given by

$$P = F\left[\frac{1}{(1+i)^N}\right] \tag{7.7}$$

Example A company desires to have $1,000 8 years from now. What amount is needed now to provide for it, if interest rate is 5%?

Answer $F = 1{,}000$, $i = 0.05$, $N = 8$; Hence the single payment present worth factor is 0.67684. The amount of investment required now is $676.84.

Annex 7.1: Some Commonly Used Interest Formulae

If the cash flow in the series has the same value, the series is called uniform series.

If the value of a cash flow varies by a constant amount G from the previous period, the series is called gradient series.

If the value of a given cash flow differs from the value of the previous period by a constant %, the series is called a geometric series.

Closed form expressions are available for these categories.

Uniform Series Compound Amount

If a uniform amount A, called annuity is deposited at the end of each period for N periods in an account earning $i\%$ per period, the future sum at the end of N periods is

$$F = A[1 + (1+i) + (1+i)^2 + \cdots + (1+i)^{N-1}] \qquad (7.8)$$

$$F = A\left[\frac{(1+i)^N - 1}{i}\right] \qquad (7.9)$$

The expression $\left[\frac{(1+i)^N - 1}{i}\right]$ is called uniform series compound amount factor.

Example If 4 annual deposits of $2,000 each are placed in an account, how much money has accumulated immediately after the last deposit, if the rate of interest is 5%?

Answer With $N = 4$, $i = 0.05$, the uniform series compound amount factor is 4.310. The future amount would be $2,000 \times 4.310 = \$8,620$.

Uniform Sinking Fund Formula

A fund established to accumulate a desired future amount of money at the end of a given length of time through the collection of a uniform series of payments is called a sinking fund.

If F is the total amount at the end of N periods, the annuity A that has to be paid is given by

$$A = F\left[\frac{i}{(1+i)^N - 1}\right] \qquad (7.10)$$

The expression $\left[\frac{i}{(1+i)^N - 1}\right]$ is called sinking fund factor.

Example How much should be deposited each year in an account in order to accumulate $10,000 at the time of the fifth annual deposit? Assume an interest rate of 5%.

Answer The sinking fund factor is 0.180975. The annuity requirement is $1809.75 (or $1,810).

Uniform Capital Recovery Formula

This calculates the amount of annuity required to accumulate to a given present investment P, with given interest rates and number of periods.
Substituting $F = P(1 + i)^N$ in Eq. 7.9 gives

$$A = P\left[\frac{i(1+i)^N}{(1+i)^N - 1}\right] \tag{7.11}$$

The expression $\left[\frac{i(1+i)^N}{(1+i)^N-1}\right]$ is called the capital recovery factor.

Example What is the size of 10 equal annual payments to repay a loan of $1,000? First payment is 1 year after receiving loan. Interest on loan is 5%/year.

Answer Here $N = 10$, $i = 5\%$, Hence the capital recovery factor is 0.1295. Hence, the annuity required is $129.5 at the end of each year.

Uniform Series Present Worth Formula

The present worth of a series of uniform end-of-period payments is given by

$$P = A\left[\frac{(1+i)^N - 1}{i(1+i)^N}\right]. \tag{7.12}$$

The expression $\left[\frac{(1+i)^N-1}{i(1+i)^N}\right]$ is known as uniform series present worth factor.

Example How much should be deposited in a fund to provide for 5 annual withdrawals of $100 each? First withdrawal is 1 year after deposit. Assume an interest rate of 5%.

Answer With $N = 5$ and $i = 0.05$, the present worth factor is 0.43295. Hence, the amount required is $432.95 (Table 7.13).

8.1 Introduction

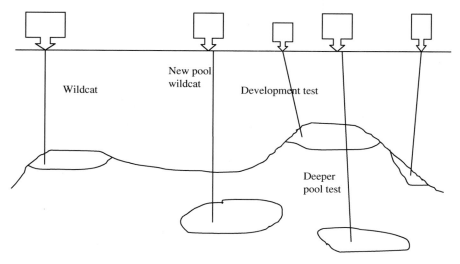

Fig. 8.2 Classification of exploratory wells

widespread and coal deposits have been found while exploring for other materials. The existence of a global market for oil and its desirable characteristics (such as high heat content and liquid form) make oil a preferred choice for explorers.

A number of different types of wells are drilled for oil and gas (see Fig. 8.2): wildcat (drilling for a new field or in a new structure which is not yet productive), new pool wildcat (drilling in an already productive structure but outside the known limits of the existing pool), deeper pool test (drilling for new resources below the known depth of the existing pool), shallower pool test (drilling as before but in the opposite direction), extension test (drilling beyond the known limits of the existing pool), and development test (additional drilling to develop a field).

Normally, the probability of success is low for wildcats (can be close to zero in many areas) but increases as for other types of wells. The success rate is generally high for development drills (70–80%).

8.1.2 Exploration Programme

The objective of any exploration programme is to maximise discovery at the minimum effort. The efficiency of the search programme would be highest if most of the reserve addition takes place quickly while it is the least when the reserve addition takes place only with very large efforts of exploration. This is schematically presented in Fig. 8.3. Any exploration programme would aim to achieve high efficiency to reduce exploration costs.

The outcome of the exploration activity is the addition to reserves. Figure 8.4 presents a typical curve showing the additions to reserves as a result of

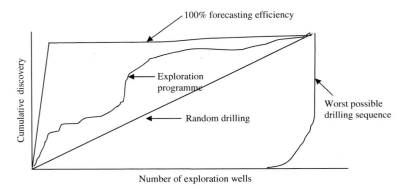

Fig. 8.3 Efficiency of exploration programme. *Source* Shell Briefing Service

Fig. 8.4 Exploration success curve (or creaming curve)

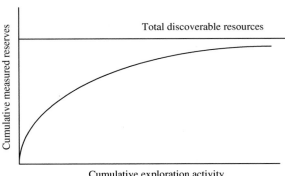

exploration. Generally, the reserve addition grows quickly at the early stages of exploration but the incremental addition to reserves slows down as exploration continues. However, the speed at which discoveries are made will affect the shape of the curve because an early discovery and faster rate of discovery will imply that the time taken to reach the discoverable resources will be shorter. This can be thought of as the efficiency of the discovery process as discussed above. However, exploration efficiency tends to depend on the field size and depth of exploration.

A common functional form used for exploration modelling is written as follows:

$$F_n = F_u\left(1 - e^{-\gamma^W}\right) \tag{8.1}$$

where F_n is the number of discovery, F_u is the ultimate number of fields, W is the exploration effort, γ is the exploration efficiency.

8.1.3 The Economics of Exploration Activities

Exploration is always for finding cheaper minerals. Any investor has to decide whether to drill within the limits of a given reservoir, or to extend it; or modestly venture into the unknown, seek new pools in the same field or look for new fields. The cost of finding oil includes cost of various surveys and studies, cost of drilling (where the possibility of finding dry holes cannot be excluded) and the cost of rentals.

Broadman (1985) argued that a number of factors are important in determining the level of exploration activities in a given area. The geological promise comes first to mind. The distribution of hydrocarbon reserves is uneven and countries with better prospects are normally expected to experience higher levels of exploration activities. Similarly, as the cost of exploration per barrel declines considerably as the field size increases, large fields are likely to attract more investments. However, even in areas of good geological prospect, the prospect risk exists.

The marginal cost of exploration bears a proportional relationship to the level of activity (see Fig. 8.5). Initially the cost tends to be low and increases as the level of exploration increases. This is because the highly prospective areas are explored first but over time, the exploration has to move to the less prospective areas which also tend to be more difficult areas to explore.

The cost of exploration varies significantly with location and geo-physical conditions. In the case of oil and gas, the cost of exploration is cheap in flat, desert countries than in offshore. Seismic costs for offshore are often lower than the onshore but drilling costs follow an inverse relationship. Tens of millions of dollars could be easily spent in an exploration activity without commercially viable discoveries.

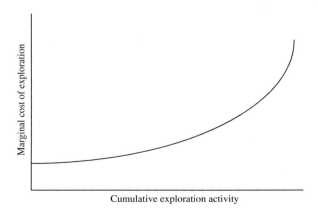

Fig. 8.5 Typical marginal cost of exploration curve

8.1.4 Investment Decision

Any investment in exploration is highly risky due to a number of uncertainties and risks. Even in areas of good geological prospect, the prospect risk exists. This is the chance of finding a dry hole (in case of oil and gas) or no resource deposits. As exploration is costly, and as the company undertaking the exploration activity bears the prospect risk, strategies are required based on an evaluation of the costs and benefits of the exploration programme.

The decision-making essentially involves a cost-benefit analysis for each exploration programme. To account for uncertainty in the entire chain, a probabilistic approach is normally used where the expected values of outcomes are considered. Thus the decision-making rule uses the expected monetary value (*EMV*) instead of normal *NPV* (Kemp 1992):

$$EMV = P \times NPV - E, \qquad (8.2)$$

where *EMV* is the expected monetary value, *P* is the probability of discoveries being made, *NPV* is the net present value of developing the discovered fields, *E* is the exploration costs.

Generally, if *EMV* is positive, the exploration activity can be undertaken.

For example, if the probability of striking oil is 15%, and the *NPV* of developing the discovered field is £100 million and if exploration costs £10 million, the *EMV* is £5 million.

At the pre-drill stage, the decision to drill or not to drill can be based on a decision-tree as indicated in Fig. 8.6. Assume that the probability of failure in drilling is 75%. The cost of drilling after tax is £10 million. In the case of success,

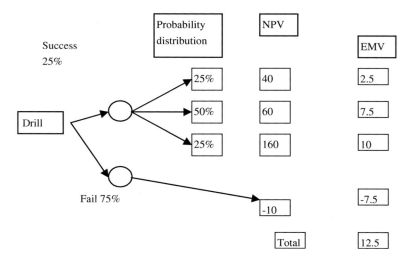

Fig. 8.6 Decision tree for exploration decision

8.1 Introduction

the probability of a low volume find or a high volume find is 25% each, while that of a regular find is 25%. The expected net present values for each case are indicated in the figure. The expected monetary value in this example is 12.5 millions, resulting in a positive outcome, thereby supporting the drilling decision.

Clearly, the odds of striking a success, the distribution of outcomes and the consequent costs-benefits influence the overall decision-making process. To obtain a better understanding of the risks, simulations involving the entire range of possibilities are also undertaken.

8.1.5 Risks in Exploration Projects

Some other factors affect exploration activities. The need for infrastructure development in general and for oil development in particular plays an important role. Lack of infrastructure can act as hindrance to exploration activities, as the infrastructure development has the characteristics of a public good, private development may not occur or may be inappropriate. This also adds to the cost of the project and affects its viability.

Even in cases where the prospect is good and infrastructure may be available, other institutional factors may hinder exploration activities. These factors include:

1) Contractual risk: A contract is required because in most countries the state is the owner of the underground resources and state's permission is required to undertake exploration. A contract provides the rules governing the allocation of risks and rewards relating to exploration. The contractual risk arises due to the possibility of changing the terms of the contract after the discovery is made, following the Obsolescence Bargain Model referred to in the literature.[1] This affects the profitability of the investment.
2) Commercial risk: The commercial risk on the other hand is related to the risk of finding less favourable commercial prospects in reality for the new found reserves than expected at the time of appraisal. This can happen due to poor geological conditions, smaller size of the reserve than initially estimated, poor quality of the output, etc. that affect the costs or benefits of the investment. Changes in the rules governing the business environment could also introduce commercial risks.

In general two basic types of legal arrangements are found in practice (Nakhle 2009):

- Concessionary systems (lease, concessions, and permits): In the first approach, permission is granted for exclusive exploration operation, development of the

---

[1] This model is widely used in the business literature to analyse the relations between the host state and the multi-national companies. See for example, Eden and Molot (2002) and Ramamurti (2001) for further details.

resources if found and disposal of the produce. The owner receives royalty in return as well as bonuses as agreed. The risk is borne by the company taking the concession.
- Contractual arrangements (production sharing arrangements, service contracts, risk-sharing service contracts and joint ventures): In the contracts, the risks and rewards can be shared between the host and the contracted company through the agreed contractual terms.

Box 8.2 provides the salient features of some of these arrangements.

Box 8.2: Alternative Legal Arrangements for Exploration

Lease: In a lease arrangement, the rights for exploration, development and production of oil and gas resources are secured from the owner (or lessor) by the lessee. The lease gives exclusive rights to the lessee to undertake the activities against payment of a consideration fee. The lease can be a negotiated agreement or decided through a competitive process. The primary terms of the lease (i.e. length of time for the initial lease period, bonus per acre and royalty percentages) are decided before an agreement is reached. This was the most common form of arrangement in the USA.

Concessions: This arrangement has been widely is used in the petroleum industry outside the USA, often under the colonial regimes. These were generally negotiated contracts, which provided for: bonus payment by companies to the producing government; a nominal rent for the designated area; and a royalty payment on the oil produced The companies in turn gained control of large areas for exploration, development and production of oil and other mineral resources for long periods.

Production sharing contracts: In this arrangement, the host enters into an agreement with the company where the company recovers the costs through cost oil and shares the profit with the host in an agreed rate. It also pays the income tax on the profit made from the operations. In service contracts, the contractor provides the service for a fee and the benefits of the activities accrue to the host. The contractor does not get any share of the consequential profit or loss.Joint venture agreements: Here risk and profit sharing arrangements are shared among a number of participants.

Clearly, any investment decision is affected by contracting arrangement and the fiscal regime of a country. This involves a trade-off:

- a country may be able to gain high short-term income by charging a higher share of the profit (or rent);
- but this tends to reduce the exploration effort over time, thereby affecting the long-term sustainability of the income.

8.1 Introduction

Thus a careful consideration is required in designing the fiscal system and the contractual arrangements to attract the investment in the industry and balance the revenue needs of the country.

In this regard, the role of government can be crucial for the level of exploration activity. In many cases, the government entity participates in profit sharing but does not take any risk of exploration. This could adversely affect the decision-making. This is because the decision-rule changes to

$$EMV = P \times NPV \times (1 - SP) - E \qquad (8.3)$$

where SP is the level of state participation without taking risk of exploration.

This reduces the available benefits and reduces the profitability of the venture.

In our example, if the state decided to have a 50% share in the project, then the *EMV* changes to −£2.5 million, making the investment unviable.

The situation changes if the state decides to participate in the project as well as in the exploration risk. In this case, the decision rule changes to

$$EMV = P \times NPV \times (1 - SP) - E(1 - SP). \qquad (8.4)$$

For the level of participation considered in the above example, the *EMV* changes to £2.5 million, making the investment profitable once again.

Thus, the state participation can influence the exploration activity. In practice the tax rules, terms of contracts, and expected market conditions all influence the decision and a more detailed economic and financial analysis is required. In addition, other decision support techniques (such as decision-tree analysis, Monte Carlo simulation, etc.) are commonly used in exploration decisions.

Normally oil price tends to influence the exploration activities. Figure 8.7 provides some information on oil rig activities in different regions of the world. The figure suggests that:

- Activity is not commensurate with the level of reserves: the level of activity is the highest in North America despite low reserves in that area,
- Activity is not particularly high in the Middle East.
- The exploratory activities reached peak after the first oil shock in the 1970s and then a general decline in the effort could be noticed, which corresponded with the low oil prices prevailing in much of the 1980s and early 1990s. Thus the long-term viability of supply could be considered to be affected by the low prices, as exploration was not found remunerative with low prices.
- Only with recent increase in oil prices, the exploratory activities have started to rise once again.

Thus there appears to be a positive correlation between prices and exploration activity levels.

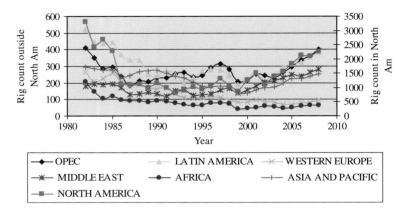

Fig. 8.7 Trend of oil exploration. *Data source* OPEC Annual Statistical Bulletin 2009

8.2 Field Development

Once a discovery is made, the next stage in the process is to decide whether to develop it or not. The crucial information at this stage relates to the size of the discovery. Various estimations are made about the size of the reserve (see Table 8.1).

The next stage is to take an investment decision based on a carefully calculated analysis. The main difference with the decision at the exploration stage is that there is no uncertainty about the existence of the resource now, although its size and other characteristics may still be inadequately known. However, any investment decision depends on the nature of resources to a great extent. Oil discovery generally attracts better industry attention because of its easy access to the global market. Gas finds and the discovery of other resources on the other hand would be treated very differently depending on its location, proximity to the market, availability of viable consumers, etc.

8.2.1 Investment Decision

Once again, the investment decision relies on the cost-benefit framework. Major costs include the investment required to develop the field or site and the operating costs. The costs are influenced by, inter alia, the environmental compliance requirements, relocation of settlements (if any), geological conditions, development schedule, additional drilling and test requirements, and local infrastructure availability for transporting the output.

For example, in order to determine the boundaries of the reservoir and to test the field conditions, appraisal drillings are carried out. These wells provide a wealth of information about the reserve characteristics, economic flow rates,

8.2 Field Development

Table 8.1 Reserve estimation methodologies

Method	Comment
Volumetric	Applies to crude oil and natural gas reservoirs
	Based on raw engineering and geological data
	Provides oil in place estimate
	Recoverable oil estimate is derived using recovery factor, knowledge of drive mechanism and spacing of wells
Material balance	Applies to crude oil and natural gas reservoirs
	More useful in estimating reserves
	Formula based on conservation of material and energy is used
Pressure decline	Applies to non-associated and associated gas reservoirs
	A special case of material balance equation
Decline curves	Applies to crude oil and natural gas reservoirs
	Used at the later stages of reservoir life
	Past production is extrapolated to determine future production and reserves
Reservoir simulation	Applies to crude oil and natural gas reservoirs
	Useful for analyzing reservoir performance
	Accuracy increases when calibrated with past pressure and production data
Nominal	Applies to crude oil and natural gas reservoirs
	Uses rule of thumb or analogy with another reservoir or reservoirs believed to be similar
	Least accurate method

nature of drive, fluid characteristics and so on. The success rate in this phase is normally higher—can be about 50% and is less expensive than exploratory drilling due to utilisation of already mobilised equipment and materials. Similar tests are required for coal fields to get additional information about the coal quality and size of the deposit.

On the other hand, the benefit side is affected by inter alia, the size of the recoverable resource, the production schedule, market price of the fuel in the future, and the tax and fiscal regime prevailing in the country. Many of these cost and benefit elements are unknown at the time of investment decision and require once again working estimates based on experience and expert judgement. The idea is to compare the costs and benefits of the investment programme and if the rate of return to be obtained from the investment is satisfactory, investment would be undertaken.

Each investor company has its own culture of investment decision-making.

- Often larger companies tend to use quite conservative rules. For example, many major oil companies still use a price of less than $20/barrel for project appraisal.
- Smaller companies (independents) tend to adopt a more aggressive approach.

A lower oil price for project appraisal makes only larger projects with cheaper production costs viable. Essentially, the investment decision requires a careful analysis of the risks in the cost and benefit streams and strategies to manage them so that the investment does not turn out to be a bad investment decision.

Figure 8.8 presents an example of cash flow for a development project. The post-tax cash flow as well as cumulative cash flow at different discount rates

Fig. 8.8 Cash flow of a development project

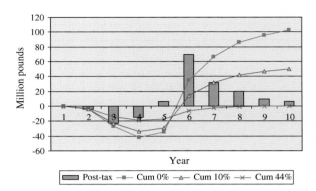

(zero, 10% and at the internal rate of return (IRR)) are presented. As is typical, the project incurs costs in the initial years, then starts earning income and finally enters into the decline phase. The project generates a positive net present value and indicates an IRR of about 44%. It pays back in about 5 years. The project shows a profitability ratio (NPV to discounted investment at a 10% discount rate) of 0.3.

As each project is unique, careful consideration is required for each of them before taking an investment decision. Some projects can involve specific issues. For example, the development of a gas field often requires a more detailed consideration. Such projects tend to cost more and benefits could be more difficult to ascertain. This is because the investment requirement for developing a gas field is generally higher than that of oil due to infrastructure-related costs and higher initial investment costs due to impossibility of phased development. The benefits are more difficult to ascertain because there is no international market price and each gas producer may be required to find its own market, find a suitable large consumer who is willing to pay a remunerative price for the gas. This can be difficult.

8.2.2 Resource Classification

A decision to develop a field creates reserves and producing capacity. Resource classification is normally done using two methods:

- the deterministic method: this approach provides a single estimate (point estimate) of the reserve;
- the probabilistic method: this approach reports reserves for different levels of uncertainty. This is more accurate but may be more difficult to understand.

The commonly used deterministic resource classification uses the McKelvy diagram shown below (Fig. 8.9), although a universally accepted set of rules does not exist. Such decisions depend on price of the fuel, cost of exploration and

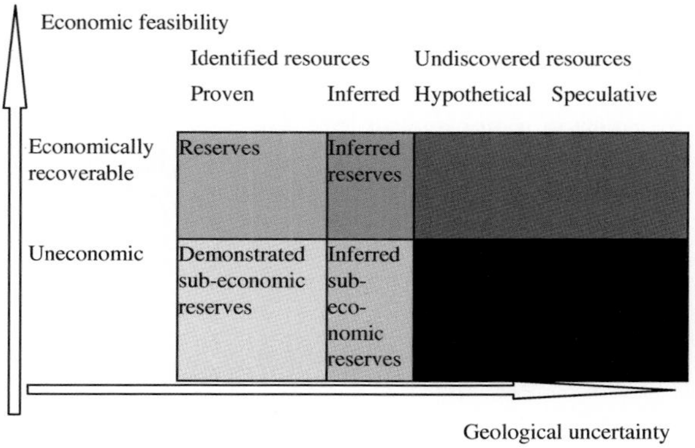

8.2 Field Development

Fig. 8.9 Deterministic classification of resources

	Identified resources		Undiscovered resources	
	Proven	Inferred	Hypothetical	Speculative
Economically recoverable	Reserves	Inferred reserves		
Uneconomic	Demonstrated sub-economic reserves	Inferred sub-economic reserves		

(Vertical axis: Economic feasibility; Horizontal axis: Geological uncertainty)

production, technological change in production and exploration, recovery methods, etc.

Figure 8.9 uses a two dimensional scheme:

- On the horizontal axis is considered the degree of geological certainty. Based on the level of geological certainty, resources can be placed under two broad categories: already identified and those not identified (undiscovered).
- the vertical axis considers the degree of economic feasibility. Resources can be grouped as economically recovered with the existing technology at the current market prices and others which are not recoverable.

Thus, four broad groups of resources can be identified:

- economically recoverable identified resources
- economically recoverable undiscovered resources,
- sub-economic but identified resources, and
- sub-economic and undiscovered resources.

Resources therefore include all naturally occurring concentrations of solids, liquids and gaseous materials that are currently or potentially extractable economically. Reserves are only that part of the resource base that is identified and extractable economically under present market conditions and with the available technology. Box 8.3 indicates the terminology used for oil.

Two things become evident immediately:

- Changes in the economic conditions would influence the size of the reserves (but not that of the resources); and
- Changes in technology would also shift the boundary between identified and undiscovered resources.

Thus there is always a tussle going on between the human ingenuity and depletion.

The problem arises because there is no universally accepted rule for shifting the boundaries of the boxes indicated above. There is no single rule to tell what is economically feasible—this is neither possible (due to diversity of the conditions and costs) nor desirable (as the risk aversion varies and all cannot be asked to have the same level of risk aversion). The problem becomes complicated because all information is not available in the public domain. This raises questions about any revision. However, it needs to be highlighted that revisions to reserve information can be perfectly legitimate as better information is available or economic conditions change.

> **Box 8.3: Terminology for Oil**
>
> The oil industry uses a somewhat different terminology. The correspondence of these terms with the above terminology is discussed below.
> Proven reserves: Generally includes measured reserves that are economically recoverable.
> Probable reserves: Generally includes indicated reserves.
> Oil in place in proven/probable reserves: Total oil in place in the reservoirs contributing to the proven/probable reserves.
> Proven/probable reserve = oil in places* recovery factor.
> Ultimately recoverable Identified resources that are economically recoverable plus an estimate of unidentified resources that might be economically recoverable.
> Proved developed resources: That portion of the proven reserves that exists in fields and or pools currently under production.
> Undiscovered oil in place: Generally undiscovered resources.

Recent resource information can be found in WEC (2009).

8.2.3 Classification of Crude Oil, Natural Gas and Coal

Once the reserve is identified, it is classified into various types based on physical and chemical properties. Briefly, these include:

a) Classification of crude oils: Based on the specific gravity crude oil is grouped into light (in the range of 30–50 degrees API) or heavy crude (less than 30 and close to 20). The API grade compares crude with the density of water using the formula below:

$$(141.5/specific\ gravity\ of\ oil) - 131.5.$$

Lighter crude yields more of lighter products which are sold at a premium in the market and hence the lighter crude is preferred to heavier crude.

Another classification is based on the sulphur content: crude with low sulphur content is called sweet crude and that with high sulphur content is called sour crude. As sulphur is corrosive and leads to acid deposition when burnt, sour crude requires special treatment before use, thereby increasing the cost of the users. Accordingly, sweet crude is sold at a premium in the market.

b) Classification of natural gas: Natural gas can be found separately from oil or in association with oil. Accordingly, gas can be associated or non-associated. Another classification of natural gas is dry and wet gas. Natural gas with large amounts of condensable hydrocarbons is called wet gas. It is called dry gas when these wet gases have been removed. High sulphur gas is called sour gas and low sulphur gas is called sweet gas.

c) Coal classification: A number of types of coals can be found in the literature and there is no universal definition for coal types.

  a. Depending on the fixed carbon content, coal is classified as anthracite (low fixed carbon content), bituminous (dark brown to black coal, most abundantly available), and lignite (a coal that has not been completely coalified).
  b. Coal is also classified in terms of ash content (high or low) and sulphur content (low or high). High ash coals require more elaborate pollution control equipment and coal preparation. Accordingly, they are less preferred by the consumers.

8.3 Production

Once a field is developed, tests and preparatory works are carried out to start production. The activities vary for coal and petroleum products.

8.3.1 Oil Production

In the case of oil and gas, this phase involves:

- Well preparation: A well needs to be cased, anchored and fitted with control mechanisms for flow control.
- Testing: Tests are carried out to determine the flow rates and possible production profile.
- Reservoir stimulation: This is used to improve flow paths and increase output. Commonly used methods are acidising and hydraulic fracturing.

Initially oil and gas flow normally due to normal pressure differences between the well mouth and the reservoir. Generally the reservoir pressure is maintained by the pressure of an underlying water aquifer or a gas cap or dissolved gas. The drive mechanism puts a physical limit to the rate of production that can be achieved without damaging the ultimate recovery of the hydrocarbon. This primary recovery of conventional oil generally allows production of some 5–20% of the oil in place.

In the case of gas, primary recovery can be much higher for non-associated gas due to better flow characteristics; but for associated gas, the decision to produce depends on the oil production schedule. As a result, associated gas is produced as a by-product and in the absence of any local or regional market gas is either re-injected into the field or flared.

To improve output, additional support to the natural drive, known as enhanced recovery techniques, is provided. These techniques allow recovery of an additional 5–25% of the remaining oil in place. Water injection and gas re-injection are more frequently used secondary recovery methods. Other enhanced recovery techniques (known as tertiary recovery) include hot water injection, steam injection, injection of chemicals, etc. See Box 8.4 for some additional details.

Box 8.4: Dive Mechanisms and Artificial Lifts Used in Oil Production

Dissolved gas drive: Lighter hydrocarbons come out in the form of gas and expands to force the oil to the well bore. Pressure declines rapidly and continuously. Wells require artificial assistance early. Recovery efficiency can be as low as 5% and as high as 30%.

Gas-cap drive: This is used where gas cap exists. Oil withdrawal reduces pressure, allowing gas cap to expand. Pressure declines more slowly than in dissolved gas case. Recovery rate can be between 20 and 40%.

Water drive: Water in the reservoir exerts pressure to move the hydrocarbons out of the reservoir. The deeper the water is, the higher the pressure. Water replaces the oil withdrawn from the reservoir. Water-drive can be bottom water drive or edge water drive. Natural drive remains effective longer. Recovery rate can be as high as 50%.

Combination drive: A combination of gas and water drive is also used.
Artificial lift
Gas injection: This is commonly used in many countries to sustain oil production and reduce gas flaring. Gas expands in the well and pushes oil up. Gas used for injection can be reused once oil is depleted.

Artificial pumping: Pump assisted lifting is also provided in some cases to enhance fluid flow.

Enhanced recovery: When natural pressure is not enough to drive oil out of the well, a decision is to be made whether to use enhanced recovery techniques or not. At the end of primary recovery as much as 75% of the oil may still remain in the well. But enhanced recovery systems add cost to production. Thus a trade off has to be reached between cost and benefits of using the enhanced recovery methods.

8.3 Production

A typical production time profile for oil fields is shown in Fig. 8.10. Production increases initially as new wells are drilled and put to production. Then the production reaches a plateau and can stay in this phase until about one-half of the recoverable reserves have been tapped from the reservoir. Production starts to decline then and is abandoned when the output falls below the economically recoverable rate. Enhanced recovery techniques are used when decline sets in due to fall in reservoir pressure and entry of water in the reservoir. The enhanced recovery techniques then extend the life of the field by pushing the decline further and ensuring higher recovery of the fuel.

Evidently, costs would play an important role in the decision making. The secondary recovery is dearer compared to natural drive and tertiary recovery is costlier compared to secondary recovery. Thus the decision to employ a particular technique depends on another set of cost-benefits analysis. The advantage in the case of oil is that this decision need not be taken at the beginning of the oil production and development phase. This modular nature of the decision-making makes oil development and production more attractive.

8.3.2 Production Decline and Initial Production Rate

As discussed earlier, production decline following exponential curve is commonly used. However such a decline would result "if the flow rate of oil is proportionate to the pressure in the reservoir and the rate at which the pressure falls is proportionate to the flow of oil." (Hannesson 1998)

The first requirement can be expressed as $q_t = k_1 h_t$, where k_1 is a constant, h_t is pressure at time t and q_t is the flow rate of oil at time t. Differentiating, we get

$$dq_t/dt = k_1 dh_t/dt. \tag{8.4a}$$

The second requirement can be written as

$$dh_t/dt = -k_2 q_t \tag{8.5}$$

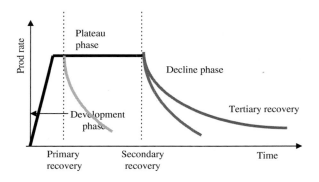

Fig. 8.10 Typical production profile of a petroleum reservoir

Combining the two, we get

$$dq_t/dt = -k_1 k_2 q_t = -k q_t, \qquad (8.6)$$

which leads to the exponential decline curve discussed above.

From the first equation, $k_1 = q_0/h_0$, where h_0 is the initial pressure in the reservoir (given by nature) and q_0 is the initial production rate. Hence the decline rate k can be written as

$$k = k_2 q_0 / h_0. \qquad (8.7)$$

This suggests that the decline rate is influenced by the initial production rate. By choosing different initial production rates, the decline can be modified. This has economic implication, as faster recovery would leave less oil in the ground while slower rate could bring future remuneration but at the risk of making it obsolete.

The initial production rate may also affect the total amount of oil that can be recovered from a reservoir. This can happen due to water flooding of the reservoir due to drop in pressure. Water may trap some oil and thereby reduce recoverable reserves. At the same time, more wells can improve draining of the reservoir, thereby augmenting production and recoverability. Oil and gas flow is influenced by permeability of rocks and viscosity of the fluid. Migration rate is less for thick oil in dense rocks. A single well cannot satisfactorily drain any reservoir but as the number increases flow improves but pressure falls. Thus there is a trade off between more wells and pressure decline. This is known as rate sensitivity.

8.3.3 Gas Production

As indicated earlier, production of non-associated gas follows a similar pattern as with oil. The exploration and development wells determine the output from a field and the rate of recovery tends to be better than oil. But once peak is reached and depletion starts, it is more difficult to improve production. Enhanced recovery of gas is complicated and expensive.

The problem for associated gas is more complex. As oil has a ready market and is highly tradable, the operator would be more interested in producing oil. Production of gas depends on oil production and the gas so obtained can be used in one of the three ways:

- Used by consumers: This depends on the availability of a market and necessary infrastructure. Lack of this can be a major constraint in many countries.
- Re-injected: The amount of gas that can be re-injected varies with well and has to decided on a case-by-case basis. Normally only a part of the gas can be re-injected to maintain the desired level of output.
- Flared: This means venting of gas. This leads to waste of a precious fuel but as high volume gas storage may not be economic and the operator may not be ready to slow down oil production to limit flaring, this waste continues,

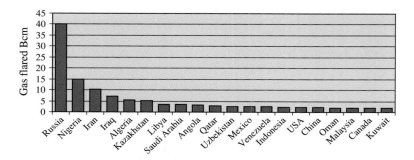

Fig. 8.11 Global gas flaring status in 2008. *Data source* GGFR website

generating environmental damage. According to the Global Gas Flaring Reduction (GGFR) partnership, about 140 Bcm of gas was flared in 2008 worldwide (see Fig. 8.11). Out of this, three countries, Russia, Nigeria and Iran, account for more than 46% of venting. Although efforts are being made to reduce gas flaring, market barriers, financing constraints and regulatory issues often deter progress in this area.

8.3.4 Coal Production

Coal production takes two different forms depending on the amount of overburden and inclination of the seam:

- Surface mining (or strip mining): In surface mining, coal is produced by removing the overburden. This is a less expensive method that can rely on high level of mechanisation, thereby increasing productivity. The recovery factor is also high in this process: can be as high as 90%.[2]
- Underground mining: When coal is located very deep, surface mining is not possible and underground mining is used. This form of mining uses either a labour-intensive "room and pillar" technique or a highly capital intensive long-wall technique. Often the former is used for shallower depths with medium to thick seams of coal. This method allows for recovery of about 50–70% of the coal in place and has a low productivity. It is also prone to higher accident risks. The long-wall technique is a modern and mechanised technique that can be used for working at large depths. However, its use is not very widespread in developing countries.

---

[2] Read more on coal extraction in Edgar (1983). See also WCI (2005).

8.4 Economics of Fossil Fuel Production

8.4.1 Field Level Economics

As coal, oil and gas production is capital intensive in nature, the fixed cost tends to be high compared to the variable cost. For any field, the fixed cost per unit falls over a range of output, showing economies of scale. The variable cost tends to increase as the field grows older. The total cost thus follows a U shape, falling first and then increasing. The operating cost increases as a field gets older. This is because output comes from a lower depth or additional support is required to bring the fuel to the ground. As a result, the fossil fuel extraction is an industry with increasing marginal costs. The marginal cost increases as output increases and the industry moves from a low cost of production regime to high cost areas as cheaper fields are exploited.

The low variable cost and high fixed cost influence the functioning of the industry. Each producer tends to operate to full capacity because of low variable cost and would be willing to supply more. This tendency leads to oversupply in the market. As long as the variable cost is recovered, the producer will continue to produce. The fixed cost is considered as sunk cost, and will not enter into operating decision.

However, oversupply would depress the price of the fuel and if this low price continues for a long time, it affects the prospects of the industry as well. Fields not able to recover costs will be abandoned earlier than expected. New costly fields will not be developed and would create problems in the future. Exploration efforts will reduce, affecting the long-term prospects of the industry.

8.4.2 Industry Level Economics

At the industry level, in a competitive market environment, output would be decided by the interaction of demand and supply curves.

- Each supplier participates in the market depending on its cost of production. Normally, the lowest cost producer is first called upon to supply, followed by the next costly producer. This process continues until demand is met.
- Those suppliers with cost below the market clearing price are called to produce. The marginal producer only recovers the operating cost while the rest would cover more than their respective marginal costs.
- In such a condition, the low cost operators should produce more while the high cost operators should provide the marginal output (see Fig. 8.12).
- Thus, even in a competitive market situation, the profit earned depends on the cost of production and demand for the output.

8.4 Economics of Fossil Fuel Production

Fig. 8.12 Industry level economics

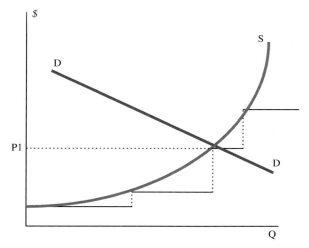

Fig. 8.13 Effect of a new low cost supplier

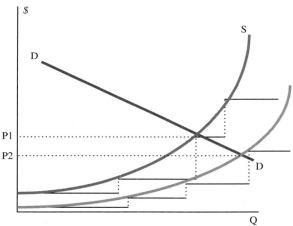

Does the above description really explain the functioning of the oil market? Based on the above description, the lowest cost producers should have higher volume of supply while high cost producers would have only marginal market share. But this does not seem to be happening in the oil market. High cost producers are producing as much or even more than the low cost producers. This is what Prof. Adelman calls oil flowing uphill. This issue will be considered in the chapter on the international oil market.

What happens when a new low cost supplier arrives? This is explained in Fig. 8.13. A low cost supplier would push the costs down and introduce a rescheduling of suppliers who would be called upon to supply the fuel. Displace the costly suppliers and some producers who were supplying earlier may find it difficult to sell their outputs. The supply curve moves outward and the prices in the market falls, which encourages more consumers to enter the market. Thus, demand increases, price falls and some producers are unable to compete and driven out of the market.

8.5 Resource Rent

We now turn to another important topic that is relevant for the supply of non-renewable energies, i.e. the aspect of economic rent. Rent is generally a category of income paid to the owner of a property to allow access to the property. Although in economics there are two commonly used concepts of economic rents- absolute rent in the sense of Marx and differential rent in the sense of Ricardo, for capital intensive energy industries, the differential rent is relevant. The differential rent arises due to differences in specific characteristics of a production unit or factor input (Otto et al. 2006). In the energy industry, four types of differential rents are found (Percebois 1989):

- mining rent due to geological conditions: those fields which could be exploited relatively cheaply compared to others located in difficult geological areas,
- technological rent: this arises due to use of a more efficient technology that reduces the costs of production;
- positional rent: proximity to markets offers added benefits to producers by reducing the cost of transports and related infrastructure
- quality rent: arises due to a favourable chemical or physical characteristics of a fuel. For example, sweet crude oil attracts a premium over the sour crude. Light crude is sold at a premium over the heavy crude oil.

In addition, rents may arise due to a non-competitive market structure (such as monopoly rent due to a monopoly market), scarcity of the resource (discussed in the next chapter) or in some cases due to changes in the market conditions or innovative practices of the firm (known as quasi-rent). Consequently, the prices for a non-renewable energy can be higher than that would prevail in a competitive market.

The economic rent for any non-renewable energy exploitation is "the returns in excess of those required to sustain production, new field development and exploration" (Kemp 1992). This manifests itself in the form of larger producer surplus (see Fig. 8.14). Sharing the rent available for energy exploitation involves a major effort in many producing countries. If the policy is not appropriate, either the state receives a low share of the rent (which can lead to dissatisfied population), or an excessive share in the short run thereby increasing uncertainty about investment and affecting the future prospects of revenue generation. Moreover, stability of the policy, its simplicity and transparency, as well as its equity effects are important considerations.[3]

---

[3] There is a well-developed literature on fiscal systems for energy and mining activities. See for example, Nakhle (2008), Tordo (2007), Otto et al. (2006) and Johnston (1998). Nakhle (2009) provides a succinct review.

8.5 Resource Rent

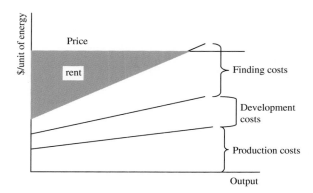

Fig. 8.14 Rent in non-renewable energy exploitation. *Source* Kemp (1992)

A number of alternative instruments are used to collect rents from energy exploitation but most common forms are the use of taxation (royalties and profit tax) and quasi-fiscal instruments (mainly the state participation). The logic for these two is discussed below.

Figure 8.15 presents the effect of royalty. Imposition of a royalty increases the price paid by the consumer, which in turn reduces the demand for the fuel and transfers not only a part of the producer surplus but also a part of the consumer surplus to the state. The price increases from p1 to p2 while the demand falls to q2. The system introduces an economic loss represented by the area E1E2B which is not captured by anybody (or known as the deadweight loss).

The royalty generates revenue for the government, is easy to administer (as the bureaucrats are quite familiar with similar fiscal measures) and involves low risk for the government. Producers loose more than what the state recovers from the producers' surplus, because of the deadweight loss. This affects their investment decision and influences the operating decisions of a field. A royalty is an additional cost to the firm and a high level of royalty can make extraction of some grades of

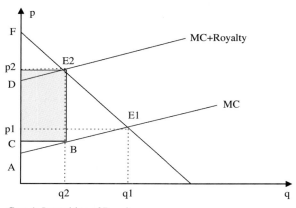

Fig. 8.15 Effect of royalty

Case 1: Imposition of Royalty

the fuel or some sizes of fields unviable. Such a tax shifts the risk on the producers and protects the government revenue better in the event of a low price scenario. Similarly, the consumers also lose but the royalty by restraining consumption leaves more resources for the future generations at the cost of the present generation.

However, deciding the correct royalty rate is not easy, especially for fuels whose prices are quite volatile (such as oil). A low royalty during high oil prices makes the government unhappy and is susceptible to unilateral changes when companies earn high profits. Conversely, a high royalty rate would leave lesser profits to the producer, thereby reducing their interest in the industry, especially during low prices. Moreover, raising a tax or royalty rate increases the revenue for the government in the short-run but affects the activities in the sector adversely in the long-term by reducing the interest in exploration and development activities (Otto et al. 2006).

The government "take"—that is the revenue captured by the state through a fiscal instrument—can ultimately be seen as the price paid by the investor to acquire access to the resource. In the petroleum industry, the government take increased after the first oil price shock in the 1970s but the trend reversed in the mid-1980s when oil prices reached very low levels. Since 2003, the trend is reversing and the government take is rising again (Van Meurs 2008). Generally, the government take can be very high (95–99%) in the case of large, low cost fields. The take is between 60 and 85% where the risks are moderate. In high risk conditions, the take ranges between 40 and 60% (Van Meurs 2008).

Many resource-rich countries have followed an alternative arrangement where the state runs the activities through a state company. The rationale behind state ownership of the rent generating industry is also to capture the rent. Instead of a royalty if a national company manages the industry and is allowed to charge monopoly prices, the same effects of imposing a royalty would be achieved (see Fig. 8.16). Prices would increase to p2 and the demand would fall. The

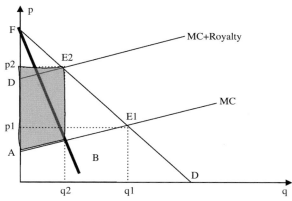

Fig. 8.16 Nationalisation as a possible method of rent capture

Case 2: Monopoly pricing

company would capture a higher share of the producers' surplus compared to the previous case (because it does not have to share with another company). As the rent ultimately accrues to the state, the state gains as well. While this logic could justify such state intervention in energy production activities, there are obvious problems as well. Lack of skills and expertise in managing production activities, political influence in decision-making, difficulty of separating company finances from that of the state, conflicts of interest in terms of regulatory and ownership roles, etc. could seriously affect the performance of the industry.

8.6 Supply Forecasting

8.6.1 Relation Between Discoveries and Production

As indicated earlier, any fossil fuel production field passes through different phases of its life. Production declines at the last phase and it is abandoned at some point in time when the variable costs cannot be recovered. This natural process is part of life of any non-renewable energy source. Normally, at any point in time, a number of different fields operate in a country and each of them would be at various stages of their life. The total output is the aggregation of production from various fields. Globally, the output can then be considered as the sum of outputs of each producing countries.

A convenient concept that is used at the national level is the reserve to production ratio. This indicates the number of years the proved reserves would survive at the current level of production. This assumes that no new reserves are added and the production remains constant at the present level. For example, the R/P ratio for coal is 204 years, while oil and gas would serve for another 40 years and 60 years respectively. However, neither R nor P remains constant over time.

If we want to maintain a certain level of R/P, new discoveries have to be made, otherwise the ratio will start to fall. This is evident from the basic identity given below (Percebois 1989):

$$R_{(t+1)} = R_t - P_t + D_t, \tag{8.8}$$

where R is the reserve, P is production and D is discovery, t is year t, $t+1$ is year following t.

The above identity can be rearranged as

$$D_t = R_{(t+1)} - R_t + P_t. \tag{8.9}$$

Let $R_t/P_t = r$ (the normal R/P ratio designated as r) and assume that the production grows at a rate n every year (i.e. $P_{t+1}/P_t = 1 + n$)

Equation 8.9 can be rewritten as

$$D_t = r \cdot P_{(t+1)} - r \cdot P_t + P_t = r \cdot P_t(1+n) - r \cdot P_t + P_t = P_t[1 + r \cdot n]. \tag{8.10}$$

Equation 8.10 provides the relation between the discoveries and production in terms of *R/P* ratio and production growth rate.

- When n is zero (i.e. there no production growth), discovery equals production. This shows that even when there is no production growth, to maintain same level of *R/P*, discovery equal to the level of production has to be made.
- In other cases, more discoveries have to be made to sustain same level of *R/P*.
  - If the industry discovers more than its production (or consumption), then the resource could be considered to be regenerated economically.
  - If the discovery is less than the production, it means the stock is being drawn down and the industry is in a phase where economic regeneration of the resource is not ensured.

Applying the above framework to the oil industry case it would appear the industry has passed through various phases:

- for a long time until late 1970s, production was less than the increment in proven reserves and during this period economic regeneration of oil was ensured.
- Then since the 1980s, the industry appears to have entered into a phase where production was higher than the reserve increment, implying that regeneration was not ensured. The first period corresponds to a situation when finding new oil is cheaper than producing from existing fields and the industry faces a decreasing long-run marginal cost of development. In the latter case, however, finding new oil is costlier and the long-run marginal cost of development increases (Percebois 1989).
- At present, we may be in the latter situation but whether this is a permanent shift or not is not at all clear.

8.6.2 Supply Forecasting Methods

In order to determine the available supply at a future date, supply forecasting is essential. A number of alternative methods are used to forecast supply.

A relatively simpler method would involve supply elasticities, which indicate the responsiveness of the supply due to changes in the driver variable, (e.g. price). The price elasticity of supply is thus defined as the percentage change in supply due to every percentage change in price. This information can be used to determine how much supply would change if fuel prices change. Similarly, supply elasticities with respect to economic activity could be used to forecast supply as economic activity changes. However, elasticity-based forecasts are not commonly used in practice.

The use of exponential decline production models is widespread for forecasting supply from an oil or gas field. Its currency in the industry arises because of

9.3 A Simple Model of Extraction of Exhaustible Resources

The basic model of the extraction of non-renewable resources was initially proposed by Hotelling (1931). The problem is to find the optimal depletion path of a firm that seeks to extract such resources to maximize its profit. There is a vast body of academic literature on this subject—see Devarajan and Fisher (1981), Fisher (1981) and Krautkraemer (1998) for further details. The basic model is based on the following assumptions: (a) the size of the resource stock is known, (b) the entire reserve is exhausted during the project life, (c) interest rate is fixed.

We define the following terms:
y_t is the quantity of resource extracted in period t;
X_t is the resource stock at the beginning of period t = fixed at $\bar{X}_0$ at time 0;
$C = C(y_t, X_t)$ = total extraction cost;
$P(y_t)$ is the inverse demand function for the resource;
r is the discount rate;
T = time horizon.
The objective is to maximize the net benefit

$$\text{Max}(y_t) \sum_{t=0}^{T} \left[\frac{1}{(1+r)^t} (p_t y_t - c(y_t, X_t)) \right] \tag{9.4}$$

S.t.

$$X_0 = \bar{X}_0; \quad X_T = \bar{X}_T \tag{9.5}$$

and

$$\frac{dX_t}{dt} = -y_t \quad \text{or} \quad X_{t+1} - X_t = -y_t. \tag{9.6}$$

The Lagrange function is given by

$$\begin{aligned} L = & \sum_{t=0}^{T} \left[\frac{1}{(1+r)^t} (p_t y_t - c(y_t, X_t)) \right] \\ & + \sum_{t=0}^{T-1} \mu_t (X_t - X_{t+1} - y_t) + \alpha(\bar{X}_0 - X_0) + \beta(\bar{X}_T - X_T). \end{aligned} \tag{9.7}$$

First order condition resulting from differentiation with respect to y_t is:

$$\frac{p_t - (\partial c/\partial y_t)}{(1+r)^t} - \mu_t = 0; \tag{9.8}$$

which can be rewritten as

$$p_t - \frac{\partial c}{\partial y_t} = \mu_t(1+r)^t = \lambda_t. \tag{9.9}$$

The net price is equal to royalty and in the special case where cost of extraction is negligible the price should grow at the rate of interest. The term on the right hand side of Eq. 9.9 is the user cost, which is directly related to the shadow price of the resource. It suggests that for non-renewable resources, the price should contain an additional element that takes care of the effect of resource depletion. This is the opportunity cost of using the resource now instead of leaving it for the future. In the special case when the cost of extraction is insignificant or zero, the price becomes equal to the rent and hence the rate of price change is just equal to the rate of interest. This is the fundamental result in the economics of exhaustible resources.

9.3.1 Effect of Monopoly on Depletion

Consider the case of pure monopoly—where one producer is functioning in the industry. The problem here is similar to the competitive market. The only difference is in the first condition of optimal depletion because the monopolist will take into account the influence of his output decision on price. The first order condition resulting from differentiation with respect to y_t is given by

$$\frac{p_t + y_t \frac{dp}{dy_t} - (\partial c / \partial y_t)}{(1+r)^t} - \mu_t = 0; \tag{9.10}$$

or MR − MC = Royalty.

Introducing price elasticity in the above equation we get

$$\frac{p_t\left(1 + \frac{1}{e_p}\right) - (\partial c / \partial y_t)}{(1+r)^t} - \mu_t = 0; \tag{9.11}$$

which can be re-written as

$$p_t\left(1 + \frac{1}{e_p}\right) - (\partial c / \partial y_t) = \mu_t(1+r)^t = \lambda_t; \tag{9.12}$$

$$p_t = \lambda_t + (\partial c / \partial y_t) - \frac{\lambda_t + (\partial c / \partial y_t)}{1 + e_p} \tag{9.13}$$

This implies that the price under monopoly would have three components: marginal cost of extraction, royalty and a monopoly rent. This third component is

9.3 A Simple Model of Extraction of Exhaustible Resources

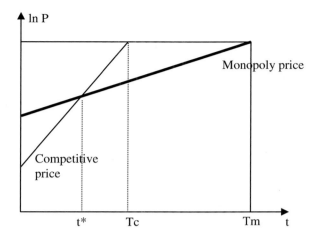

Fig. 9.2 Price path in competitive and monopoly cases

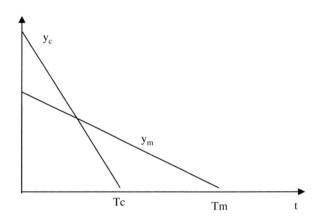

Fig. 9.3 Optimal extraction path

positive for all elasticity values greater than −1.0. In those cases, price under monopoly would be greater than the price under competition.

For a linear demand function, it can be shown that the optimal price path in the case of a monopoly is two times less rapid than that of a competitive market price path. Obviously, the two prices start at different levels and the price charged by the monopolist includes the monopoly rent. This is shown graphically in Fig. 9.2. The optimal extraction path also follows a similar path—under the competitive market situation, the resource is exhausted twice as fast as that under the monopoly in the above case (see Fig. 9.3).

Relating the above idea to the oil market would then suggest that the price change under the OPEC era in the 1970s was an adjustment process where the competitive price path was abandoned in favour of a monopolistic price path. This is shown in Fig. 9.4. Surely, this slows down the extraction and the resource will last longer in this case.

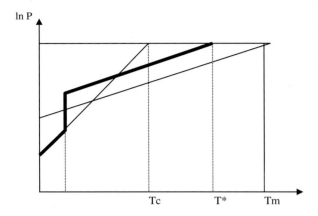

Fig. 9.4 Change in price path under OPEC after the first oil shock

9.3.2 Effect of Discount Rate on Depletion Path

As the discount rate plays an important role in the net worth calculation, the discount rate influences the decision about using non-renewable resources now or in the future. A high discount rate leads to higher rate of extraction initially but the output declines fast and therefore, the resource is exploited quickly (see Fig. 9.5). On the other hand, a lower discount rate prolongs the resource availability through a lower rate of initial extraction and a slower rate of extraction.

The price path for different discount rates again follows the similar pattern (see Fig. 9.6). A high discount rate reduces the initial price but the price path is steeper compared to a low discount rate, which in turn causes to reach the backstop prices earlier.

It needs to be mentioned here that although this application of the Hotelling principles to depletion has given rise to a large volume of academic literature, energy prices do not seem to follow the prescriptions of the theory. As shown in Fig. 9.7, the crude oil price did not follow the price path suggested by the theory, although prices have hardened in recent times. The theory relies on a number of restrictive assumptions and despite much theoretical interest, has not helped much in understanding the fuel price behaviour.

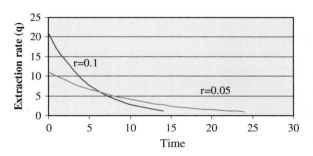

Fig. 9.5 Effect of discount rate on the extraction path

9.3 A Simple Model of Extraction of Exhaustible Resources

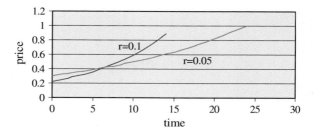

Fig. 9.6 Price path under different discount rates

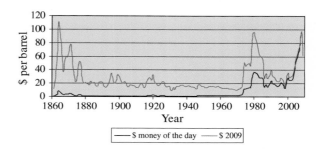

Fig. 9.7 Oil price trend. *Data source* BP Statistical Review of World Energy, 2010

Therefore, from a practical point of view, the relevance and influence of the theory has been quite limited.

9.4 Conclusion

This chapter has provided a simple and a formal introduction to the theory of exhaustible resources. The chapter has restricted itself to the basic model of the theory and did not enter into more elaborate extensions of the theory that has been suggested by various authors to relax some of the restrictive assumptions of the basic model. The outcomes of the model are at odds with the reality of the energy sector and therefore, the practical relevance of the theory remains limited.

References

Devarajan S, Fisher AC (1981) Hotelling's economics of exhaustible resources: fifty years later. J Econ Lit XIX:65–73
Fisher AC (1981) Chapter 2: Resource and environmental economics. Cambridge University Press, London
Hannesson R (1998) Petroleum economics: issues and strategies of oil and natural gas production. Quorum Books, London
Hotelling H (1931) The economics of exhaustible resources. J Polit Economy 39:137–175
Krautkraemer JA (1998) Non-renewable resource scarcity. J Econ Lit XXXVI:2065–2107

Chapter 10
Economics of Electricity Supply

10.1 Introduction

This chapter aims to provide an introduction to the economics of electricity supply by introducing the key concepts relating to this industry and indicating the specific features and simple decision-making tools that are used to make supply and investment decisions. Because of the technical nature of the industry and the influence of engineering-technical side, the tools often tend to be mathematically demanding. Emphasis is given on simple tools so as to provide a basic level of understanding, with some simple mathematical formulations but additional references are provided for those who want more advanced mathematical formulations and/or who intend to gain additional knowledge on the topic.

The electricity industry has undergone significant changes in many countries since 1990s and the industry operation and decision-making has changed from the state-dominated, planned style to the private-oriented decisions. Often the introduction of these structural changes made the decision-making more complex. As these more advanced functioning cannot be understood without the basic understanding, this chapter will generally cover the traditional style of electricity system operation and decision-making. However, indications will be given at appropriate places where restructuring of the industry has affected the decision-making.[1] Similarly, the chapter focuses on grid-based electricity supply, although off-grid electricity supply is gaining relevance in some areas, especially in remote areas of developing countries.

---

[1] Refer to Kirschen and Sadi (2004) for more on electricity system economics in the competitive era.

10.2 Basic Concepts Related to Electricity Systems

Electricity, being an energy that is difficult to store in any economically viable manner, is used by consumers at the time it is produced. Whenever a consumer who is connected to an electricity network switches an appliance, the demand is felt in the grid. The demand for an electric system at any time is the total of all consumer demand.

Although consumers can use different appliances (total of which when occurring simultaneously would give the maximum possible load or demand of the consumer), luckily most consumers do not use all appliances at the same time. Similarly, all consumers connected to the grid do not impose all their demand for electricity at the same time (see Fig. 10.1).

For example, if there are 10 consumers of 1 kW load and if all of them use electricity simultaneously, the maximum demand would be 10 kW, which is the sum of individual peak demand. As the load coincides, the demand is not diversified (i.e. diversity factor is 10/10 = 1). On the other hand, if each consumer can be made to use sequentially, the peak demand would be 1 kW only although the sum of individual maximum load is 10 kW. In this case, the load is highly diversified (i.e. 10/1 = 10).

Formally, the diversity factor is defined as the ratio of sum of maximum customer demands in a system to the maximum system load. In mathematical terms, this can be written as

$$\text{Diversity factor} = \frac{\sum \text{Maximum consumer demand}}{\text{Maximum load on the system}} \quad (10.1)$$

The diversified the load is, the lower the peak capacity requirement is. This reduces the investment need for the system. The inverse of the diversity factor is called the coincidence factor.

The demand imposed by the consumers connected to the grid varies quite significantly within a day, within a week, by season and from one year to another. The daily demand varies as the need for electricity use shows strong time dependence. The plot of demand for 24 h in a chronological order is called the daily load curve

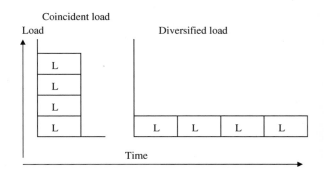

Fig. 10.1 Diversified versus simultaneous loads

10.2 Basic Concepts Related to Electricity Systems

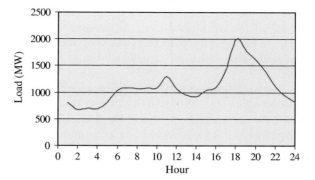

Fig. 10.2 Example of a daily load curve

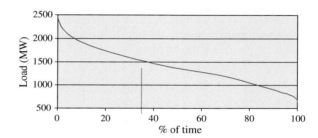

Fig. 10.3 Example of a load duration curve

(see Fig. 10.2 for an example). Such a plot indicates the variation in demand during a day. Figure 10.2 indicates that in this particular case, the peak load occurs at dusk (18 h) but the demand starts to build up from 14 h and reduces as the night progresses.

Normally it is noticed that the week-day load varies from the week-end load patterns. Similarly, the demand varies by season: normally in cold countries demand increases in winter while in tropical countries demand is more pronounced in summer. If the information of daily load curves is collected over a year, the frequency of occurrence of different loads can be determined. A plot of such a cumulative frequency distribution by load is called the load duration curve (see Fig. 10.3). For 100% of the time, the system load is in excess of a small amount of load, called the based load (which is found on the right hand extreme of the curve) while for 0% of the time the system load exceeds the highest load (or the peak load). The diagram can tell for how many hours the system experiences a load in excess of a given load. For example, the system shown in Fig. 10.3 indicates that for about 40% of the time the system load was equal to or in excess of 1500 MW.

Generally the load data is available from the electric utilities. For example, in the case of the UK, the National Grid Company provides such information on a half-hourly basis. See their website for such chronological information (http://www.nationalgrid.com/uk/Electricity/Data/Demand+Data/).

A load duration curve has typically three segments. The system faces relatively small load for most of the time. This is called the base load which exists all the time.

The system faces the highest level of demand for a shorter period. This period is called the peak period and this occurs typically for <20% of the time in a year. In between the base and peak load, the demand gradually increases to reach the peak. This period of increasing demand is known as the period of intermediate load.

The load duration curve has significance for plant operation, cost of service and system efficiency. As electricity cannot be stored in significant quantities at low cost, the demand has to be met by modulating the supply to match the demand. Therefore, for smooth system operation, three types of plants (or technologies) are required:

a. Those which would be running around the year—all the time they are available—to meet the base demand; these plants normally do not have the capability to vary the supply depending on the demand;
b. Another set of plant is required having the capability to follow the demand and vary output frequently during their operation;
c. Finally, a third set of plant is required which are suitable for running only during the peak period.

For base load plants, technologies with low operating costs are appropriate but they could have higher capital cost. On the other hand, plants with high operating costs and low capital costs are most suitable as peaking plants, because they operate only for a short period of time. The intermediate load plants incur additional wear and tear and lose some efficiency in following the load.

Consequently, different types of plants have different capacity utilization rates, called capacity factors. The capacity factor is defined as follows Eq. 10.2:

$$CF = \frac{\text{kWh produced in a year}}{(\text{capacity in kW} * 8760)} \quad (10.2)$$

The base load plants could be used almost 100% of the time, while peaking plants are used only for a very short period (usually <20% of time). If the load did not vary so widely during the year, power plants could have been used more uniformly.

Depending on the shape and size of the three elements of the load-duration curve, the overall capacity utilization is determined. This is called the system load factor (LF) and is the ratio of area under the load-duration curve to the area of the rectangle formed by the peak load for entire duration of the year (Eq. 10.3).

$$LF = \frac{(\text{kWh consumed in a year})}{(\text{peak load} * 8760)} \quad (10.3)$$

If the system load was at the peak load all the time, the load factor would be 100%. However, it is not possible to achieve such a system load factor but the closer the load factor is to 100%, the better the load profile of the system is. The objective of any utility is to improve the load profile so that the plants could be better utilized.

10.3 Alternative Electricity Generation Options

Electricity can be produced using a number of technologies employing different alternative fuels. These options could be grouped into two basic categories:

- Conventional: The conventional electricity generation options can be further grouped into two broad types: thermal and hydro. The thermal variety uses the chemical properties of fuel to generate electricity either by passing steam or in gas turbines or in combined cycle plants where both gas turbines and steam turbines are used. In the case of nuclear, the chemical properties of the fuel are used to heat another fluid (often water) which is then used to generate electricity. Hydro electricity on the other hand utilizes the energy stored in water (i.e. the potential energy) to generate electricity.
- Non-conventional: The non-conventional technologies include solar, wind, geothermal, and the like. These employ the flow of energy to generate electricity either mechanically or otherwise.

Each technology has certain features. For example, coal is a major fuel for electricity generation in many countries. Coal plants tend to offer better efficiencies as their size increases (until the size reaches around 800–1000 MW). As a result, coal plants tend to be capital intensive (see Table 10.1) but as coal is often available quite cheaply, the operating costs tend to be lower. Hydro plants and nuclear plants also tend to have very high capital costs while their operating costs are very low, see Table 10.1 for some estimates of the operating costs).

Some plants can be operated continuously while others are available intermittently. Normally those technologies which use a stock of fuel, they can be operated continuously as opposed to the flow type sources which can generate electricity as long as the flow exists (e.g. wind, tidal, solar, etc.). On the other hand, technologies using intermittent sources of energies have to be used whenever they are generating, forcing other technologies to reduce their outputs.

Similarly, each technology has its own constraints in terms operating characteristics. For example, a storage hydro plant can be brought into operation quickly and takes a few moments to reach the peak power (i.e. quick response plant). This makes such plants suitable for peak load operation. On the other hand, coal and nuclear plants take quite some time to start up and shut down and are not suitable for fluctuating loads or system frequencies. A gas plant can be brought into operation relatively quickly and its efficiency does not fall significantly as the size reduces, making modular use possible.

If a plant has to be shut down either due to outages (operational problems) or due to low demand, they cannot be brought back to operation without allowing a certain delay. In addition, the operator incurs additional costs for each start up.

Thus, generating technologies have to be operated respecting certain conditions:

a. They can only be loaded up to their maximum capacity;
b. Can be brought into operation following the manufacturer's guidelines;
c. Loads can be changed in steps that cause minimum harm to the machine;

Table 10.1 Cost and performance characteristics of electricity generating technologies

Technology	Online Year	Size (MW)	Lead time (years)	Overnight Cost in 2009 (2008 $/kW)	Variable O and M (2008 mills/kWh)	Fixed O and M ($2008/kW)	Heat rate in 2009 (Btu/kWh)
Scrubbed Coal New	2013	600	4	2078	4.69	28.15	9200
Integrated Coal-Gasification Comb Cycle (IGCC)	2013	550	4	2401	2.99	39.53	8765
IGCC with carbon sequestration	2016	380	4	3427	4.54	47.15	10781
Conv Gas/Oil Comb Cycle	2012	250	3	937	2.11	12.76	7196
Adv Gas/Oil Comb Cycle (CC)	2012	400	3	897	2.04	11.96	6752
Adv CC with carbon sequestration	2016	400	3	1720	3.01	20.35	8613
Conv Comb Turbine	2011	160	2	653	3.65	12.38	10788
Adv Comb Turbine	2011	230	2	617	3.24	10.77	9289
Fuel Cells	2012	10	3	4744	49.00	5.78	7930
Adv Nuclear	2016	1350	6	3308	0.51	92.04	10488
Distributed Generation—Base	2012	2	3	1334	7.28	16.39	9050
Distributed Generation—Peak	2011	1	2	1601	7.28	16.39	10069
Biomass	2013	80	4	3414	6.86	65.89	9451
Geothermal	2010	50	4	1666	0.00	168.33	32969
MSW—Landfill Gas	2010	30	3	2430	0.01	116.80	13648
Conventional Hydropower	2013	500	4	2084	2.49	13.93	
Wind	2009	50	3	1837	0.00	30.98	
Wind Offshore	2013	100	4	3492	0.00	86.92	
Solar Thermal	2012	100	3	4798	0.00	58.05	
Photovoltaic	2011	5	2	5879	0.00	11.94	

Source Table 8.2, Assumptions to the Annual Energy Outlook 2010 http://www.eia.doe.gov/oiaf/aeo/assumption/electricity.html

d. Loads cannot be reduced beyond a certain minimum level and for any load beyond this threshold, the plant has to be shut down;
e. If a plant is shut down, it has to be allowed a minimum cooling off period before re-starting as prescribed by the manufacturer.

In addition, at any given time all plants are never available for production. Some are taken out for scheduled maintenance while others are not working due to unplanned outages. Electricity supply is affected by the unavailable capacity due to planned and unplanned outages. As electricity has to be produced almost at the same instant as it consumed because of absence of any viable storage, any system has to be prepared to avoid any events that would cause the system frequency to go beyond a certain band or cause cascading system trips. Consequently, additional

10.3 Alternative Electricity Generation Options

capacity has to be maintained to meet the demand and any contingencies that may arise.[2]

10.3.1 Generation Capacity Reserve

The extra capacity that is required to be maintained in addition to the demand at any time is called the generation capacity reserve. This reserve allows the system to tide over any generating plant outages, errors in demand forecasting and any other faults or errors. This reserve comes in two forms:

a. Spinning reserve: This reserve comes from the plants already connected to the grid which are operating below their peak load. For example, a plant can generate up to 1000 MW but is operating at 700 MW at a time. This offers 300 MW of spinning reserve which could be used when required. The spinning reserve offers a quick response to the changes in demand and is the most reliable option. The response rate depends on the type of technology in operation and this has to be considered at the time of scheduling. Normally, it is a better strategy to distribute the spinning reserve to all committed plants instead of allocating it to one (essentially to reduce risk by not putting all eggs in one basket).
b. Quick-start reserves: These are plants which could be started up quickly to deliver the load and meet the demand. For example, some hydro plants can be brought into operation in seconds. Gas turbines can normally be operated in a few minutes. This type of reserve is less reliable than the spinning reserve because there is always a possibility that the reserve fails to start up in time.

As a consequence of the above conditions, decisions have to be made regarding choice of plants to be operated and their timing of operation (when to start and for how long), and plans for emergencies. These decisions are taken in the analysis of unit commitment.

10.4 Economic Dispatch

The objective of dispatching is to decide how much each plant should generate power so that minimum operating cost commitment results for the system. Any electricity system with many generating units has two options:

a. It can use all the plants to supply the load (kW) and the energy (kWh) all the time. In this case, many plants will run at low-loads and due to poor thermal efficiency, the operating cost would increase; or

---

[2] For a more detailed analysis of the reliability concepts related to the electricity sector and its incorporation in the system analysis, consult Munasinghe (1979).

b. It can choose those plants which would be able to meet the need. This allows all the plants to run in high efficiency region and keeps the operating cost low.

Consequently, for economic reasons only the required number of plants are normally brought on-line at any time. In order to select this list of preferred plants, a number of techniques, such as merit order, incremental cost method or optimal load flow analysis are used. We discuss the first two here for pure thermal systems.[3]

10.4.1 Merit Order Dispatch

The merit order method is the simplest one and relies on a priority list that ranks the generating units in some order of preference. The commonly used criterion is the hourly fuel cost per megawatt. Thus a list would be generated to rank all units from lowest to highest $/MWh and units required to meet the demand would be selected. If for example, there are three generating stations G1, G2 and G3 with cost characteristics such that the operating cost (OC) follows the relationship OC1 < OC2 < OC3, then generator 1 will be first loaded, followed by G2, and finally G3 will be brought in. This is a simple rule but follows a static approach and assumes that the operating costs do not change with plant output. It also does not take transmission and distribution constraints or reactive power.

The preliminary ranking can then be revised by taking into consideration the start-up costs, shut-down costs, minimum loading conditions, etc. The dispatch decision decides the short-run marginal cost of the system for each period. The short-run marginal cost is the operating cost of the costliest plant that is used to generate power in any given period. In a marginal cost-based pricing system, this is the relevant cost for price setting.

10.4.2 Incremental Cost Method

Incremental cost implies the cost of producing an additional unit of electricity. This is composed of incremental fuel cost and other cost items such as labour, materials and supplies. It is often more difficult to determine and express the non-fuel incremental costs but fuel-related costs can be better determined by relating to the fuel input required for a plant at different levels of plant output.

In the incremental cost approach, the objective is to decide plant loading to achieve minimum production costs. Assume the following:

P_i = load on plant i, $i = 1, \ldots, n$;
$F_i(P_i)$ = production cost from plant i with load P_i

---

[3] Hydro-thermal systems require somewhat more complicated analysis.

10.4 Economic Dispatch

D = demand to be met;
TPC = total production cost
We can express TPC as follows:

$$\text{TPC} = \sum_{i=1}^{n} F_i(P_i) \tag{10.4}$$

The objective is to meet the demand at lowest cost. Therefore the problem is to minimise TPC subject to

$$\sum_{i=1}^{n} P_i = D \tag{10.5}$$

The Lagrangian function can be written as

$$L = \sum_{i=1}^{n} F_i(P_i) + \lambda \left(D - \sum_{i=1}^{n} P_i \right) \tag{10.6}$$

The first order conditions are

$$\frac{\partial L}{\partial P_i} = 0 \quad \text{for all } i \text{ and } \frac{\partial L}{\partial \lambda} = 0 \tag{10.7}$$

The first condition can be rewritten as

$$\frac{\partial F_i(P_i)}{\partial P_i} = \lambda \quad \text{for all } i, \tag{10.8}$$

This implies that at the optimum loading, the incremental production costs of all plants being loaded must be equal. The second condition suggests that plants have to be loaded until the total demand is met. The optimal value of incremental cost λ^* can be obtained from the combined incremental cost curve at $\sum_{i=1}^{n} P_i = D$.

10.5 Unit Commitment

In the dispatch decision, the decision-making was based on operating costs alone. It did not consider a number of aspects related to plant operations. The unit commitment is the decision-making about economic scheduling of generating units. This needs to consider and respect unit constraints and system constraints.

Unit constraints impose restrictions on how a plant can be used. For example, a plant cannot be loaded beyond its maximum capacity. Similarly, there is a minimum load below which a plant will not operate. If a plant is switched off, it will take a minimum amount of time to return online (minimum down time). Similarly, the loading pattern of each plant varies depending on the ramp rate. Additionally, there are costs for starting up a plant from cold. Similarly, plants incur costs for shutting down as well. All these constraints have to be respected.

Similarly, the system constraints have to be respected as well. These affect more than one generating plant. The system has to meet the load demand at all times. It also has to maintain a healthy reserve to face any emergency arising out of unforeseen loss of generating plant capacity. If environmental regulation requires meeting certain environmental conditions, the scheduling has to consider such constraints as well. For security reasons, no system should violate the transmission line capacity constraint.

The purpose of unit commitment is to decide how the plants should be chosen to produce their outputs, at what point of time and to what extent so that the overall production cost is minimised satisfying all constraints. More detailed analysis of total operating costs is required to decide the units to be committed for generation. The priority list scheduling (using merit order dispatching approach) can be used as a rough guide. Proper scheduling requires cost minimization respecting the above-mentioned constraints as well as constraints related to demand—supply balancing, including provisions for reserve capacities. These require more advanced analytical tools (dynamic programming, mixed integer programming, etc.) and go beyond the scope of this chapter. See Wood and Woolenberg (1996) for further details.

Example: Consider in a system the demand for electricity is 1300, 1800 and 1500 MW between 10 PM–10 AM, 10 AM–6 PM and 6 PM–10 PM, respectively. The system keeps a spinning reserve of 10% of the current load at any period for emergency situations. There are six generating units with the characteristics given in Table 10.2. Determine which units have to be used to economise the operating costs.

The priority list in terms of operating cost is given in Table 10.3. At any time, the committed units would have to be able to meet the demand and the spinning reserve. Also, the committed units should not violate the minimum down time requirements.

After 10 PM, plant C is shut down and it is brought into operation at 10 AM. As the minimum down time is 6 h, there is no violation of downtime requirement here. Similarly, plant F is shut down at 6 PM and it only comes back to operation at 10 AM, which allows more than 5 h of down time in this case. The result is presented in Table 10.4.

The short-run marginal cost of electricity in this case is $20/MWh in the first period, $27 in the second period and $25 in the third period.

Table 10.2 Generating unit characteristics

Plants	Capacity (MW)	$/MWh	Minimum down time (h)
A	500	9	50
B	300	20	2
C	200	25	6
D	300	15	2
E	500	12	10
F	200	27	5

10.5 Unit Commitment

Table 10.3 Commitment priority list

Plant	Capacity (MW)	$/MWh	Min down time (h)
A	500	9	50
E	500	12	10
D	300	15	2
B	300	20	2
C	200	25	6
F	200	27	5

Table 10.4 Selection of generating unit

Period	Load	Load + Res	Choice	Committed capacity (MW)
10 PM–10 AM	1300	1430	AEDB	1600
10 AM–6 PM	1800	1980	AEDBCF	2000
6 PM–10 PM	1500	1650	AEDBC	1800

The price-based wholesale competition used in the initial phase of the England and Wales power sector (known as Gross power pool) followed a similar method for sorting the generators. Generators submitted price bids indicating how much they can generate and the price they wanted. This was done at a pre-determined time (say one day ahead of delivery). Based on the price bids received, the regulator (or the system operator) established the merit order dispatch schedule paying due attention to the expected demand from consumers (as submitted by the suppliers). A list of successful bidders is prepared and these generators are chosen for supplying the power. The price of the most expensive unit scheduled to be dispatched to meet forecast demand sets the system marginal price. Thus the simple idea of merit order dispatch could be used in a reformed power sector as well.

10.6 Investment Decisions in the Power Sector

Given the relatively long gestation period of power sector projects, decisions regarding capacity addition are required well in advance. As a number of technologies to produce electricity exist, it is important to decide which type of technology should be chosen for meeting future demand. In this section we discuss two simple methods: levelised bus-bar cost and screening curve analysis. More sophisticated methods exist but they require a strong mathematical (optimisation) background.

10.6.1 Levelised Bus–Bar Cost

This is a widely used approach for comparing electricity generation costs. Many studies have been published in recent years that have relied on this method (see for example Royal Academy of Engineering (2004), CERI (2004), Heptonstall (2007).

In this method the cost per kWh of electricity or the annual cost of owning and operating a generating plant is calculated to compare different technologies.

As is usual in any project analysis, the present value of investment and recurring costs is first determined. The amount is then spread into a uniform equivalent series. This is the concept of levelisation that plays an important role in the above analysis. As the life of generation technologies is normally long (15–40 years) and some components of the cost also escalate over time, cost comparisons become easier if a single, constant, present-worth equivalent value for the costs is found for each technology. The cost levelisation process does this by expressing a series of escalating annual costs into a single constant value (Fig. 10.4).

Assume that the annual cost in 2010 is 2 p/kWh and it escalates at the rate of 5% per year over the next 20 years. The present value of this escalating series is given by

$$\text{PV} = A \times \text{PVF} \qquad (10.9)$$

where PV is the present value, A is the present annual cost, and PVF is the present value factor (see Annex 1 for details).

The equivalent annual cost of the present value is obtained by multiplying the PV with the capital recovery factor (CRF). Thus,

$$\text{Levelised annual cost} = \text{PV} \times \text{CRF} \qquad (10.10)$$

The capital recovery factor is indicated in Annex 1.

In Fig. 10.4, the annual cost increases from 2 p/kWh in 2010 to 5.31 p/kWh in 2030. The uniform series that would have the same present value of the annual cost is 2.845 p/kWh (for a discount rate of 10%).

As electricity generation involves fuel, other non-fuel operation and maintenance costs and investment (or fixed costs), each of these elements is used in the levelised cost calculation. The total annual owning costs ($/year) are found out from the above components. This divided by the annual electricity generation (kWh/year) produces the unit cost of electricity ($/kWh). The following example illustrates the method by comparing two generation technologies.

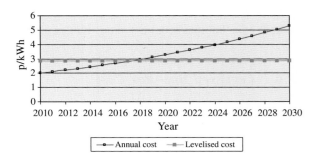

Fig. 10.4 Levelisation concept. *Source* Stoll (1989)

10.6 Investment Decisions in the Power Sector

Example: Consider two technologies: coal and combined cycle. The relevant data for the levelised cost evaluation is given in Table 10.5. Determine which option is preferable.

Table 10.5 Data for investment decision of power technologies

Details	Coal	Combined cycle
Capacity (MW)	400	400
Plant cost ($/kW)	1100	700
Fixed O and M cost ($/kW/year)	20	9
Variable O and M cost ($/MWh)	5	3
Heat rate (kcal/kWh)	2100	1950
Fuel cost ($/Gcal)	8	20
Capacity factor (%)	70	70
Fuel price escalation (%/year)	4	4
Discount rate (%/year)	10	10
Levelised fixed charge rate (%/year)	15	15
Project life (years)	20	20

Answer

Levelisation factor (using formula from Annex 2)	1.32	1.32
Annual fuel costs (M$/year)	= 400 MW * 70% * 8760 (h/year) * 2100 (kcal/kWh) * 8 ($/Gcal)/10^9 * 1.32 = 54.40 (M$/year)	126.28
Annual variable O&M charges (M$/year)	= 400 (MW) * 70% * 8760 (h/year) * 5 ($/MWh)/10 ^ 6 * 1.32 = 16.19 (M$/year)	9.71
Annual Fixed O&M charges (M$/year)	= 400 (MW) * 20 ($/kW/year) * 1.32/10 ^ 3 = 10.56 (M$/year)	4.75
Annual investment costs (M$/year)	= 400 (MW) * 1100 ($/kW) * 0.15/10 ^ 3 = 66.00	42.00
Total (levelised cost M$/year)	147.15	182.74

In this example, under the given assumptions, the coal plant has a lower levelised cost and would be preferred. However, note that the cost is influenced by the assumptions, especially those relating to the capacity factor, escalation rate and technical efficiency. Often power plants would operate at different levels of capacity factor and the assumption of a fixed capacity factor is not realistic. This problem is removed in the screening curve method.

10.6.2 Screening Curve Method

The screening curve method is an extension of the concept used in the earlier method. This is a graphical approach where the annual levelised owing cost is shown along the vertical axis and the capacity factor is plotted along the horizontal axis. Thus, the screening curve is a plot of the levelised cost for different capacity factors (see Fig. 10.5).

What does Fig. 10.5 tell? Two technologies have different cost characteristics: coal has a high capital cost but the operating cost is relatively low. On the other hand, the combined cycle plant has a low capital cost but the operating cost is higher (given by the slope of the curve). Consequently, if the new capacity is required to be operated for <30%, the combined cycle would be the preferred technology. For any higher level of capacity utilisation, the coal plant would be economical. This method can be used for a number of technology options to find out the relative merits of each technology.

The cross-over point between two technologies can be determined as follows:

Suppose that we have two different generating technologies with characteristics as follows: the fixed cost per kW of capacity is F_i and the variable cost per hour is V_i, where for the two technology case i varies between 1 and 2. We assume that the capacities are ordered in such a way that $F_{i+1} > F_i > F_{i-1}$, and $V_{i+1} < V_i < V_{i-1}$.

For any duration, h, the cost of using 1 kW of capacity of a plant is given by

$$y = F_i + V_i * h, \tag{10.11}$$

Given the cost characteristics, there exists a point where the costs of using the two types of plants are equal at the cross-over point. At this point for our two plant example, $F_1 + V_1 \cdot h = F_2 + V_2 \cdot h$, from which

$$h = \frac{(F_2 - F_1)}{(V_1 - V_2)} \tag{10.12}$$

This point indicates the number of hours during which the plant with higher variable cost shall be operated. By repeating the process for other plants, the cross-over points can be determined.

Combining the screening curve analysis with the load-duration curve gives an approximate way of determining the desired generation mix of a power system. This is shown in Fig. 10.6. The screening curve is placed on the upper half of the figure while the load-duration curve is placed on the lower half. By projecting the point of intersection from the screening curve onto the load-duration curve yields the share of generation that would come from the two technologies. In the present

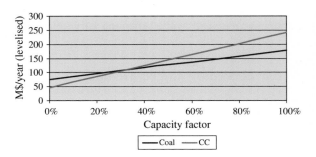

Fig. 10.5 Screening curve

10.6 Investment Decisions in the Power Sector

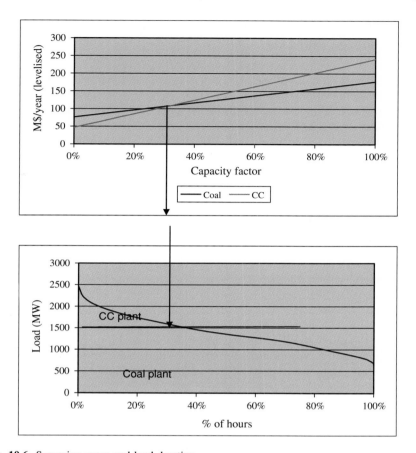

Fig. 10.6 Screening curve and load-duration

example, the combined cycle plant would have a capacity of 900 MW while the coal plant would have a capacity of 1600 MW. Thus in a generation system, different types of technologies are economic at different points in time, which forms the basis for generation mix of a power system.

The screening curve is useful in understanding the role of different technologies in the power sector. It provides a quick insight into the system optimisation strategies. This has been used in identifying the DSM options [see Koomey et al. (1990), Masters (2004) for example]. It requires limited amount of information but can capture the major trade-offs between capital costs, operating costs and the utilisation levels of various types of technologies. However, it is not adequate or a good substitute for detailed production cost analysis or system expansion analysis. It fails to capture the issues related to system reliability, resource constraints, differences in characteristics of new technologies and the old technologies, etc. More sophisticated methods are used for such purposes, which is beyond the scope of this chapter. But a brief account is given in the section below.

10.7 Sophisticated Approaches to Electricity Resource Planning

There is a large, well-developed body of literature dealing with the electricity resource planning issues that deal with the timing and optimal sizing of new capacity addition taking the demand-side solutions into consideration [see Hobbs (1995), Foley et al. (2010) for a literature review on this]. The resource planning can be considered using an integrated approach where supply, reliability, demand, resource availability, demand-side options, financial issues and rates are considered systematically. However, this tends to be too demanding in terms of computational requirement and technical capabilities. An alternative to this is to follow a modular approach where each module focuses on a specific issue and the results are then linked to find an acceptable result. However, this also requires "messaging" of outputs from each module and it is difficult or often impossible to obtain quick results (Hobbs 1995).

The simulation of electricity capacity expansion in the traditional model focused on the supply-side only. This remains the starting point in such an exercise. In the 1970s, linear programming (LP) models were formulated and used [see Turvey and Anderson (1979) for such examples]. However, an LP formulation assumed that the generation capacity is continuously variable, which in reality is not true as capacity comes in discrete sizes. Subsequently, mixed integer formulations and dynamic programming formulations have been developed.

The optimization models for resource or capacity planning generally have the following structure (Hobbs 1995):

- The objective function typically minimizes the present worth of capital costs and operating costs less salvage value.
- A set of decision variables such as capacity to be chosen in different periods that the utility aims to decide through the optimization process;
- A set of constraints that define the range within which the optimization should be performed. Such constraints include

    - The power demand constraint that indicates the demand at all times should be met.
    - Individual capacity constraint that restricts the power output of each power plant within the feasible range;
    - Hydro and other resource availability constraints;
    - Reserve margin and reliability constraints indicating the allowable range of security of supply to be built in the system.

A number of software packages such as WASP-IV of the International Atomic Energy Agency (IAEA) or EGEAS developed by the Electric Power Research Institute have been widely used around the world. Annex 10.2 provides a brief outline of WASP-IV. As the industry structure has changed in many countries from a vertically integrated one to a competitive model, the modeling techniques required significant adjustments to cater to the new issues related to competition. More sophisticated models using game theory or assuming imperfect competition (Cournot and Bertrand models), or using optimization with market equilibrium constraint have emerged (Kagiannas et al. 2004) but the use of traditional packages or simple LP models still continue in many cases for their simplicity of use.

10.8 Conclusion

This chapter has provided an introduction to the basic elements of analysing the electric power supply. It has introduced the concept of load and energy demand and their influence on the supply technologies. The chapter has then presented the decision-making challenges related to operating decisions and investment planning. Simple tools such as merit order dispatch and priority list for unit commitment were introduced, followed by discussion on simple cost analysis methods (such as levelised costs and screening curve method). Although more sophisticated approaches are used in the industry, yet the simple tools provide a quick appreciation of the issues and can help analyse problems in a systematic way. Such simple tools allow a quick evaluation of a problem and can help develop a good understanding of the economic issues of the electricity industry.

Annex 10.1: Levelisation Factor for a Uniform Annual Escalating Series

Assume that

A is the annual cost in the first year,
a is the escalation rate per year,
n is the number of years used in the analysis,
i is the discount rate,
P is the present worth of the cost series,
U is the annual levelised cost.

As the cost increases every year at the rate 'a', the cost changes from one year to the other as follows: $A, A(1 + a), A(1 + a)2,\ldots, A(1 + a)n - 1$

The present value of this cost series is given by

$$P = \frac{A}{(1+i)} + \frac{A(1+a)}{(1+i)^2} + \frac{A(1+a)^2}{(1+i)^3} + \cdots + \frac{A(1+a)^{n-1}}{(1+i)^n}$$
$$= A\left[\frac{1}{(1+i)} + \frac{(1+a)}{(1+i)^2} + \frac{(1+a)^2}{(1+i)^3} + \cdots + \frac{(1+a)^{n-1}}{(1+i)^n}\right] \quad (10.13)$$

Multiplying Eq. 10.13 by $(1 + i)$ results in

$$P(1+i) = A \left[1 + \frac{(1+a)}{(1+i)} + \frac{(1+a)^2}{(1+i)^2} + \cdots + \frac{(1+a)^{n-1}}{(1+i)^{n-1}} \right] \quad (10.14)$$

Multiplying Eq. 10.13 by $(1 + a)$ results in

$$P(1+a) = A \left[\frac{(1+a)}{(1+i)} + \frac{(1+a)^2}{(1+i)^2} + \cdots + \frac{(1+a)^n}{(1+i)^n} \right] \quad (10.15)$$

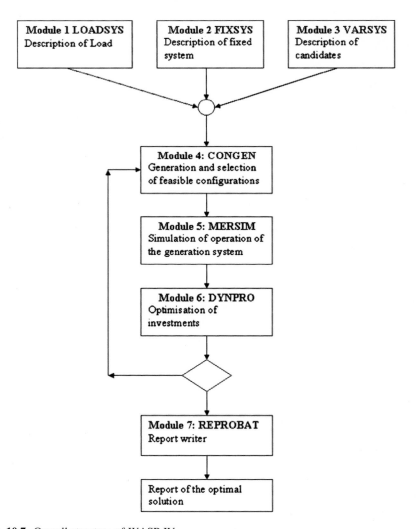

Fig. 10.7 Overall structure of WASP-IV

Annex 10.1: Levelisation Factor for a Uniform Annual Escalating Series

Subtracting Eq. 10.15 from 10.14 gives rise to the following:

$$P(i - a) = A\left[1 - \left(\frac{(1+a)^n}{(1+i)^n}\right)\right] \tag{10.16}$$

Therefore, the present worth of this annual series is

$$P = \frac{A\left[\left(1 - \frac{(1+a)^n}{(1+i)^n}\right)\right]}{(i-a)} = A \times \text{(Present value function)} \tag{10.17}$$

where present value function is

$$\text{PVF} = \frac{A\left[1 - \left(\frac{(1+a)^n}{(1+i)^n}\right)\right]}{(i-a)}$$

The annual series U that would yield the same present value as above is given by

$$U = \frac{\left[1 - \left(\frac{(1+a)^n}{(1+i)^n}\right)\right]}{(i-a)} \left[\frac{i(1+i)^n}{(1+i)^n - 1}\right] = \text{PVF} \times \text{CRF} \tag{10.18}$$

where

$$\text{CRF} = \left[\frac{i(1+i)^n}{(1+i)^n - 1}\right] \tag{10.19}$$

Note that the levelising factor is reduced to unity when there is no escalation (i.e. $a = 0$).

For the example in Fig. 10.4, a is 5%, i is 10% and n is 20 years. Using these data in Eq. 10.18 gives, $U = 1.423$.

For further details on these topics, see Stoll (1989) and Masters (2004).

Annex 10.2: A Brief Description of the WASP-IV Model

The WASP model developed by the International Atomic Energy Agency (IAEA) is a widely used tool that has become the standard approach to electricity investment planning around the world (Hertzmark 2007). The current version, WASP-IV, finds the optimal expansion plan for a power generating system subject to constraints specified by the user. The programme minimises the discounted costs of electricity generation, which fundamentally comprise capital investment, fuel cost, operation and maintenance cost, and cost of energy-not-served (ENS)[4] (International Atomic Energy Agency (IAEA) 1998). The demand for electricity is exogenously given and using a detailed information of available resources, technological options (candidate plants and committed plants) and the constraints on the environment, operation and other practical considerations (such as implementation issues), the model provides the capacity to be added in the future and the cost of achieving such a capacity addition.

To find optimal plan for electricity capacity expansion, WASP-IV programme evaluates all possible sets of power plants to be added during the planning horizon while fulfilling all constraints. Basically, the evaluation for optimal plan is based on the minimisation of cost function (International Atomic Energy Agency (IAEA) 1984), which comprises of: depreciable capital investment costs (covering equipment, site installation costs, salvage value of investment costs), non-depreciable capital investment costs (covering fuel inventory, initial stock of spare parts etc.), fuel costs, non-fuel operation and maintenance costs and cost of the energy-not-served. Overall, the structure of WASP-IV programme can be presented in Fig. 10.7.

The model works well for an integrated, traditional system but the reform process in the electricity industry has brought a disintegrated system in many countries. The model is less suitable for such reformed markets.

References

CERI (2004) Levelised unit electricity cost comparison of alternate technologies for baseload generation in Ontario. Canadian Energy Research Institute, Calgary, Canada

Foley AM, O'Gallachoir BP, Hur J, Baldick R, McKeogh EJ (2010) A strategic review of electricity system models. Energy, 35(12):4522–4530

Heptonstall P (2007) A review of electricity unit cost estimates. UKERC working paper, UKERC/WP/TPA/2007/006, UK Energy Research Centre (see www.ukerc.ac.uk)

Hertzmark D (2007) Risk assessment methods for power utility planning. Special report 001/07 March 2007; Energy Sector Management Assistance Programme of the World Bank, Washington DC

Hobbs BF (1995) Optimisation methods for electricity utility resource planning. Eur J Oper Res 83(1):1–20

International Atomic Energy Agency (IAEA) (1984) Expansion planning for electrical generating system: a guidebook, Vienna

International Atomic Energy Agency (IAEA) (1998) Wien Automatic System Planning (WASP) package: a computer code for power generating system expansion planning version. WASP-IV—user's manual, Vienna

Kagiannas AG, Askounis DTh, Psarras J (2004) Power generation planning: a survey from monopoly to competition. Int J Electr Pow Energy Syst 26:413–421

Kirschen D, Sadi, (2004) Fundamentals of power system economics. Wiley, London (see ebrary)

---

[4] Energy-not-served (ENS) or expected un-served energy is "the expected amount of energy not supplied per year owing to deficiencies in generating capacities and/or shortage in energy supplies" (International Atomic Energy Agency (IAEA), 1984).

Koomey J, Rosenfeld AH, Gadgil A (1990) Conservation screening curves to compare efficiency investments to power plants: applications to commercial sector conservation programms. In: Proceedings of the 1990 ACEEE Summer Study on Energy Efficiency in Buildings, Asilomar, CA (see http://enduse.lbl.gov/info/ConsScreenCurves.pdf)

Masters GM (2004) Renewable and efficient electric power systems. Wiley, New Jersey

Munasinghe M (1979) The economics of power system reliability and planning. The John Hopkins University Press, Baltimore

Royal Academy of Engineering (2004) The costs of generating electricity. The Royal Academy of Engineering, London (see http://www.countryguardian.net/generation_costs_report.pdf)

Stoll HG (1989) Least cost electric utility planning. Wiley, New York

Turvey R, Anderson D (1979) Electricity economics: essays and case studies, A World Bank Research Publication (see http://www-wds.worldbank.org/external/default/WDSContentServer/IW3P/IB/2001/01/10/000178830_98101911363556/Rendered/PDF/multi_page.pdf)

Wood AJ, Woolenberg BF (1996) Power generation, operation and control, 2nd edn. Wiley-Interscience, New York

Further Reading

IEA (2004) World energy investment outlook. International Energy Agency, Paris (see http://www.iea.org//Textbase/nppdf/free/2003/weio.pdf)

Marsh WD (1980) Economics of electric utility power generation. Oxford University Press, London

Munasinghe M, Warford JJ (1982) Electricity pricing: theory and case studies. The John Hopkins University Press, Baltimore

Nakawiro T (2008) High gas dependence in electricity generation in Thailand: the vulnerability analysis. Unpublished Doctoral Thesis, University of Dundee, Dundee

Park YM, Park JB, Won JR (1998) A hybrid genetic algorithm/dynamic programming approach to optimal long-term generation expansion planning. Int J Electr Pow Energy Syst 20(4):295–303

Swisher JN, Jannuzzi GM, Redlinger RY (1997) Tools and methods for integrated resource planning: improving energy efficiency and protecting the environment. UCCEE, Riso (see http://www.uneprisoe.org/IRPManual/IRPmanual.pdf)

Viscusi WK, Vernon JH, Harrington JE Jr (2005) Economics of regulation and antitrust. MIT Press, London

Chapter 11
The Economics of Renewable Energy Supply

11.1 Introduction: Renewable and Alternative Energy Background

This chapter focuses on the economics of renewable and alternative energies. The term "alternative energy" refers to any energy forms that are outside the conventional forms of energies we have considered so far. Although conventional energies can be renewable as well (such as hydropower) and can include both renewable and non-renewable sources (such as tar sand, shale gas, etc.), this chapter focuses on modern renewable energies. Most of these energies are available abundantly and the mankind has been using them for various purposes from time immemorial. The direct cost to the consumer remains low in their traditional form of use (such as drying). However, modern ways of using these energies require sophisticated conversion processes, which in turn increase the cost of supply.

The oil price shocks of the 1970s triggered new interest in renewable energy sources. Availability of easy petrodollars facilitated funding of renewable energy research and the field flourished during the periods of high oil prices in the international market. The global concern for climate change and sustainable development provided further impetus to renewable energies. Now renewable energies occupy an important place in any strategy for sustainable development in general and sustainable energy development in particular.

11.1.1 Role at Present

According to IEA,[1] around 13% of global primary energy supply in 2007 came from renewable energies. Out of 12 Giga ton of oil equivalent of primary energy

---

[1] See also Darmstadter (2003) for a review.

consumed (PEC) in 2007 globally, 1.5 Gtoe came from renewable energies. A breakdown by source of energy indicates that around 78% of the above came from traditional energies (biomass and combustible renewable wastes), another 17% came from hydroelectricity, while the remaining 5% came from modern renewable energies (solar, wind, tide, geothermal, etc.).

The share of modern renewable energies in the global primary energy supply in 2007 was about 0.6%, despite significant efforts being put in harnessing such energies. However, there are wide variations in the regional supply of renewable energies (see Fig. 11.1):

- 48% of the PEC in Africa came from renewable energies.
- 27% of PEC in Asia (excluding China) and 31% of that in Latin America came from renewable sources.
- The share is negligible in the Middle East and in the former Soviet Union countries.

However, in low income countries, biomass-based renewable energies dominate (see Fig. 11.2). The share of modern renewable energies in these areas remains low. The use of traditional energy data in the statistics overshadows the role of renewable energies. Traditional energies are often not be obtained through sustainable practices and do not rely on clean technologies. Additionally, equating the energy content of biomass to other modern energies leads to overestimation of supply and misrepresentation of shares. This is due to the low conversion

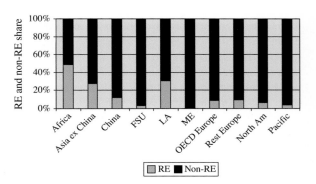

Fig. 11.1 Regional distribution of renewable energy use in 2007. *Data source* IEA (2009)

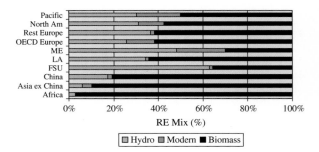

Fig. 11.2 Renewable energy use by type in 2007. *Source* Based on IEA (2009) data

11.1 Introduction: Renewable and Alternative Energy Background

Fig. 11.3 Changes in the renewable energy use by sector between 1990 and 2007. *Source* Based on IEA (2009) data

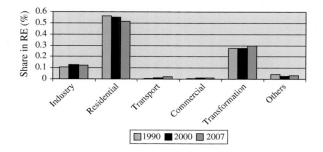

efficiencies of traditional energy technologies (often less than 10%). Despite these issues, IEA data now includes biomass as a renewable energy.

The sector-level picture indicates where renewable energies are mostly used (see Fig. 11.3). The residential sector is the most important user of these energies (51% of RE use in 2007). This is due to inclusion of biomass-based energies. It is worth noting here that the share of the residential sector is declining since 1990. Electricity generation and other energy transformation activities also represent another important use of RE (30% in 2007).[2] Inclusion of hydropower gives electricity generation an important role. Industry uses relatively small amounts of RE at present (about 12% in 2007) but this share is increasing. Transport and other activities (like agriculture) use very little amounts of renewable energies but as we will see below, the outlook is changing here as well.

Although the above background does not present an encouraging picture, the renewable energy supply has grown at an average annual rate of 2.3% over the past 30 years (IEA 2007). This rate is somewhat higher compared to the overall growth of primary energy supply. More importantly, new renewable energies like wind and solar have recorded very high growth rates, although most of this due to their miniscule base in 1971. There is a growing interest in bio-fuels as well.

Moreover, the present use of renewable energies is a small fraction of the overall potential.[3] According to de Vries et al. (2007), the estimated technical potential for wind power in 2000 was 43 PWh (peta watt-hours) and that for solar-PV and biomass was 939 and 7 PWh respectively whereas the global electricity use is 13.3 PWh. They estimate that this potential is likely to increase to 61 PWh for wind, 4105 PWh for solar PV and 59 PWh for biomass by 2050 (see Fig. 11.4 for regional potentials).

---

[2] Strictly speaking, electricity generation and other transformation do not form part of the final demand of RE.

[3] Clearly, the renewable energy potential is a function of economic factors and technological progress. There are varying estimates of potentials based on different definitions of the potential. See Verbruggen et al. (2009) for a discussion on this debate. Intergovernmental Panel of Climate Change (IPCC) has initiated a study called Special Report on Renewable Energy which aims to provide a better understanding of the renewable potentials. This report is expected to be published in 2011.

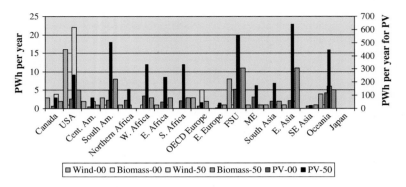

Fig. 11.4 Renewable energy potential by region. *Source* de Vries et al. (2007)

Further, EIA (2010) indicates that the use of renewable sources for electricity generation will grow at the rate of 3% per year between 2007 and 2035, making this the fastest growing source of electricity generation, ahead of coal. It also projects a rapid growth of non-conventional liquid fuels during the same period. This future outlook and the drive for a low carbon future energy path require a good understanding of the economic basis of these renewable energies. This is the purpose of this chapter.

In the rest of the chapter, we focus on renewable electricity and bio-fuels.

11.2 Renewable Energies for Electricity Generation

At present, non-hydro renewable energies occupy a very low share—around 3.6% on a global average basis—in power generation. However, the share increases to above 6% in the transformation sector due to a higher proportion of auto-producer activity in heat and power (IEA 2009). Information compiled by the American Energy Information Administration (EIA) indicates that around 161 GW of non-hydro renewable electricity capacity existed in the world in 2007 compared to an overall installed capacity of 4420 GW. According to BP Statistical Review of World Energy 2010, the installed wind capacity in 2007 was 94 GW. Geothermal capacity and solar PV account for 9.9 GW and 9.2 GW respectively. Mini- and small-hydro capacity is another large source of renewable electricity. The rest comes from biomass-based sources. Asia accounts for the highest amount of hydro-based capacity while Europe has the largest share of non-hydro capacity of renewable electricity (see Fig. 11.5).

Even within Europe, the share of renewable electricity varies widely across countries (see Fig. 11.6). Some countries in this region are well endowed with hydropower (Norway, France, Sweden and Spain for example) and accordingly, their renewable electricity essentially derives from this source. In respect of non-hydro renewable energies, Germany, Spain and Italy stand out.

11.2 Renewable Energies for Electricity Generation

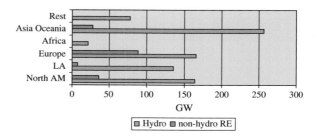

Fig. 11.5 Distribution of renewable electricity generation capacity in 2007, GW. *Data source* EIA website (http://tonto.eia.doe.gov/cfapps/ipdbproject/iedindex3.cfm?tid=2&pid=alltypes&aid=7&cid=&syid=2003&eyid=2007&unit=MK)

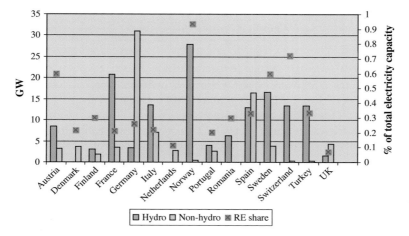

Fig. 11.6 Renewable electricity situation in selected European countries in 2007. *Data source* EIA website

Outside hydropower, electricity from wind turbines has emerged as the dominant source of renewable electricity. This technology has seen a rapid market penetration in recent terms and has recorded above 30% year-on-year growth in capacity addition between 2008 and 2009 (see Fig. 11.7). Clearly, Asia Pacific and North America has experienced a significant market expansion. China alone added more than 13 GW of wind capacity in this period, followed by the USA with about 10 GW capacity additions in 1 year. Other major markets for wind capacity

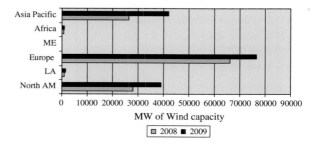

Fig. 11.7 Growth in wind electricity capacity. *Data source* BP Statistical Review of World Energy, 2010

Fig. 11.8 Average operating time of renewable electricity capacity in EU-27 in 2008. *Data source* Eurostat (2010)

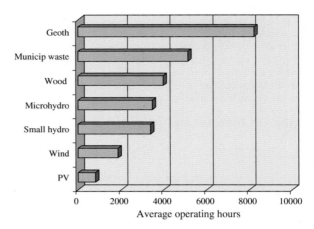

addition in 2008–2009 were Spain, Germany, India and the UK where capacity additions between 1 and 2 GW took place within a year.

However, a closer look at the output and capacity utilization information of renewable electricity capacity from EU-27 shows that some technologies like solar PV is used only less than 10% of the time in a year (see Fig. 11.8), while others like geothermal plants are operating almost at 100% capacity. Wind capacity is also used only 20% of the time, while biomass-based plants have capacity utilization rates of about 50%. The intermittent nature of wind and solar energies adds to the capacity utilisation problem and increases the cost of supply.

11.3 Bio-Fuels

Bio-fuels are the other area of renewable energies receiving current attention. Bio-fuels are produced from a variety of feed-stocks. Brazil uses sugarcane while the USA uses corn as the main source of bio-ethanol. In the European Union, countries use alternative sources of oily seeds for bio-diesel production (rapeseed, and other oil bearing seeds). In addition, animal fat and biomass can also be used to produce liquid fuel (Bomb et al. 2007). Figure 11.9 presents a schematic of alternative feed stocks and transformation processes for bio-fuel production.

According to IEA (2009), the global supply of bio-fuels in 2007 was around 51 Mt, with bio-ethanol maintaining a dominant (60%) market share (around 30 Mt) while bio-diesel accounted for 20% of the bio-fuel supply, and the remaining 20% came from other liquid fuels.[4] Although bio-fuels had just around 1.5% share in global transport fuel demand in 2007 (according to IEA (2009)), the

---

[4] IEA Statistics make this distinction where bio-fuels not used as bio-gasoline or bio-diesel are grouped as other liquid bio-fuels. This chapter follows this convention.

11.3 Bio-Fuels

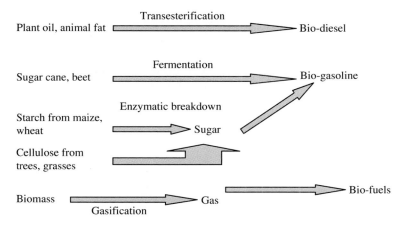

Fig. 11.9 Alternative feed-stocks and transformation processes for bio-fuel production. *Source* Bomb et al. (2007)

Fig. 11.10 Evolution of bio-fuel use in the world. *Source* IEA (2009)

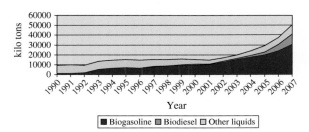

industry has grown rapidly since 1990 and has recorded an exponential growth (IEA (2009))—see Fig. 11.10. Between 2000 and 2007, bio-ethanol supply has trebled between 2000 and 2007, whereas bio-diesel production has increased many times in the same period due to a lower initial base.

The activity in the bio-fuel market is clearly dominated by three regions—North America, Latin America and Europe (see Fig. 11.11). North American market has a share of 46%, followed by the Latin American market with a 27% share, while the European market accounts for about 22% of the market share. However, the contribution of bio-fuels in the transport sector in Europe or North America is still around 2%. Only in Latin America, the share as reached 7%.

In bio-gasoline and bio-diesel segments of the business, there is now clear regional level influence: Brazil and the USA are the two main players in the bio-ethanol market. Brazil has a long experience in bio-ethanol production and bio-fuels accounted for around 15% of its transport energy demand in 2007, making it the world leader in bio-energy use in the transport sector. But the volume of output from the USA is now four times higher than that of Brazil, making USA as the world leader in bio-gasoline production (see Fig. 11.12).

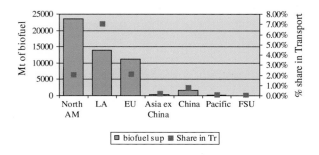

Fig. 11.11 Main players in bio-fuel market in 2007. *Source* IEA (2009)

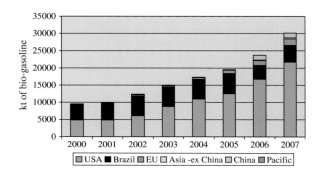

Fig. 11.12 Bio-gasoline supply around the world in 2007. *Data source* IEA (2009)

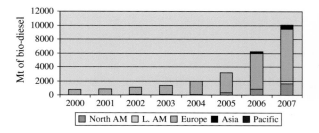

Fig. 11.13 Bio-diesel supply distribution in 2007. *Data source* IEA (2009)

In the bio-diesel market[5] on the other hand, Europe has established itself as the undisputed leader with a 75% market share (see Fig. 11.13). The supply from North America is growing but the share of other regions is pretty limited at this moment.

Although Brazil has sustained its bio-energy for over three decades now and has been successful in developing effective technologies, the size of plants tends to be small. On the other hand, American plants are exploiting economies of scale and are two to three times bigger than their Brazilian counterparts.

---

[5] See Balat and Balat (2008) for a detailed review of bio-diesel.

11.4 Drivers of Renewable Energy

Renewable energies are emerging as alternative fuels as they offer a number of advantages. Following Goldemberg (2004) these are:

a) **Reduction in CO_2 emission and mitigation of climate change**: This is the main driver of renewable energy at present. The concentration of greenhouse gases (GHG) is increasing due to fossil fuel dependence of modern economies. It is believed that the increasing concentration of GHGs has led to warming of our climate. It is forecast that without any mitigation action, the CO_2 concentration in the atmosphere would double the present level by 2050. Renewable energies being carbon free (or neutral) would help reduce the GHG concentration.

b) **Security of energy supply**: Security of energy supply has made a come-back in recent years. This is attributed to recent increases in fossil fuel prices in general and oil prices in particular; concerns for depletion of fossil fuels globally and imminent production decline in the US and UK, and consequent increase in import dependence; increasing competition for supply from emerging consuming countries; political instability in the hydrocarbon resource rich areas; and high economic impacts of energy supply disruption in the developed and rapidly developing countries.

As fuel diversification is considered as an important strategy for ensuring supply security, developing alternative energies from locally available resources can reduce import dependence and accordingly, renewable energies are being viewed favourably from this perspective.

c) **Improving energy access**: It is now believed that more than 2 billion population worldwide do not have access to clean energies. The problem is more acute in rural areas of poor countries where the supply system may be inexistent. To ensure sustainable development, it is essential to provide clean energy to these people. Renewable energies offer certain advantages in this respect—they reduce environmental and health damages, and save time in fuel collection and improve working conditions. These changes can in turn provide better opportunities for income and reduce poverty.

d) **Employment opportunities**: Renewable energy supply has the potential for employment generation, directly due to decentralised, modular structure of the technologies and local level operation of the systems. And indirectly through improved working conditions or saving in time which would otherwise be used in drudgery.

e) **Other spill-over effects**: Reliance on renewable energies would help improve macro-economic stability. The logic goes as follows: (1) Promotion of renewable energies reduces import dependence; (2) fossil fuel import being the important constituent of the international trade of importing countries, a switch over to the renewable energies is expected to reduce the trade balance; (3) this in turn reduces the possibility of economic shocks due to external factors.

In addition to the above advantages, renewable energy technologies benefited from significant cost reductions over the past decade and such a trend is expected to continue in the future. Cost reductions have made some of these technologies economically feasible and competitive (e.g. wind) with other conventional power generation technologies. In fact, Darmstadter (2003) suggests that the cost reductions were much higher than was expected by many.

Despite enjoying such advantages why are renewable energies unable to capture higher market shares?

This is essentially due to the existence of considerable barriers facing renewable energies. The literature on the subject has identified a number of barriers. Painuly (2001) provides a framework for identifying and analyzing the barriers. He suggests that the barriers can be analysed at a number of levels: first can be grouped in broad categories. Within each category, a number of barriers can then be identified. At a third level, the elements of these barriers can be identified. This disaggregated approach can provide a better clarity on the subject. Neuhoff (2005) has identified four broad categories of barriers and elements within them. These include technological barriers (related to intermittency of supply), uneven playing field (related to failure of the pricing system to internalize externalities of fossil-fuel energies), marketplace barriers (such as access to the grid, regulatory barriers, inappropriate tariffs or incentives for renewable energies, etc.), and non-market barriers (such as administrative difficulties, lack of long-term commitment, lack of information, etc.)

11.5 The Economics of Renewable Energy Supply

This section will first focus on the economics of renewable electricity and then on that of bio-fuel supply.

11.5.1 The Economics of Renewable Electricity Supply

Electricity from renewable resources has a number of technical features:

a) most common forms of renewable energies (such as solar, wind or tidal) are intermittent in nature (i.e. they are not available all the time), and
b) given that electricity cannot be stored in large quantities in a cost effective manner, these energies have to be used when they are available.

As a result of intermittency, a number of issues arise.

(1) Electricity generated from such sources cannot be dispatched following the merit-order dispatch schedule. They have to be used whenever the electricity is available. However, through better forecasting of weather conditions, more accurate assessment of local level generation can be made.

11.5 The Economics of Renewable Energy Supply

(2) As a consequence of the above, the capacity is used only for a limited time, leading to low capacity utilization. This has been indicated in Fig. 11.6 where it is noted that the average utilization of solar PV systems is less than 10% in Europe, while the average wind capacity utilization is about 20%.

(3) Consequently, such systems cannot provide reliable supplies round the clock and will require back-up capacity (or standby capacity). The standby capacity often relies on non-renewable energies and therefore, the benefits of renewable energies are not available. The standby capacity also increases the cost of supply.

In a study by Gross et al. (2006), it is estimated that the intermittency costs in Britain are of the order of 0.1–0.15 pence/kWh. This is quite substantial compared to the electricity price paid by the consumers.

In addition, renewable electricity often suffers from other biases against it. These include:

a) **Inappropriate valuation**: The value of electricity normally varies depending on whether it is used during the off-peak hours or peak-hours. The peak-period supply should fetch a higher value to the supplier; but as renewable supply is treated outside the wholesale market (being non-dispatchable), the appropriate valuation of its contribution is difficult to make. This would affect the financial and economic viability of the renewable energy projects.

b) **Inappropriate price signals**: Often such units are embedded in the distribution system and rely on net metering (i.e. considers the energy supplied less energy consumed by the unit). But unless the retail tariff is based on time-of-day pricing, the system does not provide proper signal to the consumer and the supplier. This also affects the renewable energy generation and its viability.

c) **Non-internalisation of externalities**: Renewable energies have environmental advantages compared to the fossil fuel-based electricity. Consequently, non-recognition of the external costs[6] in the pricing puts renewable energies at a disadvantage and does not allow two types of energies to be compared on the same level. This acts as a barrier to the renewable energy development.

d) **Fuel risk benefits**: Renewable energies do not face fuel price risks faced by the fossil fuels. In fact, the operating cost of renewable energies is minimal in most cases. However, the market price for fossil-fuel based electricity does not provide the correct signal to the investors and the consumers taking the premium for higher prices for fossil fuels into consideration. This has an adverse effect on the renewable energy development. Awerbuch (2003) suggested that inappropriate fuel risk and financial risk estimation renders renewable electricity costlier, which introduces a systematic policy bias against renewable electricity.

---

[6] External costs are covered in another chapter where the economics environmental damages from energy use is considered.

Any comparison of electricity supply costs should adequately capture the above differences. The basic indicator—levelised cost—is often used but it may be an in appropriate comparator as it relies only on a specific level of capacity utilisation, which varies widely across electricity generating technologies.

The screening curve approach in conjunction with the load duration curve provides a better picture as this can capture the value of energy at different stages of the load.[7] More complex simulation models are required to capture the differences in costs and technical characteristics of electricity generating techniques and their effects on the supply. This however requires more involved mathematical models, which are beyond the scope of this discussion.

11.5.1.1 Cost Features

The main elements of costs to be considered in the case of electricity supply technologies are:

a) Energy-related costs: Include those costs which are related to energy generation in a facility: costs related to fuels and variable operating and maintenance related costs. Normally, for fossil-fuel based electricity, this component is relatively high while for the renewable fuels, this element tends to be small.
b) Capacity costs: These include the cost of installing the capacity (charges to be paid in relation to installation of a capacity) and the fixed operating and maintenance costs (labour charges, stocks, etc.). For renewable energy based electricity, this is the most important cost element and could be between 50% and 80% of the overall cost of supply.
c) Other related costs: This is a broad category of cost that can include external costs due to environmental damages and climate change, costs related to standby or reserve capacity, and any other costs that should be considered to make the like-for-like comparisons.
   a. Environmental costs are higher for fossil fuels and nearly non-existent for the renewable energies.
   b. On the other hand, standby capacity costs could be important for certain types of renewable energies.
   c. Similarly, fuel price risk (or security risk) could be high for some fossil fuels and should be considered here.

Figure 11.14 presents the comparison of levelised costs of electricity supply for different electricity technologies from the Royal Academy of Engineering (2004) study.[8] Although this figure provides costs relevant for the UK market, it still provides a generic picture.

---

[7] See the paper by Kennedy (2005) for an application of this method.

[8] Heptonstall (2007) provides a review of unit cost estimates of electricity generation using different technologies.

11.5 The Economics of Renewable Energy Supply

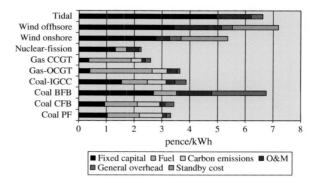

Fig. 11.14 Levelised cost of electricity generation by technologies. *Source* Based on data from Royal Academy of Engineering (2004)

The above figure suggests that most renewable-energies would be cost ineffective solutions for generating electricity even after taking environmental costs into consideration. This is because of high level of standby power costs. If standby power cost is ignored, the onshore wind power becomes quite competitive with commonly used fossil fuels like coal or gas (in an open cycle). However, tidal power and offshore wind power are still not cost effective solutions. The assumptions about fuel prices and capacity utilization rate also affect the outcome significantly. The report assumed full utilization of base load plants and 35% capacity utilisation factor for intermittent sources such as wind. As indicated before, the capacity factor of different technologies varies widely and a uniform assumption does not capture the real situation. Similarly, the fuel price assumptions were quite conservative, making the security of supply insurance premium quite small for fossil fuels.

A study by EPRI (2009) provides the levelised cost of electricity for a future date—2015 and 2025 (see Table 11.1). The message from the above discussion appears to be clear: renewable energies for electricity supply still face cost disadvantages and would require support to ensure their promotion.

Table 11.1 Levelised cost of power generation

Technology description	Cost in 2015 (2008 constant $/MWh)	Cost in 2025 (2008 constant $/MWh)
Super critical pulverized coal	66	86–101
Integrated gasification combined cycle	71	78–92
Combustion turbine combined cycle	74–89	67–81
Nuclear	84	74
Wind	99	82
Biomass circulating fluidised bed	77–90	77
Solar thermal trough	225–290	225–290
Solar PV	456	456

Source EPRI (2009)

Fig. 11.15 Feed-in tariff principle. *Source* Menanteau et al. (2003)

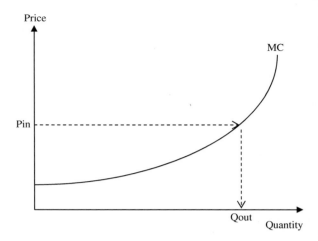

11.5.1.2 Support Mechanisms[9]

A number of intervention or support mechanisms have been used in practice to promote renewable energy based electricity to overcome barriers arising from market distortions and lack of internalisation of externalities. These include feed-in tariffs, competitive bidding process, renewable obligations, financial incentives, and taxing fossil fuels.

Feed-in Tariffs

This is an intervention by influencing the price. Here the electric utilities are required by law or regulation to buy renewable electricity at fixed prices set normally at higher than the market price. The system has evolved over time: in California, a system of standardised long-term contracts at fixed prices was initiated in the 1980s to promote renewable energies, similar to independent power project contracts. In mainland Europe, the producers were guaranteed a fixed share of the retail price and the contracts lasted for the project life (15–20 years). More recent feed-in tariffs vary by location, by technology and by plant size. The fixed price declines over time and is adjusted periodically but the tariffs are long-term in nature. The basic mechanism is explained in Fig. 11.15.

---

[9] A well-developed body of literature exists in this area covering alternative support mechanisms and their application to specific technologies or countries. See for example Menanteau et al. (2003), Sawin (2004), Mitchell et al. (2006), del Rio and Gual (2007), Bunter and Neuhoff (2004), Dincia (2006), and World Bank (1997).

11.5 The Economics of Renewable Energy Supply

In Fig. 11.15, assume that the regulatory or public authorities have fixed the feed-in tariff at P_{in}. All producers whose cost of supply is below this price will enter the market and produce an output Q_{out}. The total cost of support in this case is $P_{in} \times Q_{out}$. The important point to note here is that projects with low cost of production will earn a rent due to their locational or technological advantage. The fixed price system allows the producer to capture this rent, which provides an incentive for further innovation.

Generally the cost of subsidising renewable electricity is passed on to the electricity consumers through the electricity tariff. However, in some cases the tax payers in general or consumers in the area of utility's jurisdiction where the renewable energy development is taking place may bear the cost (Menanteau et al. 2003).

The feed-in tariff system has proved to be a successful instrument. It has been used by those who have successfully developed their renewable electricity market. These countries have often exceeded their national targets. As the producer has tariff certainty over the project life, the system reduces financing risks and facilitates financing. The system is easy to implement and if standardised, the transaction cost can be low. However, the feed-in tariff system through generous payments to producers promotes high cost supply. The long-term nature of the contract can lead to stranded investments, especially in a competitive market. Finally, it is not known in advance how much capacity addition will take place. Therefore, there is no guarantee that a given target will be achieved. If over-supply takes place, the utility has the obligation of purchasing the power, which creates a contingent liability.

Competitive bidding processes

This is a quantity restriction mechanism where the regulator or public authority mandates that a given quantity of renewable electricity would be supported but decides the suppliers of such electricity through a competitive bidding process. Interested producers are asked to submit bids for their proposals, which are ranked in terms of their cost of supply. All proposals are accepted until the target volume is reached. This mechanism is therefore an attempt to discover the supply curve through bids and can be represented in diagrammatic form as shown in Fig. 11.16.

In Fig. 11.16, for the target volume Qt, suppliers up to a marginal cost of P will be selected. However, the price paid to each supplier is limited to the bid price (i.e. pay as per bid) and not the marginal cost of the last qualifying bid. This removes the rent or producer surplus that is available in the case of a feed-in tariff. This reduces the support cost to the area under the supply curve and as a consequence, the burden on the consumers reduces. However, by removing the rent, the incentive to innovate is reduced. As the bidding system decides the quantity to be procured, there is certainty in terms of maximum volume of supply (although whether the target will be reached or not remains unknown). The price to be paid

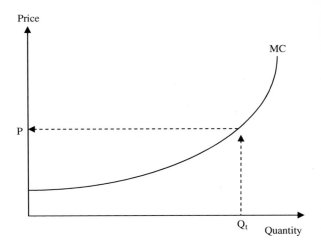

Fig. 11.16 Competitive bidding process principle. *Source* Menanteau et al. (2003)

and therefore the overall cost of support is not known ex-ante (Menanteau et al. 2003).

Renewable obligations

Renewable obligations (RO) also work through the quantity restriction mechanism where the government sets the target for renewable electricity supply and lets the price be determined by the market. The obligation is placed on the electricity suppliers to purchase a given percentage of their supply from renewable sources. The target is often tightened over time with the objective of reaching a final level by a target date. The Renewable Portfolio Standard (RPS) used in the United States of America or the Renewable Obligation system in England and Wales are the common examples of this category.

The Renewable Obligation requires the electricity supplier to supply a specific amount of renewable energy in a given year. For example, the RO in England and Wales started in 2002 with a target of 3% for 2003 but the target rises to 15.4% for the year 2015–2016. In theory, the RO is guaranteed to stay at 15.4% level until 2027—thereby guaranteeing a life of 25 years. However, in April 2010, amendments were made to extend the end date to 2037 for new projects.

A number of technologies are recognised as the eligible renewable sources (such as wind, solar energy, biomass, etc.). The producer of renewable electricity receives from the RO administrator a tradable certificate, called the Renewables Obligation Certificate (ROC), for every unit of electricity generation—either at a uniform rate for every unit of renewable electricity produced or at a preferential rate depending on the technology employed (which has been introduced in England and Wales from 1, April 2009).

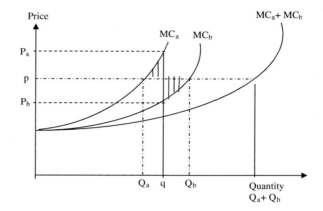

Fig. 11.17 Economic logic for certificates trading. *Source* Menanteau et al. (2003)

Generators thus have two saleable products[10]: electricity which they sell to electricity suppliers and the ROC that they can sell to electricity suppliers or traders. Certificates are tradeable and trading between suppliers and traders creates a market for these certificates. The economic logic here is that trading of certificates allows electricity suppliers to meet the target at the least cost. This is explained in Fig. 11.17.

Consider two suppliers A and B who are subjected to a renewable target of q. The marginal cost of supply for A is given by MCa while that of B is given by MCb. As A faces a steep cost curve compared to B, if it has to comply with the requirement alone, its cost will be P_a whereas B can meet the target at P_b. However, because of its cost advantage, B could easily expand its renewable supply beyond the required limit and trade the credit with A. This allows both the suppliers to benefit as the system can achieve the target at a lower price p. Thus, B produces up to Q_b while A produces just Q_a and together they still satisfy the 2q requirement set by the regulator at a lower price. This benefits the society as a whole by imposing lesser burden for promoting renewable energies.

In the English system, the suppliers can also pay a buy-out price in lieu of ROCs to meet their obligation or follow a combined approach of buying some ROCs and buy-out the rest. The buy-out price effectively sets the ceiling price for the supplier to buy renewable electricity, and acts as a protective instrument for consumers (Mitchell et al. 2006).

To prove compliance of obligation, suppliers have to redeem their ROCs with the regulator and pay the fine for non-compliance (or buy-out price if available). In England and Wales, the buy-out price is set by the regulator and the revenue so generated is recycled annually to the suppliers presenting the ROCs in proportion to their ROC holding. The market price of ROCs reflects the buy-out price and the recycle payment received by the suppliers.

---

[10] In England and Wales, the generator can also receive its share of recycled buy-out premium and payment for levy exemption certificates in the consumer is eligible for exemption under the Climate Change Levy agreements (see Mitchell et al. 2006).

11.5.1.3 Performance of Price and Quantity-Based Mechanisms Under Uncertainty and Risk

In ideal conditions of free and cost-less information, the price- and quantity-based mechanisms produce similar results. However, in reality these mechanisms do not yield same results due to incomplete information and uncertainty. Because the supply curve is not known in advance, the shape of the curve would influence the outcome considerably. If the shape of the curve is relatively flat (or elastic), the output in a price-based system will be substantially off the target when the shape in incorrectly estimated (see Fig. 11.18). On the other hand, for steep supply curves, the quantity-based systems face the risk of off-the-mark prices under supply cost uncertainties.

Assume that the regulator assumes the shape of the supply curve as indicated in MC_2 and sets a feed-in-tariff at p, expecting Q_2 as the supply to be supported. But the actual shape turned out to be MC_1, resulting in Q_1 as the supply volume. This results in an increased supply and consequently a higher volume of subsidy for support. On the other hand, for a quantity-based system, assuming the shape as MC_1, the regulator set a quantity q for renewable supply. In reality, the shape turned out to be MC_2. This leads to a significantly higher marginal price to meet the target and would facilitate entry of costly supply options. From above, the following logic can be obtained: when the slope of the marginal cost curve is gentle, the quantity-based system works better in presence of uncertainty whereas the price-based system works better when the slope is steep. In other words, a price-based approach performs poorly when the marginal cost curve is gently sloped and a quantity-based approach works poorly when the slope of the marginal cost curve is steep.

Mitchell et al. (2006) also introduce another set of risks in comparing these mechanisms. They consider price, volume and balancing risks faced by the

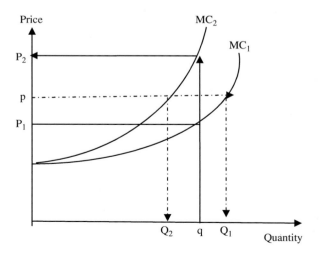

Fig. 11.18 Performance under uncertainty

investors of renewable energies under two broad types of support systems. In the case of feed-in tariffs, the electricity supplier is obligated to buy any amount of renewable electricity produced at the set price. This removes the volume risk. Similarly, the price is known in advance and the contractual arrangement facilitates financing of renewable energy projects. In the context of competitive markets, the renewable generator does not have to worry about the mismatch between predicted and actual supply in a feed-in tariff regime. It is the responsibility of the system operator to take care of the variation. There is no penalty for the mismatch.

On the other hand, the Renewable Obligations do not promise a price—this is decided by the market where supply and demand will determine the outcome. This leaves the investors with a great deal of risk and price uncertainty. Absence of a contract also affects the ability to project finance new capacity additions. Similarly, as the supply volume approaches the target, the generators face the risk that their outputs will not be purchased at the prevalent price. The suppliers would look for cheaper sources and the generator will face the volume risk. Finally, under the British system the renewable generator bears the risk of over or under-performance and faces the balancing risk. Table 11.2 summarises these risks. Accordingly, the RO appears to leave substantial risks to the generators. This can explain the slower growth of renewable electricity capacity in the U.K. However, it is important to indicate here that the British policy aimed at keeping the extra burden on electricity consumers low. The policy has succeeded in achieving this and as the technology matures, the sector and the society are expected to benefit from the prospects of declining costs of future renewable electricity.

11.5.1.4 Financial Incentives

These are fiscal measures used either to reduce the cost of production or increase the payment received from the production. Commonly used incentives include: tax relief (income tax reduction, investment credit, reduced VAT rate, accelerated

Table 11.2 Comparison of performance of support systems under risk (investor's perspective)

Risk type	Feed-in tariff	RO
Price risk	No price risk for generators	Great deal of price risk as price depends on supply–demand interactions
	Generators save money from hedging the price risk	Price likely to fall as supply approaches the target volume
Volume risk	No volume risk—obligation to buy all power produced	Exists
		Individual generators do not have any guarantee of volume
		Once target is met, no security of buying the entire output
Balancing risk	Side-stepped; no penalty for intermittent generation.	Balancing risk exists; penalty imposed for out-of-balance positions.

Source Based on Mitchell et al. (2006)

depreciation, etc.); rebates or payment grants (that refunds a share of the cost of installing the renewable capacity), and low interest loans, etc. Normally these incentives show preferences to particular technologies (hence cherry picking) and may promote capacity but not necessarily energy generation.

11.5.1.5 Taxing Fossil Fuels

The objective here is to reflect the true costs and scarcity of the fossil fuels in the prices paid by the consumers to send a clear signal. Taxing fuels for their environmental and other unaccounted for damages is one way of ensuring the level playing field. The Nordic countries are in the fore-front of such environmentally-oriented taxation. They are the pioneering countries in introducing carbon taxes (i.e. a tax on CO_2 emissions), even before the European Union launched a proposal to introduce community-wide carbon taxes in 1992 (which was never adopted although individual members have introduced some such taxes). Finland was the first country to introduce a CO_2 tax in 1990, followed by Norway and Sweden in 1992 and Denmark in 1992. Besides carbon tax, there are other taxes on energy as well—these include taxes on fuel and electricity and a tax on SO_2 emission. Despite this, it is doubtful whether the polluter is bearing the tax burden as a study by Eurostat (2003) found that the burden is shifted to residential consumers while the industry bears a relatively lower burden.

11.6 The Economics of Bio-fuels

The cost of supply of bio-fuels varies widely depending on the technology, feedstock used and the size of the conversion plant. The energy content of bio-fuels varies significantly and the energy density of bio-fuels is less compared to petrol or diesel. Generally, the plant size and feedstock cost play an important role in the bio-fuel supply cost. However, bio-ethanol and bio-diesel costs do not follow similar patterns and consequently, it is better to analyse them separately.

11.6.1 Bio-Ethanol Cost Features

Two most important cost elements for bio-ethanol production are (OECD 2006):

a) The cost of feedstock: this is the most important cost in bio-ethanol production (accounts for around 41% of the cost of supply). The choice of feedstock explains cost variation across countries to a large extent.
b) Energy and labour costs: These are also quite important in bio-ethanol production and account for about 30% of the costs.

11.6 The Economics of Bio-fuels

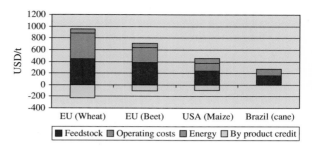

Fig. 11.19 Comparison of bio-ethanol production costs. *Source* OECD (2006)

Capital recovery can be about one-sixth of the costs while the rest is attributed to the cost of chemicals. Some credits are also obtained by selling them and this could change the economics of bio-fuels to some extent.

Brazil is the least cost supplier of bio-ethanol and produces 30% cheaper compared to the US cost and almost 2.5 times cheaper compared to the European production (see Fig. 11.19).

How does bio-ethanol compare with gasoline price? Figure 11.20 provides the comparison. Except Brazil, no other producer is yet able to produce bio-ethanol at a competitive price. The cost of ethanol from maize comes close to gasoline prices in the USA.

The cost of production however falls as the size of the conversion plant increases. In fact, it is reported that the new plants coming up in the USA are exploiting this feature to gain competitive advantage.

As the feedstock demand increases with higher fuel demand, the feedstock price will increase. Higher feedstock price would affect food prices and would encourage diversion of land and agricultural activities towards fuel feedstock supply. This could have adverse consequences for food supply, water use, and for competitiveness of bio-ethanol. In fact, this is one of the main concerns about the first generation bio-fuels.

11.6.2 Bio-Diesel Costs

The feedstock cost plays a much higher role in the case of bio-diesel—almost 80% of the operating costs (Balat and Balat 2008). An example using tallow-based

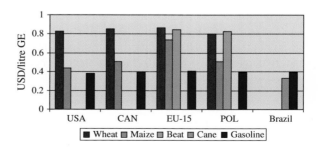

Fig. 11.20 Comparison of bio-ethanol costs with gasoline (ex tax) prices in 2006. *Source* OECD (2006)

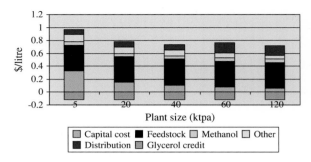

Fig. 11.21 Cost of bio-diesel production from tallow. *Source* Balat and Balat (2008)

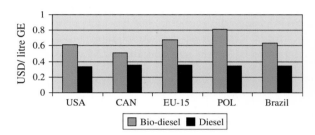

Fig. 11.22 Comparison of bio-diesel cost with diesel prices (ex-tax) in 2006. *Source* OECD (2006)

bio-diesel is provided based on Balat and Balat (2008) in Fig. 11.21. The competition from high value cooking use affects the feedstock price and the cost of production. As a result, nowhere in the world bio-diesel is yet a cost effective solution (see Fig. 11.22).

As bio-diesel or bio-gasoline is not yet competitive, support mechanisms have been developed to promote them.

11.6.3 Support Mechanisms

The generic support mechanisms are quite similar to that used for renewable electricity. The quota system (e.g. EU Directive on Bio-fuels), renewable obligation (UK Renewable Transport Fuel Obligation, RTFO), standards based system and financial incentives are commonly used.[11]

EU Bio-fuels directive: The European Union issued a directive in 2003 requiring members to ensure a minimum level of bio-fuel supply in their markets. The indicative targets set in the Directive were to supply 2% (on energy content basis) of all petrol and diesel used for transport by end of 2005, rising to 5.75% (on energy content basis) by 2010. Most of the members failed to meet the 2005 target

---

[11] For a brief review of support policies see OECD (2006, pp. 16–21). Also see Chap. 7 of IEA (2004).

and the progress towards 2010 remains limited. In 2009, the Renewable Energy Directive has set a target of 10% share of renewable energy in the transport sector.

RTFO: This is the main instrument being used by the UK to promote bio-fuels in the transport sector.[12] This obligation came in to force in 2008 and the target for 2009/10 is 3.25% renewable fuel use by volume in the transport sector. The mechanism is similar to that of the renewable obligation being used for electricity generation. Each transport fuel supplier (above a certain threshold) has a specific obligation to supply renewable fuels. They can claim certificates for renewable fuel supply and at the end of the compliance period redeem the certificates to demonstrate compliance. The supplier also has a buy-out option in case of non-compliance, set at 15 pence per litre in the first 2 years, rising to 30 p/l from the 2010/11 reporting period.

However, promotion of bio-fuels has raised concerns about food security, water scarcity and adverse effects on the poor. The competition for land for food and fuel production and the limited net energy benefits of the first generation bio-fuels have been highlighted by many, including FAO (2008) and WWI (2006). A careful analysis is therefore required before embarking on a large-scale promotion and supply of bio-fuels.

11.7 Conclusion

This chapter has provided an overview of renewable energy use and has introduced the economic concepts for analysing the developments. The levelised costs for electricity generation from renewable sources are discussed and the cost structure of bio-fuel is presented. The supporting mechanisms used by the government to promote renewable energies are also discussed to bring out the essential features and remaining challenges. Surely, renewable energies will play an important role in the energy mix in the future but many challenges remain before such energies can compete with fossil fuels.

References

Awerbuch (2003) Determining the real cost: why renewable power is more cost-competitive than previously believed, Renewable Energy World (see http://www.awerbuch.com/shimonpages/shimondocs/REW-may-03.doc)
Balat M, Balat H (2008) A critical review of bio-diesel as a vehicular fuel. Energy Convers Manag 49(10):2727–41
Bomb C, McCormick K, Deuwaarder E, Kaberger T (2007) Biofuels for transport in Europe: lessons from Germany and the UK. Energy Policy 35(4):2256–67

---

[12] See Department for Transport website http://www.dft.gov.uk/pgr/roads/environment/rtfo/.

Brown MA (2001) Market failures and barriers as a basis for clean energy policies. Energy Policy 29:1197–1207

Bunter L, Neuhoff K (2004) Comparison of feed-in tariff, quota and auction mechanisms to support wind power development. Cambridge MIT Institute Working Paper 70, University of Cambridge, UK (see http://www.electricitypolicy.org.uk/pubs/wp/ep70.pdf)

Darmstadter J (2003) The economic and policy setting of renewable energy: where do things stand? Resources for the future, (see http://www.rff.org)

De Vries BJM, Van Vuuren DP, Hoogwijk MM (2007) Renewable energy sources: their global potential for the first-half of the 21st century at a global level: an integrated approach. Energy Policy 35(4):2590–2610

Del Rio P, Gual M (2007) An integrated assessment of the feed-in tariff system in Spain. Energy Policy 35:994–1012

Dinica V (2006) Support systems for the diffusion of renewable energy technologies—an investor perspective. Energy Policy 34:461–80

EIA (2010) International energy outlook 2010, US Energy Information Administration, Department of Energy, Washington, DC

EPRI (2009) Program on technology innovation: integrated generation technology options, Technical update, 2009, Electric Power Research Institute, Palo Alto, California

Eurostat (2003) Energy taxes in the nordic countries: does the polluter pay? Eurostat, Luxembourg (see http://www.scb.se/statistik/MI/MI1202/2004A01/MI1202_2004A01_BR_MIFT0404.pdf)

Eurostat (2010) Energy—yearly statistics 2008, Publication office of the European Union, Luxembourg

FAO (2008) The state of food and agriculture 2008, bio-fuels: prospects, risks and opportunities. Food and Agricultural Organisation of the United Nations, Rome

Goldemberg J (2004) The case for renewable energies. Thematic Background paper 1, international conference on renewable energies, Bonn (see http://www.renewables2004.de/pdf/tbp/TBP01-rationale.pdf)

Gross R, Heptonstall P, Anderson D, Green T, Leach M, Skea J (2006) The costs and impacts of intermittency: as assessment of the evidence on the costs and impacts of intermittent generation on the British electricity network. UKERC, London

Heptonstall P (2007) A review of electricity unit cost estimates. Working paper UKERC/WP/TPA/2007/006, UKERC, London

IEA (2004) Biofuels for transport: an international perspective. International Energy Agency, Paris

IEA (2009) Renewables information 2009. International Energy Agency, Paris

Kennedy S (2005) Wind power planning: assessing long-term costs and benefits. Energy Policy (33):1661–1675

Menanteau P, Finon D, Lamy ML (2003) Prices-versus quantities: choosing policies for promoting the development of renewable energies. Energy Policy 31:799–812

Mitchell C, Bauknecht D, Connor PM (2006) Effectiveness through risk reduction: a comparison of the renewable obligation in England and Wales and feed-in system in Germany. Energy Policy 34:297–305

Neuhoff K (2005) Large-scale deployment of renewables for electricity generation. Oxf Rev Econ Policy 21(1):88–110

OECD (2006) Agricultural market impacts of future growth in the production of biofuels, OECD, Paris (see http://www.oecd.org/dataoecd/58/62/36074135.pdf)

Painuly J (2001) Barriers to renewable energy penetration: a framework for analysis. Renew Energy 24(1):73–89

Sawin JL (2004) National policy instruments, policy lessons for the advancement and diffusion of renewable energy technologies around the world. Thematic background paper 3, the international conference for renewable energies, Bonn (see http://www.renewables2004.de/pdf/tbp/TBP03-policies.pdf)

Verbruggen A, Fischedick M, Moomaw W, Weil T, Nnadai A, Nilsson LJ, Nyboer J, Sathaye J (2009) Renewable energy costs, potentials, barriers: conceptual issues. Energy Policy 38(2):850–61

World Bank (1997) Financial incentives for renewable energy development. Discussion paper 391 (see http://www.worldbank.org/astae/391wbdp.pdf)

World Watch Institute (2006) Biofuels for transportation, global potential and implications for sustainable agriculture and energy in the 21st century. Extended summary. World Watch Institute, Washington, DC

Further reading

IEA (2002) Renewable energies into the main stream. International Energy Agency, Paris

IEA (2007) Renewables in global energy supply: an IEA fact-sheet. International Energy Agency, Paris

Owen A (2006) Renewable energy: external cost as market barriers. Energy Policy 34:632–42

Part III
Energy Markets

Chapter 12
Energy Markets and Principles of Energy Pricing

12.1 Introduction: Basic Competitive Market Model

Any standard economics textbook starts with the theoretical world of perfect competition. In such a case, consumers maximise their utility subject to their budget constraints and producers maximise their profits subject to the constraints of production possibilities. There are numerous consumers and producers trying to transact in the market place. In a competitive market condition, all agents are price takers and there is no market power of any agent. In general, the demand for a good reduces as prices rise (i.e. inverse relationship with price) and vice versa. This gives rise to the familiar downward sloping demand curve. Similarly, producers face an upward sloping supply curve. The higher the price, the more is the supply, as at higher prices more producers become viable. The interaction of supply and demand decides the market clearing price of the good and the quantity of goods that will be sold (or purchased).

Consumers satisfy their utility (or preferences) by consuming a good. As utility is not observable, an alternative parameter for measurement of their satisfaction is the willingness to pay or accept to move from a situation to another. At any given price, consumers spend an amount equal to the price times the quantity purchased. No consumer is willing to pay for something that she does not want but some consumers may be willing to pay more than the market price. Thus the total willingness to pay at price P_0 in Fig. 12.1 is given by the area ACq_0O. But the expenditure for the good at this price is given by the area P_0Cq_0O. The difference between these two areas gives excess benefit consumers obtain, known as "consumer surplus". This is represented by the area left of the demand curve but above the price actually being charged for the good.

The sellers on the other hand incur cost for producing the goods sold and as long as the costs are recovered, they may be willing to sell for any given price. However, even at that price, some sellers will receive more benefits due to low cost production, while others will just break-even. Therefore, the benefits accrued

Fig. 12.1 Willingness to pay

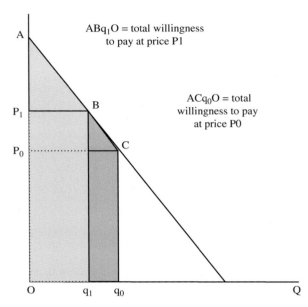

to the producers are known as "producer surplus". Total benefits to the producers then include production costs and the surplus (see Fig. 12.2).

At the equilibrium, the willingness to sell equals the willingness to pay. At this condition, the demand matches the supply. This is considered as an optimal allocation in the sense that the equilibrium cannot be replaced by another one that would produce higher welfare for some consumers without reducing welfare of others. This is depicted in Fig. 12.3.

Fig. 12.2 Willingness to sell

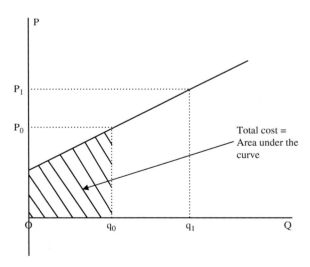

12.1 Introduction: Basic Competitive Market Model

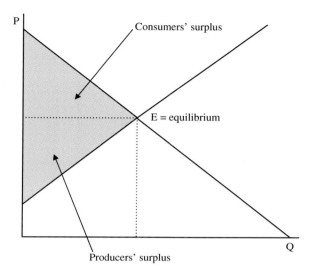

Fig. 12.3 Competitive equilibrium

Competition forces sellers to charge no more than their rivals. If one seller charges more than the market clearing price, consumers will go to others offering the same good at lower price. If someone charges less than the market price, the demand will outweigh supply, forcing a return to the market price. Individual buyers and sellers cannot affect the price. Buyers and sellers react to changes in the market price. At lower prices, some sellers will leave the market while more consumers enter it. Similarly, at higher prices more sellers are willing to offer their goods while there will be fewer consumers. The participation in the market is voluntarily and consumers or sellers are free to enter or leave the market in a perfectly competitive case. Price is equal to the marginal cost of the last supplier.

In mathematical terms, the above can be presented as follows:

The aggregate consumer surplus from consumption of a good at the prevailing price p^* is

$$CS = \int_{p^*}^{\infty} Q(p)dp \qquad (12.1)$$

The producer surplus for supplying the good having a cost function $C = C(Q)$ is

$$\pi = pQ(p) - C[Q(p)] \qquad (12.2)$$

The net economic welfare is the unweighted sum of aggregate consumer surplus and producer surplus is given by

$$W(p) = CS + \pi = \int_{p^*}^{\infty} Q(p)dp + pQ(p) - C[Q(p)] \qquad (12.3)$$

The objective is to find the price at which the welfare is maximized. This is obtained by setting the first order derivative of the welfare function with respect price to zero.

$$\frac{dW}{dp} = \frac{d}{dp}(CS) + \frac{d\pi}{dp} = 0, \text{ or}$$

$$[-Q(p^*)] + \left\{ Q(p^*) + p^* \frac{dQ(p)}{dp} - \frac{dC[Q(p)]}{dp} \right\} = 0 \quad (12.4)$$

From where we obtain, $p^* = $ MC or the price is equal to the marginal cost.

Such a market has a number of properties:

(a) the participation is voluntary—both consumers and producers enter and exit the market freely, without any compulsion.
(b) Consumers who are willing to pay the market price enter the market (which means that there could be some consumers who remain outside). Similarly, only those producers will be called to supply whose marginal cost of supply is lower or equal to the price. The marginal producer will recover only his operating costs while other producers who are called to supply would earn some additional profits (which might cover their fixed costs partly or fully depending on their cost structure). This puts pressure on the suppliers to keep their costs low to enter the market. Therefore, there is nothing wrong in a market economy to find price excluding some consumers or producers. Similarly, there is nothing wrong for some producers to earn large profits while others are barely profitable.
(c) The relevant pricing principle is essentially a short run one, with an objective of clearing the market.

However, certain basic conditions have to be satisfied to obtain such efficiency outcomes: existence of freely competitive markets, perfect and costless flow of information and knowledge, smooth transferability of resources and absence of externalities. Clearly, most of these requirements are not satisfied by the today's energy market. In addition, the energy sector is marked by certain specific characteristics such as indivisibility of capital, tradability of some products and depletion of some resources. Consequently, the basic model needs to be expanded for any meaningful analysis. We consider these aspects below.

12.2 Extension of the Basic Model

Let us consider a number of characteristics of the energy sector and see how the basic model outcome needs to be modified.

12.2.1 Indivisibility of Capital

Indivisibility of capital implies that capacity expansion takes place in discrete unit sizes of plant units, and investments are lumpy in nature. In the energy sector, this is a common feature. For example, oil fields or coal mines are developed for a particular capacity. Refineries or power plants come in particular sizes and once one unit is installed, increments are possible only in standard sizes, and not in smooth, continuous increments as is assumed in the theory. The existence of economies of scale often suggests that better cost advantages could be achieved by installing bigger sizes. The indivisibility of capital changes the shape of the supply curve, for instead of a continuous supply curve, we now have a supply curve for a fixed capacity and the addition of new capacities brings abrupt changes (or kinks) at the point where investments take place. This is shown in Figs. 12.4 and 12.5.[1]

In a fixed plant with a capacity of q_0, the output cannot go beyond the installed capacity. The marginal cost of supply is assumed to be constant at v for the entire capacity and when the capacity constraint is reached, the vertical line shows the supply schedule. Thus, at the capacity q_0, there is a rupture in the supply curve. Initially, when the demand is given by schedule D, the market clearing price is the marginal cost (v), as at this point, there is excess capacity compared to demand. In such a situation, the investor would recover his operating costs only. But as the demand shifts to D' (due to changes in income and other factors), the demand exceeds supply if the price is maintained at the short run marginal cost (i.e. v). A market clearing price would imply that the pricing mechanism would have to be used to ration demand to bring it down to the available supply level, thereby charging a price p' [which lies between v and (a + v)], thereby recovering a part of the fixed cost (but not fully yet). When the demand grows sufficiently that the price would equal (a + v), then the producer would recover his full cost of supply. But at this stage, entry would not be encouraged because of inadequate cost recovery in the past. As the demand increases further and moves to D'', the price would exceed the long run marginal cost of supply and would provide high excessive profits to producers. Sustained shortage of capacity, high prices and existence of excess capacity would encourage new entry to the market.

With new capacity, the installed capacity increases to Q_1, and brings excess capacity to the system. The intersection of demand schedule D_3 with the supply curve brings the prices down to the short run marginal cost. Thus in the process the price passes through a cycle of volatility, bringing boom and bust of the industry. This sort of inherent price instability of the energy industry is a source of major concern if the competitive market principle is applied strictly. Such instability could affect long-term investment decisions of the consumers and would increase economic uncertainties. Moreover, investors would not prefer such an environment for investment decisions. Some arrangements would be required to manage such fluctuations.

---

[1] This presentation follows Rees (1984). Also see Munasinghe (1985).

Fig. 12.4 Fixed plant case

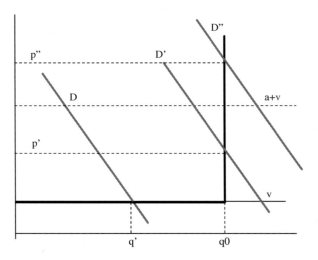

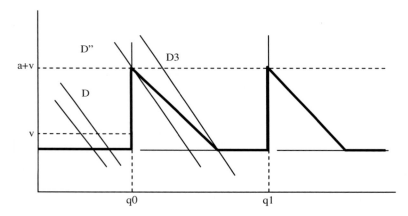

Fig. 12.5 Boom bust cycle

It is important to note here that in the literature long-run marginal cost principles are suggested a solution in such cases. As indicated above, the relevant pricing horizon is essentially a short-term one and the long-run marginal cost principle encounters practical problems in determining the cost and price. Often this requires a departure from the marginal cost principle in favour of average cost basis of some sort.[2]

---

[2] This is an area of continuous debate in the economic literature. A summary of the debate is provided in Chap. 13.

12.2.2 Depletion of Exhaustible Resources[3]

As coal, oil and gas are non-renewable resources, consumption of one unit of these resources implies foregoing its consumption at any future date. This brings in another dimension of decision-making: whether to use the resource now or later. The use decision is affected by choice of using it now or later. As discussed in Chap. 9, the price should depart from the marginal cost and include an additional item called the scarcity rent or user cost. This implies that finite resources have a value over and above their cost of production, which is due to their scarcity. Our time preference would require us to consume a bit more in period 1 than in period 2 but for this the price in period 1 has to be somewhat lower than that in period 2.

If the reserve is very large and if the prospect of export is negligible, the rent component will be practically insignificant, though theoretically it will still exist. Of the reserve is very limited, the estimation of the rent does not pose any problem either. The difference between the extraction cost of the resource and the price of the substitute fuel gives the rent cost. In all other intermediate cases, the rent can be significant and its evaluation is more uncertain and complex.

12.2.3 Asset Specificity and Capital Intensiveness

The energy sector employs highly specific assets in the sense of transaction cost economics. Assets are considered as highly specific if they have little alternative use. For example, a power generating plant has little alternative use. Similarly, investments made in an oil field could hardly be redeployed elsewhere in any other use. The asset specificity can arise because of a number of reasons—site specificity, specific investments in human capital, dedicated investment (or idiosyncratic investment) and physical (Williamson 1985). The level and nature of transaction costs depend on the frequency of transaction, the extent of uncertainty and the degree of asset specificity.

The theory of transaction costs also identifies a number of alternative arrangements for performing transactions (Williamson 1985):

- Classical contracting which includes the textbook exchanges in the market place.
- Bilateral contracting using long term contracts;
- Trilateral relationship where a third party determines the damages/adaptation following some specified procedures (such as arbitration);
- Unified governance or vertical integration that internalises the transaction with the firm.

---

[3] Please refer to Chap. 9 for further details.

Depending on the transaction attributes it is possible to identify the governance arrangements that would be most appropriate (see Table 12.1). In the energy industry given the frequency of transactions and high asset specificity, the tendency for vertically integrated arrangements prevailed. This was the case in all energy industries—oil, gas, coal or electricity but there are some differences according to the industry. In the gas industry, the trilateral contracts are more common while in the electricity industry, unified governance prevailed.

In addition to specificity, energy sector assets tend to be capital intensive as well. Often the capital cost accounts for a large part of the average cost and consequently, per unit cost falls with higher sizes, showing economies of scale.

An implication of such capital intensiveness and economies of scale is that the marginal costs tend to be low compared to the average costs and any pricing based on marginal cost would then lead to financial losses (see Fig. 12.6). But once in operation, as long as the firm is able to recover its variable costs, it would continue operating expecting to make up for the capital cost recovery at a future date. Thus, the firm would have a tendency to produce at its maximum capacity, considering fixed costs as sunk costs. This would lead to excess supply and the energy industry has an inherent tendency to be in excess supply situation. But continued oversupply situation is not beneficial for the future of any industry as no new investment would be encouraged and continued financial loss could promote premature abandonment of certain facilities.

It needs to be highlighted that a certain amount of excess capacity has to be maintained in any energy industry to cater to the unforeseen circumstances (natural calamity, disruptions, etc.), normal demand/supply fluctuations, and to ensure reliability of supply. Moreover, as storage is a problem for electricity, instantaneous supply and demand balancing is required, making the process technically demanding as well.

Table 12.1 Governance structure for transaction characteristics

Frequency of transactions	Specificity of assets		
	Non-specific	Medium	High
Rare	Classical	Trilateral	Trilateral or unified
Frequent	Classical	Bilateral	Unified

Source Williamson (1985)

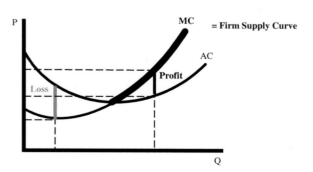

Fig. 12.6 Relevant cost curve

12.2 Extension of the Basic Model

The energy industry used two approaches to manage the problems related to indivisibility of capital and excess capacity: Oil industry used horizontal integration while the electricity and network industries used regulation. In regulation, the tariff relates to the cost of providing the service by maintaining and operating a certain mix of assets, including those required for ensuring reliability. However for a non-regulated industry (like oil), horizontal integration can work. Horizontal integration implies linking with firms at the same stage of the value chain either through merger and acquisition or through the formation of a cartel. The oil industry has seen significant merger and consolidation in the post oil-shock era, where large international companies merged together to better manage their assets. On the other hand, collusive behaviour has also been used in the oil industry to manage the problems. The major oil companies formed an effective cartel in 1928 through the As-Is agreement and froze the respective market shares until this policy became public and abandoned in the 1950s, as collusive behaviour is not legally tenable in most jurisdictions. However, the Majors found another way of influencing the market—joint ventures in the Persian Gulf, which provided them with a legal solution of perfect information exchange and thus control the market. Later when the OPEC was created, the market was controlled through production quotas and price targets in a collusive manner. But as sovereign countries are involved in these decisions, such behaviour is not illegal.

12.3 Market Failures

The competitive market model discussed above assumes a set of strong assumptions. A market failure occurs when such assumptions cannot be satisfied. Some elements of the energy sector have the technical or other characteristics that amount to the violation of the basic assumptions of a competitive market model. The common sources of market failure are discussed below.

12.3.1 Monopoly Problems

The capital intensiveness of the energy sector requires large investments and as bigger installations provide economies of scale, few large suppliers tend to dominate the market. A profit-maximising monopolist will set her price at the intersection of marginal cost and marginal revenue. But as the monopolist faces a down-ward sloping demand curve, the marginal revenue will be less than price.[4]

---

[4] The total revenue is given by $TR = P \cdot Q$, where P = price and Q = output. Marginal revenue is then $\frac{dTR}{dQ} = P + Q\frac{dP}{dQ}$, or $MR = P(1 + 1/e)$, where MR = marginal revenue and e = price elasticity of demand. As e is less than 1, MR is less than P.

Fig. 12.7 Price determination in a monopoly market

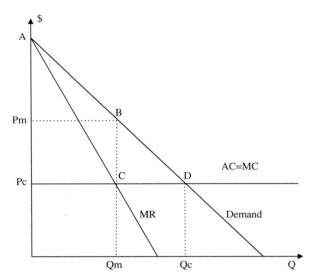

As shown in Fig. 12.7, the profit maximising output is given by Q_m, while the price charged by the monopolist is P_m.

Mathematically, the problem to maximize the profit

$$\text{Max} \, p \cdot D(p) - C(D(p)) \tag{12.5}$$

The first order condition is

$$p^* \cdot D'(p^*) + D(p^*) - C'(D(p^*)) \cdot D'(p^*) = 0 \tag{12.6}$$

$$(p^* - C') = -D(p^*)/D'(p^*) \tag{12.7}$$

But

$$e = -D'(p^*) \cdot p/D(p^*) \tag{12.8}$$

Hence,

$$\frac{(p^* - C')}{p^*} = \frac{1}{e} \tag{12.9}$$

That is, to maximize its profit, the monopolist will charge consumers inversely to their elasticity of demand. Inelastic the demand, higher the price will be.

If the monopoly results are compared with the competitive outcome, it is found that the monopolist restricts the output to Q_m compared to Q_c obtained in the competitive market. Similarly, the price paid by the consumers is P_m compared to P_c in a competitive condition. Thus the consumers pay Pm-PC as monopoly rent. The consumer surplus is reduced to APmB compared to APcD whereas the producer surplus increases to PmPcCB which was non-existent in a competitive set

Fig. 12.8 Economies of scale

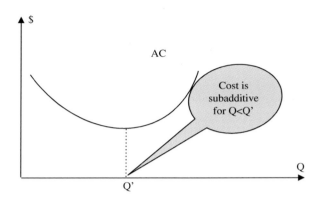

up. Monopolisation of the market leads to a deadweight loss equivalent to the triangle BCD.

In addition, following Leibenstein's observation (Leibenstein 1966), it could be indicated that a monopolist might operate in the inefficient zone of the production possibility frontier. This in other words means that a monopolist may choose the factors of production in an inefficient manner, thereby operating at a point above its theoretical cost curve. This is known as X-inefficiency. While X-inefficiency is inconsistent with profit-maximising behaviour, the inefficiency is possible given that managers may pursue their own objectives in place of owner's objective of profit maximisation.

A third source of monopoly-related inefficiency is the possibility of rent-seeking. A monopolist by charging more than the competitive market price earns a monopoly rent, which is equal to the producer surplus of PmPcCB in Fig. 12.8. The existence of such rent will set in competition among firms to seek the rent by lobbying and influencing the legislators or regulators, thus wasting resources and causing welfare loss to the society.

A fourth source of complication is the possibility of product differentiation, which allows a firm to increase its price without loosing all its sales to a competitor. In such a case, the price exceeds marginal cost, which signals misallocation of resources.

12.3.2 Natural Monopoly

This is a situation where production of a good or service by a single firm ensures least cost supply. The typical example is a single product where the long-run average cost declines for all outputs (see Fig. 12.9). As the average cost falls over the entire range of output, the marginal cost also falls. This is a case of permanent natural monopoly, because irrespective of market demand size, a single firm can produce the good at least cost.

Fig. 12.9 Permanent natural monopoly

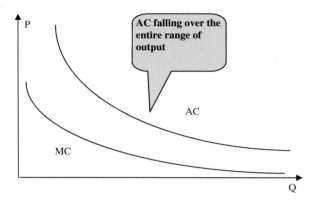

Fig. 12.10 Temporary natural monopoly

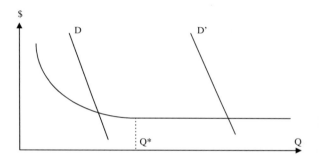

A temporary natural monopoly on the other hand exists when the average cost falls over a limited range of output as shown in Fig. 12.10. In the figure, the cost falls up to Q^* and then remains constant thereafter. Beyond Q^* level of output, a workable competitive market can develop for demand D'.

Although the economies of scale act as the driving factor for the existence of a natural monopoly, this is not a sufficient condition. Instead, the concept of sub-additivity of cost functions is used. A cost function is subadditive when it satisfies the following condition:

$$C(Q) = c(q_1 + q_2) < c(q_1) + c(q_2) \tag{12.10}$$

This implies that instead of two firms producing q_1 and q_2 quantities of a good, it is cheaper for a single firm to produce the entire quantity $(q_1 + q_2)$. For example, the cost curve shown in Fig. 12.8 suggests that the cost declines up to Q' and then starts increasing (i.e. it shows economies of scale up to Q'). The cost function is subadditive for any output up to Q'.

Next we consider what happens for outputs exceeding Q'. Figure 12.11 presents the minimum average cost curves for two firms. As for least cost production both the firms must produce at the same output rate, the second curve is obtained just by doubling the output rate for a given point on the AC curve. Thus, the minimum of

12.3 Market Failures

Fig. 12.11 Subadditivity beyond scale economies

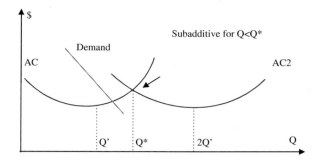

AC2 is obtained at output equal to 2 Q'. The intersection of AC and AC2 defines the range of sub-additivity. Thus it is clear that for any output less than Q^*, the least-cost option is for one firm to produce the good, even though diseconomies of scale set in at that level. Thus economies of scale are not necessary for a single product natural monopoly.

For the multi-product case, the natural monopoly obtains when the cost function is sub-additive. Economies of scale are neither necessary nor sufficient conditions for a natural monopoly for multi-product case. The concept of economies of scope becomes important as well.[5] If both economies of scale and scope exist, it is likely to lead to natural monopoly.

A related concept is the sustainability of natural monopolies. In the single product case, assume that the demand intersects the AC to the left of Q'. In this case, there is no incentive for any entrant to enter the market. However, if the demand curve intersects the AC somewhere between Q' and Q^*, the natural monopoly would be termed as unsustainable. This is because a potential entrant could enter the market and produce a part of the output. The issue of sustainability is important for entry-related decisions. In the case of sustainable monopoly, the threat of entry is not there. On the other hand, in the case of unsustainable monopoly, entry can be allowed.

The public policy dilemma in the case of natural monopoly is as follows: the natural monopoly characteristics would require a single firm to make the supply but the society would not like to suffer from the potential monopoly pricing. What alternative solutions are available to deal with such a situation? This is what we turn to now.

Marginal cost pricing: According to the economic theory, prices in a competitive market equal the marginal cost of production. Applying this principle to a natural monopolist will meet the efficiency requirement. The output will be Q_0 at price P_0. As the price is less than the average cost of production, the firm incurs a loss and is shown by the rectangular area RP_0ST in Fig. 12.12. As no private

---

[5] Economies of scope imply the potential of cost saving from joint production. This is possible because the firm can make better use of facilities and services for producing a certain mix of different outputs than leaving the production of individual products to specialty firms.

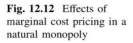

Fig. 12.12 Effects of marginal cost pricing in a natural monopoly

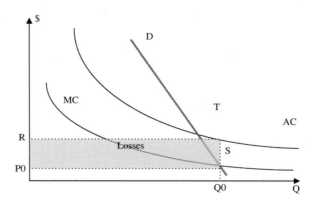

enterprise will be interested in providing a good by incurring a loss, in order to make good of this loss, a subsidy will be required. Wherefrom such subsidies will come?

The theoretical solution is to impose a lump-sum tax that does not distort other decisions throughout the economy. Such a tax is rarely used in practice. Even if such a tax is found, there are some objections to use of such instruments:

- a general tax requires non-buyers of a product to subsidise buyers. This is not fair from a distribution point of view.
- Subsidies reduce the incentive and capacity to control costs. The management and employees know that the loss will be subsidised, which can lead to inefficient practices.
- If the costs are not met, it may so happen that the benefits received by the society from the production of a good are less than the costs. In such a case, there is no justification for the production of the good. Subsidies can obscure this basic problem.
- Subsidising private firms is considered politically unacceptable in many countries.

The above suggests that the lump-sum tax may not be an appropriate solution. The pricing has to be such that the costs are at least covered. In the single-product case, balancing costs and revenues leads to pricing based on average costs for natural monopolies. This is a departure from the marginal cost principles and hence will introduce welfare losses as shown by the shaded area in Fig. 12.13.

Alternative pricing principles in such a case have been suggested. The most common methods are:

12.3.2.1 Two-Part Tariffs

A two-part tariff is a non-linear tariff system that uses a fixed fee or charge (F) and a price per unit (p) component. The price per unit can be set equal to the marginal

12.3 Market Failures

Fig. 12.13 Average cost pricing for natural monopolies

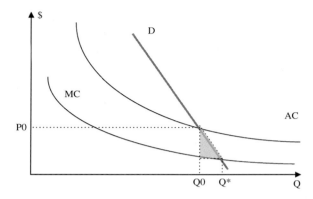

Fig. 12.14 Two-part tariff example

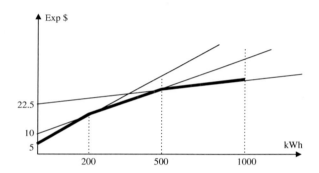

cost (MC). This, as we have discussed earlier, in the case of a natural monopoly will result in a loss. The fixed fee can be designed to cover the loss in revenue.

A simple fee could be a uniform fee for all customers. This will then be equal to total loss to be covered divided by the number of consumers using the service or the good. This is a straight forward design and is non-discriminatory. But it might cause some consumers to leave the system (especially those who consume less and have to pay a relatively high fixed fee). An alternative system is to custom design the fixed fee to suit different consumers or class of consumers. This avoids the problem of exclusion from the service but is discriminatory in nature and may be illegal in some cases.

An example of two-part tariff for electricity may be as follows:

Fixed fee per month—$5, 10 cents per kWh for consumption up to 200 kWh per month and 7.5 cents per kWh for consumption between 200 and 500 kWh per month and 5 cents for consumption beyond 500 kWh per month. This is shown in Fig. 12.14.

The curve in solid line indicates the applicable rate. It indicates that the effective fixed rate changes and variable rates change as the consumption increases.

12.3.2.2 Ramsey Pricing

Ramsey pricing is essentially a taxing method that was developed in an attempt to design a system that would lead to minimal deadweight loss. This has then been applied to the pricing issues as well as the basic problem in a natural monopoly subjected to marginal cost pricing is the recovery of losses in a less distorted manner. This principle has been analysed by a number of researcher under various conditions but Baumol and Bradford (1970) provide the most detailed and general analysis of the issue.

The Ramsey rule can be presented as follows:

$$\frac{(p_i - MC_i)}{p_i} = \frac{k}{e_i} \quad (12.11)$$

where, p is the price, MC is the marginal cost, e is the price elasticity of demand, and i indicates the product i and k is a constant.

This formula suggests that the quasi-optimal price should be more than the marginal cost and such a price that minimises the deadweight loss should be inversely proportional to the price elasticity of demand. The price shall be higher for inelastic demand and lower for elastic demand. This provides the theoretical justification for the so-called "value of services" pricing that is used by some utilities.

An alternative formulation of the Ramsey pricing is shown in Eq. 12.12

$$\Delta x_i = k x_i \quad (12.12)$$

where x is the demand of a good, Δx indicates the change in demand, and k is a constant.

This suggests that the "quasi-optimal pricing requires a proportionate change in all purchases from the levels that would be observed if prices were set at marginal costs" (Baumol and Bradford 1970). This in other words implies that the demand for all goods should be restricted by same proportion so that the total cost equals the total revenue. The implication of this is same as above, because although the output is reduced by the same amount, the prices change differently due to different elasticities of demand.

Although this deals with the revenue loss problem, the Ramsey pricing rule has equity implications. By suggesting higher prices on inelastic demand, it suggests that essential demands would be charged at higher rates. This implies that the poor may be in a disadvantageous condition if this rule is followed. Explicit accounting of equity issues has been discussed in the literature but this makes the formula more complicated and will not be covered here.[6]

---

[6] Interested readers may consult the following: Diamond and Mirrlees (1971), Feldstein (1972).

12.3.2.3 Public Ownership

An alternative solution to deal with the problems of natural monopoly is to transfer the ownership and operation rights to the government. This forms the basis for public sector involvement in production. The logic behind this is that the government will not be following the profit maximisation principle but will operate to maximise economic surplus.

The pricing rule for a public enterprise under a budget constraint was studied by Boiteux (1956). His results indicated that the pricing policy should be same as Ramsey pricing. This implies that the state monopolist should behave like a discriminating monopolist in order to reach budgetary equilibrium.

Both the options have been widely adopted in practice. In the USA, the regulatory approach was adopted while in Europe and many other parts of the world, the public ownership approach was followed.

12.3.3 Existence of Rent

As discussed in Chap. 8, the energy industry exhibits a number of differential rents arising from the differential advantages enjoyed by a production unit compared to other similar units. These rents appear as the producers' surplus and increase the profitability of the producers. In addition, the energy sector at certain times has seen monopoly rents due to the prevalent market structure. Similarly, the scarcity rent can also be applicable to non-fossil resources of energy.

In theory, the government can capture this rent without affecting supply since the company continues to receive its normal profit. This is also assumed for an efficient operation of the markets. However, in practice any fiscal measure implies an intervention of the government in the market and introduces distortions. This also implies a departure from the marginal cost-based pricing.

12.3.4 Externality and Public Goods

Energy products impose different costs on society, a part of which are supported by producers and consumers, while the rest, known as external costs, remain unaccounted for and are borne by the society. In economic terms, an externality is said to exist if any activity of an economic agent imposes positive or negative effects on the welfare of any other agent or groups of agents and when economic agents neither receive nor pay any compensation equal to the costs inflicted or the benefits conferred upon them. While this aspect will be analysed in a another chapter in detail (Chap. 25), it is important to note here that the presence of an externality introduces distortion in economic decisions and its correction requires government intervention either through taxation or through regulation.

Similarly, the provision of public goods related to the supply of energy needs to be highlighted. Public goods are those whose provision to one person or party makes them automatically available to all at zero additional costs. In the energy sector, a number of such examples can be easily cited: recreational or other benefits arising from the construction of a dam, downstream benefits as a result of upstream reforestation, etc. At the same time, ensuring adequate and secure energy supplies, adequate long-term R&D and other economic and socially desirable outcomes also share public good features. Markets may not provide these public goods left to it and as these are important issues related to energy, governments intervene.

12.4 Government Intervention and Role of Government in the Sector

The above discussion indicates that the energy sector fails to satisfy the requirements of a competitive market in a number of ways. The presence of natural monopoly and existence of rents require corrective intervention to remedy the problems. Externality, which will be considered later, is also quite pervasive in the sector, and requires further intervention. In addition, energy being of critical importance in the modern world, social, equity related and security-related issues cannot be ignored either. This so-called market failure argument is used to justify government intervention in the energy sector. Consequently, the government presence in the sector is quite widespread, both in developed and developing countries, despite waves of liberalization of the market.[7]

Governments use a wide range of instruments or measures to control the functioning of the energy sector. IEA (1996) categorises them in five following categories:

- economic and fiscal instruments;
- trade instruments;
- administration, management and ownership;
- regulation; and
- research and development (R&D).

Table 12.2 provides some examples of each category of instruments. Taxes, royalties and subsidies constitute the common form of economic instruments used in the energy sector. Although fiscal instruments can be used for various purposes including internalisation of externalities, revenue generation remains the most important motive for their widespread use. A number of trade-related instruments are used in controlling movement of energy resources and include tariffs and

---

[7] However, the market failure argument has been subjected to serious scrutiny. See Robinson (2004) for such a viewpoint.

12.4 Government Intervention and Role of Government in the Sector

Table 12.2 Main energy policy instruments

Economic/fiscal	Trade	Administration, management and ownership	Regulation		R&D
Taxes, royalties, fees	Import/export tariffs	Equity participation in or ownership of energy companies	Price and volume controls		R&D in the public sector
Tax exemptions	Import/export licences	Provision of government services	Market regulation (entry/exit, monopoly rights, anti-cartel legislation)		Funding for private sector R&D
Grants, subsidies, transfer payments	Quotas		Environmental regulations		International collaboration
Credit instruments (interest subsidies, loan guarantees, soft credits)	Selective bans/ embargoes		Technical regulations		
	Differential treatment of domestic and foreign suppliers				

Source IEA (1996)

quotas, licensing, fuel quality restrictions and political restrictions (embargoes or bans) on economic involvement in certain areas or countries or on trade.

State participation in the management, ownership and control of production and supply of energy remains quite pervasive, especially in grid-based industries. This trend was evident for much of the twentieth century in most countries and surely since the Second World War. These vertically integrated state monopolies produced reasonably satisfactory results initially, with significant growth of the sector and efficient operation in certain countries. But as sector ownership and regulation was exercised by the government and as politically motivated decision-making pervaded the sector, performance started to deteriorate, especially in developing countries. The state owned utilities suffered from poor labour productivity, deteriorating fixed facilities and equipment, poor service quality, chronic revenue shortages, inadequate investment, and serious problems of theft and non-payment (World Bank 2004).

Governments use a wide range of regulatory interventions to control the sector performance. These include price controls, competition and market access rules, private service obligations, monopoly and restrictive trade practice controls, and technical and environmental performance management. While the degree of control varies by industry, normally the networked industries are subjected to higher levels of control. The downstream side of the oil industry, at least in developed countries, is perhaps the least regulated, where a large number of wholesale and retailers compete (IEA 1996). However, the same cannot be said about the developing countries where state monopolies often supply the market.

As the oil shocks of the 1970s caught many countries unprepared and as countries struggled for effective policies and institutions to deal with energy sector problems, the government involvement in the sector rose, resulting in highly interventionist policies (such as detailed targets for the sector, price controls, support for mega projects, barriers to free trade, etc.). Many countries developed formal energy planning agencies to deal with the concerns for energy security, and protection of the economy from future shocks. But the stable energy and oil market situation since the mid-1980s and a change in the economic philosophy towards governance in certain developed countries promoted a wave towards diminishing state intervention in the energy sector. The policy of reform and restructuring of the energy sector attempted to reduce government intervention by promoting competition wherever possible and limiting regulation to core natural monopoly activities (World Bank 2004).

The World Bank and the IMF were instrumental to promote these liberalisation policies in developing countries. The "Washington Consensus" policies, as the 1989 policies for Latin America came to be known, were promoted around the world and many countries under pressure from the bilateral and multi-lateral agencies had to undertake structural adjustments to turn around their economies. The energy sector was one of the targeted sector in many countries as the sector contributed significantly to the economic distress of many countries.

However, after around two decades of persistent use of the liberalisation policies, the progress has been quite limited. Now the World Bank acknowledges that the prescription has been oversold, misunderstood and less effective (World Bank 2004). The opposition to these policies has mounted and the rate of acceptance is low. With high oil prices in the recent years, the concerns of economic downturn and security of supply are reappearing. There are calls for more intervention in the market once again. Thus a partial turn around of interventionist policies, if not total, is visible, as if the pendulum has swung back to the other side.

12.5 Conclusion

This chapter has introduced the basic economic concepts related to energy markets. Starting with the basic competitive market framework, the chapter has highlighted various specific characteristics relevant for analyzing the energy sector and indicated the implications on pricing of energy. The chapter has thus highlighted the potential for market failures in the sector, which in turn provides the basis for the widespread government involvement in the sector. The chapter ends

12.5 Conclusion

with a brief review of the cyclical nature of market-oriented and intervention-oriented developments in the global energy scene. However, the debate over the extent of state intervention in the market continues but a understanding of the critical factors will allow an informed decision-making on the subject.

References

Baumol WJ, Bradford DF (1970) Optimal departures from marginal cost pricing. Am Econ Rev 60(3):265–283
Boiteux M (1956) Sur la gestion des monopoles publics astreints a l'equilibre budgetaire. Econometrica 24:22–40
Diamond PA, Mirrlees JA (1971) Optimal taxation and public production, II: Tax rules. Am Econ Rev 61:261–278
Feldstein MS (1972) Distributional equity and the optimal structure of public prices. Am Econ Rev 62:32–36
IEA (1996) The role of IEA Governments in energy, International Energy Agency, Paris. http://www.iea.org/textbase/nppdf/free/1990/role_energy1996.pdf
Leibenstein H (1966) Allocative efficiency vs. x-efficiency. Am Econ Rev 56:392–415
Munasinghe M (1985) Energy Pricing and demand management. Westview Press, Boulder
Rees R (1984) Public enterprise economics. Weidenfeld and Nicolson, London
Williamson O (1985) The economic institutions of capitalism: firms, markets, relational contracting. The Free Press, New York
World Bank (2004) Reforming infrastructure: privatisation, regulation and competition. World Bank (see http://econ.worldbank.org/files/36237_complete.pdf). Accessed 5 Jan 2005

Further Reading

Dahl C (2004) International energy markets, understanding pricing, policies and profits. PennWell, Tulsa
Datta-Chaudhuri M (1990) Market failure and government failure. J Econ Perspect 4(3):25–39
Devarajan S, Fisher AC (1981) Hotelling's economics of exhaustible resources: fifty years later. J Econ Lit 19:65–73
Fisher AC (1981) Chapter 2. In: Fisher AC (ed) Resource and environmental economics. Cambridge University Press, London
Krautkraemer JA (1998) Non-renewable resource scarcity. J Econ Lit 36:2065–2107
Robinson C (2004) Markets, imperfections and the dangers of over-regulating energy markets. Econ Aff June:52–55

Chapter 13
Energy Pricing and Taxation

13.1 Introduction

Energy pricing represents a major instrument of the overall energy policy of any country and is used to satisfy different objectives many of which are even contradictory. Moreover, domestic energy prices are partly determined by the functioning and influences of international energy markets on the one hand and by the sociopolitical environment pf the country on the other. Additionally, since energy is an intermediate good as well as a final product, prices should distinguish between the producers and consumers. Additional criteria such as exhaustibility, capital intensiveness, and non-storability must also be taken care of where applicable. Thus pricing energy products is a complex and difficult task (Bhattacharyya 1996).

Energy supply involves a number of activities—production or procurement of primary energy from local or external sources, transformation of primary energies to usable forms, transportation of energy in bulk and distribution of energy to final consumers through retail activities. Moreover, the retail price also includes charges, duties, taxes or subsidies as imposed by the state or its agencies. Accordingly, the retail price is the end result of the combination of various cost elements involved in the entire energy value chain. A typical example is shown in Fig. 13.1.

In order to understand and account for inherent complexities of energy pricing, a two-step approach was suggested by Munasinghe (1985). In the first step, prices are considered strictly on the basis of economic principles. In the second step, economic prices are adjusted to meet other objectives, thus enabling one to know exactly the departure from the economic prices.

13.1.1 Basic Pricing Model

We start with two basic concepts of energy pricing, namely the average cost pricing and the marginal cost pricing. As these are widely used principles, the essential points are considered below.

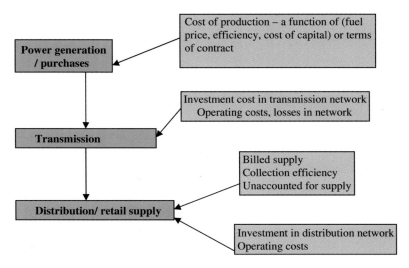

Fig. 13.1 Electricity supply value chain and cost determinants

13.1.1.1 Average Cost Pricing

The principle of average cost pricing uses the cost of production of a firm that can represent the average of the lot and such a firm neither incurs huge losses nor earns high profits. The average cost considers both the capital and operating component of costs and divides them by the output. This simplicity of the method actually lies behind its wider appeal.

From a theoretical perspective, in a competitive market under standard assumptions (of constant economies of scale, constant technologies and perfect divisibility of capital), the average cost is equal to the marginal cost at the optimal level in the long run.[1] This implies that the long-run average cost curve defines the expansion path of the firm and the average cost pricing works fine when a large number of firms are competing in the market and are producing a homogeneous good.

However, in economic terms, the average cost has certain disadvantages as well:

- It does not provide any incentive for performance improvement and allows weaker firms to co-exist with the better performing firms.
- It relies on the historic costs and does not take the cost of new capacity addition into consideration. The historic costs and the replacement costs could be very different.
- It does not provide adequate signals to the investors.

---

[1] In mathematical terms, $\frac{dAC}{dQ} = \frac{d\left(\frac{TC}{Q}\right)}{dQ} = \frac{Q \cdot \frac{dTC}{dQ} - TC}{Q^2} = \frac{Q \cdot MC - TC}{Q^2} = 0;$

13.1.1.2 Marginal Cost Pricing

The marginal cost-based approach follows from the competitive market model where prices are decided by the marginal costs of the last supplier. This was introduced in Chap. 12. This is obtained under the conditions of pure and perfect competition and such a price eliminates wasteful consumption and production of a commodity. This also assures the Pareto optimality and hence is desirable in the sense of neoclassical economics.

However, it was also indicated there that due to the specific features of the energy market, the marginal cost-based pricing may not be appropriate. The issue related to indivisibility of capital showed that following the marginal cost-based rule would lead to price volatility. But the price volatility arising from its strict application needs to be taken care of. There is a well developed literature on the consequent policy suggestion: long-run marginal cost based pricing. This tends to charge consumers for future investment in capacity addition each time consumers pay the price of a commodity. Reference was also made to monopoly and natural monopoly market structures in the energy industry. Such adjustments tend to align the pricing to average prices and hence it departs from the marginal cost concept.

In addition, there are a number of cases where the pricing may have to be adjusted to take care of other special features of energy commodities. We consider some of them below.

13.2 Tradability of Energy Products and Opportunity Cost[2]

As energy products can be traded internationally or regionally, four specific cases can arise: a country self-sufficient in energy, a country that resorts to importing so as to supplement its indigenous supply, an exporter and finally, an importer without any indigenous resources. The tradable nature of energy goods affects the supply and demand curves facing each of the above categories of countries and influences the pricing outcomes. This is explained below.

A small producer in the world market would face a demand curve containing a horizontal section corresponding to the export price. This indicates that the producers would have the opportunity to sell their goods at the international market at the export parity price (p_x). Similarly, the supply curve of a small consumer country will contain a horizontal segment corresponding to the import parity price (p_m). This implies that the country has the possibility of importing the good from the international market at the relevant price.

For a self-sufficient country the demand and costs are such that the country could meet its supply from domestic sources without resorting to imports or

---

[2] This section relies on Rangaswamy (1989). See also Bhattacharyya (1996).

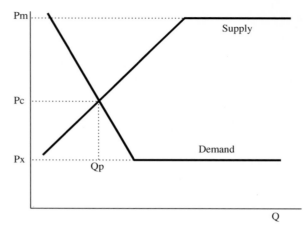

Fig. 13.2 Pricing in an energy self-sufficient country

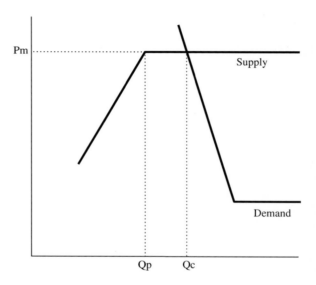

Fig. 13.3 Pricing of energy in an importing country

exports. In such a case, the relevant price will be in between p_m and p_x (see Fig. 13.2).

For an importing country, the relevant price is import parity price even for domestic production, because that price level is reached before meeting the demand (see Fig. 13.3). In this case, the domestic production would be Q_p and $(Q_c - Q_p)$ will be imported.

For a net exporter, the proper domestic price is the export value, p_x because the cost of supply is such that the cost curve intersects the horizontal part of the demand curve. In such a case, the country produces Q_p but consumes Q_c domestically, thereby leaving $(Q_p - Q_c)$ for exports (Fig. 13.4). Thus, for a tradable good, the pricing rule needs to be changed from the marginal cost principle and consumers in

Fig. 13.4 Efficient pricing for an exporting country

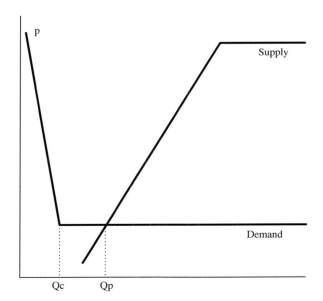

these countries may face different prices for energy depending on their demand and costs of domestic production, imports and exports. The inherent logic is that the price should be based on the opportunity cost of the commodity in question.

Arguably, the difference between the import and export price is the cost of transport and handling the good. For products with a true international market, this difference is normally small (e.g. oil). However, this difference could be quite significant for other products (like coal or gas), thereby reducing the tradability as well.

The economic prescription is that for tradable goods, the opportunity cost of the good is obtained by considering the international market prices for such goods. In practice, two methods are commonly referred to in the literature, often with reference to petroleum product pricing: import parity pricing and cost-plus pricing.

In import parity pricing, the landed cost of products is calculated by adding different charges such as CIF prices, duties and surcharges, wharfage, and landing and handling charges. The marketing and distribution margin is added to the above to obtain the selling price. However, such a pricing policy has three major disadvantages:

- First, since the freight charges are more for products than for crude oil, import parity pricing inflates the profits of the refiners when some crude oil is locally refined;
- Second, the existence of local refineries also poses problem, as import parity pricing does not pay attention to the actual cost of local production. By using the international benchmark, the local specificity is ignored but this could affect the security of supply.

- Finally, the domestic prices are subject to the same volatility as international prices, which may be difficult for consumers to accept, especially in developing country contexts.

In the case of cost-plus pricing, prices are set by adding different cost components such as cost of inputs (say crude oil for petroleum products), allowances for other operating costs, reasonable profit margin and transport and marketing costs. This is an administered pricing regime where prices are set through administrative mechanisms and is still widely used in many developing countries. However, it is difficult to arrive at a price structure that ensures a reasonable margin to suppliers as well as sends proper price signals to consumers so that the pattern of production becomes consistent with the demand pattern. Moreover, a cost-plus formula has an inherent defect in reducing the incentive to economize on costs and to inflate prices.

Moreover, in the case of some joint products such as petroleum products, only the total value of products can be related to total cost. The relative price structure would depend on the demand pattern, so that the net back value can be maximized and the imbalances can be minimized (Rangaswamy 1989). None of the above methods can satisfy this requirement.

13.3 Peak and Off-Peak Pricing

Demand of certain energy products shows significant daily and seasonal variations. To meet such varying needs, the suppliers often resort to storage facilities which can be used to balance the demand and supply. This is the case of gas or oil or coal, where the stock is built during off-peak period and the stock is drawn down during peak demand. In general terms, the use of storage option depends on the cost of storage and the cost difference between peak and off-peak production. If the cost of storage is less than the difference between the peak and off-peak period production, it makes sense to opt for storage. The issue however becomes more difficult when the economic storage possibility is limited as in the case of electricity. Here, the suppliers use different types of technologies to meet demand but the cost characteristics of these technologies are different, thereby imposing different costs of service during peak and off-peak hours of supply.

Many authors have analysed the pricing issue in such situations but a simple presentation following Munasinghe and Warford (1982, Chap. 2) is given here. In this simple version, consider two demand curves—one corresponding to the peak period (D_p) and the other corresponding to the off-peak period (D_{op}). Assume that the marginal cost curve can be simplified assuming a constant operating cost a (which is the short-run marginal cost) and the fixed cost is b (which added with the operating cost gives the long-run marginal cost)—see Fig. 13.5. During off-peak hours, there is excess capacity and the relevant price is the short-run marginal cost, a. During peak period the system feels pressure on capacity and the price

13.3 Peak and Off-Peak Pricing

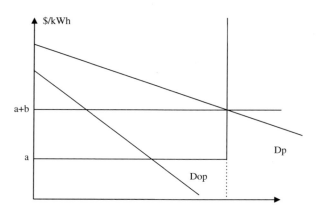

Fig. 13.5 Peak-load pricing

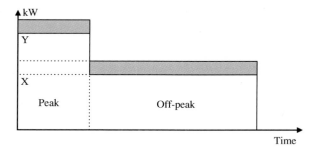

Fig. 13.6 Load duration curve for the example

would have to take into consideration the cost for adding capacity. Accordingly, the relevant price at this period is the operating cost supplemented by the capacity cost (or fixed cost). The simple rule then is that those consumers who come to the grid during peak-periods should bear the full responsibility of capacity cost and operating costs while those who use electricity during off-peak period should pay only for the short-run marginal costs.

13.3.1 Peak Load Pricing Principle

In what follows, we use a simple example to illustrate the pricing principle for peak and off-peak periods. This follows Munasinghe and Warford (1982, Appendix C).

Suppose that the annual load duration curve of the electric utility is composed of two distinct periods (see Fig. 13.6):

- An off-peak period during which the base load plants (using coal, nuclear, etc.) are used to supply the power. For simplicity it is assumed that the cost characteristics of these plants are uniform. The fixed cost per kW of capacity is a and the variable cost per hour is f. Assume also that the base capacity is given by X kW.

- A peak-period during which peaking plants are called to supplement power supply from the base load plants. The fixed cost per kW of capacity is b and the running cost per hour is g. Total load is Y kW, which implies that $Y - X$ is the peak load capacity.

It is assumed as usual that $a > b$ but $f < g$. It is also assumed that the entire capacity is fully utilised.

For any duration, h, the cost of using 1 kW of capacity of base plant is given by

$$y = a + f * h, \qquad (13.1)$$

whereas for the peak plant the cost is given by

$$z = b + g * h \qquad (13.2)$$

Given the cost characteristics, there exists a point where the costs of using the two types of plants are equal, i.e. $y = z$. This point indicates a number of hours during which the peaking plant shall be operated. By equating (13.1 and 13.2), this duration is obtained as follows:

$H = (a - b)/(g - f) = $ difference in fixed costs/difference in variable costs.

$$(13.3)$$

This is shown in Fig. 13.7.

The cost of supplying the load as shown in Fig. 13.7 can be written as:

$$C = X(a + f \cdot T) + (Y - X)(b + g \cdot H) \qquad (13.4)$$

where T is the total hours in a year (8760)

We are now going to analyse how the cost changes due to changes in peak and off-peak demand.

Case 1 Peak load demand changes by 1 kW

As the installed capacity is fully used, when the peak load demand increases by 1 kW, the utility has to install an additional peak capacity of 1 kW. This is shown by the coloured rectangle above Y. The new cost of production is given by:

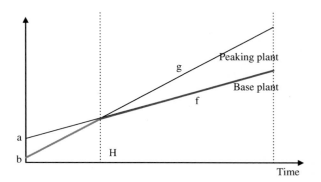

Fig. 13.7 Screening curve

$$C1 = X(a + f \cdot T) + (Y + 1 - X)(b + g \cdot H) \qquad (13.5)$$

The incremental cost is $\Delta C1 = C1 - C = b + g \cdot H$ (13.6)

This suggests that an additional demand during the peak period leads to two types of costs: the fixed cost and the running cost and the consumers should bear these costs if the tariff has to be cost-reflective.

Case 2 Demand increases during off-peak period

As the off-peak capacity is fully used, the utility has to install 1 kW of off-peak capacity. As the off-peak capacity will be available for peak load as well, the peak capacity will be reduced by 1 kW. The cost of supply can be written as:

$$C2 = (X + 1)(a + f \cdot T) + [Y - (X + 1)](b + g \cdot H) \qquad (13.7)$$

The incremental cost is given by

$$\Delta C2 = C2 - C = (a + f \cdot T) - (b + g \cdot H) = (a - b) + (f \cdot T - g \cdot H) \qquad (13.8)$$

From Eq. 13.8, $(a - b) = (g - f) \cdot H$

Replacing Eq. 13.8 in Eq. 13.7, we get

$$\Delta C2 = f \cdot (T - H) \qquad (13.9)$$

The supplementary cost is equal to the cost of running the off-peak capacity during the off-peak hours $(T - H)$. There is no fixed cost attached here and hence consumers coming to the grid during off-peak hours should pay only the running cost.

Case 3 Demand increases during the entire period

In this case, the total demand increases by 1 kW throughout. The total cost of supply is given by:

$$C3 = (X + 1)(a + f \cdot T) + (Y - X)(b + g \cdot H) \qquad (13.10)$$

Hence, the incremental cost is given by

$$\Delta C3 = a + f \cdot T \qquad (13.11)$$

This suggests that the total incremental cost of supply has to be borne by the consumers in this case.

What happens if this pricing policy is not followed? This is explained in Fig. 13.8.

Assume that the utility follows an average pricing principle (P^*) instead of peak-load pricing policy. As a result, during the off-peak period, the consumers pay more and their demand reduces (to Q^* from Q_0). This reduces the welfare of the consumers. At the same time, the consumers face lower price during off-peak period and consequently, the utility faces extra demand. To meet this demand, additional capacity is required ($K^* - K$). This could have been avoided by following the peak-pricing. In both the cases, there is some welfare loss.

Fig. 13.8 Loss due to non-peak pricing. *Source* Viscusi et al. (2005)

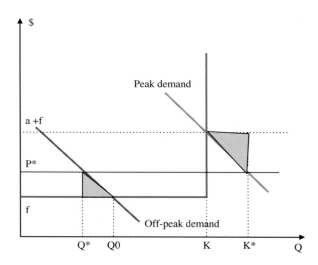

13.3.2 Short-Run Versus Long-Run Debate

The debate over the use of short-run or long-term marginal cost for energy pricing has a long history. However, the debate returned in the 1980s as the issue of stranded capacity emerged. A study by Andersson and Bohman (1985) tried to reconcile the two views by considering the studies of Boiteux (1956), Turvey (1968, 1969) and that of, among others, Munasinghe and Warford (1982). A brief summary of the debate is presented here.

The arguments in favour of pricing following the long-run marginal cost are as follows:

- When the capacity can be adjusted continuously and hence marginally and if the demand forecast is correct so that the capacity addition can be adjusted over time, the short term marginal cost is equal to the long-run marginal cost. However, if the capacity is not adapted to the demand, the pricing policy based on long-term marginal cost is preferable because the price has to stable over long term to facilitate correct investment decisions by the consumers.
- The pricing has to reflect a long-term policy because consumers can only respond over a long period of time due to lock-in effects and the response function is poorly understood.
- If the demand remains uncertain, the utility has to meet the demand in any case and therefore, the tariff has to rely on long-term pricing to take care of such uncertainties.
- The long-term marginal cost pricing coincides with the optimal resource allocation objective and allows a transfer of the burden on the consumers (Munasinghe and Warford 1982).

13.3 Peak and Off-Peak Pricing

The general criticisms against the above arguments are as follows:

- As the energy investments are indivisible, irreversible and long living ex post, the capacity cannot be varied continuously and rapidly;
- The long-run marginal cost is not easy to estimate. Schramm (1991) indicated that it is the practice in World Bank Studies to use an approximation that divides the net present worth of costs by the present worth of production but this by definition amounts to averaging and does not really reflect the marginal cost.
- If the divisibility of capital is not the concern, then the distinction between short-run and long-run costs is immaterial.
- As a forward-looking dynamic approach, the long-run marginal cost is unclear. The estimation depends on the demand forecast, and investment plan and their accuracy. Also, the future costs depend on the geopolitical situation, international energy markets and energy policies. Accordingly, the prices tend to be quite volatile and therefore, the accuracy of any forecast has been doubtful and more so for fast developing countries. In such cases, there is no guarantee that the pricing based on long-run cost will be stable.
- Munasinghe and Warford (1982) add a new dimension to the debate by introducing willingness to pay to the tariff issue. However, the marginal willingness to pay for a 1 kWh does not justify investment in a power station, the decision being guided by the total willingness to pay by the society or aggregated consumers for the entire life of the project. This in turn implies that the link between tariff and investment is more complex than it is assumed by them.

The arguments in favour of pricing based on short term marginal cost can be summarized as follows:

- The tariff-setting process is essentially a short-term phenomenon. The short term cost is well defined for a given capacity mix and available options;
- The main difference between the two concepts is that of time. In the short term, the capacity is fixed but that is not the case in the long term. But in reality, the instantaneous adjustment of capacity is not possible even in the long-term (unless technological innovation makes it possible to overcome the indivisibility of capital issue, for example using efficient but small-scale renewable technologies).
- It is strange that in the above debate the issue of the status of the company (private or public), market structure (monopolistic or not), and the nature of the transaction did not find a place. For example, Weisman (1991) remarked that the tariff decision depends on the nature of the transaction taking place between the company and its consumers. The nature of the transaction again depends on risk-sharing. In a spot market, consumers are free to choose their suppliers and therefore, the long-term marginal cost is not applicable. But in case where the supply is guaranteed by long-term contracts, consumers also share the risk. In such a case, the long-term concept becomes appropriate.

13.4 Energy Taxes and Subsidies[3]

Energy prices also include various charges, duties and taxes/subsidies which ultimately determine the price paid by the final consumers. Energy taxes are utilised for various purposes. The generation of revenue for the government is a principal objective. There are different forms of taxation that bring revenue to the treasury—excise duties on goods, royalties on domestic production of fossil fuels, and income taxes on the profit of energy companies. The relative importance of each instrument varies from one country to another but in general, indirect taxes constitute the major source of government revenue in developing countries, while in industrialized countries tax on income and profit, and the contribution for social security represent the major source of revenue.

Even for generating revenue, only the petroleum products are given more preference all over the world due to their inelastic nature of demand which provides for a stable revenue base. Consequently, more than 90% of environmentally-related tax revenue even in OECD countries comes from charges and taxes on motor vehicles and motor fuels.[4] Motor fuels are often subjected to higher taxes as they offer some attractive characteristics: (a) large, inelastic and often certain tax base; (b) easy to administer and control due to existence of a small number of economic agents; and (c) less transparent and politically sensitive compared to other taxes. In fact, as Fig. 13.9 indicates, the share of tax in the cost of a composite barrel of oil (i.e. the consumption weighted average of final consumer prices of refined products that make up a barrel) has greatly increased since 1980s.

Energy taxes are also used for demand management, macro-economic considerations and revenue redistribution to tackle equity issues. Finally, taxes on energy can also be justified for internalising externalities. Table 13.1 presents a summary of relevant factors for the determination of energy taxes. It becomes clear that the subject is complex and energy taxes, in reality, are a compromise among different objectives. Note also the value judgement aspect in the determination of relative importance of each factor depending on the country context, demand situation and energy market conditions. Therefore, the table should be considered only as an example without attaching excessive weight to the entries shown for different factors. Nonetheless, it brings to light several aspects. For instance, all factors are not equally important for all products. Similarly, the importance of different factors varies according to the sector under consideration. Clearly, all the factors are not relevant is all cases and their importance varies over time and across regions.

---

[3] This section is based on Bhattacharyya (1995, 1996 and 1997).

[4] See OECD database at http://www2.oecd.org/ecoinst/queries/index.htm.

13.4 Energy Taxes and Subsidies

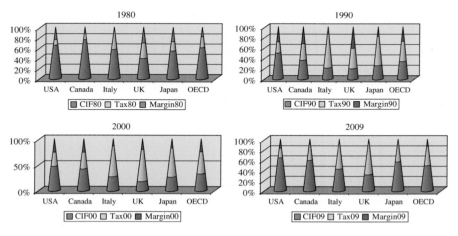

Fig. 13.9 Importance of tax in the composite barrel. *Source* OPEC Annual Statistical Bulletin 2009

13.4.1 Principles of Optimal Indirect Taxation

The economic analysis of taxation has a long history and therefore there is a well-developed body of literature on the subject[5] (See Mankiw et al. 2009 for a review. The most well-known optimal commodity tax formula dates back to 1927 when the Ramsey rule was proposed. This theory considered the problem of raising a given amount of revenue by taxing the commodities consumed by a consumer so that deadweight loss is or excess burden is minimized. In a partial equilibrium setting, assuming the demand curve DD' and a fixed producer price P and a tax t, the deadweight loss is given by shaded triangle ABC (see Fig. 13.10)

The optimal indirect tax formula requires that the compensated demand for each good be reduced by the same proportion. More precisely, if t_i is the tax rate on good i, and e_{ii} the price elasticity of demand for good i, and e_{ij} represents the cross-price elasticity of demand for i with respect to good j, then the tax rule is written for the two good case as

$$\frac{t_1}{t_2} = \left[\frac{(e_{22} - e_{12})}{(e_{11} - e_{21})}\right] \quad (13.12)$$

If cross-substitution effects are ignored, this simplifies to a proportional rule. It implies that tax rates should be inversely proportional to the elasticity of demand. Mathematically this is written as

---

[5] See Mankiw et al. (2009) for a recent review. See also Newbery and Stern (1988), Diamond and Mirrlees (1971) and Feldstein (1972).

Table 13.1 Factors determining the structure of energy taxes in different sectors

Sector	Energy	Factors to Be considered							
		Elasticity	Pollution	Charge for other services	Security of supply	Equity	Macro-economy	Demand management	Revenue
Transport	Gasoline/Diesel	+	++	++	++	+	−	+	++
Industry	Electricity	−	+	−−	+	−−	−	+	−
	Coal	++	++	−−	+	−−	+	+	−
	Natural gas	+	−	−−	+	−	−	+	−
	Petroleum products	++	++	−−	++	−−	−−	+	+
Residential and commercial	Electricity	−	+	−−	+	+	+	+	−
	Kerosene	−	+	−−	++	++	+	−	−
	Coal	+	++	−−	+	−	−	−	−
	LPG	+	−	−−	++	+	−	−−	−
	Fuel oil	+	+	−−	+	−	−	−	−
Agriculture	Electricity	+	+	−−	+	+	+	+	−
	Diesel	+	+	−−	++	−	+	+	−

Note: −− insignificant importance; − less important, + important, ++ very important

13.4 Energy Taxes and Subsidies

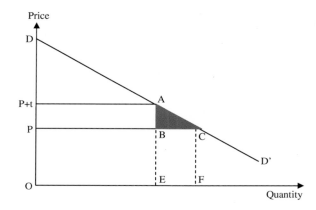

Fig. 13.10 Deadweight loss of an indirect tax

$$\frac{t_1}{t_2} = \frac{e_{22}}{e_{11}} \tag{13.13}$$

In other words, with all other factors remaining the same, tax rates should be higher on products with less than average demand elasticities. This implies that the necessities will be taxed more. Because the income elasticity of necessities is low, the "poor" will be subjected to a larger burden than the "rich", and such a tax is regressive.

The deadweight loss as shown in Fig. 13.9 may be written as follows:

$$\text{DWL} = \frac{1}{2} t \cdot \Delta Q = \frac{1}{2} \Delta P \cdot \Delta Q \tag{13.14}$$

where DWL is the deadweight loss, t is the tax (ΔP), ΔP and ΔQ are changes in the price and quantity of commodity in question after tax.

But the definition of elasticity can be used to simplify the above as follows:

$$e = \frac{\Delta Q \cdot P}{\Delta P \cdot Q}, \tag{13.15}$$

$$\text{DWL} = \frac{1}{2} \Delta P \cdot \left(\Delta P \cdot \frac{Q}{P} \cdot e \right) = \frac{1}{2} t^2 \cdot \frac{Q}{P} \cdot e \tag{13.16}$$

This is the well-known excess burden or dead-weight loss in a partial equilibrium model.

Three particular issues related to practical application of theoretical precepts can be identified. The first is related to the theoretical basis of the analysis. Any theoretical construct is based on certain assumptions. It should be ascertained whether any particular application satisfies these assumptions and what happens if they are violated. The major assumptions inherent in the above formulation are (inter alia) that there are no pure rents in the economy; that the production takes place in a competitive environment; that there are no externalities and that consumers maximize their utilities. These conditions are often violated in the energy market. Quite commonly, markets are dominated by a group of firms and collusion, rather than competition, is quite common. Unfortunately, for energy products substitutes are not absent—they need to be taken into account. Moreover, all pure rents are difficult to be taxed away and in such a case, the optimal set of taxes depends not only on the demand elasticities, but also on supply elasticities and on the share of capital in costs (Boskin and Robinson 1986). Moreover, for exhaustible resources, the pattern of taxes over time is very important for intertemporal consumption decisions. This calls for, ceteris paribus, the permanent levying of taxes to keep the present value of tax payments constant, which in turn, necessitates commitment of present and future governments to the same tax rates—a demand difficult to meet in reality. Additionally, since energy is an intermediate input for production as well as a final product for consumption, the elasticities of substitution between capital and energy may be important. Above all, these so-called optimal tax theories consider only efficiency without paying any attention to equity, and are thus vulnerable to criticism.

13.4.2 Equity considerations

No democratic governments can possibly ignore the importance of equity considerations in pricing. It is well known that policies that improve economic efficiency frequently have distributional impacts. Changes in energy prices affect income in different ways (Kumar 1985): directly through a change in real purchasing power and through changes in prices of other goods that use energy, and indirectly through changes in macroeconomic forces, like inflation, exchange rate, and employment. The main concern for equity derives from the fact that the poorer section of the population spends proportionately higher amounts on energy than price than their richer counterparts, and therefore is adversely affected by an energy price increase. Often the goal of equity is to minimize the adverse effects on the poorer households.

Some economists prefer to disregard equity issues either from the feeling that equity is a subjective criterion or from the belief that over long term the positive effects and negative distribution effects on any group average out (Griffin and Steele 1980). Others are sympathetic to the equity issue and advocate subsidies for the fuels used directly by poorer households. For any policy maker, at least three issues complicate the problem further: targeting the poorer sections of the population and then reaching them, avoiding unintended consequences, and providing subsidies at low cost. Rarely does any subsidy policy meet all these criteria, and this fact strengthens the anti-subsidy campaign.

The initial development of optimal commodity tax theory did not take the distribution aspect into consideration. Ramsey considered only a single individual.

13.4 Energy Taxes and Subsidies

It is Diamond and Mirrlees (1971) and Feldstein (1972), who have explicitly incorporated the distributional equity in the analysis. The Feldstein rule is quite similar to the original Ramsey rule. It can be expressed as follows:

$$\frac{t_1}{t_2} = \frac{[e_{22}(R_1 - L) - e_{12}(R_2 - L)]}{[e_{11}(R_2 - L) - e_{21}(R_1 - L)]} \quad (13.17)$$

where

$$t_i = \frac{(p_i - m_i)}{p_i} \quad (13.18)$$

p_i is sales price of good i, m_i is marginal cost of good i, e_{ii} own price elasticity of good I, e_{ij} is cross price elasticity of good i for a change in the price of good j, $i \# j$, R_i is the distributional characteristics of commodity i, and L is the shadow price of the budget constraint.

The value of R_i will be greater for a necessity than for a luxury. The higher the income elasticity of demand for a good, the lower the value of R_i. The relative optimal prices will thus depend on three factors: price elasticities of demand, distributional characteristics and budget constraint. Note that in the special case in which the distributional characteristics are irrelevant, the formula yields the Ramsey rule. When cross elasticities of demand are zero, Eq. 13.17 reduces to the following:

$$\frac{t_1}{t_2} = \frac{[e_{22}(R_1 - L)]}{[e_{11}(R_2 - L)]} \quad (13.19)$$

This ratio of optimal tax rates is the product of an efficiency factor (the Ramsey ratio of price elasticities) and a distributional equity factor. If the goods had equal demand elasticities, then in the absence of distributional considerations, they would have the same proportionate mark-ups over the marginal cost. When distributional considerations are taken into account, the good with the higher distributional characteristic will have the lower mark-up. For any good the markup is lower the higher the value of R. If $R > L$, then $t < 0$, and the price will be below the marginal cost (Rees 1984). The above formula requires information on distributional characteristics of each product and the shadow price of the budget constraint, and the computational difficulty increases.

13.4.3 Issues Related to Numerical Determination of an Optimal Tax

The determination of a numerical value for a tax that satisfies different objectives is a major problem. First, although the Ramsey rule allows us to compare different taxes, it leaves unresolved the problem of tax, as the formula leads to infinite solutions (Nan 1995). This formula needs to be modified to include explicitly the budget constraint of the government in order to be a useful guide for taxation purposes. However, this is part of the whole issue. For example, in order to take into account factors like pollution or national security or other macroeconomic aspects, the cost due to each factor needs to be determined. This is an involved task and information is not easily available in many cases. Moreover, the question of arriving at a unique solution for a tax requires that values determined for all these factors are to be combined at different proportions and there cannot be any unanimity in this regard. The question of double counting also arises. For instance, if a tax is imposed for budgetary purposes, it takes care of pollution, demand management or national security to a certain extent. Whether there should be a supplementary pollution tax (or any tax to recover other damages) becomes a major question. All these factors render the tax issue opaque, and it appears to be impossible to arrive at a consensus on all issues. This is so because all factors are important to various degrees and it depends on policy makers to fix a tax rate.

The structure of energy taxation for different products faces some other problems due to specificities of energies. For instance, kerosene and diesel are not very different in terms of quality and may be used as substitutes. Normally these products are destined for different consumer groups. Yet, if the difference is price is large, there exists possibilities of illegal use. Similarly, the difference between diesel and gasoline prices also favour motor conversions, dilution of gasoline with kerosene, and a rise in the stock of diesel-driven vehicles. Thus differential pricing of products poses serious problems.

Moreover, it becomes evident from Table 13.1 that theoretical prescriptions for different factors may be in contradiction or may act in different directions. For example, in a coal-producing country, the considerations for security of supply may prescribe subsidies on coal. On the other hand, coal being a highly polluting fuel, will shoulder a heavy tax on pollution grounds. The case for other factors or other fuels is similar. Thus, tax and subsidies based on different factors make the final outcome less transparent.

In addition, problems exist in relation to use of taxes (or subsidies) for correcting market failures. The valuation of environmental costs is a case in point that is fraught with many difficulties. As discussed in a subsequent chapter (Chap. 25), the case of internalisation of external costs faces a number of challenges. The damage functions or private benefit functions require information on output of firms, pollution created by firms, long-term accumulation of pollutants, monetary evaluation of the cost of damage and marginal private benefit derived from the output. The impact of pollution on health and ecology still remains a hotly debated issue with wide diversity in opinion. Great uncertainties regarding data, technology characterisation and atmospheric modelling adversely affect the valuation process. Lack of transferability of results, asymmetry in willingness to pay versus willingness to accept, aggregation of damages and benefits for different types of impacts, and the assumption of perfect competition undermine the valuation.

Thus although the issue of energy taxation may appear quite simple at first glance, a more careful analysis shows that larger issues loom. The inherent subjectivity of

13.4.4 Energy Taxes in Nordic Countries: An Example

Nordic countries are generally perceived as pro-active in the matter of environment protection. They are the pioneering countries in introducing carbon taxes (i.e. a tax on CO_2 emissions), even before the European Union launched a proposal to introduce community-wide carbon taxes in 1992 (which was never adopted although individual members have introduced some such taxes). Finland was the first country to introduce a CO_2 tax in 1990, followed by Norway and Sweden in 1992 and Denmark in 1992. Besides carbon tax, there are other taxes on energy as well—these include taxes on fuel and electricity and a tax on SO_2 emission.

The importance of energy taxes in the gross domestic products of the Nordic economies can be seen from Table 13.2 (Eurostat 2003). The importance is quite similar in all four countries with minor variations. Except in Norway, where the tax has tended to be lower than 2% of GDP, the tax share in the GDP has not changed much since 1990.

The relationship between energy consumption and tax payment by different users reveals interesting information (see Fig. 13.11). The pattern of energy consumption is quite varied in the four countries as is their energy sector. Sweden and Norway rely heavily on hydropower for electricity generation while Finland and Denmark depend on thermal power. Consequently, a significant share of their energy is used in the energy sector. Manufacturing plays an important role in these countries except in Denmark while the service sector is an important user of energy in Sweden and Norway. Households consume around 20% of energy in these countries. However, when it comes to payment of energy taxes, close to 60% of the burden is borne by households and another 30% falls on the service sector. The manufacturing sector and the energy sector, despite being major polluters contribute little to the pollution tax. The above information does not corroborate the idea of the polluter pays principle.

Often taxes on energy products are decided on revenue considerations. But CO_2 and SO_2 taxes in the Nordic countries were introduced to reduce CO_2 and SO_x emissions. Accordingly, the tax and the emission should have close relationship if the instrument is to be used effectively. Figure 13.12 provides the information while Table 13.3 provides the tax burden on different categories of polluters (Eurostat 2003).

Table 13.2 Taxes as % of GDP, 1999

% of GDP	Sweden	Norway	Finland	Denmark
Energy taxes (excl. CO_2 taxes)	2.1	1.5	2.2	2.2
CO_2 taxes	0.7	0.6	0.4	0.4
SO_2 taxes	0.0	0.0	0.0	0.0
Total energy related taxes	2.8	2.1	2.6	2.6

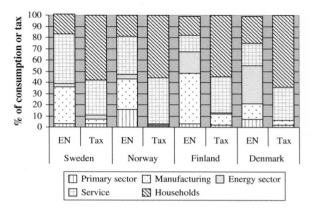

Fig. 13.11 Who pays energy taxes in the Nordic countries?. *Source* (Eurostat 2003)

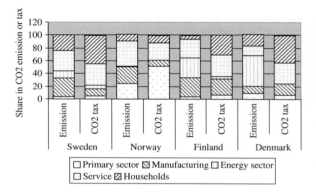

Fig. 13.12 CO_2 emission and carbon tax in Nordic countries in 1999. *Source* (Eurostat 2003)

13.4.5 Who Bears the Tax burden?

The imposition of a tax increases the price of a good in the market and consumers who consume the goods ultimately bear the effect. This might suggest that consumers ultimately bear the burden of tax. However, both the suppliers and the consumers share the burden but the level or degree of burden sharing depends on the elasticity of demand.[6]

The sharing of the burden depends on the elasticity of demand. When the demand is inelastic, changes in prices will not affect the demand for the good. This results in transferring the burden to the consumers. On the other hand, when the demand is elastic, consumers will switch to other products and the demand will be

---

[6] See Pearce et al. (1994) for more details.

13.4 Energy Taxes and Subsidies

Table 13.3 Effective CO_2 rate in 1999 (euro/tonne CO_2)

Category	Sweden	Norway	Finland	Denmark
All	23	16	8	10
Households	43	17	46	23
Agric and fishing	36	13	16	15
Mining	14	40	12	1
Manufacturing	9	5	6	14
Energy sector	13	7	1	0
Financial sector	43	218		107
Services	39	25		59
Construction	44	21	17	13
Transport	15	9	6	9
Trading	43	11	14	42

Source Energy taxes in the Nordic Countries: Does the polluter pay? Eurostat, 2003 (see http://www.scb.se/statistik/MI/MI1202/2004A01/MI1202_2004A01_BR_MIFT0404.pdf)

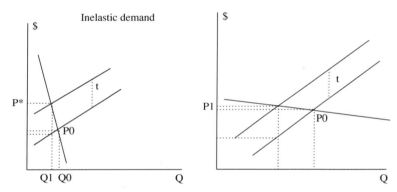

Fig. 13.13 Tax burden sharing

affected substantially. In this case, the producers will bear a larger share of the burden. This is shown in Fig. 13.13.

13.4.6 Subsidies

Subsidies can be defined as the difference between the price that would exist in a market in absence of any distortion or market failures and the price faced by consumers at a given time. If market distortions/failures exist, instead of market price some reference price has to be used, correcting the problem. In case of traded goods, relevant border prices are considered as appropriate reference prices in absence of externalities.

Subsidies normally have a number of perverse consequences: they send wrong price signals to consumers and promote over-consumption, often inefficiently; they divert scarce financial resources at the cost of depriving other needs; they hinder growth of alternatives and act as a trade barrier. Subsidies take various forms. Subsidies to producers help to lower the cost of production, while subsidies to consumers lower prices faced by them. Fossil fuel subsidies in developed countries tend to support particular indigenous fuels such as coal (as in Germany or in UK) to protect employment. Nuclear energy also receives significant subsidy in many countries. Consumer subsidies take the form of support to lower income groups but other consumers often pay higher taxes to compensate for the revenue losses. On the other hand, developing countries provide much extensive levels of subsidies, often across the board. Price controls by the governments remain the most commonly used method of intervention. But being non-targeted, the effectiveness of the subsidies is questionable, as the benefits do not reach the desired groups.

Subsidies for fossil fuels are pervasive in both developed and developing countries and have emerged as a major theme in international discussions and negotiations aimed at promoting sustainable development. According to Morgan (2007), global energy subsidies cost between USD 250 to USD 300 billion per year net of taxes. This is equal to 0.6% to 0.7 of world GDP (see Table 13.4). Developing countries pay subsidies to lower consumer prices while in developed countries subsidies often go to the producers. However, Table 13.4 makes it clear that the amount of subsidies in the developing world dominate the global subsidy scene.

But not all subsidies have negative consequences. When the social benefit or environmental improvement exceeds the cost of subsidies, a positive effect can result. Examples can include subsidies for promoting renewable energies.

Subsidy programmes can be designed in different ways: through price-reduction systems, through a gifts and voucher scheme, and through an all-or-nothing system. The choice of any system depends on the objective of the policy maker. For example, if the objective is to ensure a minimum level of consumption of a commodity, subsidies acting through the price mechanism are preferable to direct cash transfers. If, however, the objective is to improve general welfare, purchasing

Table 13.4 Size of energy subsidies

Fuel	OECD	Non-OECD	World
Oil	n.a.	90–110	90–130
Natural gas	n.a.	70–90	n.a.
Coal	5.8–6.7	10–13	16–23
Electricity	n.a.	55–70	n.a.
Nuclear	4 (R&D only)	n.a.	16
Renewables	1 (R&D only)	n.a.	16
Total	20–30	220–280	240–310

Source Morgan (2007)

power transfers are better than price reductions. Moreover, there seems to be a conflict of interest between donor and recipients. Taxpayers wishing to minimize the cost of a assuring a target level of consumption of a particular good by recipients will prefer the minimum cost all-or-nothing subsidy to the price-reduction subsidy. The price-reducing subsidy, in turn, is superior to cash grants. On the other hand, recipients will prefer cash grants to a price-reduction subsidy for a given total subsidy cost to the taxpayer, and they will prefer a price-reducing subsidy to the minimum cost all-or-nothing scheme. This suggests that choice of any system can always be subject to criticism.

13.5 Implications of Traditional Energies and Informal Sectors in Developing Economies for Energy Pricing

Developing countries face a dilemma in the case subsidies as the environmental benefits of subsidy removal may be offset by increases in traditional energy utilisation. Traditional energies play a crucial role in the energy sector of developing countries. While developing economies transit from traditional energies to modern energies as they climb up the income ladder, the speed at which countries move varies and consequently, around 1.4 billion will lack access to electricity in 2030 while 2.7 billion will continue to use traditional energies for cooking unless new policies are undertaken (Birol 2007).

Simultaneously, the informal sector plays an important role developing economies. These are mostly unorganised, isolated, decentralised activities and mostly localised in rural and urban peripheries. The major client of the services of this sector is the households, especially those in the low-income groups. A special feature of these activities is that non-monetised transactions (in kind payments or barter transactions) co-exist with a presumably growing monetised subsector. The presence of informal sector introduces non-optimal choices (Shukla (1995) and Pandey (2002)) and leads to violation of the basic assumptions of the neoclassical paradigm (Bhattacharyya (1995)) because of incomplete markets, costly information and transaction costs in developing countries.

Since traditional energies are mainly used in the residential and commercial sectors, pricing of modern energies for those sectors needs to take certain additional factors into account. While prices of traditional energies do not affect the consumption of modern fuels, the reverse is not true. Generally, energy prices often play a minor role in any changeover to modern fuels. The decision to switch to commercial energies are largely determined by income, regularity of paid employment, availability of financing for appliance purchases through formal and informal credit markets, exposure to and knowledge of the qualitative attributes of commercial energies and so on (Bose (1993) and Bhattacharyya (1996)). Any substitution from traditional energies to commercial ones (and hence to a monetized activity) would mean that households would need access to adequate and continuous streams of income, which they would then be willing to spend on commercial fuels rather than on other goods or savings. This implies that socio-economic conditions involving poverty, unemployment, seasonal employment, in-kind wage payment, or prospects for the future could act as impediments to a transition to commercial energies.

The need to modify modern fuel prices to take into account traditional fuels may be justified for two reasons: first, it is often considered that a reduction in prices of modern fuels will enable a switch from traditional to modern fuels; second, this is justified on environmental grounds. In many countries, heavy reliance on traditional fuels leads to chronic health problems. One way to avoid or reduce this is to shift to modern fuels or use traditional energies in an efficient way. In order to induce consumers to change their consumption pattern, incentives such as subsidized fuel are often used.

If traditional and commercial energies are considered as substitutes, increases in commercial energy prices could promote switching to traditional energies, as the households require a minimum amount of energy to meet their needs, whatever the source of energy may be. Such a substitution may have adverse environmental consequences, as the efficiency of traditional energy use is much lower than that of using modern energies. The social cost of such a substitution could be significant as well.

13.6 Conclusion

This chapter has provided a tour of energy pricing and taxation issues. The basic principles of energy pricing are first presented and then departures from those principles due to various factors are considered. The chapter has then introduced the economic logic behind taxing or subsidizing energy products. One main theme of this chapter has been the complexity and subjectivity of some of the issues involved. This happens because of dependence on a single instrument to cater to various problems. Consequently, the subject remains controversial and highly debated, which, in turn, makes it difficult for policymakers to agree on a common or harmonized tax systems in a region or across countries. This assumes importance in the context of climate change where global actions are required.

References

Andersson R, Bohman M (1985) Short- and long-run marginal cost pricing: on their alleged equivalence, Energy Econ October:279–288
Bhattacharyya SC (1995) Internalising externalities of energy use through price mechanism: a developing country perspective. Energy Environt 6(3):211–221
Bhattacharyya SC (1996) Domestic energy pricing policies in Developing countries: why are economic prescriptions shelved? Energy Sources 18:855–874
Bhattacharyya SC (1997) Energy taxation and environmental externalities: a critical analysis'. Int J Energy Dev 22(2):199–223 Spring 1997

References

Birol F (2007) Energy economics: a place for energy poverty in the agenda? Energy J 28(3):1–6
Boiteux M (1956) Sur la gestion des monopoles publics astreints a l'equilibre budgetaire. Econometrica 24:22–40
Bose S (1993) Money, energy and welfare: the state and the households in India's rural electrification policy. Oxford University Press, New Delhi
Boskin MJ, Robinson MS (1986) Energy taxes and optimal tax theory. Energy J, Special Issue, 1986, pp 1–16
Diamond PA, Mirrlees JA (1971) Optimal taxation and public production II: tax rules. Am Econ Rev 61:261–278
Feldstein MS (1972) Distributional equity and the optimal structure of public prices. Am Econ Rev 62:32–36
Griffin JF, Steele HB (1980) Energy economics and policy. Academic Press, New York
Kumar MS (1985) Socio-economic goals in energy pricing policy: a framework for analysis. In: Siddayao CM (ed) Criteria for energy pricing. Graham & Trotman, London
Mankiw NG, Weinzierl M, Yagan D (2009) Optimal taxation in theory and practice, NBER Working paper no. 15071, The National Bureau of Economic Research, Cambridge, MA, USA (see http://www.nber.org/papers/w15071)
Morgan T (2007) Energy subsidies: their magnitude, how they affect energy investment and Greenhouse gas emissions, and their reform, UNFCCC Secretariat, Geneva (see http://unfccc.int/files/cooperation_and_support/financial_mechanism/application/pdf/morgan_pdf.pdf)
Munasinghe M, Warford JJ (1982) Electricity pricing: theory and case studies. World Bank, Washington D.C
Munasinghe M (1985) Energy pricing and demand management. Westview Press, Boulder, Colorado
Nan GD (1995) An energy Btu tax alternative. Resour Energy Econ 17(3):291–305
Newbery D, Stern N (eds) (1988) The theory of taxation for developing countries. Oxford University Press, Oxford
Pandey R (2002) Energy Policy modeling: Agenda for developing countries. Energy Policy 30:97–106
Pearce D, Turner K, Bateman I (1994) Environmental economics: an elementary introduction. John Hopkins University Press, Baltimore
Rangaswamy V (1989) Domestic energy pricing policies, Energy Series Paper 13, World Bank, Washington DC. World Bank, 2004, Reforming infrastructure: privatisation, regulation and competition, World Bank
Rees R (1984) Public enterprise economics. Weidenfeld and Nicolson, London
Schramm G (1991) Marginal cost pricing revisited. Energy Econ 13(4):245–249
Shukla PR (1995) Greenhouse Gas Models and abatement costs for developing countries: a critical assessment. Energy Policy 23(8):677–687
Turvey R (1968) Optimal Pricing and investment in electricity supply. Allen and Unwin, London
Turvey R (1969) Marginal Cost. Economic Journal 79(314):282–299
Viscusi WK, Harrington JE, Vernon JM (2005) Economics of regulation and antitrust. The MIT Press, Massachusetts
Weisman D (1991) A note on first-best marginal cost measures in public enterprise. Energy Economics, October, pp 250–253

Further Reading

Barnes D, Halpern J (2000) The role of energy subsidies, Chapter 7. of Energy Services of the world's poor, World Bank, Washington DC (see http://www.worldbank.org/html/fpd/esmap/energy_report2000/ch7.pdf)
Datta-Chaudhuri M (1990) Market Failure and government failure. J Econ Perspect 4(3):25–39

Energy Charter Treaty (2007) Putting a Price on Energy: international pricing mechanisms for oil and gas (see http://www.encharter.org/index.php?id=218)

Fattouh B (2005) The causes of crude oil price volatility. Middle East Econ Surv 58(13). See http://www.mees.com/postedarticles/oped/v48n13-5OD01.htm

IEA (1999) World Energy Outlook, Looking at energy subsidies: getting energy prices right. Int Energy Agency, Paris

RWE (2005) World Energy Report 2005, Determinants of energy prices, Essen, See http://www.rwe.com/generator.aspx/konzern/property=Data/id=266754/worldenergyreport-2005.pdf

UNEP (2008) Reforming energy subsidies: opportunities to contribute to the climate change agenda, UNEP, Genenva (See http://www.unep.org/pdf/PressReleases/Reforming_Energy_Subsidies.pdf)

Chapter 14
International Oil Market

14.1 Introduction

This chapter focuses on the international oil market and presents an overview of the developments in this industry by looking at the resource positions, production and consumption patterns. It also traces the changes in the organisational pattern of this industry over time and highlights the nature of market interactions in these industries. The purpose of the chapter is to capture the essence of the changes in the industry without entering into an elaborate analysis or discussion, which is outside the scope of this chapter. This chapter is organised as follows: first a brief history of the evolution of the oil market is presented by considering two important phases of development—pre-OPEC era and OPEC era. This is followed by an analysis of some key aspects of the market.

14.2 Developments in the Oil Industry

Oil was discovered by Colonel Drake and William A. Smith in 1859. The oil industry has undergone four distinct phases between 1859 and 1960, when OPEC was formed. Here a brief description of the pre-OPEC and post-OPEC era is given. Detailed discussions can be found in, among others, IFP (2007).

14.2.1 Pre-OPEC Era

The four phases of this period are: the period of gold rush, the phase of Standard Oil domination, the internationalization of the industry and the rise of the Seven Sisters. Each phase is described below.

14.2.1.1 Phase 1: Oil Rush and Intense Competition (1859–1870)

The discovery of oil triggered an oil rush in America as fortune seekers rushed to the site to buy land and construct oil derricks. As the American law confers the ownership of the underground resources to the landowner, there is an enormous incentive to pump oil from the ground as fast as possible, so as to surprise the neighbour. With a low recovery rate (about 5% at that time), this rush led to excessive drilling and a considerable wastage. Moreover, every time a significant new field is discovered, prices fell and output soared. Prices sometimes varied even 100 times during this period (see Fig. 14.1).

Oil had limited use for lighting purposes—kerosene was produced by refining crude and sold to light homes and businesses. Horse-drawn wagons and railroad were primarily used for transporting crude oil to refineries. The industry was essentially an American industry.

14.2.1.2 Phase 2: Monopoly of Rockefeller company (Between 1870 and 1911)

During this phase, John D. Rockefeller dominated the oil industry. He entered the refining business in 1863 in Cleveland Ohio but carefully avoided the production segment because of high risk involved in it. His strategy was to gain control of the industry by controlling the bottleneck facilities such as the refining, transportation and distribution segments of the industry. Rockefeller and his assistant Henry Flagler set up the Standard Oil Company in 1870. A new era began with the establishment of this company.

Standard Oil took advantage of pipeline transport and managed steep railroad rebates. They gained control of 90–95% of the refining in the United States through aggressive mergers between 1870 and 1880. Standard Oil became larger than its competitors through economies of scale and could influence the producer prices and the output prices using its market power.

However, Standard Oil's dominant position came under threat from a number of developments.

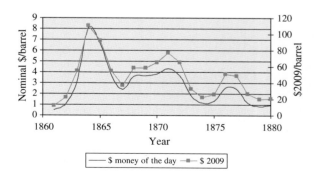

Fig. 14.1 Crude oil prices between 1860 and 1880. *Data source* BP Statistical Review of World Energy 2010

(a) Oil discoveries in other countries, especially in the Baku region of Russia and in the Dutch East Indies were a source of competition. Nobel brothers acquired oil properties in Baku in 1872 and expanded the activities to transport Russian oil to the West. Russian oil production increased rapidly during this period. Similarly, the Royal Dutch Company and Shell started operations in now Indonesia in the last decade of the 19th century and the merger of these two companies to form the Royal Dutch/Shell in the first decade of the 20th century gave impetus to their global ambitions. The new company started operations in Romania (1906), Venezuela (1910), Egypt (1911), Russia (1912), US (1912), Trinidad (1913), and Mexico (1913).
(b) New large discoveries in Texas in 1901 and the emergence of new companies like the Texas Oil Company and the Gulf Oil Company emerged as new threats on the domestic front. By the 1920s both Texaco and Gulf had established themselves as integrated global oil companies.
(c) But the changes to the legal and regulatory frameworks proved to be the most important threat to Standard Oil. The passage of Sherman Antitrust Act of 1890 and the subsequent proceedings against the company ultimately led to the dissolution of Standard Oil. In accordance with the Court order, Standard Oil was divided into several separate entities, including, among others:

    a. Standard Oil of New Jersey, with almost half of the total net value was the largest of the lot and was renamed as Exxon 1972.
    b. Standard Oil of New York with 9% of the net value was another large company. Later it became known as Mobil Oil Corporation before its merger with Exxon.
    c. Standard Oil California which later became Chevron Corporation in 1984. Chevron also acquired Gulf Oil in 1984.
    d. Standard Oil of Indiana which later became known as Amoco Corporation in 1985. Amoco merged with BP in 1998.
    e. Standard Oil of Ohio—now the American arm of British Petroleum since 1987.

This is an important phase in the history of the oil industry that recorded the rise and fall of a major player. But the newly created companies from the old giant proved strong enough to emerge as giants (or Majors) in due course in their own right.

14.2.1.3 Phase 3: Internationalization of Oil Industry (1911–1928)

During this period, oil started to displace coal as the dominant fuel in the world economy. The maturity of the automobile industry spurred demand for gasoline and the break out of the First World War fuelled oil demand for military and other services. Oil emerged as a strategic commodity for the first time and the state intervention in the oil business started with the British Government deciding to acquire 51% of the Anglo-Persian Oil Company (renamed as the British Petroleum afterwards).

The company acquired a vast concession for exploration from the Shah of Persia (now Iran).

Around the same time, in 1911, the beginning of the Mexican Revolution created an uncertain political situation in Mexico. Further, the new constitution (of 1917) gave ownership of subsoil resources to the state. These changes affected investor confidence in a country where foreign investors controlled the majority of the oil operations at the time of the revolution. Although the country continued to produce and supply, the long-term sustainability was damaged and from 1927 onwards, Mexico gradually lost its ability to compete in the world oil market, particularly against Venezuela, because of higher production costs, increasing taxation, and the exhaustion of existing fields.

Industrialised economies realized their increasing dependence on oil after World War I and even there was concern about rapid depletion of the American oil reserves. As a response, oil companies started to invest in new crude oil production, which in turn led to a glut in world oil supplies by 1928 and a drastic fall in prices. A part of this production came from low cost fields, which drove the marginal producers out of the market. In an attempt to stabilize the market, three leading oil companies, namely Standard oil of New Jersey, Royal Dutch/Shell and Anglo-Persian, decided to minimize competition and organize the market on an "as is" basis. This is known as the Achnacarry Accord of 1928. The American market was excluded from these agreements.

One of the developments as a result of the Achnacarry Accord was the application of the Gulf-plus pricing scheme (see Fig. 14.2). The method considered fob price from the US coast of Gulf of Mexico plus the cost of transportation. These prices were also known as world parity prices for oil. Mathematically, at any point

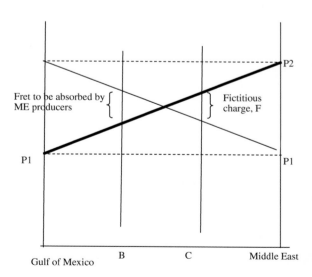

Fig. 14.2 Gulf-plus pricing system

in the world (say C on Fig. 14.1), which is at a distance of D_{1j} from the Gulf of Mexico, the CIF price will be given by

$$P_j = P_1 + a \cdot D_1 \quad (14.1)$$

where a is the unit cost of transport, D_{1j} is the distance between Gulf of Mexico and point C.

Even if the supply comes from the Middle East, the same price will be charged but this is equivalent to an uplift charge F over the price that would apply if the cargo started from the Middle East. Middle East producers benefited from a fictitious fret for any delivery close to source but supported extra cost for supplies closer to Gulf of Mexico. Under the system, American production was protected and prevented expansion of markets for new sources of supply.

14.2.1.4 Phase 4: Between 1928 and 1960: Rise of the Seven Sisters

During this period, a number of major developments took place that changed the face of the oil industry. These include:

- Greater assertion of the host governments in the oil affairs as exemplified by (1) the cancellation of the Anglo-Iranian Concession in 1932 as a result of fall in royalty income due to the Great Depression of 1930s and (2) Nationalization of oil industry of Mexico in 1938 and the establishment of a national oil company—Petroleos Mexicanos (PEMEX). This is the precursor for the subsequent nationalization movement in the industry.
- During this period, the Middle East started to emerge as the focal point of all attention due to its huge reserve potential. Although under the "as-is" agreement, oil companies decided to protect their spheres of influence in the Middle East, they failed to regulate the market because of factors beyond their control. There was new entry into the Middle East in search oil and production outside these two regions also continued to grow—Venezuela became a major producer in South America. Major oil companies operating in the Middle East worked a joint operating arrangement
- The claim for a bigger share of the oil profit started to emerge in oil exporting countries. They also started to exert claims for sovereignty over the oil beneath the ground. As foreign companies possessed the technical know-how, financial power and the network for distribution, the states were unable to operate without the multinational support. But the pressure for better deals succeeded in 1943, when Venezuela obtained a deal with 50–50 split of profits. Here, the companies would pay a lump sum royalty to the host country plus a 50–50 split in profits (i.e., selling price minus production cost). This became the industry norm within a few years time.
- The importance of oil continued to grow and in the Second World War, oil played an important role.

Fig. 14.3 Double basis point pricing

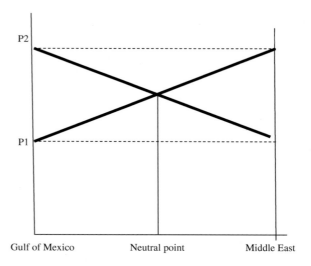

- Seven oil companies[1] dominated the oil scene during this period—these international, integrated oil companies controlled the entire supply chain and had influenced the market.
- More importantly, the USA became a net importer of oil in this phase and this required a change in the pricing policy as the objective now shifted to ensuring competitive oil supplies from the Middle East as opposed to protecting the American export market outside.

As a first step, the single base point pricing system lost its appeal during this period as production from the Middle East grew. Consequently, the Arabian Gulf was accepted as the second base point but the price was equated to the FOB price of Gulf of Mexico. There is a cross-over point where the transport cost becomes equal to that from the other reference point. In other words, if D_{1j} is the distance of a point j from reference point 1 (say Gulf of Mexico) and D_{2j} is the distance from reference point 2 (say Middle East), then there is a point where $a_1 \cdot D_{1j} = a_2 \cdot D_{2j}$ where a_1 and a_2 are the unit transport costs from two sources. If the unit costs are same, then $D_{1j} = D_{2j}$ (i.e. halfway between the two reference points). This protected the American production and allowed market segregation depending on the distance over which oil was transported (see Fig. 14.3). This also provided a better remuneration to the Middle Eastern producers. The method was followed until 1948 when the USA became a net importer.

---

[1] These were Standard Oil Company of New Jersey (later Exxon, now ExxonMobil), Standard Oil Company of New York (later Mobil, now ExxonMobil), Standard Oil of California (now Chevron), Texas Oil Company (now Chevron), Royal Dutch Shell, Anglo-Persian Oil Company (now BP) and Gulf Oil (now part of Chevron and BP).

Fig. 14.4 Double base point pricing with double posting

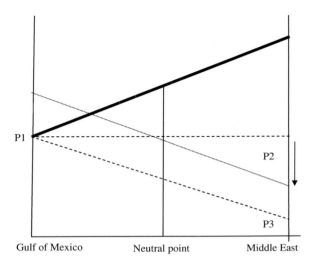

A further adjustment to the pricing principle became necessary in 1948 to allow an expansion of the Middle Eastern supply. This was done by introducing a second posting price using Middle East price. As the cost of production in the Middle East was lower, the second posting price allowed exporting of Middle East oil further and eventually competing with the American supply in America itself (see Fig. 14.4). The increase in the cost of production in the USA and its change of status prompted the above shift in pricing. But the policy of market expansion resulted in a lower price for oil in the international market and affected the revenue income of host governments from royalty payments. The system continued until 1959. However, the competition from the Middle East also brought new issues—American administration became concerned with the rising import dependence and the Texan producers feared loss of market as a result of external competition. The American government reacted by imposing import quotas in 1959.

However, since 1956 international companies started to apply a new system—the posted price. Here, prices for crude and products were posted: crude postings were fob well head excluding gathering and collection costs whereas product postings were fob refinery. These are buyer-set prices (i.e. set by the refineries who bought crude for refining) and were used as the reference price for tax and royalty determination. Yet, realized prices (or market selling prices) were different from posted prices as various discounts on posted prices—of the order of 20–30%—were offered. Such discounts included cash discounts, long term commitment discounts—to reflect reduced risk on the producer and large order discounts—to reflect reduced handling charges. There was a growing sentiment that the pricing policy eroded the value of the Middle Eastern oil and served the interests of the importing countries. The producing countries therefore decided to work together to protect their own oil interests, as discussed below.

14.2.2 OPEC Era

The Organization of the Petroleum Exporting Countries (OPEC) was established in Iraq in September 1960 by five leading oil producing states (Iran, Iraq, Kuwait, Saudi Arabia and Venezuela) in an attempt to co-ordinate petroleum policies of member states so as to secure a fair and stable remuneration for their outputs. At the peak of its time, OPEC had 14 members[2] but now has 12 full members (Indonesia suspended its membership in 2009 and Gabon terminated its membership in 1995).

The organization was established at a time when the Middle Eastern production was rising but the price in the market was falling (or stable) in real terms. As discussed in Chap. 12, in a capital-intensive industry with a low operating cost, this is an essential feature as the producers continue near full-capacity operation to minimize operating losses, which in turn depresses the price. As the host countries were dependent on royalty incomes, such low prices affected their oil revenue significantly (see Fig. 14.5). Simultaneously, this period also saw emergence of new independent states through decolonization and one of the aspirations of such states was to exert control over their resources to be able to control the future course of their economic and social development. This phase of development can be split into a number of phases

14.2.2.1 First Phase: 1960–1973

During this initial phase, OPEC members were disunited and the organization played a cautious role and achieved moderate gains. It focused on three areas of activity: tax system changes, production control and steps towards nationalization of concessions. Until OPEC gained price fixing power, short term objective was to maximize the share of producers' surplus by reducing tax deductions. This was achieved by disallowing marketing allowance, expensing royalties and increasing tax reference price. Members agreed not to recognize prices below August 1960 as reference, achieved freezing of posted prices. Production control program started in 1965 but was abandoned in 1967. A system of maximum annual growth rates in export of members was fixed but did not work. The Manifesto of 1968 provided the blue print for nationalization of concessions. The 1968 Manifesto encouraged creation of national oil companies to develop oil reserves, and national participation in concessions by purchasing of operating concessions as well as development of relinquished concessions.

As the producing countries had to rely on the international companies to sell their oil, an arrangement of buy back by oil companies was designed. The buy-

---

[2] These are Qatar (1961), Indonesia (1962), Libya (1962), United Arab Emirates (1967), Algeria (1969), Nigeria (1971), Ecuador (1973), Gabon (1975) and Angola (2007).

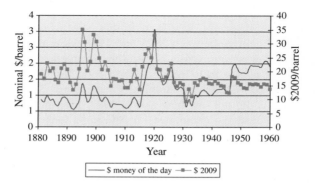

Fig. 14.5 Crude oil prices between 1880 and 1960. *Data source* BP Statistical Review of World Energy, 2010.

back price was the price at which companies purchased producers' share of oil. This was a negotiated price but often this was lower than the posted price.

Towards the end of this period, a number of political events took place. These include Algerian independence and attempts towards nationalization, Israeli victory in the 1967 war, overturning of the Libyan monarchy in 1969 and higher price for oil sales in 1970. In 1971, OPEC opened negotiations with companies on a 5 year pact and forced a price increase of 21% for Saudi light, an increase in tax rate from 50 to 55% and an escalation of 2.5% in prices per year for inflation. Although the pact remained in force until 1976, the market price changed quite significantly during this period, and the second phase of OPEC history started.

14.2.2.2 Phase 2: 1973–1975

This period coincides with a more assertive role of OPEC. Following the Yom Kippur war in October 1973, Arab producers imposed oil embargo against USA, Holland, Portugal and South Africa. They also decided to cut oil production by 25% (5 million barrels per day). OPEC also declared unilaterally tax reference price in October 1973 that forced a price increase from $3.01/bbl to $5.12/bbl. Market price increased due to demand increase as a result o panic buying, fear of import problems and production reduction. Prices increased to close to $12 per barrel by end of 1974, causing the first oil price shock. Nationalization of oil assets also continued and by 1975, OPEC members completed the process. However, they still needed international oil companies to explore, find, develop, transport and market oil. The posted price system was abandoned by the end of this period and OPEC started its official price mechanism.

The first oil shock was an eye-opener for importers. This marked the era of cheap oil and energy issues started to gain importance in international arena. Importers started to look for alternative options and the use of domestic fuels gained currency. Coal in many cases was favoured and energy efficiency and demand management options were considered for the first time. Renewable energies also received some attention.

14.2.2.3 Phase 3: 1975–1981

OPEC policies between 1975 and 1978 were aimed at demand stabilization and moderate price increase. There were divergences among the members in terms of pricing policy. In 1976, for a short period, a policy of double official price regime was used—$12.09 for Arabian producers and $12.70 for other OPEC members. But this did not last long and in 1977, the single official price system returned. OPEC created a committee for devising a long term strategy.

Between 1979 and 1981, historical events influenced OPEC policies significantly. Iranian revolution and subsequent Iran–Iraq war dominated the international scene. Oil production stopped in these countries, leading to a second wave of price rise that sent shock waves around the world. Prices rose to $24/bbl in 1979, to $32/bbl in 1980 and to $34/bbl in 1981. The difference between the official OPEC price and the market price started to grow and it became difficult to control OPEC members to take advantage of high market prices. Consequently, 25% of OPEC oil was sold in the spot market and even in some cases members annulled long term contracts for this.

14.2.2.4 Phase 4: 1981–1986

The reaction to the second oil shock was quite dramatic. Importing countries reduced their consumption and started to search for alternatives. Production of oil from outside OPEC received greater attention and the share of non-OPEC oil in the international market started to rise. OPEC had to deal with declining market share as cheap oil remained underused while costly oil became viable. OPEC opted for price stability and decided to fix $34 per barrel price. To achieve this, a 10% reduction in production in 1981 was initiated and Saudi Arabia decided an upper limit of 8.5 Mbd for Aramco in 1982. Production quota was introduced for the first time in March 1982 but disagreements surfaced in 1983 for new quotas. Saudi Arabia decided to act as the swing producer at this time to control prices but the global economy was under a severe recession. In addition, the developing world entered into a debt crisis due to spiraling interest rates. OPEC was struggling to manage its revenue but decided to defend a price of $29b in 1983. However, this was not sufficient to arrest price fall and Saudi Arabia as a swing producer had to reduce its output. The market share of OPEC members declined rapidly (see Fig. 14.6) and reached close to 20% from a high of above 50% in 1973. From 4.9 Million barrels per day (Mbd) in the second quarter of 1984 Saudi reduced its output to 2.3 Mbd in the third quarter of 1985, while non-OPEC producers filled the market. However, against this background of falling market share and declining prices, discontent increased among members. OPEC members decided to regain market share and a price war began. This is known as the third oil shock or counter shock, when Saudi Arabia decided not to support any production cut any further. Prices started to fall to $15, $10 and to $7 in July 1986 (Fig. 14.7).

14.2 Developments in the Oil Industry

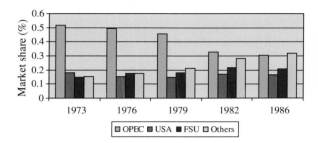

Fig. 14.6 Declining OPEC share between two oil price shocks. *Data source* BP Statistical Review of Energy Statistics, 2010

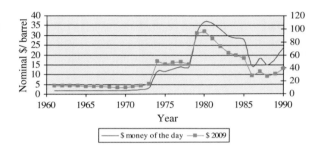

Fig. 14.7 Crude oil price between 1960 and 1990. *Data source* BP Statistical Review of Energy Statistics, 2010

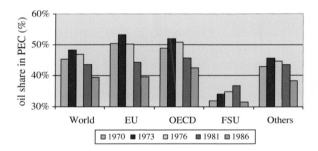

Fig. 14.8 Declining importance of oil in the primary energy mix. *Data source* BP Statistical Review of Energy Statistics, 2010

A major consequence of the oil price spikes in the Seventies was that importing countries became more concerned about their import dependence and many promoted programmes to replace oil and promote alternative energies and energy saving. These actions started to produce results by the middle of 1980s when the developed world's dependence on oil reduced quite considerably (see Fig. 14.8). Consequently, the share of oil in the global commercial energy demand fell below 40% for the first time in many decades.

14.2.2.5 Phase 5: OPEC in the 1990s

In the post-1986 period, the glut in the market continued. The return of the cheap oil era adversely affected the viability of costly oil production in non-OPEC

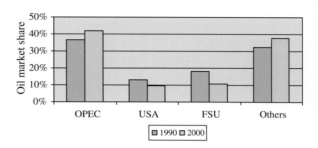

Fig. 14.9 OPEC market share in the 1990s. *Data source* BP Statistical Review of Energy Statistics, 2010

countries and other initiatives related to energy diversification through alternative energies. This situation created a long-term effect in the oil industry by depressing the investor interest and creating a permanent scar in the minds of the oilmen.

A few major events marked the international oil market situation during this period.

a. The wave of liberalization and market restructuring influenced the market-oriented operations in many economies and the oil sector also saw some deregulatory efforts.
b. The collapse of the Soviet Union affected the Russian oil industry greatly, resulting in a significant loss of output during the first half of the 1990s.
c. The Iraqi invasion of Kuwait in 1990 and the subsequent war saw an assault on the oil infrastructure that resulted in a significant loss of oil supply capacity. This event also implanted the seed of a greater international operation involving Iraq in the following decade. Oil price soared for a few weeks but other producers such as Saudi Arabia and Venezuela reacted quickly to make up for the supply shortfall.
d. The Asian economic crisis in 1997 and its contagion effect in the rest of the world severely dampened the growth of oil demand and oil prices return to its $10 per barrel level once again.

During this period, OPEC worked through quota adjustments and supply adjustments. This helped maintain a reasonable price level for oil (between $15 and $20 per barrel). Without such control, prices would have collapsed to its marginal cost level. The organization was effective is managing this phase of excess capacity and its market share improved although non-OPEC share continued to dominate the supply (see Fig. 14.9). Such a situation encouraged members to produce beyond their quotas and OPEC was not able to ensure strict adherence to its quota policy.

14.2.2.6 Phase 6: Return of High Prices

At the turn of the century, things started to look differently. Prices showed greater volatility and high prices were sustained over a number of years (see Fig. 14.10). Prices have been increasing since January 2002 after staying at very low levels

Fig. 14.10 Recent high oil prices. *Data source* BP Statistical Review of Energy Statistics, 2010

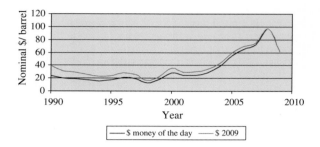

since 1997 when the financial crisis hit Asian economies. The oversupply of oil due to poor demand drove the prices down. But oil prices started to rise in 2000 and in October–November averaged three times the level that existed in February 1999. This level of price was highest in the previous 15 years, except the Gulf war period (IMF 2000). The prices fell sharply after the "9/11" event and started to pick up again in 2002.

A closer look at the recent time (see Fig. 14.11) indicates that the prices maintained a steady upward movement since early 2004 and by the end of September 2005 prices in nominal terms reached a monthly average of 60 US dollars per barrel. Since then, prices have risen to reach $145 a barrel in July 2008. This is a very high price by any standard and there is a sentiment that the changes may not be transitory in nature, implying that a part of such high prices may become a permanent feature of the global economy (ESMAP 2005).

The market has seen greater price volatility due to a number of factors:

- The demand has grown significantly from non-OECD countries and especially from China and other fast developing countries. For example, China's oil demand has almost doubled between 1999 and 2009, whereas India' demand has increased by 50% during the same period. Consequently, the share of developing countries in the world oil demand has increased from 33% in 2000 to 42% in 2009. This represents a rapid growth in the developing country share

Fig. 14.11 Daily price movements in recent times (nominal prices). *Data source* EIA website (spot prices)

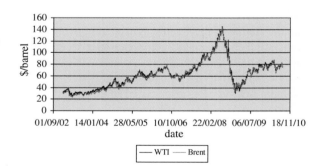

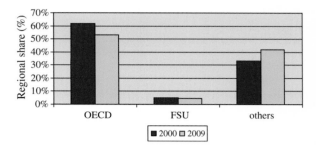

Fig. 14.12 Changes in regional oil demand mix in recent times. *Data source* BP Statistical Review of Energy Statistics, 2010

(see Fig. 14.12). This rapid growth was poorly anticipated by the market due to data issues.
- Simultaneously, a part of the production capacity became available due to natural calamity (tropical storms and hurricanes), industrial action by workers (in Venezuela for example) and political unrest in some producing areas (such as Iraq, Nigeria). Consequently, the available spare capacity reached low levels, which in turn fuelled concerns for supply security and supply disruptions. At certain times, there was hardly any spare capacity available, and spot market prices reacted to such situations, by adding a risk premium to oil price. The value of the risk premium is a matter of empirical analysis but experts believe this to lie between $5 and $15 per barrel of oil (IFP 2007).
- Speculation by traders is also considered to have played some role in this respect. A report by the US Senate (2006) suggested that billions of dollars worth of speculative investment in oil futures contracts is partly influencing crude oil price rise.

During the first part of this phase, when the spare capacity was limited, OPEC had hardly any instrument to regulate the market. The market was in "auto" mode during this period, when above normal prices have attracted investment in the sector, even in the costly conventional and non-conventional oil resources on the belief that the change is a permanent one, and the era of cheap oil is over. However, prices collapsed as the financial crisis in the Western banking and financial sector deepened after the collapse of one of the most prestigious Wall Street player, Lehman Brothers, when it filed for bankruptcy protection in September 2008. Subsequent run for cover by the banks on both sides of the Atlantic and the frenzy of the Central Banks and governments for injecting money into the panic-stricken financial sector caused concerns about the prospects of a deep recession in major economic powers. The speed with which events unfolded caught most analysts by surprise, although the problems of investing in toxic financial instruments have started to emerge since the sub-prime crisis hit the US economy in late 2007. Subsequently, the OECD economies entered into a deep recession and energy demand in general and oil demand in particular fell sharply. The spare capacity reappeared and OPEC once again had regained some market controlling power.

14.2.3 Commoditisation of Oil

Although spot transaction for oil is as old as the industry itself (Razavi 1989), the commmoditisation of oil started in a big way in the 1980s. According to Razavi (1989), by 1985 about 80–90% of internationally traded oil was spot-traded. This represented a new phenomenon in the oil market which was hitherto dominated by contract sales.

The spot market is an alternative mechanism where oil is exchanged on a day-to-day basis instead of exchanges through long-term contracts. The spot market developed due to a number of reasons, including among others, de-integration of petroleum industry in the 1970s that required a balancing mechanism, increase in production outside OPEC who needed an alternative transaction mechanism, diversification of sources of supply and consumption, and spot-related sale by OPEC members.

The spot market performs important functions: (a) it provides pricing information as it provides market clearing price of crude and petroleum products; (b) it is sharing or transferring risks as speculators take the risk; and (c) it provides an alternative channel of oil trade. Information is processed quickly and disseminated instantaneously. There are no institutional barriers to distribution of information.

The spot market is an informal worldwide network of contacts carrying out cargo-by-cargo sales and purchases of oil and products. The main participants are

a. Major oil companies who buy products from spot and acquire crude supplies on a spot basis. They are actively involved since 1979.
b. Independents played an important role as they depend on spot sales and they are often affected by the price fluctuations.
c. Traders take up positions (i.e. contracts to buy or sell real cargoes of oil and is responsible and liable for the cargo). They take risk of price fluctuations.
d. Brokers—they hold no title to the cargo but facilitate in discovering the needs or availability for a commission.

A complimentary development in this regard is the futures market. Futures market is one involving promises of sale and purchase of a petroleum product at a future date but for a price fixed immediately. The futures market serves two important purposes: it provides an organized forum for hedging the price risk and it helps in price discovery in the oil market. Two benchmark crude oil types dominate the futures—WTI (West Texas Intermediate) which is being traded on the New York Mercantile Exchange (NYMEX) since 1983 and Brent Blend on the International Petroleum Exchange in London since 1988.

The basic distinction between the spot and the futures market is that in the former real commodity transactions take place where actual goods are bought and sold whereas the latter deals with standardized contracts (or papers) with no immediate transfer of ownership of the commodity. These contracts can be exchanged and cancelled out prior to the delivery month and therefore they need not involve any real delivery. In general, only 2% of such contracts end up with a real delivery.

In recent times, the oil futures market has deepened. In addition to the traditional players such as oil producers and refineries, investment banks, pension and hedge funds have also entered the market. As a consequence, there is a growing disconnect between the physical and the futures market (IMF 2005). This increase in the speculative activity is believed to fuel price volatility in the spot market.

14.3 Analysis of Changes in the Oil Market

In this section, we analyse a few specific elements that influence the international oil market.

14.3.1 Evolution of Oil Reserves, Oil Production and Oil Consumption

The oil industry has been successful in improving its proven reserves over the past three decades. This has been possible through the use of better technologies for exploration and through exploratory activities (Fig. 14.13).[3] However, as is well known, more than two-thirds of these reserves are found in the Middle-East, whereas the remaining reserves are distributed in the rest of the world.

Within OPEC, five Middle Eastern members, namely Saudi Arabia, Iran, Iraq, Kuwait and United Arab Emirates accounted for about 70% of the reserves in 2009, with Saudi Arabia alone holding close to 26% of the OPEC reserves. This highly skewed distribution of oil reserves makes the industry heavily dependent on the Middle East for the security of supply. However, the oil operations in most of these members are carried out through the respective national oil companies. This restricts the access of other oil companies, especially the majors to prospective oil reserves. It is also worth noting that the OECD share of the reserve is fast declining: it had about 7% share of the global oil reserves in 2009, down from 16% in 1980.

However, the global oil production follows a different pattern (see Fig. 14.14), which results from the market regulation role played by OPEC. Consequently, non-OPEC production accounts for about 60% of global oil supply at the moment. Figure 14.14 also clearly indicates that the OECD makes an important contribution until now—about 22.5% of global oil production in 2009 came from this region despite its low reserve position. Evidently, the North American production has dominated the OECD output so far with a 75% share.

---

[3] Excludes non-conventional oil reserves.

14.3 Analysis of Changes in the Oil Market

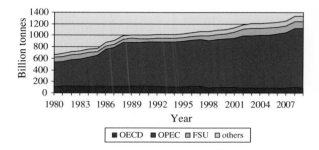

Fig. 14.13 Oil reserve distribution. *Data source* BP Statistical Review of Energy Statistics, 2010

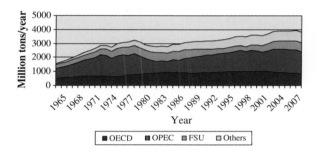

Fig. 14.14 Oil production history. *Data source* BP Statistical Review of Energy Statistics, 2010

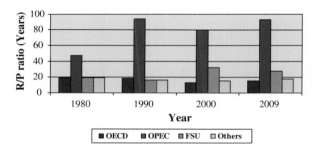

Fig. 14.15 Changes in the oil R/P ratio. *Data source* BP Statistical Review of Energy Statistics, 2010

But this practice of "sweating out" of the reserves in the OECD region leaves the countries vulnerable in terms of their ability to supply in the future. The reserve to production (R/P) ratio for oil is the lowest for the OECD region (see Fig. 14.15). The preference for short-term gains by private companies compared to the societal preference for long-term benefits drives such a development.

As indicated earlier, oil demand traditionally originated from the OECD countries (see Fig. 14.16). More than 70% of oil demand came from this region in 1965 but the share has been falling as the demand from developing countries started to pick up in the 1980s. Although OECD demand still accounts for more than 50% of global oil demand, the developing country share has reached above 40% in 2009.

Fig. 14.16 Oil demand trend. *Data source* BP Statistical Review of Energy Statistics, 2010

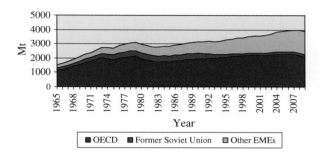

Fig. 14.17 Changes in oil trade movements

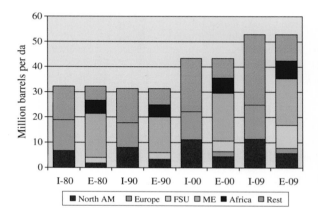

China has emerged as the second largest oil consuming country in the world after the USA, while India became the fourth largest oil consuming country in 2009. There are now indications that the oil demand in the industrialized world has past its peak and is in the decline phase. The average growth rate of demand between 2000 and 2009 was −0.7% in the OECD region, as opposed to a growth of 3.5% in the rest of the world (excluding the Former Soviet Union countries). China's oil demand has grown at an average rate of about 7% during this period, showing clear indications of a major shift in the centre of attention in terms of global oil requirements.

As a consequence of the regional demand–supply imbalances, the trade volume has been growing over time (See Fig. 14.17). While Europe and North America (mainly the USA) remain major importers, the growth in trade since 1990 is originating from the other areas (mainly Asia–Pacific). The level of oil import has more than doubled in this region between 1990 and 2009. In terms of sources of supply, the return of the Former Soviet Union supply to its normal level and a greater participation in international trade is clearly evident. The share of the Middle Eastern supply in the trade did not change significantly, which implies that a greater diversification of sources and trading partners has occurred over the past two decades.

14.3.2 Constrained Majors

The history of the 150 year-old oil industry has been dominated by the international oil companies (IOC). Only since the emergence of OPEC and the subsequent nationalization of the oil industry, the national oil companies (NOC) became relevant. Although NOC depended on various services by the IOC and still in many countries the co-operation continues, the canvass has changed quite dramatically. Over time, the role and the power of the NOC became more important and according to Jaffe and Soligo (2007), "14 out of 20 top upstream oil and gas companies in the world are national oil companies" in terms of reserves holdings. In the 1970s and 1980s, when OPEC was pursuing the policy of market control by restricting its output, IOC have invested heavily in non-OPEC countries and benefited from OPEC market regulation which reduced the market uncertainty.

However, as the old fields deplete and concerns for long-term supply security start to bite, IOC start to face stiff competition for acquiring non-OPEC opportunities. The aggressive expansion of China in acquiring overseas reserves, coupled with similar strategies by other developing countries has reduced access for IOC. Chinese success followed a different strategy where the oil company or the Chinese government entered into strategic alliances with the host government for wider economic development of the host country, thereby giving it a special advantage compared to IOC offers. At the same time, the high oil prices of the new millennium have also revitalized resource nationalism especially in Venezuela and Russia. Consequently, IOC while still healthy in financial terms, are finding it difficult to replace their reserve to ensure future sustainability (see Fig. 14.18).

Moreover, the production is already declining in the case of a number of majors and their R/P ratio in recent times is precariously low—ranging between 6 and 13 years (see Fig. 14.19). Contrast this with the average R/P ratio of OPEC of above 90 years in 2009—clearly, the effect of limited access to reserves on the Majors' activities becomes evident.

Oil Major's low R/P ratio along with the precarious R/P ratio of OECD countries clearly justifies their concern for long-term oil supply security. No wonder that the debate about peak oil has a strong developed country bias—and

Fig. 14.18 Declining Major's reserve *Source* OPEC Annual Statistical Bulletin 2003, 2008 and 2009

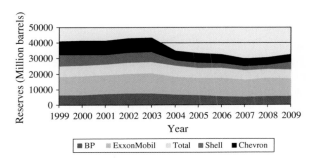

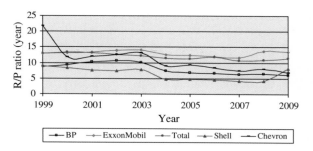

Fig. 14.19 R/P ratio for oil of selected Major oil companies. *Source* OPEC Annual Statistical Bulletin 2003, 2008 and 2009

support of the Majors. After all, the future of major oil companies looks very uncertain if their exclusion from the oil-rich region continues.

14.3.3 Analysis of the OPEC Behaviour

There is a vast literature analyzing the OPEC behaviour and strategies (see for example Slant 1976; Percebois 1989; Greene 1991; MacAvoy 1982; Griffin 1985; Dahl and Yucel 1991, Alhajji and Huettner 2000a, b and Ramcharran 2001). As usual in such an area, there is no consensus about how best OPEC can be described. This difficulty arises because OPEC has followed different strategies at different times to determine prices and production levels (Fattouh 2007). In this section, a simple, diagrammatic presentation of the models analyzing OPEC behaviour is presented.

The models on OPEC behaviour can be categorized into broad groups of models: (a) cartel models such as the dominant firm model or (b) non-cartel models such as target revenue model, and the competitive model. Only a few models were statistically tested and results have been contested by others due to model weaknesses.

14.3.3.1 Cartel Model

A cartel occurs when a group of firms or organizations enter into an agreement to control the market by fixing price and/or limiting supply through production quotas. A cartel may work in a number of ways: as if there is a single monopoly producer, or with market-sharing agreements. The objective is to reduce competition and thereby generate higher profits for the group. In the absence of any agreement, the competitive market conditions will prevail and the price p_c and quantity q_c will be obtained in Fig. 14.20. However, if the producers enter into an agreement to enforce a monopoly price (p_m) in the market, they will have to agree to reduce supply to q_m in such a way that the marginal revenue equals the marginal cost. Each member of the cartel then receives a higher price for the output but any

14.3 Analysis of Changes in the Oil Market

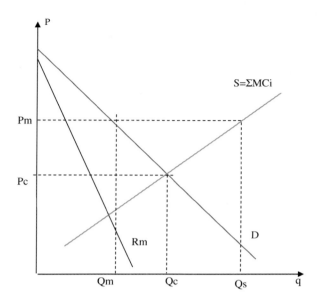

Fig. 14.20 Cartel as a monopolist

producer will be interested to participate only if it can extract more benefits compared to a competitive environment. As long as this condition is satisfied, members will be happy to support the collusive behaviour.

But, each member would have the tendency to increase its output based on its marginal cost of supply so that its individual profit is maximized. This tendency to cheat will lead to an overproduction (q_s) and the market will see a return of the competitive market price. This represents a natural threat for internal cohesion of any cartel.

Any cartel thus faces a number of problems: in most jurisdictions it is illegal for firms to enter into such a collusive behaviour. OPEC as a group of sovereign nations escapes from this argument. The tendency to cheating by members for individual gains by undermining the collective position is another major threat. Finally, in order to control the market, the group must have accurate information about the shape of the demand and supply curves, elasticity of demand, and actual production by members. Often this information is not readily available although some generic idea may be available.

Stability of any cartel then depends on a number of factors:

a. Group size: A small group is better placed to have a tighter control than a large group.
b. Group characteristics: Homogeneous group members acting on a product with a captive demand or inelastic demand is more likely to succeed than a heterogeneous group.
c. Dispersed, large number of buyers: Widely dispersed consumers will have little chance of colluding with each other. This is an essential requirement for a cartel.

d. Member gains: Each member of the cartel must benefit from the action—otherwise, there is no incentive to join the group.
e. Group discipline: A group that is committed to play by the rules of the cartel is also an important condition.
f. Policing: Any cartel being vulnerable to cheating would need an effective policing mechanism to detect cheating.

Clearly, these requirements are difficult to satisfy in reality.

14.3.3.2 Cartel with a Leader (Dominant Firm Model)

Because of the inherent issue of cohesion, a cartel needs to ensure that the group is able to maintain the market power even if some members are cheating. A cartel with a leader is such a cartel where one of the members can regulate his behavior to maintain the group coherence and can make the group agree to its proposals, thereby protecting leader's interest. A leader should have an important market share, high flexibility in capacity utilization, low financing requirement, and be less sensitive to changes in energy markets.

Consider that the leader knows the market demand D_T well and understands the supply curve of other cartel members (S_O). Based on these, the leader determines its own demand curve D_L such that $D_L = D_T - S_o$. Note that the leader will have no demand when the price rises to $p1$ and it faces the entire demand when the price falls to $p2$ level. The leader then decides its output to maximize its profit and would impose the price p_L (Fig. 14.21) on the cartel. Rest of the members produce $q_T - q_L$. The leader is the price maker and the rest are price takers. The elasticity of

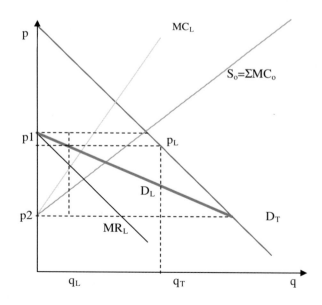

Fig. 14.21 Cartel with a leader

14.3 Analysis of Changes in the Oil Market

demand facing the leader is a key determinant of price. Saudi Arabia is such a leader in OPEC.

14.3.3.3 Limit Pricing Model

Limit pricing model examines the effect of changes in demand for cartel. Competition can arise from non-cartel producers as well as from other fuels. Producers outside the cartel affect the demand and supply.

Consider that So represents the supply curve of other suppliers while MCc presents the cartel supply. At price p1, the producers outside the cartel are not able to produce; cartel faces the total demand. At price $p2$, they would be able to supply the entire demand, the cartel faces no demand. The cartel demand curve is obtained from these facts (see Fig. 14.22). The cartel then decides its price-output combination to achieve its profit maximization objective—that is by equating the marginal revenue to marginal costs. The output of cartel would be q_c at price p_e. Producers outside cartel would produce q_0 such that $q_t = q_0 + q_c$.

The above model can be used to analyse cartel strategies. Two general strategies have been considered: an offensive strategy where the cartel declares the price war and another defensive strategy where the cartel conserves its resources leaving non-cartel producers freedom and space to operate in the market.

In the price war strategy, the cartel will try to drive the competitors out of the market. As the cartel benefits from cost advantage, it can forgo the market control strategy and let price drop down to the competitive level. At this point, costly producers who were benefiting from the price protection offered by the cartel will

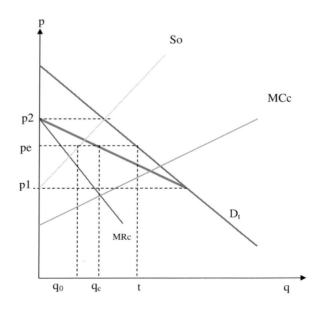

Fig. 14.22 Limit pricing model

Fig. 14.23 Price war

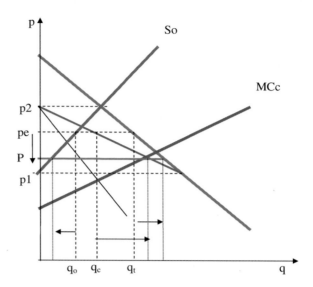

become non-competitive and will be displaced by cartel output. Hence, the cartel will see its market share increase but the price will reach the competitive market levels (see Fig. 14.23). The overall market supply will increase as well.

The cartel can adopt a defensive strategy when it faces competition from other substitutes that threaten the demand of the commodity under cartel control. Generally, such substitutes are viable when the price reaches a certain level where it becomes profitable for alternatives to appear. In such a case the cartel can decide to set the price below this threshold level where the profit for the cartel may not be maximized but it prevents entry of new substitutes. A related strategy would be to continue with the equilibrium price p_e indicated above but use the threat of price war as an effective deterrent. This avoids profit loss for the cartel but keeps substitutes at bay. OPEC has used both the strategies to ensure its control over the oil market.

14.3.3.4 Target Revenue Model

The Target Revenue Theory was developed by Ezzati (1976). According to Ezzati (1976), OPEC production decisions are made with reference to national budget requirements. The budgetary need of OPEC members is a function of economies capacity to absorb productive investments. Investment projects can be ranked in order of decreasing rate of return and projects with return above market rate of return would be undertaken. The investment requirement can thus be determined. Isorevenue schedules can be drawn for different rates of oil and quantities of oil producing export revenue equal to investment requirements. If the share allotted to the country is not enough to meet the investment demand, the country would cheat or seek an increase in share. If share is more than that required to meet investment demand, the country may voluntarily reduce output. Only members who are marginal in oil resources would have tendency to cheat. Rich members may not prefer to leave oil to ground as the return may not be remunerative. Small producers may like to defer production. The behaviour of OPEC production in the mid-1970s was consistent with the Target Revenue Model (Ramcharran 2001).

A significant amount of research work has since been done to verify, extend and refine the initial theory. For example, Cremer and Salehi-Isfahani (1980) retained the target revenue hypothesis but modeled OPEC in a competitive framework. Teece (1982) also analysed the OPEC behaviour using the competitive framework with a target revenue constraint. But his analysis differed from Cremer and Salehi-Isfahani (1980) in a number of respects. Teece (1982) considered investment and expenditures as fixed whereas Cremer and Salehi-Isfahani (1980) considered them as endogenous variables. Griffin (1985) used quarterly data for the period between 1973 and 1983 and adopted an ordinary regression model to test the target revenue hypothesis but did not find strong support to the idea. Dahl and Yucel (1991) tested two variants of the competitive model and found no support for the competitive hypothesis. Alhajji and Huettner (2000a) modified Griffin's model by using static and dynamic econometric models but the static models did not give good estimates due to auto-correlation problems. Even the dynamic models did not find support to the target revenue hypothesis. Using a longer set of data Ramcharran (2001) examined the production behaviour of OPEC and non-OPEC countries and found some support for the target revenue hypothesis.

The above shows the differing views on the subject. The results often reflected the choice of the model, data set used and the econometric method used. Empirical evidence did not provide any conclusive outcome on the issue.

Irrespective of the approach used in analyzing the OPEC behaviour, it is important to note that the group represents the interests of major oil producers. Given the size of their reserves, they will remain an important player in the oil market and the group cannot remain idle to any challenge that tries to destroy the captive oil demand arising from the transport sector.

14.3.4 A Simple Analytical Framework of Oil Pricing

To end this chapter, a simple diagrammatic framework is presented based on Stevens (1995), (1996) that combines the supply and demand curves of oil and can be used to explain the oil price movements. The framework is captured in Fig. 14.24.

The demand curve has three segments—highly inelastic for a wide range of prices, with some elastic segment at very low and very high prices (shown as D in the figure). This arises due to capital intensive nature of the appliances used for consuming oil. In the short-run, only some adjustments in the capacity utilization

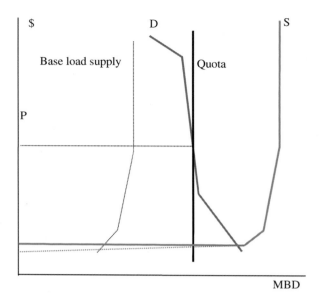

Fig. 14.24 A diagrammatic framework for oil prices

of the appliance is possible, making the demand inelastic over a certain price range. At very high prices, substitutes will appear and make demand elastic. Similarly, at very low prices oil would replace other fuels and therefore would have a greater elasticity of demand.

Similarly, the supply curve has a low cost segment, followed by an increasing cost segment. The horizontal segment represents the low marginal cost of oil supply. This is assumed to be same throughout the world but the argument does not change even if an increasing marginal cost argument is used (as shown by the dotted line). The vertical segment of the supply curve represents the change in the marginal cost due to the fixed capacity (or the capacity constraint) at any given time.

Two groups of suppliers are considered—base load and residual suppliers. The base load suppliers are price takers and supply to capacity for a given price. The residual suppliers are price makers and try to regulate the price by controlling their output. If the price regulation is not used, the supply and demand in the market will decide the price and the market clearing price will be the marginal cost-based. For a target price, the residual producers are then striving to set a quota that yields the desired result. However, this act requires accurate information about the demand and supply. If the information is imperfect, "between a wide range, any price could be regarded as an equilibrium price which 'cleared the market'" (Stevens 1996).

This simple demand–supply based framework can be used to explain the price movements in the oil market. For example, in 2008 when the demand moved outwards, the supply became capacity constrained. Consequently, the prices reached very high levels but the economic consequences of such high prices resulted in demand destruction and the demand curve moved inward to result in a sharp price drop.

14.4 Conclusion

This chapter has presented a brief history of the oil market and its development through different phases under the international companies and the control of OPEC. It has also presented the current supply and demand positions and considered the changed taking place in the market. The economic logic behind the developments is also explained using a simple diagrammatic approach. These help explain the strategies of the actors and help understand the oil market growth and issues related to it.

References

Alhajji AF, Huettner D (2000a) The target revenue model and the international oil market: empirical evidence from 1971 to 1994. Energy J 21(2):121–144
Alhajji AF, Huettner D (2000b) OPEC and world crude oil markets from 1973 to 1994: cartel, oligopoly, or competitive? Energy J 21(3):31–60
Cremer J, Salehi-Isfahani D (1980) A competitive theory of the oil market. What does OPEC really do? Working Paper No. 80–84, University of Pennsylvania
Dahl C, Yucel M (1991) Testing alternative hypothesis of oil production behaviour. Energy J 12(4):117–138
ESMAP 2005 The impact of higher oil prices on low income countries and on the poor, Energy Sector Management Assistance Programme, World Bank, Washington DC
Ezzati A (1976) Future OPEC price and production strategies as affected by its capacity to absorb oil revenues. Eur Econ Rev 8:107–138
Fattouh B (2007) The drivers of oil prices: the Usefulness and limitations of non-structural models, the demand and supply framework and informal approaches. Oxford Institute for Energy Studies, WPM 32
Greene DL (1991) A note on OPEC market power and oil prices. Energy Econ 13(2):123–129
Griffin JM (1985) OPEC behaviour: a test of alternative hypothesis. Am Econ Rev 75(5):954–963
IFP, 2007 Oil and gas exploration and production: reserves, costs, contracts. Editions Technip, Paris
IMF 2000 The impact of higher oil prices on the global economy. International Monetary Fund, Washington, DC
IMF 2005 The structure of the oil market and causes of high prices. International Monetary Fund, Washington, DC (see http://www.imf.org/external/np/pp/eng/2005/092105o.htm)
Jaffe AM, Soligo R (2007) The international oil companies, The changing role of national oil companies in the international energy markets, The Baker Institute, Rice University, Houston (See http://www.rice.edu/energy/publications/docs/NOCs/Papers/NOC_IOCs_Jaffe-Soligo.pdf)
MacAvoy P (1982) Crude oil prices as determined by OPEC and market fundamentals. Ballinger Publishing Company, Cambridge
Percebois J (1989) Economie de l'Energie. Economica, Paris
Ramcharran H (2001) OPEC production under fluctuating oil prices: further test of the target revenue theory. Energy Econ 23:667–681
Razavi H (1989) The new era of petroleum trading: spot oil, spot-related contracts and futures markets, World Bank Technical Paper 96, The World Bank, Washington DC
Slant S (1976) Exhaustible resources and industrial structure: a Nash Cournot approach to the world oil market. J Political Economy 84(5):1079–1093

Stevens P (1995) The determination of oil prices 1945–1995: a diagrammatic interpretation. Energy Policy 23(10):861–870

Stevens P (1996) Oil prices: the start of an era? Energy Policy 24(5):391–402

Teece DJ (1982) OPEC behaviour: an alternative view. In: Griffin JM, Teece DJ (eds) OPEC behaviour and world oil prices George Allen, Unwin (Publishers) Ltd, UK pp 64–93

US Senate (2006) The role of market speculation in rising oil and gas prices: A need to put the cop back on the beat, Staff Report, prepared by the Permanent Sub-Committee on Investigations of the Committee on Homeland Security and Governmental Affairs, United States Senate, June 2006, Washington, DC (See http://hsgac.senate.gov/public/_files/SenatePrint10965MarketSpecReportFINAL.pdf)

Chapter 15
Markets for Natural Gas

15.1 Introduction

Although natural gas been around for a long time,[1] it has come to prominence over the past 30 years and the industry has seen a rapid growth in the late 1980s—early 1990s when the environmental concerns emerged as a major global issue. In the new millennium when the energy supply security concerns emerged, the attention on natural gas increased due to its better distribution of reserves. By 2009, natural gas accounted for about 24% of the global primary energy demand on average. The industry has seen major developments during the course of its history in terms of market structure, technological changes, as well as economic and political dimensions. Although oil and gas industries are often considered in a similar manner, there are significant differences and therefore, it is important to study these industries separately. This chapter aims to provide a general understanding of the natural gas market from an economic perspective.

The chapter is organised as follows: first we present the advantages of natural gas and review the specific characteristics of natural gas and its distinctive properties compared to oil. Then a status report of the development of the industry in terms reserves, demand and supply is presented. The economics of piped gas and LNG is then considered. Finally, the issues related to pricing, market development and internationalisation of the gas industry is discussed.

---

[1] The Chinese are believed to have used natural gas around 500 BC for salt water desalination. Natural gas seepages were discovered in the United States in the seventeenth century and the first gas well was dug in 1821.

15.2 Specific Features of Natural Gas

15.2.1 Advantage Natural Gas

Natural gas is a different hydrocarbon—not a by-product of oil. Its resource base is, therefore, not linked to that of oil. The geological conditions for gas are much less severe than those for hosting oil. Oil in liquid form can be found up to a certain depth and at higher depths oil dissociates into gas. But for gas there is no such depth limits, which implies that gas can be found in conditions different from that of oil. Therefore, gas is more abundant and widespread (Rogner 1989).

Natural gas is a more diversified resource than oil in terms of availability. Although Middle East and FSU countries together account for about 70% of the gas reserves, gas has been found in all continents and new findings are reported even from Europe and elsewhere. The diversified availability of gas ensures better security of supply. Moreover, since the second oil price shock, an unexpected growth in oil reserves was noticed. But for each major oil discovery, at least twice that amount of gas was discovered using the standard technology. As advanced technologies develop for exploration of gas at greater depths, even better results could be expected. Gas is now available at places which were earlier considered as unlikely locations for gas traps. This confirms that gas may be available more widely and more abundantly.

Gas is environment friendly compared to other fossil fuels. For example, natural gas emits 56.1 tCO_2 per TJ while the emissions from oil and coal are 73.3 tCO_2 and 94.6 tCO_2, respectively (BP Statistical Review of World Energy 2010). Therefore, natural gas emits 30% less CO_2 compared to oil and almost 70% less compared to coal for an equivalent amount of energy. As the reduction of greenhouse gas emissions is expected to gain momentum in the future, natural gas will surely strengthen its position.

15.2.2 Gas Supply Chain

The gas industry consists of a number of technically demanding and capital intensive functional activities. These include production, gathering, storage, pipeline transportation, distribution and supply to end-users (Teece 1996). Production of natural gas involves a set of operations such as exploration, drilling and production which are required to deliver the gas at the wellhead. Producers incur substantial start-up costs, most of which is fixed and often risky, before they can start producing gas. This is because of the technical nature of exploration as well as risk involved in the process. Normally, bigger fields are found first and the success rate declines as the area is intensively explored. Similarly, development of fields and production are also costly activities. Their cost depends on the number of wells to be developed, well locations, reservoir condition and the surface infrastructure required (Julius and Mashayekhi 1990). As a result, normally producing companies tend to be large, although they may be small compared to the market size.

Transportation facility establishes the link between the producer and the city gate using high pressure pipelines (similar to transmission network in electricity). These facilities form an essential and unique part of the industry and have high asset specificity (i.e. no or limited alternative use). Investment in transport facility tends to be lumpy and huge. The investment cost depends on the size (diameter) of the pipeline, which in turn depends on the length and peak demand. Consequently, the average cost of transportation tends to fall over a large range of output, indicating that the transmission system has the characteristics of a natural monopoly. This implies that the market may be better served by a single pipeline company. However, once a pipeline is developed, increases in demand may allow the possibility of other pipelines to emerge. Thus in sufficiently large markets (as in the US or Germany) it is quite common to find many transmission pipelines (IEA 2000).

Distribution involves delivery of gas from the city gate to end consumers using low pressure pipelines (similar to distribution networks in electricity). Investment in distribution is normally based on the peak demand in the system and depending on the size and pattern of customer demand the cost would vary (Julius and Mashayekhi 1990). It is normally considered that distribution systems offer both scale and scope economies and hence have natural monopoly characteristics.

Supply (which can be at the wholesale or retail level) is a trading activity. Traders and suppliers need little up-front investment and hence the size of a trader tends to be much smaller than the market size. This limited economy of scale makes this segment potentially competitive (Juris 1998a).

Continuity and integrity of supply security are main concerns in the gas industry. Disrupting supply even briefly and re-establishing the supply can lead to gas leaks, fire and explosion (Newbery 1999). Like electricity, gas demand also varies daily and seasonally. Newbery (1999, p. 353) indicates that the daily peak demand in Britain is typically 1.5 times the average demand and 4 times the minimum demand. The demand variation can, however be handled through storage or by varying the extraction rate from the wells. As a result, natural gas industry, like electricity, requires close co-ordination and co-operation of system activities to ensure smooth operation of the pipeline system. These include pressure control, load balancing, gas rerouting during line work, storage and gas mix (Teece 1996). Generally a system operator performs the scheduling and central dispatching functions. Importance of this function depends on the number of pipeline users and the industry structure. A disintegrated industry requires more co-ordination while more consumers require better co-ordination. A set of rules for system operation and balancing thus becomes necessary (Juris 1998b).

15.2.3 Specific Features

The physical attributes of natural gas adds to its specificity. First, its gaseous state makes transportation difficult and costly. It also makes the fuel more risk prone to accidents. Second, its heat content per unit of volume is another source of problem. Oil will have 7,000–8,000 times higher heat content than natural gas in a cubic metre, thus making gas very weak in terms of energy density. Moreover, unlike oil, gas often finds little use where it is produced. In order to use gas, often physical links have to be created between the producer and the consumer by developing necessary transportation infrastructure. This in turn requires huge capital investments in inter-related facilities (Julius and Mashayekhi 1990). As a result, natural gas industry, unlike oil industry, is not international in character. The markets have developed in different regions where adequate supply and demand could be met and where adequate infrastructure is available for supply.

It is also a non-renewable resource, implying that gas not produced today can be used tomorrow. This adds a dimension of scarcity rent. Similarly, gas has to compete with other fuels at the end-use level, which essentially determines the value it can fetch. Depending on this value, gas production may yield a resource rent as well (Newbery 1999, p. 344). As the investments in transportation and distribution networks are high, inflexible, durable and sunk, investors seek for long-term commitments from consumers. Thus, pipelines need a long-term commitment from the distribution network to buy gas while the distribution networks need a similar commitment from the end consumers. Similarly producers are not willing to invest in production in the absence of secure long-term contracts (Newbery 1999, Viscusi et al. 2005). All these features tend to restrict gas industry to high value users, at least at the initial stage of gas industry development, to ensure finances for investment in infrastructure and production facilities.

Additional specific features for gas that make it different from oil include (Davidson et al. 1988):

(a) Markets for gas are not often ready and need to be developed. Due to specific characteristics of gas, large consumers are required to develop viable projects. Because of asset specificity and nature of gas, firm commitments are required between sellers and buyers.
(b) Gas transportation is costly and requires a physical connection between supply and demand centres. Infrastructure development is an additional task. This makes global gas market difficult.
(c) Market structure: As the industry requires costly bottleneck infrastructure, there is a greater tendency to integrate supply and demand sides, thereby creating a more integrated structure at least at the initial stage of the market development.
(d) Project development: Unlike oil, gas development is difficult to phase out. This requires higher initial outlay than oil, which also acts as a barrier.
(e) Attitude problem: companies consider gas as a problem because of risk, lack of market, lack of commensurate rewards, etc.

The specific characteristics of gas impose some pre-requisites for the gas market development and include (Percebois 1986):

(1) A critical mass of potential users: as a physical link is required to connect gas fields to the users, unless there is sufficient potential demand for gas in an area, the investment is difficult to justify. This feature makes gas penetration easier in developed countries compared to developing countries.
(2) A technical solution to non-correspondence of demand and supply centres. This normally takes two forms: pipeline transports or liquefying gas and then transporting it. However, a network is required to supply gas to the final consumers, even in this case.
(3) A competitive end-user level price: natural gas lacks any captive market, although it can compete with a number of fuels (coal, heavy fuel oil, and electricity). This lack of indispensability requires gas to rely on economic virtues of relative prices for market penetration. Environmental considerations have helped natural gas to expand its market but otherwise, gas market was developed where it could compete effectively with other fuels.
(4) A remunerative wellhead price: competition at the end-user level effectively decides the viability of gas production. Unless sale of gas fetches remunerative prices to the producer, no gas will be produced. Thus fields which could profitably be developed at any given market price may be limited in number, leaving other fields unexploited.

15.3 Status of the Natural Gas Market

This section presents the dynamism of the gas industry by considering the reserves, production, consumption and trade.

15.3.1 Reserves

Natural gas reserves have more than doubled between 1980 and 2009 (see Fig. 15.1). The proven reserves in 1980 stood at 81 TCM while in 2009, it reached almost 188 TCM. About 70% of the reserves are located in the Middle East and in the Former Soviet Union states. The Middle Eastern reserves have grown faster than other regions during the above period: in 1980, the region had about 25 TCM of gas reserves but in 2009, this has increased to more than 76 TCM. But the reserves of the rest of the world have also seen such a tremendous growth during the period—thereby improving gas resource distribution around the world. Note that OECD countries hold only less than one tenth of the global gas reserves.

Three countries, namely the Russian Federation, Iran and Qatar, held more than 50% of the global natural gas reserves in 2009. But there are 18 other countries in

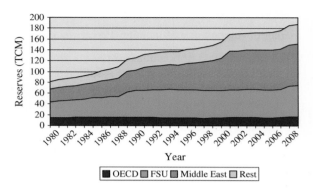

Fig. 15.1 Proven reserves of natural gas. *Data source* BP statistical review of world energy 2010

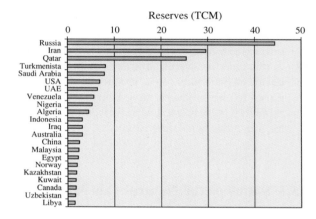

Fig. 15.2 Countries with at least 1.5 TCM gas reserve in 2009. *Data source* BP statistical review of world energy 2010

the world with at least 1% of the global gas reserves or about 1.8 TCM of gas reserves. These countries are found in all continents, which make gas reserves more widely distributed than oil reserves (see Fig. 15.2).

15.3.2 Production

However, as common with oil, the regional distribution of natural gas production is skewed towards countries with limited gas resources (see Fig. 15.3). OECD countries despite holding less than 10% of the global gas reserves have produced 38% of global gas output in 2009. As the figure indicates, the share was as high as 75% in 1970, but the share has declined as production from other countries has risen. But most of the growth came from outside the main gas reserve holding areas. Even in 2009, just 37% of the global production came from the Former Soviet Union countries and the Middle East. This shows the preference for oil in these countries and the market specificities of natural gas that constrain the market

15.3 Status of the Natural Gas Market

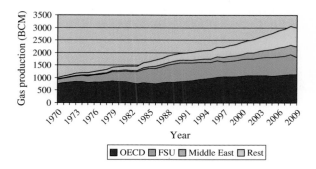

Fig. 15.3 Natural gas production trend. *Data source* BP statistical review of world energy 2010

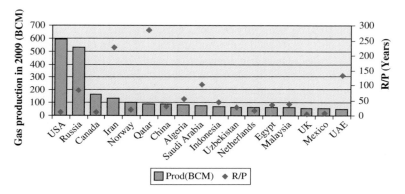

Fig. 15.4 Production and R/P ratio of major gas producers in 2009. *Data source* BP statistical review of world energy 2010

development. However, natural gas production grew at an average rate of 2.8% per year between 1970 and 2009, which is higher than the growth rates for other fossil fuels, implying the rising importance of natural gas in the global energy scene.

A careful look at the production data indicates that five countries (USA, Russia, Norway, Iran and Canada) produced about one half of the global production and the rest came from a large number of countries. However, some of the large gas producers do not have high reserve endowments and are operating with a low reserve to production ratio (see Fig. 15.4).

This clearly indicates that countries with limited reserves are generally exploiting their resources more intensively than those with large reserves. Accordingly, the reserve to production ratio for natural gas varies widely: from about 5 years for the United Kingdom in 2009 to close to 300 years for Qatar. On a global average, at the present level of gas consumption, the available reserves will last for about 63 years.

Figure 15.4 also shows that there are 12 other producers with outputs between 50 and 100 BCM per year. Together they produce about 28% of the global production. Thus, 17 major producers of gas account for about 80% of the global gas production.

15.3.3 Consumption

The demand for natural gas has a strong developed country bias (see Fig. 15.5). The OECD countries accounted for 75% of the global gas demand in 1965 and even in 2009 their share was about 49%. These countries represent the most important market for natural gas.

Natural gas has always played an important role in the Former Soviet Union countries. This region now accounts for about 20% of the global gas demand. But there has been a spectacular growth in gas demand in the developing world since late 1980s and the phenomenal growth in demand (at above 5% per year on average between 1990 and 2009) has catapulted this region to the second most important demand centre globally.

However, the striking similarity between Figs. 15.4 and 15.5 also suggests that gas is consumed close to its production source (Fig. 15.6). In fact, out of five major producers, only Norway has a low domestic demand. The US domestic gas supply is insufficient to meet its own needs while Russia consumes about two-thirds of its production domestically. Iranian and Saudi Arabian gas feeds into their local markets, mainly due to political or policy constraints, while Canada consumes about 60% of its local output. Similarly, the output of China, UK, Mexico and the UAE is generally inadequate to meet their domestic needs, with UK leading this group with a demand 40% higher than its local production in 2009. This leaves a handful of major gas producers with exportable gas—notably Qatar and Algeria.

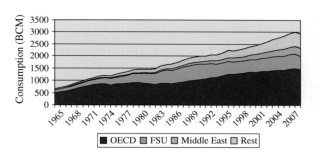

Fig. 15.5 Evolution of natural gas demand. *Data source* BP statistical review of world energy 2010

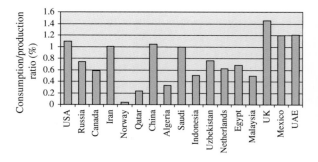

Fig. 15.6 Producers as consumers of gas. *Data source* BP statistical review of world energy 2010

15.3 Status of the Natural Gas Market

In terms of regional supply–demand disparities, it becomes apparent from Fig. 15.7 that Europe and Asia-Pacific represent two major regions where demand exceeds the regional production. Out of these two, Europe (or more specifically industrialised Europe) is the most important demand centre. As a consequence of this imbalance and due to geographical proximity of some producers to these distinct demand centres as well as perceived opportunities for selling surplus gas, three regional gas markets, namely the North American market, the European market and the Asia-Pacific market, have emerged. However, each market has its own characteristics, which is captured in the trade sub-section below.

In terms of use of gas, the electricity generation has emerged as the major use in all regions (see Fig. 15.8). However, the Pacific region (Japan) is the leader in this area with a share of 55% of its gas supply being used in the power sector, whereas the share is close to 30% in North America and in Europe. The emergence of combined cycle gas turbines in the 1980s and the deregulation of the electricity market in the 1990s have brought a sea change in the technology preference for electricity generation. The climate change agenda has also helped this transformation and natural gas has become the preferred fuel for electricity generation worldwide. Its environmental appeal, lower capital cost, shorter gestation period, higher efficiency and the modular technology challenged 'the bigger the beautiful' notion of the past (Thanawat and Bhattacharyya 2007). This trend has started with

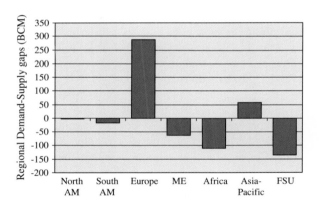

Fig. 15.7 Regional demand–supply imbalances in 2009. *Data source* BP statistical review of world energy 2010

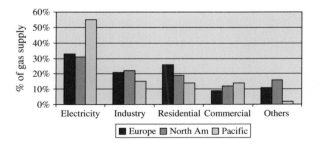

Fig. 15.8 End-use of gas in 2007. *Data source* IEA energy statistics

the 'dash for gas' in the English and Wales system after liberalization of the electricity market and has been followed in many other jurisdictions.

Industry and the residential sectors are other major consumers with around 20% share of gas supply in each sector. The rest is used in other activities such as in the commercial sector, in non-energy uses (petrochemical or fertilizer industries) and in agriculture.

15.3.4 Gas Trade

As a result of supply–demand imbalances, trade in natural gas developed. But because of the specific features of the commodity, the traded volume of gas was limited to 30% of gas consumption in 2009. Two modes of gas transportation are commonly used—pipelines and transportation upon liquefaction as liquefied natural gas (LNG). The overall volume of gas trade in 2009 was 877 BCM, of which piped gas trade accounted for 72%. The regional distribution of gas trade shows some particularities of regions (see Fig. 15.9). Europe had a 45% share in the gas trade in 2009, followed by Asia-Pacific and North America. These three regions account for more than 80% of global gas trade. LNG is the dominant mode of supply in the Asia Pacific where 89% of the import takes place in the form of LNG but in the rest of the world, piped gas transport is the common mode of supply.

Both these modes of gas trade are discussed below in the context of three main consuming regions.

15.3.4.1 North American Market

As indicated above, this region is a major producer and consumer of natural gas. Canada, Mexico and the USA produce significant quantities of gas and except for Canada, the other two countries have demands greater than their domestic supplies. Consequently, gas trade is within the region and outside the region has

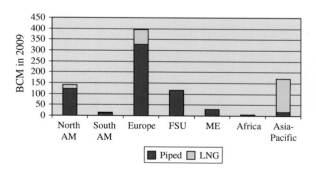

Fig. 15.9 Regional distribution of gas trade in 2009. *Data source* BP statistical review of world energy 2010

developed. Yet, trade volume is much lower compared to oil and the share of gas traded in the region is only 17% of total gas used in the region.

Because of geographical advantage and the historical nature of gas market development, the pipeline mode of trade of transport is well developed in this market. This is due to geographical reasons of proximity of producers to the demand centres and also due to the historical development of the gas market in the USA where cross-country pipelines played an important role in connecting the suppliers and consumers. Consequently, gas trade through pipelines dominates the market (see Fig. 15.10), accounting for about 88% of the overall gas trade in the region. USA and Canada are main trading partners—Canada exporting about 92 BCM of gas in 2009.

The LNG trade on the other hand plays a minor role at present and mostly restricted to the US market. The source of LNG supply is quite diversified but Trinidad and Tobago is the dominant supplier (see Fig. 15.11) with a 44% market share in the region.

Egypt and Nigeria are two other main players—with almost 29 and 18% shares, respectively. Three LNG suppliers therefore account for 90% of the LNG market of the region.

The recent developments of shale gas in North America have brought some changes in the business. The availability of domestic gas nearer to the demand centres has reduced the demand for imported gas and is likely to affect the market dynamics substantially.

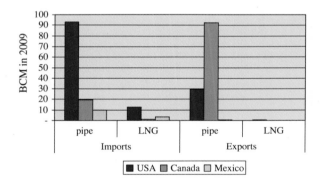

Fig. 15.10 Gas trade in the North American market. *Data source* BP statistical review of world energy 2010

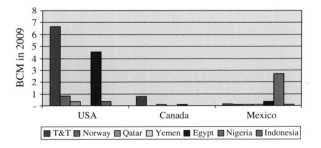

Fig. 15.11 LNG suppliers to North America in 2009. *Data source* BP statistical review of world energy 2010

15.3.4.2 European Market

As indicated before (Fig. 15.7), the European market is the most important market for gas trade. Although most of the European countries participate in gas trade, the market is dominated by a few major players (see Fig. 15.12): Germany had a 22.5% market share in 2009, followed by Italy (17.5%) and France (12.5%). Three other countries, namely Spain, UK and Turkey, had about 10% market share each. These six countries accounted for more than 80% of the gas trade in the region. In all countries except Spain, piped gas was the dominant mode of supply but LNG had a market share of 17% in this region, which was better than that of North America.

In terms of sources of supply, the market is dominated by two major suppliers, namely Russia and Norway (see Fig. 15.13). Russia supplied 26% of imported gas while Norway provided another 25%. Algeria and the Netherlands supply for another 13% each. These four countries thus accounted for more than 75% of the European imported gas in 2009. Most of these supplies come through pipelines—only supplies from Trinidad and Tobago, Nigeria and Libya were in LNG form while the Algerian supply comes in both forms.

Although almost one half of the gas supply in Europe comes from other European countries (such as Norway or the Netherlands), the maturity of European supply and the growing demand in the region is making Europe heavily dependent on foreign gas supply. As most of the gas travels long distances through pipelines,

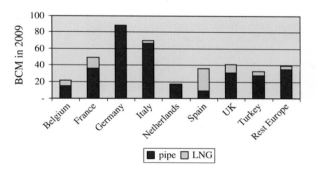

Fig. 15.12 Major gas importers in Europe in 2009. *Data source* BP statistical review of world energy 2010

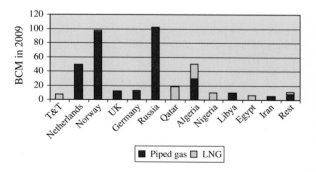

Fig. 15.13 Major gas suppliers to Europe. *Data source* BP statistical review of world energy 2010

the region faces the risk of supply disruption due to periodic episodes of disputes between gas suppliers and the transit countries as well as possible security threats to the gas transport infrastructure. This has emerged as an important issue for the region.

15.3.4.3 Asia-Pacific Market

The Asia-Pacific market is the second largest gas market in the world and is the only market where LNG dominates (see Fig. 15.14). However, this market is dominated by Japan which had a 50% market share in 2009 but it is the oldest LNG market in the world. South Korea is the other major player in this market with a share of 20% in 2009. The other emerging players in this market are India, Taiwan and China. These five players account for about 88% of this market and are active players in the LNG market. The rest of the market is devoted to pipeline gas trade within South-East Asia.

The Japanese policy of promoting natural gas for power generation was initiated in the aftermath of the first oil crisis when the country was highly vulnerable due to excessive dependence on imported fuel oil for electricity generation. The availability of South East Asian natural gas offered a solution through the LNG option as the pipeline option was not viable due to the long distance between the source and the demand centre. The same logic was followed by South Korea in the 1990s and subsequently by other importing countries of the region. Importation of the piped gas is limited to some members of the ASEAN group of countries. Although there are talks about pipeline projects in the region, not much has been achieved in terms of infrastructure development and consequently, the trade in piped gas remains limited.

The main suppliers of gas are mostly from the region and from the Middle East. Indonesia, Malaysia and Brunei are long-time players in this market with long-term sales agreements with Japan. Qatar and Australia have also acquired a significant market share (see Fig. 15.15) in the LNG sales in this region. These five countries supplied more than two-thirds of the imported LNG used in the region in 2009. The rest comes from a variety of sources. The piped gas supply on the other

Fig. 15.14 Asia-Pacific gas market. *Data source* BP statistical review of world energy 2010

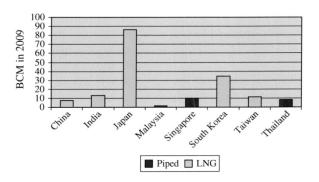

Fig. 15.15 Major gas suppliers to Asia-Pacific in 2009. *Data source* BP statistical review of world energy 2010

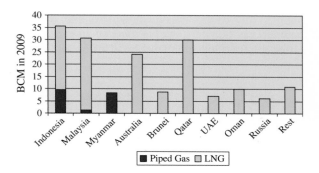

hand came essentially from Myanmar and Indonesia for use in Thailand and Singapore.

As gas pricing and transportation costs play an important role in gas trade and gas use, we turn to these aspects below.

15.4 Economics of Gas Transportation

Natural gas is costly to transport and the mode of transportation is often decided by the cost. Because of high cost of transport, only a small fraction of gas is internationally traded and the rest is used where it is produced or flared. Although gas could be used in different forms from the source to transport to the consumer,[2] the most common options are through the pipelines and through LNG. The economic drivers of pipeline transport and the LNG mode are different and accordingly, it is important to consider them separately.

15.4.1 Economics of Pipeline Transport of Gas

Transportation of natural gas through a pipeline results in the economies of scale. As the diameter is increased, the volume increases following $\pi r^2 L$ formula where r is the radius (or half of the diameter) of the pipeline and L is the length but the material required for the pipeline increases following $2\pi r L$ formula. However, as the length of the pipeline increases, the gas pressure falls due to friction and therefore, compressors are required to raise the pressure. The resistance due to friction increases as the pipe diameter reduces. There is therefore some trade-off possible between the pipeline size and the line pressure drop to minimize the cost of gas transport for a given length of distance.

---

[2] Thomas and Dawe (2003) indicate following other options: gas to liquids, compressed natural gas, gas to solids, gas to wire and gas to commodity.

15.4 Economics of Gas Transportation

The economic literature on pipeline transport is somewhat limited. However, Banks (1987) and more recently Yepez (2008) and Massol (2009) have considered this issue by following and extending the seminal work of Chenery (1949). The presentation here follows Yepez (2008).

Let us define the variables first.

Q = pipeline flow (in millions of cubic feet per day)
T_b = base temperate in Rankin scale (degree R, which is equal to 460 + temperature in degree F)
T = mean flowing temperature in degree R
P_b = base pressure in pounds per square inch absolute (psia)
P_1 = initial pressure in the pipeline in psia
P_2 = terminal pressure in the pipeline in psia
$R = P_1/P_2$
G = gas specific gravity, which varies between 0.59 and 0.70 (with air = 1)
h = difference in elevation between the inlet and outlet of the pipe (in feet)
L = pipeline length between compressor stations (miles)
f = friction coefficient (dimensionless)
D = pipe inside diameter (inches)
Z = compressibility factor at average conditions (dimensionless)
H = compressor horsepower per million cubic feet of gas
R = compression ratio, P_1/P_0
P_0 = suction pressure, psia
P_1 = compressor discharge pressure, psia
E = mechanical efficiency of the compressor
K = ratio of specific heats

According to Yepez (2008), the general flow equation can be written as

$$Q = C_0 D^{8/3} \sqrt{\frac{(R^2 - \alpha)}{L}} \tag{15.1}$$

where

$$C_0 = 433.5 \times 10^{-6} \frac{T_b}{P_b} \sqrt{\left(\frac{1}{GTZ}\right) P_2} \sqrt{\frac{s}{(\alpha - 1)}} \tag{15.2}$$

$$s = 0.0375 \frac{Gh}{T} \tag{15.3}$$

$$\alpha = e^s \tag{15.4}$$

The power needed to compress gas is given by the following

$$H = C_1(r^\beta - 1) \cdot Q \tag{15.5}$$

where

$$C_1 = 3.0325 E \frac{P_b T Z}{T_b} \left[\frac{1}{\beta}\right] \qquad (15.6)$$

$$\beta = \frac{k-1}{k} \qquad (15.7)$$

If it is assumed that $P_0 = P_2$, and $R = r = P_1/P_2$, then Eq. 15.1 can be reduced to

$$Q = C_0 D^{8/3} \sqrt{\frac{(r^2 - \alpha)}{L}} \qquad (15.8)$$

By eliminating r from Eqs. 15.5 and 15.8, we obtain Eq. 15.9.

$$F(D, H, Q) = \frac{L \cdot Q^2}{C_0^2 D^{16/3}} + \alpha - \left(\frac{H}{C_1 Q} + 1\right)^{2/\beta} = 0 \qquad (15.9)$$

This gives the engineering production function of gas transport pipeline.

To determine the cost function, the costs related to the pipeline and those related to the compressor have to be considered. It is important to note that the main costs in any transmission network in the capital cost. Yepez (2008) considered a log-linear function for these costs and suggested that the total annual cost of the line per mile can be written as the sum of capital costs and the operating costs.

$$C_D = a D^\alpha \tau^\rho + b D^\delta \qquad (15.10)$$

where C_D = total annual pipeline cost for diameter D, a, b are regression coefficients, D = diameter of the pipeline (inches), τ = pipeline thickness (inches), α, β, and δ are elasticity parameters.

Yepez (2008) provides an estimate of the above equation empirically as follows.

$$C_D = 7,144.59 D^{0.881} \tau^{0.559} + 317.61 D^{0.809} \qquad (15.11)$$

Similarly, the cost function for the compressor system can be written as

$$C_H = c \cdot H^\varepsilon + d \cdot H^\psi \qquad (15.12)$$

where c and d are coefficients estimated through regression and ε and ψ are elasticity parameters.

Yepez (2008) provided an estimation of Eq. 15.12 and is shown in Eq. 15.13.

$$C_H = 1,256.33 H^\varepsilon + 6,145.177 H^{0.4523} \qquad (15.13)$$

Therefore the total cost for the pipeline and the compressor system is given by

$$L \cdot C_D + C_H \qquad (15.14)$$

15.4 Economics of Gas Transportation

where L is the length of the pipeline.

For any firm, the objective is to minimize the annual total pipeline cost subject to the gas flow constraint. This leads to the minimization problem that can be written as follows:

Minimise $L \cdot C_D + C_H$
s.t. $F(D, H, Q) = 0$

The Lagrangian for the minimization problem is

$$\tilde{\lambda} = L \cdot C_D + C_H + \lambda \cdot F(D, H, Q) \tag{15.15}$$

The solution to the above minimization problem gives the optimal combination of pipeline diameter and compressor horsepower. The first order condition indicates that the ratio of the marginal costs of pipeline and the compressor system should be equal to the ratio of marginal productivity of pipeline diameter and compressor horsepower. This ratio has to be satisfied for optimal combinations of diameter and the compressor power. However, given the nature of the equation, numerical solution is generally required.

Pipelines require costly investments—between $1 billion to $1.5 billion per 1,000 km for a large diameter pipeline. Moreover, long distance cross-border pipelines often generate conflicts between the gas producer and transit countries (ESMAP 2003). The recent incidences between Russia and Ukraine have created security concerns of gas supply to European markets. In addition, such pipeline projects have long gestation periods and new pipeline projects can years from their conception to the successful completion. For consumers far away from the sources of supply and at places far removed from the existing pipeline networks, alternative transportation arrangements are required. This led to the development of LNG mode of transport.

15.4.2 Economics of LNG Supply

Liquefied natural gas is obtained by cooling clean natural gas (i.e. devoid of impurities) to -162 degrees centigrade that changes the physical state of natural gas to liquid. In the process, the volume reduces to 1/600th of its original volume at room temperature and atmospheric pressure. LNG is then transported in liquid form in specially designed vessels to the users where the liquefied gas in stored and converted back to gaseous form before use. Thus the supply chain involves three distinct phases (apart from natural gas feed supply)—liquefaction, transportation and storage and re-gasification. Thus the economics of LNG supply depends on three inter-related elements of the supply chain.

The liquefaction plant is the first component of the capital intensive LNG supply chain. This segment of the business benefits from economies of scale, implying that the unit cost falls as the size increases. A report by the Energy Charter Secretariat (2009) indicates that in the 1990s, the train size was limited to 2.5 million tons due to compressor size limitations. But technological innovation has allowed now "super sizing" of trains with Qatar installing a very large train close to 8 million tons in 2010. Jensen (2004) indicated that a 25% reduction in cost can be achieved by doubling the plant size from 2 million tons to 4 million tons. This scale economy effect was expected to bring the capital costs per ton down to $200 from a level of $800 prevailing in the 1990s. But costs have been rising in recent times due to high demand for LNG plants that has overloaded the engineering contractors for such works and escalated the cost of materials required for such plants. However, the global economic crisis of 2008 led to a cancellation of a number of Greenfield projects. Moreover, the demand for LNG was affected by the discovery and exploitation of shale gas. Consequently, the viability of new liquefaction plant is being questioned again.

The shipping segment of the business establishes the link between the producer and the consumers. These are specially designed ships that carry cryogenic liquids and consequently, the cost of transportation is much higher compared to oil. The tanker size has seen some improvements over time: in the 1980s, the average size was below 125,000 cubic meters. By 2007, the average size has increased to 151,000 cubic meters (Energy Charter Secretariat 2009). This however masks the recent trend towards super tankers. Noble (2009) indicates that the Q-Flex ships having a size between 200,000 and 250,000 cubic meters and Q-Max ships of over 250,000 cubic meters capacity are now operating. Table 15.1 presents the fleet distribution by size in 2009.

The ship building industry is concentrated in Asia–Japan is the oldest centre for LNG tank building. South Korea entered the market next, followed by China. However, the high demand for ships in the first few years of the new millennium exceeded the building capacity. Consequently, there was an increase in the cost and the orders took longer to deliver. However, recent reduction in LNG demand has changed the market again. This "boom-bust" cycle is a common feature of the highly capital intensive ship building industry.

The shipping industry has changed as well in terms of ownership of ships. Traditionally, major energy companies used their captive fleet to transport LNG to ensure adherence to contractual requirements for supply destinations. Independent ship owners used to play a minor role. However, this trend has changed now with many new ships being owned by independents. The traditional owners have also divested their shipping business to focus on the core activities.

The shipping element can easily account for between 25 and 40% of the delivered LNG cost. The shipping rate (or freight rate) is a daily fee that is charged

Table 15.1 LNG fleet by size in 2009

Type	Size range (cubic meters)	Number
Q-Max	>250,000 cubic meters	4
Q-Flex	200,000–250,000 cubic meters	20
Standard	100,000–200,000	244
Small	<100,000 cubic meters	30
Total		298

Source Noble (2009)

15.4 Economics of Gas Transportation

Table 15.2 An example of LNG costs

Element	Capital cost ($ billion)	Cost of service ($/million Btu)
Field development	$1.3	$0.8
Liquefaction	$1.6	$1.22
Tankers	$1.6	$0.98
Regasification	$0.5	$0.39
Total	$5.0	$3.39

Source Jensen (2004)

for hiring the ship. The rate varies considerably depending on the demand, contract duration, etc. The average charter rate in 2008 was between $40,000 and $50,000 per day.

The last element in the LNG supply chain is the regasification terminal where LNG is stored and re-gasified before injection into the gas pipeline. The cost of such a terminal is generally site specific but generally the storage facilities account for 40–45% of the costs while processes and utilities account for another 40% of the costs. The storage tanks also benefit from economies of scale and large scale tanks of 150,000 cubic meters are reducing costs of re-gasification. This element accounts for about 15% of the total LNG supply costs.

The following example (Table 15.2) from Jensen (2004) illustrates the capital cost and delivered gas cost distribution of LNG for a Greenfield plant[3] of 2 trains of 3.3 Million. Although the actual cost at any time varies depending on the source, market conditions, etc. this relative distribution does not change much.

15.4.3 LNG Versus Pipeline Gas Transport

The cost structure of the two modes of transport as described above suggests the possibility of a trade-off. Generally, LNG requires higher initial investments compared to pipelines but the cost increases proportionally with distance whereas LNG transportation cost rises slowly with distance. Using the traditional break-even analysis therefore a threshold point can be identified below which pipeline transport becomes cheaper compared to LNG. An illustrative diagram is shown in Fig. 15.16. It shows that the LNG option becomes viable for distances above 3,000 miles.

However, the economic trade off depends on various factors—such as utilisation rate, geographical conditions (terrain, sub-sea conditions), technological developments, economic, political and regulatory factors affecting costs. Generally, the cost drops as the utilization rate increases—however, if due to market conditions, the level of utilization falls, the cost per unit increases. The economic

---

[3] The following assumptions were used in Jensen (2004): 2 trains of 3.3 million tons, 6,200 nautical miles of transport, and 10 tankers each costing $160 million. The cost of service is based on a regulated tariff using the traditional regulation.

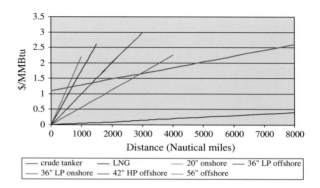

Fig. 15.16 Illustrative cost comparison of LNG and pipeline transport. *Data source* Jensen (2004)

factors vary from one region to another—generally in the Asian devel-oping countries or in Africa, the cost of labour tends to be lower than that in the developed world and therefore, the labour component in the infrastructure development cost or in operation can be lower (Okimi 2003). Also, any cost reduction due to economies of scale or technological development affects the trade off.

15.5 Gas Pricing

As natural gas does not have a captive market, to penetrate any market natural gas has to compete with other fuels. But such a price has to provide adequate incentives to the producers as well. These inherent conflicting objectives require a balancing act, which is a difficult task. As a result, depending on the market condition, either the consumers' or the producers' perspective tends to dominate in the pricing decision and a number of alternative pricing mechanisms have emerged in the market. This section provides a brief introduction to various natural gas pricing mechanisms.

15.5.1 Rules of Thumb

Brown and Yucel (2007) provide an account of three rules of thumb used in the gas industry. These rules are simple and can give a reasonable value in normal conditions.

(a) Simple rule of thumb—The 10-to-1 rule suggests that the natural gas price is one tenth of crude oil price. If the price of WTI crude is $60 a barrel, the price for natural gas would be $6 per million Btu (see Fig. 15.17). Another simple thumb rule is the one-sixth rule that tries to take the heat content into account.[4]

---

[4] The heat content of one barrel of WTI is 5.85 million Btu, so one-six of this produces 1 million Btu approximately.

15.5 Gas Pricing

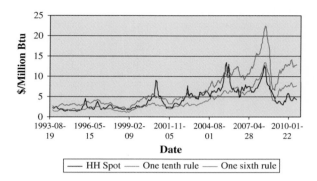

Fig. 15.17 Henry-Hub gas price and rule-based approximation. *Data source* ALFRED

These thumb rules tend to provide a good approximation during periods of stable prices but the one-tenth rule appears to under-estimate the price while the one-sixth rule overstate gas prices.

(b) Burner-tip parity—This is a more complex rule that captures the competition between gas and other fuels at the consumers' end. Generally, for heating purposes natural gas competes with fuel oil and hence the burner-tip parity would imply that natural gas price will adjust to the required level as oil price changes. As the heating value of a barrel of fuel oil is 6.287 million Btu, and because fuel oil is historically priced at 95% of a barrel of crude oil, the parity rule is

$$P_{HH} = 0.1511 \times P_{WTI} \tag{15.16}$$

where P_{HH} is the gas price at Henry Hub in \$/million Btu while P_{WTI} is the price of crude oil in \$/barrel.

Brown and Yucel (2007) however suggested a correction to reflect the differences in transportation costs. This is given as

$$P_{HH} = -0.5 + 0.1511 \times P_{WTI} \tag{15.17}$$

If oil price is \$60 per barrel, the gas price as per this rule will be \$8.66 per million Btu. This thumb rule appears to trace the historical price well (Fig. 15.18) until 2006 but since then it is clear that the burner tip thumb rule is over-estimating the actual price. Note that this rule will vary from one market to another and it is difficult to generalize such a rule in the absence of a common reference point.

(c) Simple regression: The third rule suggested by Brown and Yucel (2007) is a relation obtained by simple regression of WTI price and Henry Hub Spot price. They used the weekly price information to arrive at the following relation

$$P_{HH} = -0.1104 + 0.1393 \, P_{WTI} \tag{15.18}$$

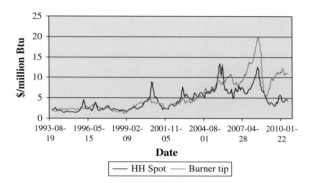

Fig. 15.18 Henry Hub spot gas price compared with burner tip rule. *Data source* ALFRED

This formula is very similar to the burner-tip parity rule and has similar problems. It is surely possible to improve on the above relationship but the simplicity of the rule will be lost.

15.5.2 Parity and Net-Back Pricing

Two commonly used pricing mechanisms for trading natural gas, namely the parity pricing and net-back value pricing, try to link gas prices to crude oil or petroleum product prices. The parity pricing principle tries to work out the price starting from the upstream side whereas the net-back pricing principle follows a reverse path (i.e. starting from the down-stream or end-user price parity, it finds the well-head price). Thus the two mechanisms are similar in nature.

The parity pricing is favoured by the exporters as it passes the costs of transportation and other infrastructure on to the consumers. The opportunity cost to the exporter is the free on board (FOB) price and at the point of export, the exporter should be able to charge a price identical to that of a competing fuel. As natural gas is more environment friendly compared to its competitors, it is claimed that such parity is appropriate and fair. The exporter will therefore have adequate remuneration for exploiting its gas reserves and supply sufficient quantities of gas. This was the principle used in the Algerian gas export to France and Belgium in the late 1970s and early 1980s and led to huge debates in those countries as gas prices soared.

The net-back argument on the other hand suggests that for consumers the relative price of a fuel is the most important factor. As natural gas does not have a captive market where it can exercise its monopoly power, gas has to compete with other fuels. Therefore, in order to ensure competitiveness of gas, the parity has to be established at the consumer end and not at the well-head. All costs related to transportation and infrastructure use are then deducted from the end-use parity to arrive at the price to be paid to the exporter. This moves backward to find the well head price and hence it is called the net-back pricing. Clearly, such a price favours the importers perspective to keep imported gas competitive. This is the principle used in the continental European market since 1980s.

Clearly, two positions lead to conflicting results and a comprise solution is achieved in the gas purchase contracts by sharing the available rent. For the producer, the minimum acceptable price is a price that allows it at least to cover the cost of production and local transportation. At any price lower than this, the producer will either flare her gas or re-inject (if such an opportunity is available). On the other hand, an importer will not be interested in buying any gas that costs more than the burner-tip parity of other competing fuels. As the cost of transport in the case of oil is much cheaper than that of gas, the importer will not be interested in any gas whose transport differential adjusted cost exceeds the FOB price of oil or oil products. This can be written as (based on Percebois (1986)):

$$C_p + C_{gt} <= C_{ot} + C_{o(FOB)} \quad (15.19)$$

where C_p = cost of gas production and local transport by the producer; C_{gt} = cost of gas transport; C_{ot} = cost of oil transport; $C_{o(FOB)}$ = FOB price of oil; All costs are in \$/million Btu.

The above equation can be reformulated as

$$C_p + C_{gt} - C_{ot} <= C_{o(FOB)} \quad (15.20)$$

Suppliers with cheaper cost of production or transport costs will have an advantageous position compared to those with costly production or those located at a long distance from the importer.

Most of the long-term contractual arrangements have linked gas price to oil price or oil product prices. They also include an indexation mechanism by which prices are adjusted as economic or market conditions change. The aim was to ensure the competitive advantage of natural gas compared to competing fuels while providing the producer with an opportunity to earn a reasonable return on its investments. The specific formulation used in the industry varies from country to country but in most cases the variables in the index referred to readily available information.

However, depending on the market, natural gas faces competition from different fuels. For example, in the electricity sector, the competitor can be coal or fuel oil depending on whether it is for base load or peaking use. In the industrial sector, fuel oil or coal would be the main competitors. For the residential and commercial sectors, the competition will be with coal or electricity. Therefore, the netback value will vary from one use to another even within a given country and the value of gas sold to different countries would be different depending on the market share of competing fuels, and pricing mechanism in use for those fuels. As the market does not give a correct signal to account for the external effects, and given that the taxation policies are often socially or politically motivated, it is difficult to believe that the netback pricing or parity pricing results in an efficient outcome. See Stern (2007) and Miyamoto and Ishiguro (2009) for a detailed analysis of the issue.

15.5.3 Spot Prices of Natural Gas

As the organization of the industry has changed from an integrated business to a more competitive one in Europe and North America, spot markets have emerged. Gas trade became more transparent and developed around inter-connected grids called hubs. Since the 1990s, two such hubs emerged as the leaders—the Henry Hub in the North America and the national balancing point (NBP) in the UK, which is also the main hub of Europe. Many other hubs have also come to existence but in terms of liquidity and volume of transactions, they still play a local role.

The price at a hub is determined by supply and demand and hence is competitive in nature. Through organised spot trading on an exchange, the price signal is generated, which provides timely information on the working of a natural gas market and facilitates decision-making for investment purposes. However, because the spot market is more concerned with the short-term pricing, any change in the underlying demand–supply fundamentals can affect the prices, thereby rendering the prices more volatile. Moreover, active participation of traders and speculators in the market has also increased the price volatility in recent terms.

Yet, spot prices continue to be "anchored in a long-term relationship with crude oil prices" (Brown and Yucel 2007). Although factors like weather, seasonality, supply disruption risk, or storage capacity condition influence the short-term prices, but in the long-run the spot prices appear to adjust to the changes in the crude oil prices.

15.6 Natural Gas in the Context of Developing Countries

While natural gas is more widely used in the developed countries, and receives more attention in the academic literature, the context in many developing countries is quite different. As indicated before, natural gas reserves are more widely distributed around the world but the development of natural gas faces a number of constraints in the developing world. Gas field and infrastructure development requires huge capital investment—often involving billions of dollars. Many developing countries do not have either the investment climate or the economic strength to attract or invest such quantities of capital. Although in many cases gas is found by foreign oil companies, the absence of any ready market for natural gas in the country often prevents development of such gas fields. If the gas is associated with oil, in such cases the company attaches priority to oil production and any gas will be re-injected to the extent possible and the rest will be flared.

To justify investment in a gas field and transportation facilities, a core group of large consumers is required with sufficient demand and paying capacity so that the investment in the entire supply chain can be justified. In general, the power sector is considered to be a candidate industry in most countries. But in many developing countries, the power sector may be quite small in size and the demand for power may not itself be sufficient to support a field development. Moreover, the power sector is generally financially weak due to poor pricing policies and in such a case, selling gas in bulk to financially weak companies does not ensure economic or financial viability of the gas supplier. An alternative could be to identify large industrial consumers with sufficient demand and develop a power generating company to cater to industrial and more affluent consumers. However, this involves entering into a business outside the core competence of the gas company. Moreover, the regulatory environment of the country may not allow such a development as the power sector in many cases is still heavily state controlled.

In addition, the regulatory system is weak in many developing countries and often politically motivated. The system is generally non-transparent and in the case of gas industry, the legal provisions may be insufficient to support the entire chain of activities. Because gas is often found while exploring for oil, the upstream regulation for oil exploration generally applies. However, this generally gives inadequate guidance for developing a viable gas industry or its governance mechanisms. Investors do not feel confident enough to risk their money in such business environments. Moreover, the domestic pricing of gas is an issue in itself—the desire to maintain a low-cost supply to promote economic development often leads to unviable business propositions. Consequently, gas market development remains painfully slow in the domestic context of developing countries.

This domestic challenge encourages gas producers to look for export options. However, one of the main factors here is the availability of sufficient gas reserves to support a dedicated export supply agreement. For example, to develop two trains of 3.3 million ton of LNG facility over 20 years of contract period will require about 280 BCM of reserves (Jensen 2004). Large fields of sufficient capacity are not widespread and accordingly, gas deposits may not qualify for export-oriented development. Wherever they qualify, reaching the financial closure and putting in place all pieces of the investment jigsaw takes a long time—in many cases running into decades. This makes the gas industry very different from the oil industry and highlights the dilemma about whether to develop or not to develop the gas market (see Box 15.1 for an example from the Republic of Congo).

Box 15.1: To Develop or Not to Develop the Gas Market?

The Republic of Congo or Congo-Brazzaville[5] is a sub-Saharan African state lying between Gabon and the Democratic Republic of Congo (DRC); thus forming part of the West Central African Region.[6] With an area of

---

[5] Which should not to be mistaken for the Democratic Republic of Congo (DRC) or Congo-Kinshasa formerly known as Zaire.
[6] The members of the West Central African region are Cameroon, Central African Republic, Chad, Congo, DRC, Equatorial Guinea, Gabon, and Sao Tome and Principe.

342,000 km² and a population of only about 3 million[7] Congo can be regarded as a small country.

The majority of the Congolese population (about 70%) is located in the 3 principal cities of the country: Brazzaville the capital city; the harbour and industrial city of Pointe-Noire which is the economic capital and where the petroleum activities are concentrated and export companies located because of its port; and finally Loubomo which is a major agricultural area.

GDP was estimated in 2006 at $5.1 billion with a real GDP growth rate of about 6% for the same year. The oil sector is the major source of foreign exchange supplying around 90% of total export revenues and accounting for roughly 65% of the country's gross domestic product (GDP).[8] However, three successive and intense civil conflicts in the 1990s (1993, 1997, and 1998–1999) destroyed physical capital, displaced thousands of individuals, and weakened institutions.

Congo with proven oil reserves of 1.6 billion barrels is the fifth largest oil producer in Sub-Saharan Africa.[9] Moreover, probable oil reserves are estimated at about 15 billion barrels. Around 80% of the oil production is located offshore.

The national petroleum company SNPC[10] develops PSAs with each company operating in the country in order to ensure a constant minimum flow of revenue for the government. Under PSAs, companies carry out exploration and development for an agreed period of time, while financing all investment costs which are recovered after the start of production. Tax breaks and a royalty system are also offered under those agreements. The SNPC is responsible for selling the government share; about one-third of the oil produced goes directly to the government, which provides about 70% of its annual revenues.[11] Primary foreign companies are Total (France) and Eni (Italy), and wells operated by Total account for roughly 47% and those operated by Eni for 22% of Congo's total oil production.[12] Additional smaller oil producers include Perenco (United-kingdom), Congorep (Perenco-SNPC consortium), and Likouala S.A. (private domestic company).

Over the past 20 years, Congo's crude production has quadrupled, from 65,000 bbl/day in 1980 to an average of 280,000 bbl/day in 2000. Since that

---

[7] July 2005 estimates; CIA the World Factbook 2005, http://www.cia.gov/cia/publications/factbook/geos/cf.html.

[8] Congo-Brazzaville country analysis brief, p1. Available on-line at http://www.eia.doe.gov/emeu/cabs/Congo_Brazzaville/pdf.pdf.

[9] Congo-Brazzaville country analysis brief, p2. Available on-line at http://www.eia.doe.gov/emeu/cabs/Congo_Brazzaville/pdf.pdf.

[10] It was set up in April 1998 by law No 1/98 and it has a capital of 900 millions CFA.

[11] AfDB/OECD op. cit., p. 166.

[12] IHS Energy as of February 2007 cited in EIA op. cit. p. 2.

peak period, oil production steadily declined up to 2005 with an average of 227,000 bbl/day. This was due to drop-off in production at mature fields, and delays in bringing new fields online. The national oil production has rebounded over 2006 with an average production of 245,000 bbl/day. In 2006, Congo consumed about 6,000 bbl/day and exported about 234,000 bbl/day of crude oil mainly to the Asian markets.

The indigenous hydroelectric resources form the basis of the Congolese power generation. In 2001 for instance, electricity production was estimated to be about 358.1 million GWh. Of this production 0.3% was from fossil fuel sources, and 99.7% was from hydro. However, the electricity consumption in the same year was about 633 millions kWh. The installed electrical capacity of Congo is in fact supposed to be 109 MW, but the effective production is only 56 MW, and the electrification rate is no more than 50% in urban areas and 5% in rural ones.[13]

The electricity sector is characterised by a huge, unrealised hydro potential of 3,000 MW. The country faces the problem of capacity and energy shortage, mainly because of the inefficiency (or lack) of the national network.

The country has a proven gas reserve of 4 TCF in 2005, but the majority (2/3) of these reserves is associated gas. Most of the fields are of medium size but due to lack of suitable infrastructures and the long distance between the oil fields and the exploitation sites, the associated gas is flared in the fields. Indeed, 97% of the annual gas production is flared or re-injected.

The use of natural gas for power generation would represent the most profitable and operational use due to the constant inadequacy between supply and demand for energy in the country. But the hydro potential of the country and the import option from DR Congo could impede with natural gas use in the power sector.

Only a small 20 MW power plant has been built up recently based on open-cycle gas turbine process as a demonstration project. SNPC through its gas subsidiary has developed the network for gas gathering and supply to the power plant. The power plant has been set up through a joint venture that works along the lines of an IPP. The venture received gas at no cost to it on a 10 year contract and sells the power generated to the electricity company through a power purchasing agreement. But the utility seems unhappy about the cost of the gas-based power and does not want to pay anything more than its benchmark cost of buying hydropower.

A cement plant of 1 Mt/year capacity has been installed recently and a Magnesium alloy plant will be set soon which will require approximately 120 megawatts of electrical power and up to 120 million cubic meters of natural gas annually. Also the economic hub of the country has a

---

[13] Data from the "Mission économique de Brazzaville" available at http://www.izf.net/izf/EE/pro/congo/5020_electricite.asp.

> concentration of small and medium industries and commercial centres which could use some gas if the network is developed.
>
> The gas reserves are not sufficient to engage in large-scale LNG export but its known natural gas reserves are well in excess of the local market's needs. For exports of 300 MCF/d as pipeline gas or LNG over a period of 20 years, about 2.5 TCF of NG reserves would need to be dedicated to the project. However, some of its neighbours (such as Equatorial Guinea) have developed export oriented LNG and such a model can be looked into.
>
> The primary problem is to find a sufficiently lucrative market to cover the very large capital investment in pipelines that is required to bring the gas to its market. Such a market would be willing to pay the price for gas and would have enough demand. The government is willing to develop the local gas market and needs your help in finding out the best way forward. There is no proper gas market structure in the country. There is no gas specific law either.
>
> *Source* Ibata (2009).

15.7 Conclusion

Although natural gas has emerged as a preferred fuel in the developed world and in emerging developing countries, the market still remains fragmented regionally. The capital intensiveness of the entire supply chain of the industry and the possibility of opportunistic behaviour by the consumer or the owner of the bottleneck facility or both creates particular hazards for large investments. Moreover, competition from other fuels poses a continuous threat to the industry. Consequently, the developments have been slow compared to the oil industry. But the threat of climate change and the search for low-carbon fuel has brought new opportunities for the gas industry and the industry has seen unprecedented growth in recent times. Through an account of the demand–supply and trade of natural gas and a discussion of the regional markets and prices, this chapter has provided an understanding of the changes in this market.

References

Banks FE (1987) The political economy of natural gas. Croom Helm, London
Brown SP, Yucel MK (2007) What drives natural gas prices? Working paper 0703, Federal Reserve Bank of Dallas, Dallas, Texas, USA
Chenery HB (1949) Engineering production functions. Q J Econo 63(4):507–531
Davidson A, Hurst C, Mabro R (1988) Natural gas: Governments and oil companies in the third world, OIES. Oxford University Press, London

References

Energy Charter Secretariat (2009) Fostering LNG trade: developments in LNG trade and pricing, Energy Charter Secretariat, Brussels, Belgium

ESMAP (2003) Cross-border oil and gas pipelines: problems and prospects, energy sector management assistance programme. The World Bank, Washington, DC, USA

Ibata B (2009) To develop or not to develop a natural gas market in the Republic of Congo? Unpublished doctoral thesis, University of Dundee, Dundee

IEA (2000) Regulatory reform European gas, IEA, Paris

Jensen JT (2004) The development of a global LNG market: is it likely? If so, when?. Oxford Institute of Energy Studies, Oxford, UK

Julius DeAnne, Mashayekhi A (1990) The economics of natural gas: pricing, planning and policy. Oxford University Press, London

Juris A (1998a) The emergence of natural gas markets. World Bank working paper, WPS 1895, World Bank, Washington, DC

Juris A (1998b) Market developments in the United Kingdom's natural gas industry. Policy research working paper 1890, World Bank, Washington, DC

Massol O (2009) Cost function for the natural gas transmission industry: further considerations, Cahiers de recherche 09.09.86, CREDEN. University of Montpellier, Montpellier, France

Miyamoto A, Ishiguro C (2009) A new paradigm for natural gas pricing in Asia: a perspective on market value, NG 28. Oxford Institute for Energy Studies, Oxford, UK

Newbery DGM (1999) Privatisation, restructuring and regulation of network utilities. MIT Press, Mass

Noble P (2009) A short history of LNG shipping 1959–2009. SNAME, Texas section (see www.sname.org)

Okimi H (2003) Comparative economy of LNG and pipelines in gas transmission. Paper for world gas conference 2003, Japan. See http://www.igu.org/html/wgc2003/WGC_pdffiles/10392_1045815366_9772_1.pdf

Percebois J (1986) Gas market prospects and relationship with oil prices. Energy Policy 14(4):329–346

Rogner HH (1989) Natural gas as the fuel for the future. Ann Rev Energy 14:47–73

Stern J (2007) Is there a rationale for the continuing link to oil product prices in Continental European long-term gas contracts, NG 19. Oxford Institute for Energy Studies, Oxford, UK

Teece DJ (1996) The uneasy case for mandatory contract carriage in the natural gas industry, in New Horizons in Natural Gas Deregulation. Praeger, London

Thanawat N, Bhattacharyya SC (2007) High gas dependence for power generation in Thailand: the vulnerability analysis. Energy Policy 35(6):3335–3346

Thomas S, Dawe RA (2003) Review of ways to transport natural gas energy from countries which do not need the gas for domestic use. Energy 28(14):1461–1477

Viscusi K, Vernon JM, Harrington JE Jr (2005) Economics of regulation and antitrust. MIT Press, Mass

Yepez RA (2008) A cost function for the natural gas transmission industry. Eng Econ 53(1):68–83

Chapter 16
Developments in the Coal Market

16.1 Introduction

Coal is the oldest form of fossil fuel that is used in large quantities even today. Coal first displaced traditional energies to bring the first energy transition globally (see Chap. 18) in the nineteenth century and reigned supreme for about one hundred years when it was displaced to the second position by oil. Environmental concerns over coal use have restricted its growth from time to time but its abundance and the allure of secure and affordable supply helps it return to the limelight. At present it accounts for more than a quarter of global primary energy demand. In addition, the technological developments have also made coal use safer and less damaging to the environment.

This chapter provides a brief overview of the coal industry.

16.2 Coal Facts

Coal is abundantly available around the world and is widely distributed in 70 countries (World Coal Institute 2005). According to the World Coal Institute (2005), the estimated coal reserve (including probable reserves) is 984 billion tons while according to the BP Statistical Review of World Energy 2010 the proved reserves stood at 826 billion tons in 2009. The regional distribution of coal reserves (see in Fig. 16.1) shows that three regions, namely North America, Former Soviet Union and the Asia Pacific, account for about 83% of the proved coal reserves. Five countries are endowed with 78% of the proved reserves—US 28.9%, Russian Federation 19%, China 13.9%, Australia 9.2% and India 7.1%. At the current rate of production, coal reserves will last for 119 years at the global average level but the R/P ratio varies at the regional and country levels (see Fig. 16.2). Thus coal offers far greater volumes of reserves amongst fossil fuels.

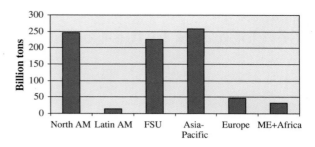

Fig. 16.1 Coal reserve distribution in 2009. *Data source* BP Statistical Review of World Energy 2010

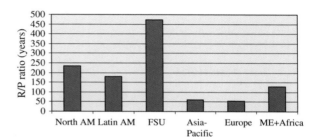

Fig. 16.2 R/P ratio of coal at the regional level in 2009. *Data source* BP Statistical Review of World Energy 2010

The global coal production in 2009 was 6.9 billion tons. The ranking changes to some extent. China is the undisputed leader in global coal production with a share of 45.6% of global production in 2009. The US comes next with a 15.8% share while Australia and India produced just above 6% of the global output each. Indonesia, Russia and South Africa individually produced between 4 and 5% of the global output in 2009. Taken together, seven countries accounted for just over 87% of the global coal production. The hard coal production was 5.99 Bt while the Brown coal production was 0.9 billion tons (IEA 2010).

It is worth mentioning here that China has recorded a spectacular growth in production since the new millennium (see Fig. 16.3). In the 1990s, China's production ranged between 1100 and 1300 Mt per year. However, since 2000, there has been a step-change in Chinese production and in 2009, China has produced more than 3 billion tons of coal, thereby more than doubled its output within a period of nine years.

Simultaneously, the over the past three decades, some countries have lost a significant amount of their outputs. Germany and Poland are two such cases: German output has declined from about 500 Mt per year in the 1980s to close to 200 Mt in recent times. Similarly, Polish output has fallen from 260 Mt in the mid-1980s to a low of about 140 Mt now. High cost of operation in Germany and market reform policies of the Polish government are responsible for such changes.[1]

In terms of quality of output, bituminous coal dominates the production (61% in 2007). Lignite comes next with a 14% share, followed by coking coal with a 12%

---

[1] See Zientara (2007) for further details on the Polish coal industry.

Fig. 16.3 Trend of global coal production. *Data source* BP Statistical Review of World Energy 2010

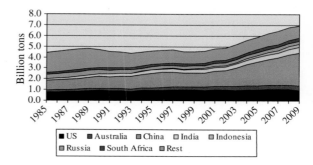

Fig. 16.4 Coal demand trend. *Data source* BP Statistical Review of World Energy 2010

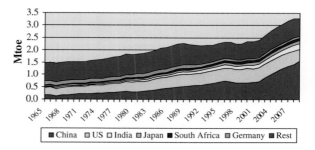

share. However, the output mix varies considerably across regions. Underground mining still continues to dominate in coal production with about 60% of coal being produced using this method but some major producers like Australia and the US are relying more on open-pit mining (World Coal Institute 2005).

Coal is mostly consumed in the country of its production. Over the years, coal has found an ally in the electricity industry. About 41% of global electricity in 2008 came from coal (IEA 2010). Almost 70% of coal output is used in electricity plants, combined heat and power (CHP) plants and in heat plants, while a major share of the rest is used in industries. Steel-making industries uses about 13% of hard coal output and 70% of the steel making industry relies on coking coal (World Coal Institute 2009).

Asia and the Pacific is the most important market for coal: 65% of global coal demand arose from this region in 2009. China is undoubtedly the most important player, with a 47% share in global demand for coal. The US with a 15% share and India with a 7% share come second and third in terms of coal demand. The trend of coal demand shows a steady growth (see Fig. 16.4). In fact, between 2000 and 2009, China's demand has grown almost at 10% per year. Consequently, coal demand has grown faster than all other fuels during the same period globally.

Coal is traded at a relatively lower rate globally—about 13–14% of global coal primary supply is traded. The cost of transport and the difficulties in transporting coal effectively reduces coal trade. Although countries that import coal have inadequate domestic coal supplies, but major producers like China, India and the

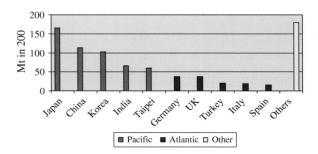

Fig. 16.5 Major coal importers in 2009. *Data source* IEA (2010)

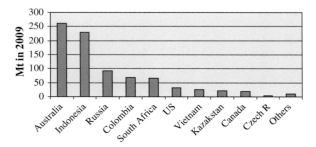

Fig. 16.6 Major coal exporters in 2009. *Data source* IEA (2010)

US also import some amount of coal of specific grades and in some regions where importing coal can be a cost-effective option due to logistical problems.

Because of transport-related constraints, coal trade has developed around two regional markets—the Atlantic market consisting of the West European importers such as the United Kingdom, Germany, Italy and Spain, and the Pacific market consisting of Japan, South Korea and Chinese Taipei. The Pacific market is the dominant market at present—accounting for 62% of the global hard coal trade in 2009 (see Fig. 16.5). The Atlantic market is comparatively small—essentially due to environmental restrictions on coal use in power plants. However, this can change if the carbon capture and storage technology takes off.

Similarly, the major coal exporters are located in different regions (see Fig. 16.6). Australia and Indonesia are two major exporters, who together account for about 50% of the total export in 2009. The rest comes from a variety of other sources—such as Russia, Columbia, South Africa and the US. Most of the coal trade involves steam coal and the coking coal accounts for about a quarter of the total coal trade. More than one half of the coking coal for trade is supplied by Australia.

The trend of coal price is shown in Fig. 16.7. As can be seen, coal price has dramatically increased since 2001 and continued to increase until 2008. The price of coal has followed a similar path as that for crude oil. However, generally, the price in the Japanese market was higher compared to the American or European market. But the spread has reduced in recent times. The rapid rise in prices was partly attributable to high demand for coal in Asia and higher cost of input (labour and materials) for coal production due to increases in oil prices.

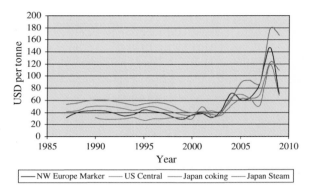

Fig. 16.7 Coal price development. *Data source* BP Statistical Review of World Energy 2010

It needs to be mentioned that on an energy content basis, coal still remains the cheapest fossil fuel. For example, the average import price for steam coal paid by the European Union members in 2007 was $3/MBtu while the price for natural gas import via pipeline was $7.5/MBtu. The import price for high sulphur fuel oil was even higher at $8.8/MBtu (WEO 2008).

The cost of coal supply varies considerably depending on the geological condition of mines, mining technique used, size of the mine, etc. The cost can even vary within a particular country as well. Generally, the cost of production varies between $20 and $50 per tonne. Comparatively low cost of production of some exporters like Australia, Colombia and Indonesia leave them with sufficient profit margins when export prices are high, as was seen between 2005 and 2008. High prices also attract new investments in the industry but the emergence of excess capacity leads to low prices subsequently.

According to the forecasts by the International Energy Agency, coal will continue to play its role as the second most important source of global primary energy until 2030. WEO (2008) indicates that the global primary demand for coal will rise to 4908 Mtoe by 2030, and the share of coal in the global energy mix will increase to 29% by that date. Most of the additional demand will arise from Asia, especially from China and India, due to their continued reliance on coal-fired power generation. Steam coal will remain the dominant type of coal due to increasing demand for this variety and account for about 80% of the global coal output. However, this assumes that adequate investments will be made on time in developing new mines to add adequate supply capacity. However, concerns are emerging in India that its coal reserves may not be sufficient to support adequate supply in the long term and that the quality of the output may deteriorate at a fast rate.

According to WEO (2008), inter-regional coal trade will not see any major change between now and 2030. India is likely to emerge as a major coal importer during this period, while Japan and South Korea will continue to import large quantities of coal. Australia will remain as the major exporter and Russia is likely to expand its export considerably during this period.

However, due to higher demand and use of coal, CO_2 emissions from coal use are likely to increase considerably. WEO (2008) suggests that CO_2 emissions from coal will reach 18.6 Gt in 2030 from a level of 11.7 Gt in 2006. This represents 42% of global energy-related emissions in 2030. Initiatives for mitigating the climate change will therefore require concerted efforts in this area.

16.3 Changes in the Coal Industry

The global coal industry has undergone a metamorphosis (Martin-Amouroux 2008). At a time when constraining carbon emission appears to be one of the most serious issues facing the world, coal has improved its position in the overall primary energy supply and even this trend will continue until 2030 as reported in WEO (2008). This contradiction, although intriguing, highlights important changes and developments that have taken place in the global coal industry. Tracing the history of the coal industry, Martin-Amouroux (2008) indicated that three major developments are note worthy:

- An east-ward shift of the coal industry's centre of gravity from a more developed country phenomenon of the past. China's rapid growth as a major coal producer has brought a dimensional change to the industry.
- Migration of production to new areas and the emergence of a group of coal exporters led by Australia. New members are being added to the list from across the world, which is broadening the industry's sphere of influence.
- Large to very large companies are now managing the operations of the coal industry. Coal producing nations now rely on big firms capable of extracting between 50 and 200 Mt per year. Similarly, the international market is dominated by four major firms: "BHP-Billiton, Anglo, Xstrata/Glencore and Rio Tinto control the majority of global exports: South Africa (86%), Colombia (82%, if we add Drummond), Australia (67%), Indonesia (38%) and even Russia (40%), through commercial agreements. They also dominate production (50% and more) in the top three exporting nations" (Martin-Amouroux 2008).

Looking at the Chinese coal industry, Zhu and Cherni (2009) provide an account of the transformation and confirm that the coal industry in the country is also moving towards the west of the country in search of new coal prospects. The government has launched the Giant West Development Programme to promote coal mining in that region. They also highlight the following characteristics of the Chinese coal industry:

- large mining companies co-exist with small-scale entities and the performance varies significantly. The average mine size of a small local firm is just 40 kt per year whereas large firms operate mines having on average size of 10 Mt.
- The small firms still contribute a significant share of country's coal output, despite government intentions of closing their operations. Their low cost of

operations and high demand for coal make their operations viable despite their small size.
- The environmental impact of the coal industry has been significant in terms of damages to the local environment.

Martin-Amouroux (2008) found that the metamorphosis of the industry was supported by the re-organisation of the industry. The integration with down-stream activities and consolidation of the industry through mergers and industrial reorganisation has created more efficient entities that are well placed to manage business risks. For example, steel makers have acquired stakes in coal mining to ensure adequate supplies of coking coal while the coal industry received capital for expansion and modernisation of their business. This synergy was exploited throughout the world and some companies have even looked at overseas investment opportunities (such as the Brazilian company Vale investing in Mozambique, Chinese company Shenhua investing in Indonesia WEO (2008)).

Simultaneously, the reorganisation of the industry and infusion of capital has allowed the industry to exploit the economies of scale. Newer technologies such as longwall method of extraction are being employed as well. The productivity of the industry is improving as a result.

16.4 Technological Advances and the Future of Coal

Despite it large resource base and affordable price, coal suffers from the environmental externality. It is much more carbon intensive compared to other fossil fuels and in a world where the global attention is towards limiting carbon emissions, coal faces the greatest challenges. However, a number of new technologies have arrived to offer some respite.[2]

a) The Integrated Gasification Combined Cycle is one such option that is more efficient compared to the conventional pulverised fuel or supercritical pulverised coal plants. An IGCC plant reduces SO_x and NO_x emissions and almost completely eliminates particulate emissions. However, an IGCC plant is a costlier option compared to the standard coal plants for power generation.
b) Another much talked about option is the carbon capture and storage technology. This is a technology in the demonstration phase where either existing coal-fired power plants are retrofitted with carbon capture facilities or new capacities are installed with the option. The captured CO_2 is then stored in secure sites thereby preventing their release to the atmosphere. The commercial application of this technology faces the investment challenge and will increase the cost of electricity production significantly. Also, the net carbon benefit is reduced by the additional energy requirement for the process. But coal can benefit from

---

[2] See IEA (2008) and EC (2007) for further details.

this technology if costs reduce and the energy requirement reduces with further development of the technology.
c) Underground coal gasification is another option being considered whereby coal can be converted to gaseous form and extracted more efficiently compared to the conventional mining of coal. WEO (2007) estimated that a conservative estimate of the size of underground coal gas could be almost similar to the current size of conventional natural gas reserves. However, the commercial development of this technology is not expected in the near future.
d) Coal to liquid—This technology has been in existence for a long time and has been commercially used in South Africa. Recent high oil prices have also renewed the interest in this technology. Here coal is converted to commercially valuable liquid fuels such as diesel or pure liquid cooking fuels for households. Experiments are being in China and Australia to commercialise this technology but generally the technology can be competitive only at high oil prices.
e) Coal bed methane—Coal mines release methane that can be captured for energy use. The amount of methane available from a mine varies with the depth of the mine and rank of coal. As methane build up in a mine has to be controlled for safety reasons, capturing it for energy purposes can be a viable option, especially for local use. Developing countries also can benefit from the Clean Development Mechanism credits for such projects. This has led to a flurry of activities in China.

The present growth pattern of coal, especially in fast growing Asian economies, and the relative abundance of coal could allow its continued use for a long time in the future. Whether it will regain a dominant position is matter of empirical analysis but its cheap availability will surely ensure its position in the power sector for a long time, unless the climate-related policies turn very hostile to coal.[3] The increased level of pollution at the local level could also affect its growth but the use of advanced technologies would hopefully mitigate some of the problems.

16.5 Conclusion

This chapter has briefly introduced the coal markets. Because of its characteristics, coal does not enjoy a global market and is mostly used in countries where it is produced. However, this old industry has shown great resilience in weathering out major challenges and has continued to grow at a fast rate in recent times. In a carbon-constrained energy future, coal faces further challenges and it remains to be seen how the industry faces them to retain its prominent position.

---

[3] See Gordon (2009) and Martin-Amouroux (2008) for more details.

References

EC (2007) World energy technology outlook 2050, WETO-H2. European Commission, Brussels
Gordon RL (2009) The prospects for coal in the twenty-first century, chapter 19. In: Evans J, Hunt LC (eds) International handbook on the economics of energy. Edward-Elgar, Cheltenham
IEA (2008) Energy technology perspectives—scenarios and strategies to 2050. International Energy Agency, Paris
IEA (2010) Key world energy statistics. International Energy Agency, Paris
Martin-Amouroux JM (2008) Coal—the metamorphosis of an industry. Int J Energy Sector Manage 2(2):162–180
WEO (2007) World energy outlook 2007. International Energy Agency, Paris
WEO (2008) World energy outlook 2008. International Energy Agency, Paris
World Coal Institute (2005) The coal resource: a comprehensive overview of coal. World Coal Institute, London
World Coal Institute (2009) Coal facts 2009 edition. The World Coal Association, London (see http://www.worldcoal.org)
Zhu S, Cherni JA (2009) Coal mining in China: policy and environment under market reform. Int J Energy Sector Manage 3(1):9–28
Zientara P (2007) Polish government policies for coal. Int J Energy Sector Manage 1(4):273–294

Chapter 17
Integrated Analysis of Energy Systems

17.1 Introduction

We have looked into different components of the energy system. We started with the demand-side and considered demand analysis and forecasting tools. We have looked into the supply-side of the system by specifically analysing the investment and operating decisions of fossil fuel supply and electricity supply. In between we have also considered the demand-side management options and economic analysis of investments.

This chapter pulls all these elements together to present an overall picture of the energy system. The objective of this chapter is to introduce the tools that could be used to analyse the energy system in an integrated manner for present and future decision-making. In the process we also briefly consider the economic and environmental aspects related to the integrated energy system. As before, the focus is on simple tools but we provide hints and directions for additional reading for more complex and practically useful methods.

17.2 Evolution of Energy Systems Models[1]

Since the early 1970s, when the energy system came to limelight because of sudden price increases, a wide variety of models became available for analysing energy systems or sub-systems (such as the power system). These models were not developed for the same purpose—they were concerned with better energy supply system design given a level of demand forecast, better understanding of the present and future demand–supply interactions, energy and environment interactions, energy-economy interactions and energy system planning. According to Hoffman

---

[1] This is based on (Bhattacharyya and Timilsina 2009, Bhattacharyya and Timilsina 2010). See also Urban et al. (2007), Jebaraj and Iniyan (2006), and Nakata 2004.

and Wood (1976), "Energy system models are formulated using theoretical and analytical methods from several disciplines including engineering, economics, operations research, and management science." As a consequence, these models apply different techniques. Even in the 1970s, Hoffman and Wood (1976) identified a number of techniques: "mathematical programming (especially linear programming), econometrics and related methods of statistical analysis, and network analysis." The list has grown in recent times.

The data requirement for the models vary as well: some technologically explicit models require a huge database, most of which is not readily available in the developing country contexts. The skill requirement and computing requirement for some models can be too onerous for developing countries where the pool of skilled human resource may be in short supply. Most of these models were developed in the industrialised countries to analyse a specific issue or a problem in a specific context. Some of these models have been applied to the developing country contexts but such a transfer of modelling technologies is fraught with difficulties. A relatively few set of models are found in the literature that are developed in the developing countries but often such models did not cross national boundaries to generate a wider developing country portfolio of modelling tools.

17.2.1 Historical Account

As an energy balance provides a simple representation of an energy system, the energy accounting approach is one of the frameworks used in energy system analysis. Hoffman and Wood (1976) describe the initial efforts in this area and suggest that this consistent and comprehensive approach has been used since 1950s in the US. The accounting framework of analysis is very popular even today and models such as LEAP or MEDEE/MAED essentially employ this framework.

A natural extension of the energy balance framework was to use a network description of the energy system to represent energy flows. This development took place in the early 1970s and has found extensive use until now. The reference energy system (RES) captures all the activities involved in the production, conversion and utilisation of energy in detail by taking the technological characteristics of the system into account. This approach allows incorporation of existing as well as future technologies in the system and facilitates analysis of economic, resource and environmental impacts of alternative development paths. This approach was developed by Hoffman and Wood (1976) and has set a new tradition in energy system modelling.

Although the pictorial presentation becomes complex with addition of more technologies and resources, the advantage of this approach derives from the ease of developing an optimisation or a simulation model based on the RES to analyse complex problems. The fundamental advantage of this approach was the ability to apply optimisation techniques to analyse alternative forms of system configuration using alternative technologies and energy sources, given a set of end-use demand.

Thus from the early stage of RES development, the linear programming models were used. One of the well-known applications of the early days was the BESOM model (Brookhaven Energy System Optimisation Model) that was developed for efficient resource allocation in the US. The first version of the model was implemented at the national level for a snap-shot analysis of a future point in time. A number of other versions were developed subsequently, that extended the capabilities of the model, including a macro-economic linkage through an input–output table (Hoffman and Jorgenson 1977). Similarly, multi-period or dynamic models have emerged and in fact, one of today's best known energy system models, MARKAL, is indeed a derivative of the BESOM model.

Munasinghe and Meier (1993) indicate that many countries followed the BESOM example and developed their own model or adapted the BESOM model. Examples include TEESE model for India, ENERGETICOS for Mexico, etc. In addition to country specific models, more generic models for wider applications also came into existence. EFOM and MARKAL models come under this category. For developing countries, RESGEN was widely used (Munasinghe and Meier 1993).

In the US, Hudson and Jorgenson (1974) pioneered the tradition of linking an econometric macroeconomic growth model with an inter-industry energy model. The input–output coefficients of the inter-industry model is endogenously determined, and the macro-model allowed a consistent estimates of demand and output.

While most of the above initiatives were at the national level, the pioneering works of large-scale global modelling started with the efforts of Jay Forrester for his World Dynamics and its application in Limits to Growth by Meadows et al. (1972). As is well known now, the doomsday prediction of this report fuelled a fierce debate about resource dependence for economic growth and the issue of sustainability. Despite its limited representation of the energy sector and the limited following of the report, this initiated a new trend of global modelling. At a collective level, the efforts of the Workshop on Alternative Energy Sources (WAES 1977), of US Energy Information Administration (EIA 1978) and of International Institute for Applied System Analysis (IIASA) [in Haefele et al. (1981)] stand out.

One of the major developments during 1973–1985 was the investigation and debate about the interaction and interdependence between energy and the economy. In a simple aggregated conceptual framework, Hogan and Manne (1979) explained the relationship through elasticity of substitution between capital and energy, which consequently affects energy demand. Berndt and Wood (1979) is another classical work in this area which suggested that capital and energy may be complimentary in the short-run but substitutable in the long-run. In contrast, Hudson and Jorgenson (1974) used a disaggregated study using the general-equilibrium framework to analyse the effects of oil price increases on the economy.

The other major development of this period is the divergence of opinion between top-down and bottom-up modellers. While the traditional top-down approach followed an aggregated view and believes in the influence of price and markets, the bottom-up models stressed on the technical characteristics of the energy sector. Despite attempts of rapprochement the difference continues until now.

The high prices of oil in the 1970s emphasised the need for co-ordinated developments of the energy systems and led to a number of modelling efforts for strategic planning. IAEA developed WASP for the electricity sector planning in 1978. This model has been used extensively and modified over the past three decades to add various features. Electricity related models often tend to rely on optimisation as the basic approach. Hobbs (1995) identifies the following as the main elements of their structure:

a. an objective function where often cost minimisation is considered but financial and environmental goals can also be used;
b. a set of decision variables that the modeller aims to decide through the model;
c. a set of constraints that ensure the feasible range of the decision variables.

The concept of integrated planning received attention at this time and efforts for integrated modelling either by linking different modules or by developing a stand-alone model multiplied.

At the country level, we have already indicated the developments in the US. A set of alternative models was developed in France, including two widely used models, namely MEDEE and EFOM. India relied on an input–output model for its planning purposes and included energy within this framework. Parikh (1981) reports an integrated model for energy system analysis. This was a sort of hybrid model that had a macro-economic element connected with a detailed end-use oriented energy sector description. The focus shifted to energy-environment interactions in the mid-1980s. This is the time when deregulation of the energy sector also started. The energy models incorporated environmental concerns more elaborately and the practice of long-term modelling started at this stage. Later, TEEESE (Teri Energy Economy Environment Simulation Evaluation model) was used in India for evaluating energy environment interactions and in producing a plan for greening the Indian development (Pachauri and Srivastava 1988).

In the 1990s, the focus shifted towards energy-environment interactions and climate change related issues. Most of the energy systems models attempted to capture environmental issues. For energy models this was a natural extension:

- the accounting models could include the environmental effects related to energy production, conversion and use by incorporating appropriate environmental coefficients;
- the network-based models could similarly identify the environmental burdens using environmental pollution coefficients and analyse the economic impacts by considering costs of mitigation;
- energy models with macro linkage could analyse the allocation issues considering the overall economic implications.

Markandya (1990) identified four approaches that were used for the treatment of environmental issues in electricity planning models as follows:

a. Models that includes environmental costs as part of energy supply costs and to minimise the total costs;

b. Models that include environmental costs in the supply side but minimises costs subject to environmental constraints;
c. Models that aim for cost minimisation but also include an impact calculation module that is run iteratively to evaluate alternative scenarios;
d. Models not based on optimisation but analyses the impacts of alternative power development scenarios.

During this period, the effort for regional and global models increased significantly and a number of new models came into existence. These include AIM (Asian-Pacific Model), SGM (Second Generation Model), RAINS-Asia model, Global 2100, DICE, POLES, etc. At the same time, existing models were expanded and updated to include new features. MARKAL model saw a phenomenal growth in its application world wide. Similarly, LEAP model became the de-facto standard for use in national communications for the UNFCCC reporting. As the climate change issue required an understanding of very long terms (100 years or more), modellers started to look beyond the normal 20–30 years and started to consider 100 or 200 years. However, the uncertainty and risks of such extensions are also large and the validity of behavioural assumptions, technological specifications and resource allocations becomes complex. This has led to incorporation of probabilistic risk analysis into the analysis on one hand and new model development initiatives on the other (e.g. VLEEM initiative of the EU).

17.3 A Brief Review of Alternative Modelling Approaches

We consider the following models: (1) Bottom-up optimisation models, (2) bottom-up accounting models, (3) econometric models, and (4) hybrid models.

17.3.1 Bottom-up, Optimisation-Based Models

17.3.1.1 Model Description

This is a widely used approach in energy modelling. Even today this remains the most popular tool for energy system analysis. Examples of such models include RESGEN, EFOM, MARKAL and TIMES.

RESGEN, a model developed by the Resource Management Associates, was a widely used model in the 1990s for energy planning in developing countries. It relies on the RES approach and uses linear programming as the solution technique. It allows three different types of demand structures: econometric specifications, industry/project specific demands and process models. For the electricity sector, plant specific dispatching is permitted using a linearised load duration curve.

More recently, this was used in RAINS-ASIA model for generating energy scenarios for a large number of Asian countries.

EFOM was initially developed in the 1970s by Finon (1974), (as indicated by Sadeghi and Hosseini (2008)) and was then widely used in the European Union and other countries in Asia (Pilavachi et al. 2008). It is a multi-period system optimisation tool based on linear programming that minimises the total discounted costs to meet the exogenously specified demand of a country. The model can be used to analyse a specific sector (single sector mode of analysis) or for the overall energy system planning exercise (multi-sector mode). The electricity industry is extensively covered by the model. To increase the environmental capability of the model, the model was modified and a new version called EFOM-ENV came into existence. This is normally considered as a sister model of the MARKAL family of models.

EFOM employs the network representation of the activities in the form of a RES. Being an end-use driven model, it is also technologically rich and covers both supply and end-use technologies. Its optimisation approach allows identification of marginal costs, and accordingly, the results have intuitive and economic appeal.

The MARKAL (Market Allocation Model) model is the most widely used and best known in this family of optimisation models (Seebregts et al. 2001). The model uses the linear optimisation technique to generate the least cost supply system to meet a given demand given the energy system configuration (technical aspects including the efficiency), energy resource availability specified by the users. The model identifies the optimal feasible configuration that would ensure least-cost supply of energy to satisfy the demand.

The model covers the entire energy system—from energy resources to end-uses through energy conversion processes. Like other bottom-up models, the model provides a detailed technological representation of the energy system and can be used to analyse the environmental effects as well. The building blocks of the standard model are indicated in Fig. 17.1 below.

The original model has been extended in various ways and now a family of MARKAL models exists. The assumption of exogenous demand specification of the standard model has been overcome in some extensions to make demand price-responsive. This produces a more realistic solution than the standard model under the tax policies or emission constraints.

The TIMES model is the new avatar of the MARKAL and EFOM models where the features of the two widely used models have been integrated to produce a powerful analytical tool using the optimisation technique (Loulou et al. 2005; Vaillancourt et al. 2008). The model produces the least-cost solution as MARKAL or EFOM but also considers the investment and operating decisions and can be applied for the entire system or a specific sector.

The demand-side of the model uses exogenous assumptions about demand drivers and the elasticities of demand with respect to these drivers and prices. Through these elasticities, the model can capture the effects of policy changes (price or tax or environmental constraints) on demand. This is an enhanced capability of the model compared to standard MARKAL model.

17.3 A Brief Review of Alternative Modelling Approaches

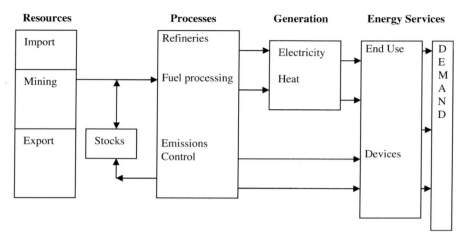

Fig. 17.1 MARKAL building blocks. *Source* Seebregts et al. (2001)

The supply-side consists of a set of supply curves representing the potential available resources. The model accepts multi-stepped supply curves, with each step representing the potential corresponding to a given cost. The model seeks to optimise the total surplus (consumers and producers surplus) and leads to partial equilibrium solutions.

The model is a multi-period model that can be applied to a large number of regions and can capture trading options. This is another additional feature of this model that was not available in the MARKAL model.

17.3.1.2 Mechanics

The basic formulation of an optimisation model for an energy system can be explained using a simple example given below. Consider that there are four end-use demand categories in the country for simplicity, namely water heating and cooking, lighting, industrial operations and transport. Assume also that there are four types of energy resources available in the country: hydropower, coal, oil and natural gas. The following additional assumptions are also made:

- all fuels can be used for cooking;
- electricity and oil can be used for lighting;
- industrial activities are electricity operated;
- transport uses only petroleum products;
- losses in the refinery or in the transportation/distribution system are ignored.

The reference energy system can be presented as in Fig. 17.2.

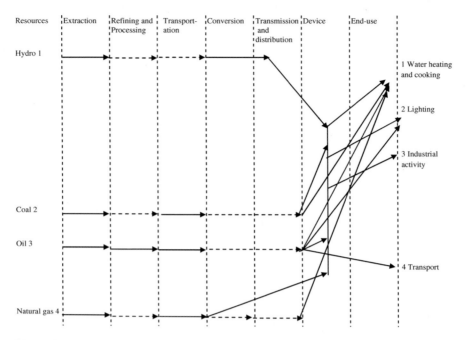

Fig. 17.2 RES for the example case

Let us define the notations for mathematical formulation first.

X_{ij0} = quantity of primary energy of type i supplied to demand category j in the form of secondary energy (such as electricity),
X_{ij1} = quantity of primary energy of type i supplied to demand category j without any transformation,
U_{ij0} = efficiency of utilising device in demand category j using energy type i in the form of secondary energy,
U_{ij1} = efficiency of utilising device in demand category j using energy type i without any transformation,
V_i = conversion of generation efficiency of energy type i into electricity,
S_i = maximum quantity of available energy of type i,
D_j = useful energy requirement of demand category j,
C_i = per unit cost of using energy type i,

The objective is to minimise the total cost of supplying energy to meet the needs of different end-use categories. This can be written as:

$$\text{Min}[C_1 \cdot (X_{110} + X_{120} + X_{130}) + C_2 \cdot (X_{210} + X_{220} + X_{230} + X_{211}) \\ + C_3 \cdot (X_{310} + X_{320} + X_{330} + X_{311} + X_{321} + X_{341}) \\ + C_4 \cdot (X_{410} + X_{420} + X_{430} + X_{411})] \quad (17.1)$$

17.3 A Brief Review of Alternative Modelling Approaches

The minimisation problem is subject to the following constraints:

(a) Demand constraints
The energy needs for cooking and hot water supply have to be met. This can be written as

$$U_{110} \cdot V_{110} \cdot X_{110} + U_{210} \cdot V_{210} \cdot X_{210} + U_{310} \cdot V_{310} \cdot X_{310}$$
$$+ U_{410} \cdot V_{410} \cdot X_{410} + U_{211} \cdot X_{211} + U_{311} \cdot X_{311} + U_{411} \cdot X_{411} = D_1 \quad (17.2)$$

The lighting need has to be satisfied, which can be written as

$$U_{120} \cdot V_{120} \cdot X_{120} + U_{220} \cdot V_{220} \cdot X_{220} + U_{320} \cdot V_{320} \cdot X_{320} + U_{420} \cdot V_{420} \cdot X_{420}$$
$$+ U_{321} \cdot X_{321} = D_2 \quad (17.3)$$

Similarly, the energy needs for industrial activity and transport have to be met and are written below.

$$U_{130} \cdot V_{130} \cdot X_{130} + U_{230} \cdot V_{230} \cdot X_{230} + U_{330} \cdot V_{330} \cdot X_{330}$$
$$+ U_{430} \cdot V_{430} \cdot X_{430} = D_3 \quad (17.4)$$

$$U_{341} \cdot X_{341} = D_4 \quad (17.5)$$

(b) Supply constraints
For each type of energy source, a specific constraint indicating its availability has to be added.
For hydropower,

$$(X_{110} + X_{120} + X_{130}) \leq S_1 \quad (17.6)$$

Similarly for other sources, the constraints are written as:

$$(X_{210} + X_{220} + X_{230} + X_{211}) \leq S_2 \text{ [for coal]} \quad (17.7)$$

$$(X_{310} + X_{320} + X_{330} + X_{311} + X_{321} + X_{341}) \leq S_3 \text{ [for oil]} \quad (17.8)$$

$$(X_{410} + X_{420} + X_{430} + X_{411}) < S_4 \text{ [for natural gas]} \quad (17.9)$$

(c) Non-negativity constraints
All decision variables are non-negative. That is

$$X_{ijk} \geq 0 \quad (17.10)$$

This completes the mathematical formulation of the simple problem. The linear programme can be solved for specific values of parameters (efficiencies), and end-use demands and supply availabilities. The solution will provide the optimal supply strategy to minimize the supply cost. The shadow price or the opportunity cost of using the resources can also be obtained from the dual solution.

This is a flexible tool that can capture the technical features of an energy system in great detail. It can also include new and emerging technologies as well as new demands. Alternative pathways can be easily considered to see how they perform in the optimal solution. Generally, the linear programming models are sensitive to data and a minor change in the parameter or constraints can result in substantially different results. While the method can capture the technical details and energy system structure in detail, the data requirement is very high to populate a representative system. This also requires a good understanding of the energy sector and high numerical skills.

17.3.2 Bottom-up, Accounting Models

This is another category of highly popular energy system models. These models follow the accounting framework (discussed in Chap. 2) to generate a consistent view of energy demand (and supply) based on the physical description of the energy system. They also rely on the scenario approach to develop a consistent storyline of the possible paths of energy system evolution. Thus for the demand forecasting, these models do not optimise or simulate the market shares but analyses the implications of possible alternative market shares on the demand.

The supply-side of the model does not try to find the least cost solution or system configuration as in the optimisation model but uses accounting and simulation approaches to provide answers to "what-if" type of analysis under alternative possible development scenarios. Such models can be developed in spreadsheet like tools and are flexible enough to consider various data requirements. The framework, based on a popular model LEAP (Long-range Energy Alternative Planning Model), is presented in Fig. 17.3.

Well-known tools of this category include LEAP, MEDEE (Model for Evaluating Demand for Energy) family of programmes and MAED (Model for Analysis of Energy Demand). The Long-range Energy Alternatives Planning (LEAP) is a flexible modelling environment that allows building specific applications suited to particular problems at various geographical levels (cities, state, country, region or global). As an integrated energy planning model LEAP covers both the demand and supply sides of the energy system. MEDEE and MAED belong to the same family of models developed by Chateau and Lapillonne (1978, 1990). The difference lies in the improvements made in the revisions to the models to enhance flexibility and capability. As they are easy to use, end-use models have been widely used both in developed and developing countries.

There are two main concerns about this approach: (1) such models are highly data intensive and require a well planned database; (2) their assumptions may be internally inconsistent. These models use assumptions about the economic structure

17.3 A Brief Review of Alternative Modelling Approaches

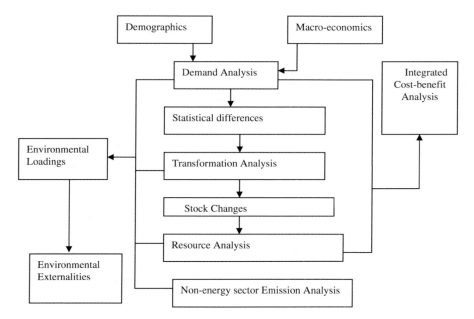

Fig. 17.3 LEAP framework. *Source* Heaps (2002)

and other policy variables but whether or not such assumptions are consistent with the economic realities of a country is difficult to ascertain.

17.3.3 Top-down, Econometric Models

This tradition of modelling follows directly from economic theories and hence is sympathised by the economists. Here a set of econometrically estimated relationships are used to describe energy supply and demand of a country. They include price and economic activity as the prime drivers and depending on the sophistication of the model, additional features can also be found. An example of such a model is the tool used by the British government agency for its energy forecasts and future carbon emission estimations. The erstwhile Department of Trade and Industry (DTI) relied on an econometric model that covered both supply and demand sides but the demand was fairly elaborate. The demand model contained 150 econometric relationships to determine the demand in various sectors of the economy. The model followed the Error Correction Modelling approach and uses price and economic activity as main variables although time trends are used in some sectors. The model had 13 final users who are then grouped into four major sectors, namely industry, transport, services and domestic. Each final user sector is

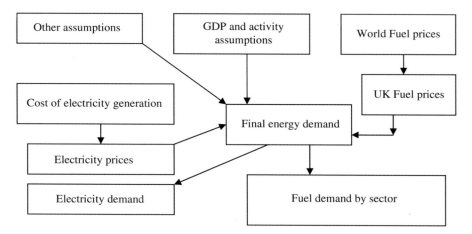

Fig. 17.4 DTI energy model overview. *Source* Based on DTI (See http://www.berr.gov.uk/files/file26611.ppt)

further disaggregated by fuels. The model structure of the model is shown in Fig. 17.4.

The supply side of the model considered the electricity supply system in detail. It captures the diversity of the capacity mix, technological differences and characteristics, and determines the cost of generation and operation of the system to meet the demand. The supply and conversion of other fossil fuels are taken into consideration as well.

17.3.4 Hybrid Models

This approach, as the term indicates, relies on a combination of two or more methods—often in an attempt to bridge the difference between the top-down and the bottom-up approaches. These models have become very widespread now and it is really difficult to classify any particular model into a specific category. For example, econometric models now adopt disaggregated representation of the economy and have internalised the idea of detailed representation of the energy-economy activities. Similarly, engineering-economy models use econometric relationships at the disaggregated levels thereby taking advantages of the econometric estimation method. The end-use approach heavily relies on the scenario building approach to enrich itself.

A number of hybrid models are widely referred to in the literature. These include—NEMS (National Energy Modelling System) used by the US Department of Energy, POLES (Prospective Outlook on Long-term Energy Systems) model

17.3 A Brief Review of Alternative Modelling Approaches

used by the European Union, and WEM (World Energy Model) used by the International Energy Agency for its World Energy Outlooks.

NEMS is a model of energy-economy interaction that is used to analyse the functioning of the energy market under alternative growth and policy scenarios. The model employs a technologically rich representation of the energy sector and covers the spatial differences in energy use in the US. At the same time, it retains the behavioural analysis of top-down models.

POLES is a recursive, disaggregated global model of energy analysis and simulation. This is embedded in the end-use tradition but uses calibrated relationships similar to econometric models. It also uses energy price as one of the drivers as well unlike most end-use models. It is meant for long-term energy policy analysis. The model has four main modules: final energy demand, new and renewable energy technologies, conventional energy transformation system and fossil fuel supply. Accordingly, the model captures the entire energy system.

While the regional and country level analyses generate the respective energy balances, they are horizontally linked through an energy market module which is used to clear the market. For oil, a single global market is considered while for coal three regional markets have been used. For gas, bilateral trade flows are considered. This price-driven formulation of the model makes it different from others of its kind (i.e. accounting, end-use models).

Like POLES, WEM is a global energy market model. The model has evolved over time but the basic model has four main components: a final demand module, power generation module, fossil fuel supply and emissions trading. The model considers oil and gas supply in detail by taking OPEC, non-OPEC and non-conventional oil production. For gas supply, net importers and exporters are considered separately and the regional nature of the gas market is taken into account. Coal supply is not explicitly modelled but is included in the supply system.

Despite retaining its general structure, the model has undergone significant changes over time. In recent times, the access issue has been considered and the residential sector has been modified considerably. Similarly, the industry and transport sectors details have been improved and in its latest version, the model was linked to a macro model to ensure macro-economic consistency of model assumptions.

17.3.5 Some Observations on Energy System Modelling

A review by Bhattacharyya and Timilsina (2009) concluded that the purpose built models of national focus generally lack transferability and are not suitable for wider applications. This is especially true for models of the econometric tradition and hybrid modelling approach. On the other hand, end-use models tend to be more generic in nature and require relatively less skills due to their accounting approach.

There is no guarantee that complex models necessarily lead to better results. Moreover, the developing countries have certain specific characteristics which are not adequately captured by models originating from the developed countries.

The problem is more pronounced with econometric and optimisation models than with accounting models. The level of data requirement and the theoretical underpinning of these models as well as their inability to capture specific developing country features such as informal sectors and non-monetary transactions (Shukla 1995) make these models less suitable. The accounting type end-use models with their flexible data requirements and focus on scenarios rather than optimal solutions make them more relevant for developing countries. The global models also suffer from the same problems—as the developing countries are given limited focus in such models and the modelling approach is not modified for developing countries.

In addition to lack of adequate data and skills, the issues like existence of informal sector, non-monetary transactions, shortages of energy supplies and the transition from traditional to market economic systems require careful consideration. Often such non-price related issues cannot be captured in the econometric or econometric-style relations. In contrast, inherent simplicity of use and the ability to capture alternative structural and socio-economic conditions in end-use models make them popular for policy analysis purposes.

17.4 Energy Economy Interactions

So far, we have considered the energy sector alone, without considering its relation with the other components of the economy. Yet, the energy sector does not exist in isolation. As energy is an important input for economic activities changes in the availability and prices impose adverse effects on the economy and can be a source of concern.

Energy supply requires other factor inputs which come from the households and firms in the country or from the external sector. Energy supply may depend on imports (or lead to exports for an exporter), which could affect the balance of payments and the external sector of the economy significantly. Depending on the importance of the imports (exports), the country could be exposed to risks of price changes in the traded energy commodities. This risk often translates into exposure to higher inflation because of the impact on prices of other factors of production. As this reduces the purchasing power of the salaried labour, the demand for wage adjustments would arise. If wages are revised upwards due to inflationary pressure, a wage-price spiral results.

Changes in relative prices influence any substitution of energy by other factors of production or by other types of energies. While price influences the inter-fuel substitution, the derived nature of the demand also implies that the joint decision about appliance choice also influences the substitution process and the time for adjustment. Such a substitution is also influenced by the supply situation and the linkages of that fuel supply chain within the economy. For example, if an increase in oil price makes coal relatively cheaper to use, the demand for coal would increase. The increased demand for coal would encourage suppliers to supply more coal, which would be met initially through increased production from the existing fields. This would require higher inputs to the coal industry in terms of labour, capital, equipment, and transport facilities. If there is scarcity of any of these inputs, the supply would be constrained, which in turn would affect the demand as well. If the demand is sustained and prices provide enough remuneration to the coal producers, new fields would be developed to ensure future supplies as well.

In addition, changes in energy prices affect energy-labour and energy capital substitution processes. As the wage rate increases, labour intensive processes become unviable, permitting substitution by labour-saving technologies and processes in the energy sector as well as in other sectors. This can have important bearing on the labour use, especially in labour affluent countries, thereby bringing new issues into focus. In addition, labour-saving processes being capital intensive, the investment needs increase.

Energy price changes also affect the economy indirectly through transportation of goods and services where energy input cost can be important. This effect can have an urban bias and would be more pronounced in a large country. The mode of transport and distance would also influence the effect.

Similarly, due to capital intensive nature of the investments, the sector imposes high demand on the financial resources available for investment and thus competes with other investment demands. If a part of the investment comes from outside the economy, the indebtedness of the economy would increase. The investors expect to recover at least the opportunity cost of their investment, which then influences the supply price. But unless the price reflects the cost of supply, the demand and supply would not match and the profitability of the investment would not be ensured. Thus the investment and price link plays an important role (Fig. 17.5).

The remuneration received by the suppliers of capital and other inputs (labour, materials and land) would be partly used for purchasing energy from the suppliers. A part of the remuneration may also go outside the economy and represent the foreign debt service payments. The conditions of the markets in those inputs would affect the cost of supply while the high level of demand in the energy sector in turn influences their market price as well.

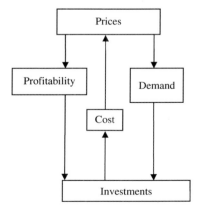

Fig. 17.5 Price–investment link

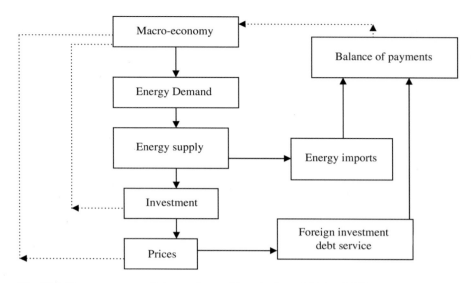

Fig. 17.6 Energy economy linkages. *Source* Munasinghe and Meier (1993)

On the other hand, the sector outputs are demanded partly by the consumers and partly by outsiders (for export). As we have seen previously, the demand is influenced by prices, supply situations and the economic condition of the country. In any properly functioning economy, adequate energy supply to meet the demand is ensured through correct pricing and adequate investments in supply. However, this is not the situation in many developing countries where the supply demand gap is a chronic problem. The economy bears some cost for inadequate supply due to loss in production, cost of alternative resource utilisation and damage to the equipment and to the environment.

The above linkages to the economy are referred to as forward and backward linkages. The forward linkage captures the macroeconomic impacts on the energy sector while the backward linkages capture the impacts of the energy sector on the national economy. Figure 17.6 indicates this in a simple diagram where the firm lines indicate the forward linkage while the dotted lines indicate the backward linkage.

17.4.1 Modelling Approaches

The analysis of energy economy interactions requires specific analytical tools. In Chap. 5, the input–output model was introduced. Input–output models are used to provide a consistent picture in inter-sectoral flows due to changes in demand and/ or prices. To analyse the price effect, the dual model associated with the standard Leontief model is used with the assumption that the price depends on the cost of production but not on the level of production. This allows price estimation without

17.4 Energy Economy Interactions

linking it to the activity level. Various multipliers (such as output multipliers, income multipliers and employment multipliers) are used in such works.

17.4.1.1 Analysis Using SAM

A related technique uses the Structural Accounting Matrix (SAM) that relies on the national accounting information. The SAM is normally considered as the basic starting point for any multi-sectoral modelling exercise. The SAM provides a consistent picture of the circular flow of funds within an economy. It is presented in a square table form (i.e. with same number of rows and columns) and ensures consistency of data that can be used in multi-sector models. It also allows an analysis of impacts following techniques similar to the input–output methods and can be used to capture effects of exogenous price changes on the economy.

SAM has a flexible structure and the level of disaggregation and details can be decided according to the modelling needs. The table is always presented in a square matrix where the rows and columns have the same order of arrangement. The total of the rows is always equal to that of the columns and the entries are in monetary unit (Taylor 1990).

Table 17.1 provides an example of a simple SAM. Producers buy inputs T11 and factors of production T21 to produce output Y1. T41 represents all spending for taxes and imports. Households provide the labour and employ their capital to receive a remuneration T32 and income from outside T32. Households spend a part of their income T13 for consumption goods and the rest is used for saving and tax payments (T43).

A SAM is prepared from the input–output table, national accounts information and other information about households. It is important to ensure data coherence as the definitions used in different sources may vary. Similarly, the level of disaggregation and higher accuracy of analysis has to be traded off against the data availability, model complexity and analytical difficulties. As any analysis using SAM only requires matrix manipulation, a higher level of disaggregation can be easily incorporated.

The accounting properties of SAM also allows for more detailed analysis of economic issues. Assume that the price depends on the cost of production but not on the level of production. Assume that group 1 of Table 17.1 is an endogenous

Table 17.1 A simple SAM

Income → Expenditure ↓	I	II	III	IV	V
I Production	T11	0	T13	T14	Y1
II Factors of production	T21	0	0	T24	Y2
III Households	0	T32	T33	T34	Y3
IV Rest of the World	T41	T42	T43	T44	Y4
V Total	Y1	Y2	Y3	Y4	

Source Roland-Host and Sancho (1995)

variable while the rest are exogenous variables. If p_i is the price of activity i, reading column 1 of Table 17.1, we can write the following formula:

$$p_1 = p_1 A_{11} + \bar{p}_2 A_{21} + \bar{p}_4 A_{41} = (\bar{p}_2 A_{21} + \bar{p}_4 A_{41})(I - A_{11})^{-1} = v_1 M_{11} \quad (17.11)$$

where v_1 is the exogenous cost vector, A_{ij} is the normalised coefficient matrix obtained from T_{ij} of Table 17.1, and M_{11} is the multiplier.

Yet, the above equation ignores the links between households, producers and the factors of production assuming them exogenous. By endogenising these variables, the links can be better considered. For example, if we assume that only the rest of the world is exogenous, the rest are endogenous to the model, then, we can write the following equations:

$$p_1 = p_1 A_{11} + p_2 A_{21} + \bar{p}_4 A_{41}$$
$$p_2 = p_3 A_{32} + \bar{p}_4 A_{42} \quad (17.12)$$
$$p_3 = p_1 A_{13} + p_3 A_{33} + \bar{p}_4 A_{43}$$

If $A = \begin{bmatrix} A_{11} & 0 & A_{13} \\ A_{21} & 0 & 0 \\ 0 & A_{32} & A_{33} \end{bmatrix}$, the matrix containing different accounts, where

- the sub-matrix A_{11} represents inter-industry transactions;
- the sub-matrix A_{21} reflects the income for factors of production;
- the sub-matrix A_{13} records households' propensity to consume different outputs;
- sub-matric A_{32} consistes of incomes received by households for supplying factors of production; and
- sub-matrix A_{33} captures the inter-household transfers.

Suppose that $p = (p_1, p_2, p_3)$ be the price vector for the endogenous sectors of the SAM and that $v = p_4 A_{(4)}$ be the exogenous costs of elements A_{41}, A_{42} and A_{43}. In matrix notation,

$$p = pA + v = v(I - A)^{-1} = vM \quad (17.13)$$

where M is the multiplier matrix.

Pyatt and Round (1979) and Roland-Host and Sancho (1995) have shown that the multiplier M can be decomposed into three economically significant elements: the first element is known as transfers multiplier that captures the effect of exogenous transfers on any account. The second element captures the inter-group effects while the third element is called the circular or closed-loop effects. The details are provided in Box 17.1.

17.4.1.2 Computable General Equilibrium Models

A more sophisticated method is known as the Computable General Equilibrium (CGE) modelling technique which captures the relationship of energy industries

17.4 Energy Economy Interactions

Box 17.1: Decomposition of Price Multiplier

In mathematical terms, the decomposition of the price multiplier can be written as follows:

$$p = pA + v = vM = vM_1 M_2 M_3 = M_3^* M_2^* M_1^* v \qquad (17.14)$$

where

$$M_1^* = \begin{bmatrix} (I - A_{11})^{-1} & 0 & 0 \\ 0 & I & 0 \\ 0 & 0 & (I - A_{33})^{-1} \end{bmatrix} \qquad (17.15)$$

$$M_2^* = \begin{bmatrix} I & A_{21}^* & A_{32}^* A_{21}^* \\ A_{13}^* A_{32}^* & I & A_{32}^* \\ A_{13}^* & A_{21}^* A_{13}^* & I \end{bmatrix} \qquad (17.16)$$

$$M_3^* = \begin{bmatrix} (I - A_{13}^* A_{32}^* A_{21}^*)^{-1} & 0 & 0 \\ 0 & (I - A_{21}^* A_{13}^* A_{32}^*)^{-1} & 0 \\ 0 & 0 & (I - A_{32}^* A_{21}^* A_{13}^*)^{-1} \end{bmatrix}$$

$$(17.17)$$

$$A_{13}^* = A_{13}(I - A_{33})^{-1}; \; A_{21}^* = A_{21}(I - A_{11})^{-1}; \; A_{32}^* = A_{32} \qquad (17.18)$$

The matrix M_1^* captures the effect of an exogenous change on the account itself through direct effects. It contains only the diagonal elements. The first element of the matrix captures the effect of an exogenous increase in the cost on the productive activities through the inter-industry linkages. The second element is an identity matrix because of absence of transfer among factors of production. The third element captures the multiplier effect originating from direct transfers among institutions.

The matrix M_2^* captures the inter-group effects. The first column indicates the effect of an exogenous increase in the cost on the factors of production and on the households.

The matrix M_3^* represents circular effect in the sense that it captures the effect of an exogenous increase in the cost through all accounts. For example, the first column of the matrix M_3^* captures the impact of the increased cost first on the households, then the factors of production and finally on the productive activities.

Source Roland-Host and Sancho (1995).

along with other economic activities. These models are applied ones that find numerical solutions to a given problem (and hence the term computable). The follow the theoretical tradition of general equilibrium analysis where all markets are considered to be clearing and such market clearing actions decide the price. A wide range of styles have been used in this area. A brief introduction is provided below.[2]

In the tradition of multi-sectoral analysis, the most important tool is the computable general equilibrium models. This is a logical extension of the input–output models where the substitution possibilities in production and demand are incorporated. These models are applied models designed to undertake a quantitative analysis of a given economic problem. In a CGE model, the markets clear and a equilibrium is reached.

There are different traditions of CGE modelling: for example Jorgenson followed an econometric approach to develop the relationships; Taylor (1990) created a tradition of structuralist modelling whereas the neoclassical approach was generally followed by many others. In Box 17.2 a simple example is provided just to explain the modality of the approach.

Box 17.2: A Simple Example of a CGE Model

Consider a simplified example with a single productive sector, a single product and a representative household. The productive sector produces a good Y using two factors F and E, where F is the primary inputs and E represents energy. A CES production function (Constant elasticity of substitution) with constant returns to scale combines these inputs to outputs as follows:

$$Y = f(F, E) = A\left[\alpha F^{(\frac{\sigma-1}{\sigma})} + (1 - \alpha)E^{(\frac{\sigma-1}{\sigma})}\right]^{\frac{\sigma}{(\sigma-1)}} \quad (17.19)$$

where

$\alpha = $ is the share parameter;
$\sigma = $ substitution elasticity;
$A = $ scale parameter.

The first order condition of profit maximisation gives the following:

$$\frac{\delta Y}{\delta F} = A^{\frac{(\sigma-1)}{\sigma}} \cdot \alpha \cdot \left(\frac{Y}{F}\right)^{\frac{1}{\sigma}} = \frac{w}{p}; \quad (17.20)$$

where

$w = $ price of F;
$p = $ price of product Y;

---

[2] Interested readers can refer to Bhattacharyya (1996), Wing (2009), Dervis et al. (1982) and Shoven and Whalley (1984).

17.4 Energy Economy Interactions

$$\frac{\delta Y}{\delta E} = A^{\frac{(\sigma-1)}{\sigma}} \cdot (1-\alpha) \cdot \left(\frac{Y}{E}\right)^{\frac{1}{\sigma}} = \frac{r}{p} ; \qquad (17.21)$$

where

$r =$ energy price.

Rearranging Eqs. 17.20 and 17.21 we get,

$$\frac{F}{Y} = A^{\sigma-1} \alpha^\sigma \left(\frac{p}{w}\right)^\sigma \qquad (17.22)$$

$$\frac{E}{Y} = A^{\sigma-1}(1-\alpha)^\sigma \left(\frac{p}{w}\right)^\sigma \qquad (17.23)$$

Combining Eqs. 17.22 and 17.23 we get

$$\frac{E}{F} = \left(\frac{1-\alpha}{\alpha}\right)^\sigma \left(\frac{w}{r}\right)^\sigma \qquad (17.24)$$

It is assumed that the households own the factors of production and receive remunerations for them. It is also assumed that they spend their entire income to consume good Y. The demand is equal to the supply, which can be written as:

$$Y = \left(\frac{wF + rE}{p}\right) \qquad (17.25)$$

The market clearing price is obtained from the equilibrium condition, which takes the following mathematical form:

$$p = \frac{wF}{Y} + \frac{rE}{Y} = A^{-1}[\alpha^\sigma w^{1-\alpha} + (1-\alpha)^\sigma r^{1-\sigma}]^{\frac{1}{1-\sigma}} \qquad (17.26)$$

This system of equations does not provide a unique solution. In effect, if F and r are assumed to be fixed, we still have three unknowns (E, w and p) for two equations. Therefore, a numeraire has to be chosen to solve the problem.

A CGE model offers three main advantages: first it is based on a solid macroeconomic foundation and uses the basic elements of the neo-classical theory of optimizing behaviour of producers and consumers. This strong economic basis allows a transparent model development and helps analyzing the results using economic logic. Second, a CGE model allows an integrated analysis of a problem taking all main drivers into consideration. The third advantage relates to the numerical solution of the problem. Whereas the analytical solution using differential calculus is valid for a small change, the problem is avoided in the CGE by adopting numerical solutions (Borges 1986).

However, this tradition also has a number of constraints. First, the model becomes complex and intractable as the level of disaggregation increases. The use of representative consumers in many CGE applications reduces its practical appeal, especially where distributional aspect can be important. The strict application of the neoclassical principles also is an issue where the markets are not competitive or markets are incomplete. Further, the technological specification of such models is not given much importance, which can also limit the practical importance of the results. Finally, these models are inherently more complex and require high skills for their application.

17.5 Conclusion

This chapter has attempted to integrate the learning from the previous chapters on demand and supply and provided an integrated approach to deal with the energy sector issues. The chapter has presented a review of alternative models that have tried to address this issue. While the sector level tools are useful for energy sector analysis, the impact of the sector level decisions and actions on the economy cannot be analysed using the sector-level tools. Possible energy-economy interactions are also presented and a brief introduction to the tools to deal with economy-wide impacts is given.

References

Bhattacharyya SC (1996) Applied general equilibrium models for energy studies: a survey. Energy Econ 18(3):145–164
Bhattacharyya SC, Timilsina GR (2009) Energy demand models for policy formulation: a comparative study of energy demand models. World Bank Policy Research Working Paper 4866, World Bank, Washington, DC
Bhattacharyya SC, Timilsina GR (2010) A review of energy system models. Int J Energy Sect Management 4(4):494–518
Borges AM (1986) Les modeles appliqués d'equilibre general: une evaluation de leur utilite pour l'analyse des politiques economiques. Revue Economique de l'OCDE 7:7–47
Brendt ER, Wood DO (1979) Engineering and econometric interpretations of energy-capital complimentarity. Am Econ Rev 69(3):342–352
Chateau B, Lapillonne B (1978) Long-term demand forecasting—a new approach. Energy Policy 6(2):140–157
Chateau B, Lapillonne B (1990) Accounting and end-use models. Energy 15:261–278

References

Dervis K, De Melo J, Robinson S (1982) Computable general equilibrium models for development policy. Cambridge University Press, Cambridge, UK

EIA (1978) An evaluation of future world oil prices, EIA Analysis Memo EIA-010121/4, Energy Information Administration, Washington, DC

Finon D (1974) Optimisation model for the French energy sector. Energy Policy 2(2):136–151

Haefele W et al. (1981) Energy in a finite world. International Institute for Applied System Analysis, Austria

Heaps C (2002) Integrated energy-environment modelling and LEAP, SEI Boston and Tellus Institute, Boston, http://www.energycommunity.org/default.asp?action=42

Hobbs BF (1995) Optimisation methods for electric utility resource planning. Eur J Oper Res 83:1–20

Hoffman K, Jorgenson DW (1977) Economic and technological models for evaluation of energy policy. Bell J Econ 8:444–466

Hoffman K, Wood DO (1976) Energy system modelling and forecasting. Ann Rev En 1:423–453

Hogan WW, Manne AS (1979) Energy-economy interactions: the fable of the elephant and the rabbit. Advances in economics of energy and resources 1:7–26

Hudson EA, Jorgenson DW (1974) US energy policy and economic growth: 1975–2000. Bell J Econ 5:461–514

Jebaraj S, Iniyan S (2006) A review of energy models. Renew Sustain Energy Rev 10(4):281–311

Loulou R, Remne U, Kanudia A, Lehtila A, Goldstein G (2005) Documentation for the TIMES model, Part 1, ETSAP, http://www.etsap.org/Docs/TIMESDoc-Intro.pdf

Markandya A (1990) Environmental costs and power system planning. Utilities Policy 1(1):13–27

Meadows DH, Meadows DL, Randers J, Behrens WW III (1972) The Limits to Growth, Vol. 1. Universe Books, New York

Munasinghe M, Meier P (1993) Energy Policy Analysis and Modelling. Cambridge University Press, Cambridge, UK

Nakata T (2004) Energy-economic models and the environment. Progress in Energy and Combustion Science 30(4):417–478

Pachauri RK, Srivastava L (1988) Integrated energy planning in India: a modelling approach. Energy J 9(4):35–48

Parikh J (1981) Modelling energy demand for policy analysis. Planning Commission, Government of India, New Delhi

Pilavachi PA, Dalamaga Th, Rossetti di Valdalbero D, Guilmot J-F (2008) Ex-post evaluation of European Energy Models. Energy Policy 36(5):1726–1735

Pyatt G, Round J (1979) Accounting and fixed price multipliers in a social accounting matrix framework. Econ J 89:850–873

Roland-Host D, Sancho F (1995) Modelling prices in a SAM structure. Rev Econ Stat LXXVII:361–371

Sadeghi M, Hosseini HM (2008) Integrated energy planning for transport sector—a case study of Iran with techno-economic approach. Energy Policy 36(2):850–866

Seebregts AJ, Goldstein GA, Smekens K (2001) Energy/environmental modelling with MARKAL family of models, in Operations Research Proceedings 2001—Selected Papers of the International Conference on Operations Research (OR2001)

Shoven J, Whalley J (1984) Applied general models of taxation and international trade: an introduction and survey. J Econ Lit XXII:1007–1051

Shukla PR (1995) Greenhouse gas models and abatement costs for developing nations: a critical assessment. Energy Policy 8:677–687

Taylor L (1990) Structurally relevant policy analysis: structuralist computable general equilibrium models for the developing world. The MIT Press, Cambridge, MA

Urban F, Benders RMJ, Moll HC (2007) Modelling energy systems for developing countries. Energy Policy 35(6):3473–3482

Vaillantcourt KML, Loulou R, Waaub J-P (2008) The role of nuclear energy in the long-term energy scenarios: an analysis with the World TIMES model. Energy Policy 36(7):2296–2307

WAES (1977) Workshop on Alternative Energy Sources, Global Prospects 1975–2000, McGraw-Hill, New York

Wing IS (2009) Computable general equilibrium models for the analysis of energy and climate policies, Chapter 14. In: Evans J, Hunt LC (eds) International handbook on the economics of energy. Edward Elgar, Cheltenham

Part IV
Issues Facing the Energy Sector

Chapter 18
Overview of Global Energy Challenges

18.1 Introduction[1]

This chapter provides a stylised overview of the issues and challenges faced by the energy sector. It starts by taking stock of the historical developments and goes through various energy issues by identifying the drivers behind them and discussing some possible policy options available to deal with them. This chapter would set the scene for the subsequent chapters of this part of the book and would help you understand them better.

The issues of and concerns about the energy sector are multi-faceted, often international, spatially differentiated and dynamic. This is due to the pivotal role of energy in any individual's day-to-day life as well its importance as a key input to the production processes that transform inputs to goods and services. In addition, the sector has a multi-dimensional strategic importance in terms of macro-economic influences, geo-political implications, and environmental concerns. Often these interactions and mutual dependencies create complex problems (as indicated in Chap. 1).

Yet, we find that a common set of issues having a common appeal across the board tends to emerge at any given time. Although security of supply related issues are dominating recent discussions and drawing public attention, they do not constitute the only challenges facing the global energy sector. While "energy haves" are concerned about the future prospects, billions of others are struggling to get access to affordable, reliable, and acceptable energy services. Access to energy has been identified as a major challenge for achieving sustainable development worldwide. At the same time, fingerprints of unsustainable practices abound: unprecedented demand growth with demand arising from new centres of growth in contrast to traditional centres; globalisation of wasteful consumption patterns; supply concentration in politically unstable regions; financially bankrupt state entities perpetuating supply, etc. Governance issues, restructuring and reform

---

[1] This chapter is based on Bhattacharyya (2007a and b).

of the sector and environmental concerns remain relevant. This complex set of challenges would shape the future of the energy sector and consequently, the sector would perhaps charter a new path where the sector activities are organised and performed differently.

Concurrently, in a dynamic world where profound political, social, economic, technological and even ideological changes shape our present and future lives and living conditions, changes in the energy scenario are quite natural and inevitable. In the past, two grand transitions have shaped the developments of the global energy system by bringing profound changes in energy demand and supply, in the functioning of the energy industry as well as in the organisation and conduct of economic activities. Despite the uncertainty about the timing and nature of the next energy transition, such a change will also bring profound changes to the energy sector and to the economic activities globally. But the cost and pain associated with adjustment and adaptation required for such changes and the fear of being caught unprepared make investigations into this subject interesting.

Similarly, not so long ago we tended to think that the markets are the solutions for all evils in the sector and market-oriented policies were promoted. The energy industry has changed significantly and the way business is carried out has changed—in some cases beyond recognition. Still it is dawning on us that perhaps markets are not catering to all our needs the way we would have thought them to. Security of supply concerns mentioned earlier, investments in socially desired areas, protection of the environment and the climate and the like cannot all perhaps be left alone to the market.

The objective of this chapter is to put the challenges in their global context and proffer an overview of possible policy responses. It provides a panoramic view of the energy sector issues along with the drivers behind them and the key policy responses that can be used to address them.

18.2 Grand Energy Transitions

Historically, energy demand patterns have undergone significant changes since mankind started using energy. Prior to the Industrial Revolution, energy was mostly derived from the natural energy flows and human and animal power. Energy usage was limited due to restricted availability and mechanical energy sources were limited to animate energy, water and windmills. Energy conversion also had a limited scope—conversion from chemical energy to heat and light and the chemical energy was derived from natural sources—animal or plants. Energy consumption typically did not exceed 0.5 toe per capita per year (Nakicenovic et al. 1998).

Since then, two grand transitions have shaped energy systems at all levels (see Fig. 18.1). The first transition involved a shift towards coal, which was made possible by a radical technological innovation of steam engines powered by coal. This innovation allowed conversion of fossil energy into work and made the

18.2 Grand Energy Transitions

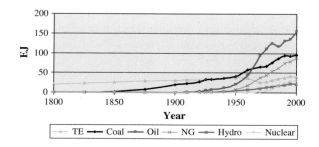

Fig. 18.1 Energy transitions of the past. *Data source* Grubler (2008)

supply of energy site independent, as coal could be transported and stored as needed. Stationary steam engines found first application in water lifting from coal mines, which improved productivity. Latter, they were introduced to factories and led to an entirely different form of production organization: the factory system. Mobile steam engines, on locomotives and ships, brought a transport revolution. Railway networks were laid in various countries making supply of energy and productive resources easier and less site dependent. Ships got converted from sail to steam engines. The steam age came in the middle of the 19th century and replaced traditional non-fossil energy sources in the industrialized countries. Energy consumption increased as well and was about 2 toe per capita per year in this period.

The second transition was triggered by a number of innovations that resulted in a diversification of supply options and end-use technologies. Electricity was the most important innovation of this period. It allowed easy conversion of energy to light, heat or work at the point of use. Another important innovation was the internal combustion engine, which revolutionized individual and collective mobility through the use of cars, buses and aircraft. Availability of new technologies allowed new energy supply feasible and petroleum, which was discovered in the mid-19th century found new application. Oil emerged as the dominant energy form due to the technological progress.

In both the transitions, scarcity of energy in the sense of physical exhaustion did not play any role. Biomass and wood fuel were and are available in quantities equivalent if not more than that were used during pre-industrial revolution. Coal provided an alternative to the traditional form of energy use and it gained acceptance because of higher energy density, flexibility and mobility. Similarly, the second transition did not happen due to the threat of scarcity of coal. Coal was available abundantly and is available even now. But it could not compete with oil in terms of ease of use, versatility and universal appeal. Oil is the dominant form of energy source even now and has maintained its position for over four decades.

Two grand transitions also resulted in far-reaching structural changes in terms of industrialization and urbanization. Economies started to move away from agriculture towards industry and manufacturing. The present trend is to move away from basic industries to services and information-based industries. Urbanization has also led to profound changes in the economies in terms of economic activities, life styles, social values, relocation from rural to urban areas. Two grand transitions brought profound changes in energy demand and supply. Industrialization, urbanization and increasing monetisation of economic activities led to a shift towards commercial energies away from traditional energies. Demand for flexible, convenient and cleaner energy forms increased. This also resulted in higher per capita energy demand but decrease in energy intensity.

Yet, the benefits of grand transitions did not reach everybody, as about a third of the world population still relies on traditional energies to meet their demand. The poor, particularly in the rural areas, are disadvantaged in this regard and the poor access imposes undue burdens on the children and the women. At the same time, uncontrolled urbanisation, rapid population growth, changes in the economic and social structures, and imitation of developed country life styles are imposing additional burden on energy. But politically motivated pricing policy, revenue generating tax policies, inefficient functioning of the energy market due to inappropriate market structure or collusive behaviour and poor performance of the firms in many areas bring additional problems to the sector. Moreover, restructuring efforts to fix some of the performance related problems did not progress well in most parts of the world and new issues are arising in places where it did progress well. Activities related to supply and consumption of energy created various environmental damages at local, regional and global dimensions. While billions of people are deprived of modern energies, countries around the world flare around 140 BCM of natural gas per year (equivalent to more than 100 million tons of oil), wasting a valuable resource and damaging the environment.

Any future transition is expected to emerge from the present day issues and challenges. High reliance on oil centralised in a single region has led to the concerns of scarcity and security of supply. Unprecedented demand growth, skewed regional distribution of demand, profound divergence in consumption level, 2 billion people without access to clean energies, and multi-dimensional concerns of environmental damage due to energy use—in short the symptoms of unsustainable energy practices would continue to haunt us in the future. It is obvious that the energy sector requires a paradigm shift to return to a sustainable path. This implies that the sector has to organise differently and to change practices and policies at various levels.

The next question that arises relates to the future transition. What will such a transition involve and when will this happen? This is a matter of speculation. Some believe that oil will continue its domination for another three four decades while others believe oil will face a faster decline. It could be replaced by natural gas. Some even imagined the possibility of reversal of fortunes of coal[2] which has maintained a high share in the electricity generation could re-emerge as the dominant fuel if peak oil concerns prove correct. Such a reversal of fortune could initiate a new trend in the energy studies. Despite these uncertainties, the future energy system would look different from the present and managing the change to ensure smooth transition is a main challenge.

---

[2] Martin-Amoroux (2005).

18.2 Grand Energy Transitions

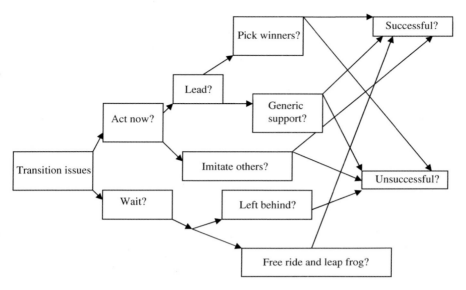

Fig. 18.2 Transition management challenge

Any future transition is expected to emerge from the present day issues and challenges. But such a transition will lead to a large number issues Does it make sense to act now or is it better to wait and see? If one decided to wait, will it be disadvantageous pushing it lag behind others? Or will waiting provide the opportunity for free-riding and leapfrogging? If waiting is not desirable, what sort of action will be desirable? Should it take a leading role or follow others? If a leading role is preferred, should it be for any specific types of technologies (i.e. should the winners be picked?) or should there be a generic support to all technologies so that a level playing field is provided for development. Will the chosen technologies be socially acceptable, environmental friendly and economically viable? If not, what would be the fall-back option? What happens if the chosen winners turn out to be economically unviable or are displaced by more favourable technologies? Should government play the role of facilitator or should it take a more active role? Figure 18.2 captures these issues related to transition management challenges.

As the sector faces both micro-level operating issues which are short-term in nature as well as those involving the medium and long-term future, and because of specific characteristics of the energy sector (discussed before in other chapters), the decisions need to be taken well in advance for the future and the present greatly shapes the future outcomes, although with a greater level of uncertainty. Moreover, the specifics of the decisions will vary depending on the circumstances (e.g. resource rich or resource poor country), economic conditions (developed or developing country), time dimension, and the like. Accordingly, the sector management issues are multi-dimensional in nature and a simplified typology is outlined in Fig. 18.3.

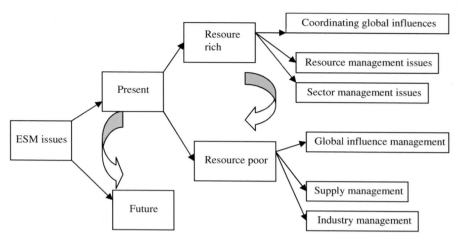

Fig. 18.3 Typology of energy sector management issues

In the following sections, some of these issues are discussed.

18.3 Issues Facing Resource-Rich Countries

Countries rich in energy resources face a different set of issues compared to those of resource-poor countries. As indicated in Fig. 18.3, three issues of importance for this group of countries are: coordinating the influences of other producer countries, resource management issues and sector management domestically. The first two elements are discussed here while the third element is presented jointly for both resource-rich and resource-poor countries towards the end of this chapter.

18.3.1 Co-Ordination of Global Influences

As is well known, energy resource endowment is not uniform across regions, with oil and gas highly concentrated in the Middle East and Russia while coal is somewhat more evenly distributed (see Fig. 18.4). Similarly, despite emerging economies' gaining importance, dominance of the OECD countries in energy demand is still evident (see Fig. 18.5). Consequently, energy trade at the regional and international level plays an important role in closing the mismatch in domestic supply and demand.

A number of issues are emerging in respect of global energy supply. From the energy transition perspective the pertinent question is the timing of the next

18.3 Issues Facing Resource-Rich Countries

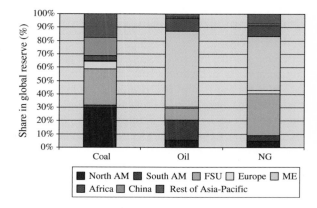

Fig. 18.4 Regional distribution of resource endowment in 2009. *Data Source* BP Statistical Review of World Energy, 2010

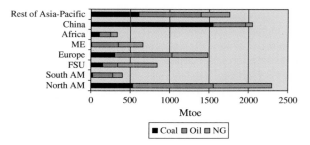

Fig. 18.5 Regional composition of Primary Energy Demand in 2009. *Data source* BP Statistical Review of World Energy, 2010

transition and the nature of the next dominant fuel. Recent outpours of information about impending oil scarcity (see among others Deffeyes 2001) suggesting end of oil domination and the concentration of oil and gas in two regions—Middle East and FSU, and the smallest reserve to production (R/P) ratio for oil in general (a ten year R/P ratio in North America and an even quicker depletion possibility of European oil) have fuelled speculation of global oil depletion in recent times.

Oil scarcity has a short term and a long-term dimension. High oil prices in the recent past may be a short term phenomenon arising out of lack of investment in exploration, production and refining capacities to replace old and aging facilities and to ensure surplus capacity to meet future demand. The low variable cost of oil production and the application of by-gone rule encourage the oil industry to be in an inherent state of overproduction, thereby bringing price crashes and consequent losses to investments. Restricted access of international oil companies to prime oil acreages since the nationalisation of oil industry in the 1970s, the market co-ordination policy of the OPEC by leaving the market to non-OPEC producers with high cost oil and erosion of excess capacity due to war, conflicts and unprecedented demand growth contributed to lower capacity expansion in the past decade. Demand management through price mechanism in a capacity constrained situation leads to price spikes as is experienced recently but the expectation of such a situation will lead to capacity addition, bring prices down to low levels again, which in turn reinforces the need for global co-ordination of supply activities. In line with the oil supply co-ordination, the gas producers are also working to develop a market influencing mechanism through the Gas Exporting Countries Forum (see Wagbara (2007) and Hallouche (2006)). Despite the differences in the oil and gas markets and producer behaviours, any international co-operation or collusion would change the energy market significantly.

The longer-term problem is more complex. Depletion of oil in other producing areas would mean that increasing amounts of supply would come from the resource-rich regions in the future. Despite the fact that both reserve and resource estimates are essentially guesses, and conservatism reigns supreme in such areas, there is still some uncertainty whether enough resources are really there to meet the long-term demand. If enough oil is not there in the ground, then is relying on oil a logical strategy? Lack of correct information and transparency are two major barriers in this area. Although new initiatives have been made to remedy both the above barriers, a lot remains to be done.

A careful look at the global energy evolution indicates that the substitution of oil in electricity generation has continued since the oil shocks of the 1970s and oil could not regain its lost market (Fig. 18.6). Domination of oil at the final energy consumption level arises because of nearly total oil dependence of the transport sector (see Fig. 18.7). Thus appearance of any viable substitute in the transport could end oil-era. High oil prices provide necessary encouragement to try and find alternatives. Already hybrid fuels, bio-fuels and derived fuels are making desperate attempts to break-even. It is not necessary that only a single universal alternative is required—a combination of substitutes in separate niche markets can be sufficient to ensure economic abandonment of oil. This would be in line with the lessons from previous transitions. Whatever oil is left is then, in Adelman's words "unknown, unknowable and totally uninteresting".

A viable substitute for oil in the transport will bring a sea change in the energy markets: importing countries would be liberated from the security concerns while the exporting countries will discover that the golden goose is no more laying golden eggs. Countries sitting on huge oil reserves and hoping wind-fall gains in the future when oil becomes scarce in other areas would be the worst hit. The transition could be violent and could make the world a lot more unsafe place. What could the resource-rich countries do to prevent such a disastrous scenario to develop? Does co-ordination of producing country activities hinder the long-term

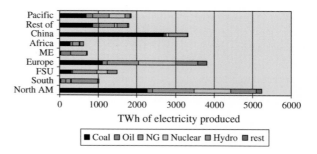

Fig. 18.6 Regional variation in electricity generation mix in 2007. *Data source* IEA website

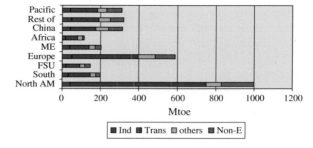

Fig. 18.7 Final oil consumption by sector and region, 2007. *Source* IEA website

viability of the industry? What is the most preferred way of ensuring long-term sustainability of resource demand?

18.3.2 Resource Management Issues

Revenue from the energy resources often constitutes the main source of income of many resource-rich developing countries. Although natural resources could drive economic growth of a country, a negative correlation between economic growth and resource abundance is often found, which is variously termed in the literature as 'paradox of plenty' or 'resource curse'.[3] Sudden change of an economy upon discovery and production of these resources coupled with absence of diversified economic structure, poor institutional endowment and arrangements, and poor management of new-found riches often hinders economic development. Consequently, resource management issues assume great importance for the resource-rich developing countries.

As the price of traded energy resources is quite volatile, resource exporting countries receive windfall gains when prices increase (and financial distress during low prices). The size of the windfall is often large compared to the national output, but lack of adequate production linkages (i.e. the forward and backward linkages in the production system) in a high rent economy and higher propensity of importing goods for consumption (i.e. adverse consumption linkages) could act as hindrance to growth and development. Use of windfall through direct consumption by expansion of public services or transfer of revenues could create distortion. Moreover, expanded service becomes a pain at the time of economic downturns. Any investment in infrastructure in anticipation of demand to spur growth can be a deficiency-correcting measure and not a solution for long-term growth. In addition, reliance on a few decision-makers for use of windfall tends to promote large-scale, prestigious projects, making the country dependent on a few key investments instead of a diversified portfolio of investments, and thereby making the country vulnerable. Promotion of economically unviable projects or politically motivated

---

[3] There is a vast body of literature on this theme. See Stevens (2003); Davis and Tilton (2005) for recent reviews.

projects may not contribute to the growth of the economy. Rent-seeking, corruption and personal enrichment instead of general development of the country also accompany such developments.

This brings to the major development strategy issue: what should be the desired development policy for a resource-rich country? If the absorptive capacity of the local economy is poor and the development of the resource does not encourage economic growth, why should the country develop such resources when leaving it in the ground for future use remains a viable alternative? Or will it require scaling down the projects and develop in phases for ease of revenue management? Will such projects be able to exploit scale economies and will such a policy be acceptable politically, socially and economically? Would such slow down of resource developments produce institutional arrangements to attenuate negative impacts of resource developments? Will private owners accept such investment logic or will this disadvantage some countries compared to others? If slowing down of resource development is considered as the desired policy, would the resource supply be enough to meet the demand? Will the consequent price promote substitution of other energy forms and threat long-term valuation of the in-ground resources? If the long-term prospect is affected, will then producers rush to develop the projects? All these then lead to a vicious circle of issues (see Fig. 18.8).

18.4 Issues Facing Resource-Poor Countries

The set of issues facing the resource poor countries is somewhat different. As this group includes both the developing and developed countries, there is wide

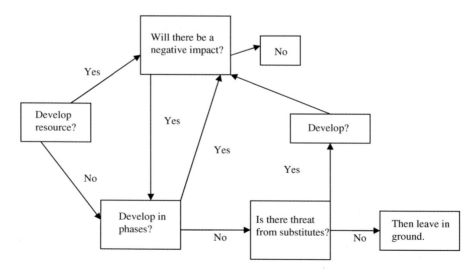

Fig. 18.8 Resource management issues

18.4.1 Managing Global Influence

The most important influences faced by resource-poor countries are those related to vulnerability of their economies due to price volatility of imported energy and ensuring security of energy supply.

18.4.1.1 Managing Effects of Price Shocks

Rise in oil prices in the first half of this decade has brought the vulnerability issue to the fore once again. Based on a sample of 97 oil importing countries, ESMAP (2005a) finds that the effect of a $10/barrel increase in oil price would be felt most severely by the poorest countries while the effect on GDP will be much lower in the industrialised countries (see Table 18.1).

A variety of inter-linked effects can then be expected (IMF 2000):

1. Higher oil prices result in a fall in oil demand, as the consumers with limited budget try to reach an alternative equilibrium position.
2. The cost of production of goods and services rises, which puts pressure on profits of the firms. The effect depends on the energy intensity of production: normally developed countries with lower energy intensity are expected to face lower pressure than the developing countries.
3. Higher costs of goods and services put pressure on general price levels, fuelling inflation.
4. Higher costs and inflation, and lower profit margins would put pressures on demand, wages and employment, affecting the economic activities.
5. Effects on economic activities influence financial markets, interest rates and exchange rates.
6. Finally, depending on the expected duration of price increases, consumer and producer behaviours would change. Producers may invest in new capacities and developments while consumers may tend to economise.

Different economic sectors are expected to be affected differently as a result of oil price shocks. Energy-intensive production is expected to be worst hit as the cost

Table 18.1 Effect of $10/barrel rise in oil prices on GDP of oil importing countries

Per capita income (US$)	% change in GDP
<300	−1.47%
>300 but <900	−0.76
>900 but <9000	−0.56
>9000	−0.44

Source ESMAP (2005a)

of production would rise significantly. Consumer goods industry, where the goods tend to be non-essential (i.e. demand is elastic) also face a falling demand. In contrast, industries providing essential goods are not expected to suffer great loss in demand. Managing these effects for the economy and for the sector remains a major challenge for these countries.

ESMAP (2005b) estimated the vulnerability of various regions in 1990 and 2003 and found that Sub-Saharan Africa and East Asia were the most exposed regions in both the periods. High import dependence and high share of oil in energy mix in Sub-Saharan Africa aggravated the situation while low energy intensity partly offset the exposure. As resource endowment is unlikely to be changed and these countries are likely to experience higher oil intensity as income grows, the vulnerability of this region to oil price shocks is expected to remain high. How to manage the external effect due to energy price volatility in a cost effective manner and without imposing undue burden on the economy?

18.4.1.2 Energy Supply Security

Another concern is related to ensuring adequate energy supply in an affordable and reliable manner to meet the future demand. Although concerns for energy security first started in the aftermath of the first oil shock in the 1970s, easing of oil prices and supply constraints in the later periods shifted the energy policy focus to environmental and industry restructuring issues. There was a belief that the markets would be able to solve the problems of the energy sector and that no specific attention needs to be paid to energy security concerns. However, concerns about peaking of oil supply and supply capacity constraints to match the demand have brought back an era of sustained high oil prices, staging a come-back of energy security concerns to the limelight.

Although the definition of energy supply security varies, it is commonly defined as the "reliable and adequate supply of energy at reasonable prices" (Bielecki 2002). Reliable and adequate supply refers to uninterrupted supply of energy that is able to meet the demand of the global community. As adequate supply has direct links with energy demand at any given time, this segment of the definition points to both supply and demand aspects related to security. Similarly, supply adequacy and reliability is not a matter of external dependency alone, although the control over external supply can be limited in most cases. Most of the literature on energy security focuses on dependence on external supply, although in many developing countries the internal sources of supply could equally be problematic.

The security of supply problem has a number of components (Toman 2002): (a) exercise of market power by suppliers to raise prices, (b) macroeconomic disruption due to energy price volatility, (c) threats to infrastructure, (d) localised reliability problems, (e) environmental security, etc. Accordingly, the security concern has a physical as well as an economic dimension. Moreover, there is a time dimension of it: in the short-term, the main concern relates to the risks of disruption to existing supplies essentially due to act of god, technical or political problem; in the long-term, the risks related to future energy supply also arise.

While developed countries participate in an emergency response programme through the International Energy Agency (IEA),[4] there is no such arrangement for the developing countries. Although any country benefits from the reaction of the market due to stock release, the issue of global co-operation and unified response mechanism for all importing countries has received little attention at the international level. Instead, countries who are not members of the IEA try to develop their own individual strategies but given the costs, stockpiling strategies may not be efficient for smaller countries. Thus the pertinent issue here is whether it is possible to develop an international arrangement of cooperation and response co-ordination to provide an insurance policy publicly against future supply shocks. The issues related to reserve sizing, timing and method of stock utilisation and the institutional mechanism for stock use would require attention. If such a global mechanism is not feasible, how could this sort of global influence be managed so that even the small countries would not be disadvantaged?

18.4.2 Issues Related to Supply Provision

A number of issues arise in this area including adequate supply arrangement, energy access issues, etc. These are discussed below.

18.4.2.1 Investments

Two major changes in demand pattern can be noticed from the evolution of demand: a shift towards more reliance on electrical energy by final users and a geographical shift of demand towards the Asian region. These two changes would require more capacity addition in the conversion processes and facilities closers to demand centres.

The global energy demand is projected to grow at an annual rate of 1.6% between 2006 and 2030, thereby requiring about 1.48 times the energy demand of 2006 in 2030 (WEO 2008). The share of non-OECD countries would increase to 63% of global primary energy demand in 2030 and the Asian developing countries will emerge as the second most important demand centre in the world with a 38% share of global demand (WEO 2008). According to WEO (2008), an investment of $26 trillion (in 2007 dollar) is required for the energy sector between 2007 and 2030, of which around. 52% of this investment ($\sim$ $14 trillion) would be required

---

[4] IEA member countries hold a stock of oil equivalent to 90 days of net imports in the previous year. The emergency measures apply when the supply disruption exceeds 7% of IEA or any member country supply.

in the electricity industry while oil and gas industries would need 45% of the overall investment (~$12 trillion). About 65% of the above investment demand would arise in developing countries. Mobilising such huge investments in a timely manner at places where required is a major challenge. The issue becomes particularly challenging in developing countries which would need to step up investments manifolds. The recent slump in private investment in the energy sector is not an encouraging indication.

The problem aggravates due to the following as well:

1. Financial crisis of 2008 that has constrained the credit flow and affected the global capital market adversely.
2. Financial difficulties of the national energy companies, especially the electric utilities, due to poor pricing policies, poor management and political interference in decision-making;
3. Budget constraints of the governments, forcing them to commit less funds to energy sector projects and higher levels of competition for funds due to refocusing of state activities and social needs;
4. Unfinished reforms of the energy sector in many countries create a prolonged transition state, where the state withdraws from the sector expecting private investment to come, while private investment does not materialise due to unstable environment. Investments cannot wait until reforms are finished but being a political process any reform depends on the political stability and willingness to reform.
5. Heightened security concerns in many countries for energy installations due to political or other conflicts fuel risk premium. It is reported that such concerns have cost billions of dollars for Saudi Arabia to protect its oil installations in recent years and international oil prices now bear a risk premium of $10–15 per barrel, much higher than the cost of oil production in many areas.

The global phenomenon of the 1990s was the emphasis placed on three R's: energy market Reforms, Renewable energy technologies and Right prices. Energy sector reform has not been a great success in many countries and the state funding for energy has deteriorated, without any concomitant participation from the private sector. Depending on reforms for solving the energy access problem does not appear to be a logical approach. Policies to avert the problem could involve a return of the state-led investment in major energy projects, especially those involving longer-term financial commitments and strategic importance. As restructuring and reform is failing to deliver in many countries, state presence is becoming more important, especially in developing countries who did not benefit from the inflow of private capital in the energy industry and where private capital did not develop nationally desired energy sources such as hydro, nuclear and the like. Similarly, instead of wasting decades on reforms that can be hardly implemented, it may be better to focus on better management of state agencies through enterprise level changes.

Right pricing for energy and promotion of renewable energies have been on the policy forefront for quite sometime. Removal of price distortions has proved to be

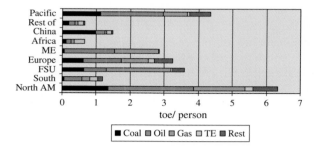

Fig. 18.9 Regional distribution of per capita primary energy consumption in 2007. *Data source* IEA website

a difficult task in both developed and developing countries alike. Distortions in the energy markets, distorting taxes for revenue generation, and price distortion for social or other reasons militate against correct price signals. Relying on such a non-attainable goal for enhancing energy security does not make much sense. Instead removal of trade barriers and flooding of markets with low-cost, efficient appliances appear to work better.

18.4.2.2 Search for Alternative Strategies for Energy Access Problem

The disparity in primary energy consumption per capita is now a well-recognised fact (see Fig. 18.9). Consumption in Africa and Asia excluding China is just 40% of global average consumption, estimated at 1.65 toe per person in 2002. China and Latin America also have low per capita energy consumption, between 60 and 65% of world average.

When consumers in Africa meet their energy needs by consuming 111 kgoe of coal, 140 kgoe of oil, 86 kgoe of gas and 310 kgoe of traditional energies, their counterparts in North America consume 1.4 toe of coal, 2.5 toe of oil, 1.5 toe of gas and one toe of other energies (including traditional energies). Thus coal consumption per person in North America is 12 times higher, while oil and gas consumption is about 20 times higher.

Despite past growth and development, a large section of the world population does not have access to clean energies. The most commonly cited figures on the lack of access to energy indicate that about 1.4 billion (i.e. about 22% of the global population) is without access to electricity (IEA 2009). It is also believed that there are about 2 billion people without adequate access to clean cooking energy.[5] According to IEA (2009), 42% of those lacking electricity access reside in South Asia and another 40% reside in Sub-Saharan Africa, while East Asia contributes another 15% of such population. Providing access to clean, affordable energy is a major challenge.

Sustainable, long-term solutions for energy access problem can neither rely on subsidized supply of clean energies nor on piece-meal solutions that address only a part of the problem. What is required in the long term is to ensure adequate supply

---

[5] Similar figures are quoted in WEHAB (2002) and DfID (2002).

of monetary resources to households to sustain a life style that relies on clean energies and other monetized inputs. Thus the energy access issue joins here the problem of ensuring economic development, which in turn calls for an integrated approach of combining various development efforts at a decentralised level (DfID 2002 and WEC 1999) as opposed to treating electrification or energy supply issues in an isolated manner. Given the diversity in terms of energy use, resource availability and other conditions, appropriate local solutions have to be found instead of universal or global solutions to the problem. The policy objective should be to promote innovative solutions rather than prescribing templates for adoption. Thus each decentralised unit will have to search for own solutions.

1. Focusing on the creation of opportunities for higher income generation in monetary terms, as opposed to in-kind income. Unless money flow increases to the poor, commercial energies stand little chance of competing with traditional energies.
2. by developing local energy markets taking into account the specificities of local energy situation (resources, needs, capacities, strengths and constraints) and adopting appropriate supply mechanisms and organizational structures to cater to the local needs.
3. by selective and judicious use of market interventions to make energy supply affordable but ensuring financial viability of energy supply. Unless the supply is financially viable, it cannot be sustained.
4. by ensuring local community participation in the decision-making and policy implementation process.

Such a bottom-up policy is inherently multi-dimensional and necessarily complex. It comes as a contrast to the existing policies which are top-down in nature and essentially imposed on the population. Implementation of such a policy would require development of a common framework that can be adapted to each situation, creation of a organizational set up to carry out the policy, building organizational capacity, adequate funding arrangements, and above all a complete review and perhaps an overhaul of the mode of functioning of the government, existing organizations and the economic activities to facilitate decentralized mode of functioning of the economy.

18.5 Other Sector Management Issues

18.5.1 Management of Environmental Issues of Energy use

As different activities in the energy system (production, conversion and utilisation) lead to various environmental impacts, the future growth of energy demand and continued reliance on fossil fuels to meet the demand raise the environmental concerns of energy use. The environmental concern has a social dimension as the poor, women and children are adversely affected due to low access to clean energy

18.5 Other Sector Management Issues

discussed earlier. Use of biomass and coal in inefficient stoves and in indoor conditions without adequate ventilation at times when most of the young and vulnerable members are present leads to significant health hazards, and significant social costs, which has received attention only recently. Given that people tend to move up the ladder of higher fuel quality with higher income, the issue of local level environmental damage control rejoins the issue of ensuring equitable distribution of the benefits of economic development. Sector-specific intervention through promotion of subsidised fuels or technologies does not seem to work without leaving a heavy revenue burden on the government or the supplier.

At the same time, the urban areas of developing countries are expected to face major environmental problems. About 70% of world urban population lives in developing countries of Asia, Africa or Latin America (UNEP 2002) and most of these urban areas suffer from poor air quality. 12 out of 15 cities with highest levels of particulate matters and 6 out of 15 cities with highest levels of SOx are found in Asia (GEO-2, UNEP 2000). Use of traditional energies, reliance on coal and other fossil fuel energies and inefficient combustion techniques have degraded the environmental quality. With high level of urbanisation and more energy consumption, the situation can deteriorate unless active protection measures are taken through technological, economic, legal and behavioural changes.

The climate change issue which has received a disproportionate level of attention in recent times has proved to be a challenging task given its global nature, long-term time frame (involving inter-generational issues), public good dimension, north–south divergence of opinion and even the dispute about the scientific veracity of the problem. Consequently, the debate over whether to slow or not global warming remains inconclusive and passionate, with each party sticking to its own position. Despite some progress in the 1990s, no long-term solution appears to have emerged.

While both resource-rich and resource-poor countries have adopted varying degrees of measures to protect the environment, increase in supply to meet the future demand poses challenges for the environment both for the producing and consuming nations. How to meet the demand without adversely affecting the environment and imposing social burdens remains a major issue. The possibilities of energy transition and the emergence of new energy carriers would bring additional challenges in terms of security and environmental management. The debate over the choice of appropriate technological fixes, regulatory intervention and economic instruments would continue but everyone would have find the locally desirable solutions for which generic templates of solution does not exist.

18.5.2 Renewable Energies and the Management Challenge

Renewable energy sources attracted attention of energy policy analysts and researchers just after the oil price shocks of the 1970s. Availability of easy petrodollars, the global concern for climate change and sustainable development and the concerns for fossil-fuel dependence and energy scarcity provided necessary impetus to renewable energies. Now renewable energies occupy an important place in any strategy for sustainable development in general and sustainable energy development in particular.

Renewable and decentralised energies are based on 'small is beautiful' approach as opposed to scale and scope economy dominated energy sector. Our very poor knowledge about how to manage small energy systems and how to organise the activities around them act as the main barriers. Managing such decentralised systems is different from the traditional approach of providing energy through centralised supply systems. The skill set required for such new types of activities may not be available locally and there is a need to focus on such capacity development. Second, there appears to be a market failure in some areas as private investors are interested in lucrative, high cost markets, while there is little investor interest in low marginal cost of supply (such as for solar cooking). The commercial profitability of low marginal cost ventures being low, some alternative form of supply is required, which has not yet emerged. Traditional economics would suggest that government should intervene in such a case but governments do not have any special skills for managing the activities of decentralised energy supply systems. Most governments work in a centralised manner and tend to impose its decision at all levels. There can however be a case for governments providing capacity building facilities and opportunities for undertaking such activities, at least initially.

The issues related to renewable energies are similar to that of transition management as most of the technologies are still under development and possibilities of new breakthroughs imply that preference would be given for low lock-in effects and weak path dependence, so that if required new courses could be easily adopted. In such a case, is it logical to use the poor as guinea pigs?

18.5.3 Reform and Restructuring

The history of evolution of energy industries[6] around the world shows a highly competitive beginning, followed by an integrated monopolistic development of the industry and the emergence of reforms and restructuring thereafter. Government intervention in the energy sector has been promoted either through ownership or through regulatory practices or due to the existence of a public good[7] argument (greater societal benefits compared to costs for certain technologies such as nuclear, hydroelectric energy, broader social benefits of energy investments,

---

[6] See IEA (1999) for a more detailed account. Also consult EIA (1996) and Newbery (1999).

[7] A public good is characterised by non-exclusivity and non-rivalry. Non-exclusivity implies that it is difficult to exclude others from using the good or service without withholding the good or service. Non-rivalry on the other hand means that consumption by one does not reduce its potential to be used by others.

long-term view of public investments, etc. (Jaccard (1995)). Consequently, the energy sector is dominated by state-owned companies and the market is often regulated—either independently or otherwise.

The wave of reform, restructuring and privatisation that swept the global economy since 1980s did not spare the energy industry. Many developing countries embraced restructuring and sector reform under the influence of the Breton Woods duo and other donor countries. Poor performance, due to monopolistic market, state-ownership of the utility, lack of transparent regulation and political interference in the sector—all prevented the utility to function commercially, and resulted in an abysmal financial health of the utilities, poor service quality, inefficient system operation and poor economic growth. The general prescription then was restructuring the sector through unbundling of the state-owned utility, creation of an independent regulatory body and privatisation of the utilities.

However, as the World Bank (2004) report states, "for much of the 1990s privatisation was heralded as the elixir that would transform ailing, lethargic state enterprises into sources of creative productivity and dynamism serving the public interest". But "the privatisation was oversimplified, oversold, and ultimately disappointing—delivering less than promised" (World Bank 2004). The same report indicates that scepticism and hostility to reform has grown in size and dissatisfaction with reform and privatisation is fuelled by loss of jobs, price rise, high profits of firms, etc. (World Bank 2004). It is now considered that the reform had significant distributional impacts in many cases and the poor and the vulnerable were affected disproportionately.

Given that the scorecard of reform remained unimpressive and that some of those who embraced the reform has had disastrous experience with competitive power markets, new issues are appearing. The system expansion need to cater to growth, low regulatory capacity for ensuring proper functioning of the unbundled system, path dependence introduced by the reform, sequencing of reform and privatisation are some such concerns. The search for alternative viable options continues as the standard reform falters and fails to deliver in an increasing number of cases. A logical approach is to strive at achieving improved performance of the state-owned enterprises through appropriate regulation and governance. While this area has received poor attention so far, it promises to be an area of increased focus in the future.

18.6 Conclusion

This chapter has presented the multi-dimensional interactions of the energy sector and analysed the multitude of sector management issues. Despite uncertainty about the timing and nature of the next transition, managing the transition and preparing strategies for such a change remains important challenges. Such issues cover a wide range from possibilities, including whether to act now or not, act in a specific area or not, whether the chosen action would be successful or not, and whether left out options would turn out to be viable or not. While oil depletion has generated a passionate debate about the future energy transition, the lessons learnt from energy transitions suggest that an alternative will emerge before physical depletion of oil. A viable substitute to oil in the transport sector will have a serious effect on the economies of oil rich countries, which could unleash a price war to regain market share. Both the price-war strategy and the development of a substitute for oil in transport would bring dramatic changes to the energy scene in the future. At the same time, experience suggests that the experts' consensus view never worked in the energy sector. Nuclear was the consensus fuel of the future in the aftermath of oil shocks in the 1970s but gas turned out to be a more important energy carrier in reality. Renewable energies did not succeed in penetrating the energy scene while traditional energies did not retreat either. There is no reason to believe that this time around the experts won't be wrong.

The sense of energy scarcity has brought the resource-rich countries to limelight. The windfall gains from the resource development have also brought the revenue management issues to foster economic growth. Being poorly endowed with institutional capacities and economic diversities, they face the challenge of properly utilising their revenues avoiding negative impacts. If the resource development hinders economic development, the challenge is to mitigate the undesirable effect through acceptable solutions. The search for such solutions has not ended yet but if this slows down the resource development, new challenges arise for the producing and consuming countries. Producers may face resource obsolescence while the consumers face the prospect of high prices. Managing the impact of energy price volatility and ensuring adequate supply at reasonable prices becomes the priority for the consumers.

Similarly, the energy access problem begs new initiatives and thinking. Alternative strategies based on overall economic development, selective intervention, free trade practices and access to low cost capital and efficient technologies could work better. At the same time, with symptoms of unsustainable energy practices abound and a history of sustained market failures influencing the sector policies, the energy sector requires a paradigm shift to return to a sustainable path. This implies that the sector has to organise differently and to change practices and policies at various levels. Finding those solutions to manage the sector in a sustainable manner remains the most important management challenge.

References

Bhattacharyya SC (2007a) Energy sector management issues: an overview. Int J Energy Sector Manage 1(1):13–33

Bhattacharyya SC (2007b) Policy responses for the future of energy: global overview and perspectives. In: Boscheck R (ed) Energy futures. Palgrave Macmillan, Hampshire, UK

Bielecki J (2002) Energy security: Is the wolf at the door? Q Rev Econ Finance 42:235–250

Davis GA, Tilton JE (2005) The resource curse. Nat Resour Forum 29(3):233–242

References

Deffeyes KS (2001) Hubbert's peak: the impending world oil shortage. Princeton University Press, Princeton, NJ

DfID (2002) Energy for the poor: underpinning the Millennium Development Goals. Department for International Development, UK

EIA (1996) The changing structure of the electric power industry: an update. Energy Information Administration, Washington, DC

ESMAP (2005a) The impact of higher oil prices on low income countries and on the poor. Energy Sector Management Assistance Programme, World Bank, Washington DC

ESMAP (2005b) The Vulnerability of African countries to oil price shocks: Major factors and policy options—the case of oil importing countries. Energy Sector Management Assistance Program, World Bank, Washington DC

Grubler A (2008) Energy transitions. In: Cleveland CJ (ed) Encyclopedia of earth. Environmental Information Coalition, National Council for Science and the Environment, Washington, DC

Hallouche H (2006) The gas exporting countries forum: is it really a Gas OPEC in the making? Oxford Institute for Energy Studies, Oxford

IEA (1999) Electricity market reform: an IEA Handbook. International Energy Agency, Paris

IEA (2009) World energy outlook. International Energy Agency, Paris

IMF (2000) The impact of higher oil prices on the global economy. International Monetary Fund, Washington, DC. See http://www.imf.org/external/pubs/ft/oil/2000/

Jaccard M (1995) Oscillating currents: the changing rationale for government intervention in the electricity industry. Energy Policy 23(7):579–592

Martin-Amoroux JM (2005) Coal Phoenix-like. Oil, Gas Energy Law Intell, 3(3). Internet journal

Nakicenovic N, Grubbler A, McDonald A (1998) Global energy needs: past and present, in global energy perspectives. Cambridge University Press, London

Newbery DMG (1999) Privatisation, restructuring and regulation of network utilities. MIT Press, ISBN 0262640481

Stevens PJ (2003) Resource impact: curse or blessings, A literature survey. J Energy Lit 9(1):3–42

Toman MA (2002) International Oil security: Problems and policies, RFF Issue Briefs. Resources for the Future, Washington, DC. See http://www.rff.org/rff/Documents/RFF-IB-02-04.pdf

UNEP (2000) Global environment outlook 2000. United Nations Environment Programme, Nairobi

UNEP (2002) Global environment outlook 3: past, present and future outlook. United Nations Environment Programme, Nairobi

Wagbara O (2007) How would the gas exporting countries forum influence gas trade? Energy Policy 35(2):1224–1237

WEC (1999) The Challenge of rural energy poverty in developing countries. World Energy Council, London

WEHAB (2002) A framework for action on energy. WEHAB Working Group, UN

World Bank (2004) Reforming infrastructure: privatisation, regulation and competition. World Bank, Washington DC

Chapter 19
Impact of High Energy Prices

19.1 Introduction

The importance of energy in the global economy was recognised in the aftermath of the First oil price shock in the 1970s. Since then energy issues received significant government attention. But in the 1990s, the focus shifted to liberalisation of markets and climate change issues on the assumption that markets would be sufficient to provide reliable supplies at reasonable cost. With high energy prices back in the first half of this decade, the issue of their impacts has resurfaced. The objective of this chapter is to introduce to this important issue and discuss how economics can help analyse the effects. The focus is on oil price increases, although similar arguments could be made for other energies as well.

High oil prices affect both oil exporting and importing countries, albeit differently. The impact varies across countries and the issues could be quite different from one country to another. Despite this difficulty of generalisation, the problem is better analysed considering exporting countries and importing countries separately.

19.2 Recent Developments in Energy Prices[1]

The oil price trend since 1970s indicates that in nominal dollar, prices recently have reached higher levels compared to the oil shocks of the past (see Fig. 19.1). In real terms, prices were quite stable in the post price war period (i.e. between 1986 and 1996) and even reached very low levels in 1997 after the Asian financial crisis. The oversupply of oil due to poor demand drove the prices down. But oil prices started to rise in 2000 and after prices fell sharply after the "9/11" event,

---

[1] There is a well-developed literature on this topic. See Krichene (2006), Koyama (2005), Fattouh (2005), IMF (2005) and IMF (2000) among others for a review of issues.

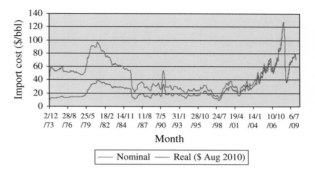

Fig. 19.1 Oil price trend in nominal and real terms (2007, CPI adjusted). *Source* EIA website

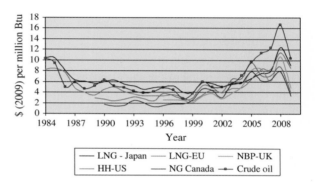

Fig. 19.2 Natural gas price trend. *Data source* BP Statistical Review of World Energy 2010 and EIA website

they started to pick up again in 2002. This period of high oil prices continued for about 6 years.

A closer look at the period between 2002 and 2009 indicates that the prices maintained a steady upward movement since September 2003 and the real prices remained above 60 US dollars per barrel for about 30 months continuously. This represents a significant increase in prices compared to the previous periods and there is a sentiment that the changes may not be transitory in nature, implying that a part of such high prices may become a permanent feature of the global economy (ESMAP 2005a). The impact of such higher prices on the global economy then becomes an important energy-economy issue.

High oil price has also influenced the price of other fuel prices. For example, both natural gas and LNG prices followed a similar path as that of crude oil and prices increased since 2002 to reach the peak in 2008 (see Fig. 19.2). Natural gas prices during the period increased three-four folds and added further to the economic distress of consumers.

Similar is the case with coal (see Fig. 19.3)—prices increased by a factor of three during the same period. This again shows the link between crude oil price and other fossil fuels and supports the claim that oil still hold's the driver's position in the international energy scenario.

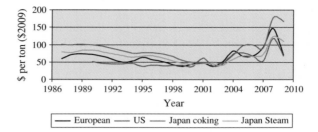

Fig. 19.3 Coal price trend. *Data source* BP Statistical Review of World Energy 2010 and EIA website

19.3 Impacts of Energy Price Shocks: Case of Importing Countries

As sudden changes in prices affect various elements of the economy, it is often useful to analyse the issues at three levels (ESMAP 2005a):

- at the micro-economic level, the reaction of the consumer could be analysed;
- at the intermediate level (or meso-economic level), the analysis could focus on how the micro-economic decisions are reflected in the factors such as oil dependency and energy intensity; and finally,
- at the macro-economic level, the effect of energy price shocks on macro-economic variables could be analysed.

19.3.1 Consumer Reaction to Oil Price Increases

When an increase in oil price occurs, how would the market respond to such non-marginal price changes? This is explained in Fig. 19.4. A consumer with a given income and hence budget could acquire a certain amount of energy and other goods and services. The relative prices of energy (PE1) and other goods and

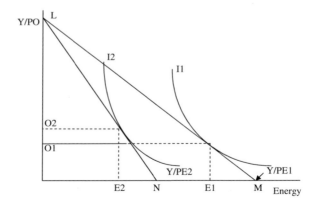

Fig. 19.4 Consumer's reaction to price shock

services (PO) decides the initial equilibrium position. This equilibrium is obtained where the budget line is tangent to the indifference curve I1, allowing the consumer to select E1 amount of energy and O1 amount of other goods and services.

When the energy price increases, her budget will permit her to procure lesser quantities of energy if she spends the budget only on energy but she could still procure the same quantities of other goods and services with her budget. This is reflected by shifting the budget line from LM to LN. The rise in price forces the consumer to an inferior indifference curve, reflecting loss of consumer welfare. A new equilibrium will be reached where the ratio of prices and the marginal rate of substitution will be equalised. The consumer will be consuming O2 quantities of other goods and services and E2 quantities of energy.

The aggregation of such consumer-level reaction will give rise to the markets response of demand consequent to price changes. This is shown in Fig. 19.5. As the price changes to p2 from p1, the demand falls but the reduction tends to be relatively small largely because of the inelastic nature of energy demand in the short-run. In a real world where the ceteris paribus assumption does not hold, the adjustment takes place in a dynamic manner. The dynamic response is influenced among other factors by changing incomes, expanding economies, lock-in effects, changing technology, and changing expectations. This makes the dynamic response more complicated.

As energy is used both for final use and as a factor input of production and services, the effects of price shocks affect the activities. Facing higher energy prices, the producers would follow a similar thought process as the households and would try to achieve the least-cost combinations of inputs for the output. This would involve substitution of factor inputs subject to the constraints faced in real life situations such as rigidity of the labour market, locked-in appliances, and so on. Given the limited possibility of substitution in the short-run, the firms would face higher cost of production for goods and services, which makes their products less attractive to consumers. Moreover, facing budget constraints, consumers would continue their adjustment process by substituting goods and services and in

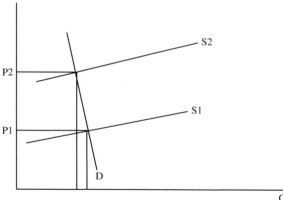

Fig. 19.5 Market reaction to price shock

general, would be able to afford less quantities of the goods and services. Thus the demand for other goods and services would fall subsequent to higher energy prices. This can be considered as the direct effect of oil price increase.

19.3.2 Transmission of Reactions to the Economy

The response of consumers and firms would then form the reaction of the economy. As a result of changes in the demand for goods as well as output of production activities, the demand for factors of production would be affected. As producers substitute factors to minimise costs, demand for certain factors would increase. This substitution effect is not limited to energy alone, as factors of production for goods and services could provide substitution possibilities. As a consequence of such direct effects, indirect factor demand could change as well. For example, higher oil prices could lead to higher demand for coal, which in turn would fuel demand for coal-miners (Codoni et al. 1985).

Similarly, changes in the demand for factor inputs would encourage adjustments in prices to bring supply–demand parity. This could result in lower prices for some factors of production, notably wages for labour. But in a world of regulated markets, organised labour is likely to resist any decrease in real wages, and labour substitution policies. This would push the production costs upwards and prices would rise, fuelling inflationary pressures. The normal policy intervention in such circumstances is to tighten the monetary policy by raising the interest rates.

However, as higher prices erode the real disposable income of consumers, the output of companies falls, leading to lower economic growth and increased unemployment. The inflationary pressure starts to ease off as well at this stage. To promote economic growth, interest rates are reduced to ease borrowing and reduce cost of capital. However, uncertainties about the timing of policy interventions, magnitude of reaction and interaction with the external sector remain and would affect the second round of effects. Figure 19.6 captures the above interactions between monetary policy and oil price rise diagrammatically.

The effect of reduced economic activities (i.e. lower GDP growth) and tight monetary policies affect the income levels and income distribution in a country. Normally the poorer section is expected to be worse hit due to lack of employment and above average share of energy-related expenditure compared to income. The induced effect on the economy works through the economic links with various sectors and activities. However, such induced effect may manifest slowly and over time.

Thus, a variety of inter-linked effects can then be expected (IMF 2000):

1. Higher oil prices results in a fall in oil demand, as the consumers with limited budget try to reach an alternative equilibrium position (as discussed above).
2. The cost of production of goods and services rises, which puts pressure on profits of the firms. The effect depends on the energy intensity of production: normally

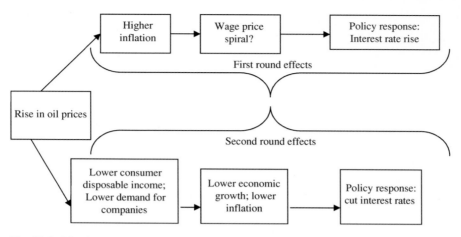

Fig. 19.6 Oil price rise and monetary policy interactions. *Source* RBS (2004)

developed countries with lower energy intensity are expected to face lower pressure than the developing countries.
3. Higher costs of goods and services put pressure on general price levels, fuelling inflation.
4. Higher costs and inflation, and lower profit margins would put pressures on demand, wages and employment, affecting the economic activities.
5. Effects on economic activities influence financial markets, interest rates and exchange rates.
6. Finally, depending on the expected duration of price increases, consumer and producer behaviours would change. Producers may invest in new capacities and developments while consumers may tend to economise.

Different economic sectors are expected to be affected differently as a result of oil price shocks. Energy-intensive production is expected to be worst hit as the cost of production would rise significantly. Consumer goods industry, where the goods tend to be non-essential (i.e. demand is elastic) also face a falling demand. In contrast, industries providing essential goods are not expected to suffer great loss in demand.

19.3.3 Linkage with the External Sector

The macro-economic impacts of oil price increases for any oil importing country first manifest in the form of higher import bills for oil and oil products and consequent shrinkage of non-oil imports in import share (and perhaps in absolute terms) depending on the foreign exchange reserves/balances, import dependence of the country, oil content in imports, etc. The restraint on import demand is achieved through a reduction in domestic demand for consumption and investment, often by imposing trade restrictions (such as imposing quotas and import duties). At the same time, a reduction in demand for goods and services discussed earlier reduces demand for exports, which in turn affects economic activities and economic growth. ESMAP (2005a) reports that for a USD10 per barrel increase in oil price would cause a 0.5% fall in GDP of the OECD countries. It was noticed during previous oil shocks that developing countries with relatively lower level of participation in international trade faced severe balance of payment problems. This led to the policy of structural adjustments at the behest of the World Bank and the IMF.

The balance of payment problem is also related to the exchange rate policies. In dollarised economies, the impact of higher oil price will be immediately felt on the balance of payments. In other cases, the effect is somewhat offset by exchange rate adjustments.

However, the overall trade position is affected indirectly as well, as higher oil price influences by other factors as well. Some such factors are (RBS 2004):

1. Overall oil dependence—industrialised countries are now much less dependent on oil imports compared to 1970s (oil import accounts for 1% of GDP in OECD now compared to 3% in 1970s (RBS 2004). The situation varies by country, but the US economy is more oil dependent than many European economies, while Japan is fully import dependent for oil, making it more vulnerable. Oil importing developing countries are affected more adversely because of higher energy and oil intensities.
2. Energy content of non-oil imports—as oil prices affect the cost of goods produced from or using oil, countries that rely on import of such oil-based products are also exposed to the effects of high oil prices.
3. Energy tax: consumers face less dramatic changes in prices in countries with a higher level of tax on oil products than where taxes are low. Moreover, governments in high tax countries have the possibility of lowering the tax levels (at the cost of the exchequer) to offset some of the welfare loss due to higher oil prices. However, this option does not exist in low tax regimes.

In addition, three other elements could affect the balance of payment positions: remittances from workers employed in oil-exporting countries, exports to oil-exporting countries and concessional aid oil-exporting countries (Munasinghe and Meier 1993). All these elements tend to offset higher import bills but the flow of such funds may occur with a time lag and may escape the official banking system in part (thereby manifesting in the black market) or may be channeled in the form of consumer/valuable goods instead of cash. The exchange rate and import duty policies tend to influence these decisions to a large extent.

Munasinghe and Meier (1993) offer a simple way of measuring the impact of oil price shocks described above. First, consider that oil-related payments leave a gap G, which can be expressed as:

$$G = I - W - Y - L \qquad (19.1)$$

where I = net oil imports; W = worker remittances from oil-exporting countries; Y = exports to oil-exporting countries; and L = net financing from oil-exporting countries.

As we are interested to find out how the above terms adjust in response to oil price changes, we focus on the change from one period to another, which can be written as:

$$\Delta G = \Delta I - \Delta W - \Delta Y - \Delta L \qquad (19.2)$$

As the oil import bill depends on the import price of oil and the quantity of oil imported, the change in the import bill can be decomposed into two terms to reflect the impact of price change and the impact of volume adjustment. This can be written as:

$$\Delta I = \Delta P + \Delta Q, \qquad (19.3)$$

where ΔP = price effect; ΔQ = quantity effect of imports.

Rearranging we get the decomposition of the response to oil price shock as

$$\Delta P = \Delta G + \Delta W + \Delta Y + \Delta L - \Delta Q \qquad (19.4)$$

The response can be measured by determining the appropriate value for the above elements using national statistics.

19.4 Energy Price Shocks and Vulnerability of Importers

ESMAP (2005a) suggests that the direct impact of oil price increases on the GDP can be estimated by a simple relationship

$$\% \text{ change in GDP} = \% \text{ price rise} * (\text{share of oil imports in GDP}) \qquad (19.5)$$

This relationship is based on the assumption that if the price elasticity for oil and oil products is zero, then for any change in oil price, GDP will have to adjust to the same level as the change in the net value of oil imports. ESMAP (2005a) suggests that the above formula can be used to get a quick measure of the severity of oil price shocks. Based on a sample of 97 oil importing countries, the reports finds that the effect of a $10/barrel increase in oil price would be felt most severely by the poorest countries while the effect on GDP will be much lower in the industrialised countries. Table 19.1 provides the results from ESMAP (2005a) study.

The term "vulnerability" means different things to different users of the term. In general it is used to describe "the risks of being negatively affected by shocks" (UN 1999). In this chapter the term has been used from an economic perspective and implies the cost to the economy of being exposed to external or internal shocks. This idea has been used to analyse the exposure of energy importing

Table 19.1 Effect of $10/barrel rise in oil prices on GDP of oil importing countries

Per capita income (US$)	% change in GDP
<300	−1.47
>300 but <900	−0.76
>900 but <9,000	−0.56
>9,000	−0.44

Source ESMAP (2005a)

countries, especially oil importing countries to energy price shocks (World Bank 2005) and oil supply disruptions (USGAO 1996).

Babusiaux et al. (2007) and Percebois (2007) suggest that a number of indicators could be used to measure vulnerability in energy supply. These include: energy interdependence, import concentration, energy intensity, and net energy import bill. ESMAP (2005a, b) have analysed the vulnerability of countries to an oil price shock. The oil import bill as a percentage of GDP is the key variable that captures the vulnerability. A high value for the above ratio would suggest greater exposure to the oil price shock and vice versa. Vulnerability ratio of depending on a particular fuel can be expressed in a multiplicative form using as follows:

$$\frac{\text{Fuel import bill}}{\text{GDP}} = \frac{(\text{Fuel price} * \text{import volume})}{\text{GDP}}$$
$$= \text{Fuel price} \frac{(\text{Fuel import volume})}{\text{total fuel use}} \frac{(\text{total fuel use})}{\text{total energy use}} \frac{(\text{total energy use})}{\text{GDP}}$$
$$= \text{Fuel price} (\text{Fuel import dependency})(\text{fuel dependence})(\text{energy intensity})$$
(19.6)

Thus vulnerability to oil price shock can be analysed by considering policies related to oil import dependency, oil dependence in the energy mix and the intensity of energy use for economic activities.

1. Oil import dependency can be changed by discovering and producing more locally. Since the first oil shock, the number of oil producing countries has increased significantly and many countries have become important producers. By encouraging self-sufficiency in oil supply, vulnerability could be reduced.
2. Oil dependence for energy supply could be reduced by promoting oil substitution where possible or by diversification of fuel mix. The relative price of fuels is an important factor in the substitution decision-making. If substitute prices also align with that of oil, substitution becomes less likely. Gas prices have shown a tendency in the developed countries to follow oil prices, reducing the possibility of oil to gas substitution. Coal still commands price advantages but the environmental concerns of coal use acts as an impediment. Resource availability however is the main factor for reducing oil dependence and poorer countries may have a tendency to increase reliance on traditional energies facing oil price shocks.

3. Energy intensity evidently would affect exposure to oil shocks. Efficient use of energy for conducting economic activities would reduce vulnerability while higher GDP intensity of energy increases the exposure to risks. Energy intensity can be reduced through structural changes (i.e. by promoting less energy intensive activities in the economy), by adopting efficient technologies or less energy intensive technologies (e.g. dry process of cement making rather than wet cement making), and through behavioural or other changes by reducing energy use. As can be observed from IEA (2004), there is wide disparity in energy intensity across countries, which suggests significant potential for policy intervention in this area.

ESMAP (2005b) estimated the vulnerability of various regions in 1990 and 2003 and found that Sub-Saharan Africa and East Asia were the most exposed regions in both the periods. High import dependence and high share of oil in energy mix in Sub-Saharan Africa aggravated the situation while low energy intensity partly offset the exposure. As resource endowment is unlikely to be changed and these countries are likely to experience higher oil intensity as income grows, the vulnerability of this region to oil price shocks is expected to remain high.

Many other studies have now been reported in the literature trying to analyse the fuel dependence and vulnerability of countries (e.g. WEC 2008; Bhattacharyya 2009; Nakawiro and Bhattacharyya 2007). For example, Bhattacharyya (2009) has analysed the vulnerability of dependence on natural gas and coal for electricity generation in some European countries. Figure 19.7 presents the results of vulnerability due to gas dependence for electricity generation (i.e. gas bill for electricity generation as a ratio of GDP) for selected European countries.

The indicator used here is not from an import dependence point of view but from the perspective of the end-users' risk exposure. Italy, which used to spend around 0.1 percent of its GDP towards gas bill for power generation in the mid-1990s, has also seen its gas bill per unit of GDP rise almost five times during the period. In fact, Italy has recorded the highest growth in the vulnerability indicator of the five countries.

On the other hand, Spain remained the least vulnerable gas dependent system as it spends the least on gas consumption for its electricity generation. It was

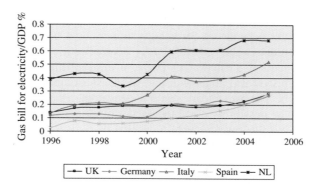

Fig. 19.7 Vulnerability due to gas dependence in electricity generation. *Source* Bhattacharyya (2009)

successful in maintaining this position through out the period of study except for the terminal year. It also has faced a gradual increase in its gas vulnerability—essentially since 2001. This therefore confirms that Spain's well diversified electricity system ensures low exposure to fuel price shocks.

It is commonly believed that the adjustment to high oil prices involves three phases (Munasinghe and Meier 1993): in the first stage, income is transferred from consumers to producers, leading to windfall gains while the general prices in consuming countries increase. In the second phase, oil-exporting countries begin to increase imports from oil-importing countries. This encourages energy producers in importing countries to search for fuels themselves or expand their production facilities. Finally, in the third phase, energy consumers fully pay for the higher prices through a transfer of real resources—reflected in higher exports to foreign producers of energy.

19.5 Impact of Higher Oil Prices: Case of Oil Exporting Countries

The effect of higher oil prices on the oil exporting countries is quite different as the income is transferred from consumers to producers, leading to windfall gains. Gelb (1988) indicated that each oil price shock of the 1970s and 1980s resulted in a transfer of US$300 million a day from oil importing countries to exporting countries. Based on a crude export of around 53 million barrels per day in 2009, the windfall for every $10 increase in price of oil (per barrel) amounts to USD 530 million per day. As most of the revenue accrues to the treasury, the balance of payment situation improves and the governments would have extra room for maneuverings.

19.5.1 Windfall Gains

The oil export revenue of major MENA countries show a clear pattern (see Fig. 19.8a, b) of close links with oil price movements. Three periods can be easily identified:

- the sharp revenue fall up to 1986;
- an extended trough between 1986 and 2000 where a minor revenue recovery is observed; and finally
- income growth after 2000 that continued to 2008.

Two countries in the sample dominate the picture—Saudi Arabia and Iran—each having almost similar levels of oil export revenue in 2006, although in 1980 Iran had less than a quarter of Saudi revenues (because of Iran–Iraq conflicts).

Fig. 19.8 **a** Oil export revenue trend of some MENA countries. **b** Oil export revenue trend of Iran and Saudi Arabia. *Source* Bhattacharryya and Blake (2009)

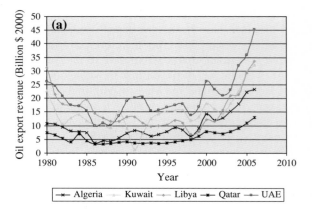

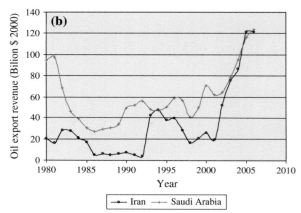

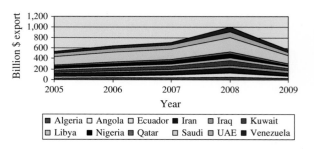

Fig. 19.9 Oil export revenue of OPEC countries. *Data source* OPEC Annual Statistical Bulletin, 2009

Oil export revenues of the Middle East region have seen a significant growth in the recent past as oil prices soared until 2008 (see Fig. 19.9). Between 2005 and 2008, oil export revenue of the group almost doubled in nominal terms. Similar was the case with other oil exporters. As a result of higher export earnings, OPEC members have seen significant improvements in their current account balance (see Fig. 19.10) and OPEC as a group had reported a 70% increase in the trade balance

19.5 Impact of Higher Oil Prices: Case of Oil Exporting Countries

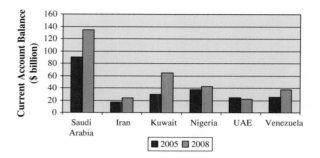

Fig. 19.10 Improvements in current account balances of some OPEC members. *Data source* OPEC Annual Statistical Bulletin, 2009

between 2005 and 2008. This has resulted in a higher economic growth during the period. For example, the GCC countries averaged a GDP growth of 8% per year between 2002 and 2007 (Saif 2009). However, the economic crisis of 2008 has reversed the situation and the oil revenue of OPEC returned to their 2005 levels in nominal terms by 2009.

Clearly, these countries face a number of challenges: the high volatility of oil price in the international market makes the investment decision difficult—especially during low oil prices. Given the captive market for oil in the transport sector and because there is a limited likelihood of loss of this market in the foreseeable future, it may make sense to leave the oil in the ground when prices are low. The price volatility also affects the export revenue of these countries and because of the resource curse issue that is associated with the export dependence of a high-value natural resource like oil, the producing countries would always be worried about the misuse of windfall gains from the volatile market. At the same time, the domestic energy use may not be as efficient as in other countries and with higher income, there may be a larger tendency of wasting more energy, which in turn can put pressure on the oil export surplus. Clearly then an efficient domestic energy utilisation policy will leave more exportable surplus and improve the export revenue in the future.

The vulnerability of oil exporters due to dependence on oil revenue can be analysed using in the vulnerability analyses proposed in ESMAP (2005a, b) and Bacon and Kojima (2008). While those studies considered the effect of oil imports on the gross domestic product, here for exporters the focus changes to oil export revenues as a ratio of GDP. This ratio can be calculated using data in local currency or in a common currency depending on the data.

The components of this oil export dependence can then be identified using a Kaya (1990) type identity as follows:

$$\frac{OER}{GDP} = \frac{OER}{OEV} \times \frac{OEV}{POS} \times \frac{POS}{PEC} \times \frac{PEC}{GDP} \qquad (19.7)$$

where OER = oil export revenue (in constant US dollar terms, million); GDP = gross domestic product (in constant US dollar terms, million); OEV = oil export volume, Mtoe; POS = primary oil supply (Mtoe); PEC = primary energy consumption (supply) (Mtoe).

Equation 19.7 identifies four drivers of oil export dependency (oil export/GDP) as follows:

The first term (oil export revenue to oil export volume) captures the effective export price, on average in constant US dollar per toe of export.

The second term (oil export volume to primary oil supply) captures the importance of oil export compared to domestic oil use. Normally exporters are expected to have a high ratio, implying a greater importance of exports compared to oil use in the economy.

The third term (primary oil supply to primary energy supply) refers to oil dependency of the economy (toe/toe). High oil dependence is likely to be inversely related to the second term, thereby affecting the overall export revenue potential.

The last term (primary energy supply as a ratio of GDP) indicates the primary energy intensity of the economy (toe/$). This, being a measure of effectiveness of energy utilisation in the economy, influences the overall local demand and thereby affects the volume of exportable oil.

Accordingly,

$$\text{Oil export dependency} = (\text{Effective price}) \times (\text{oil export importance}) \\ \times (\text{oil dependency}) \times (\text{primary energy intensity})$$

(19.8)

By considering these components and performing a cross-country comparison, it is possible to identify the best practices for each factor and to analyse the effect of adopting the best-practice policies on other countries using a what-if type analysis.

Based on Bhattacharryya and Blake (2009), the oil export dependence of some MENA countries is presented in Fig. 19.11.

For most countries in the above case, oil export dependency remained high—between 30% and 70% of the GDP. All the countries have seen a significant level of volatility in the oil export revenue dependence during the period—as oil price fluctuated.

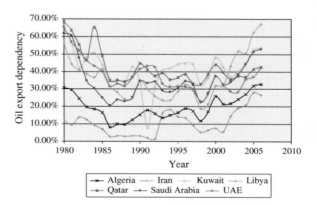

Fig. 19.11 Dependence of MENA countries on oil revenue. *Source* Bhattacharryya and Blake (2009)

19.5.2 Effect of Windfall Gains

There is no consensus among analysts whether development and growth are furthered or hindered by windfall gains. Gelb (1988) provides a detailed analysis, which is beyond the scope of this chapter. The issue of management of windfall gains assumes importance in this context. The additional revenue could be used for current consumption, offsetting past debts, undertaking policy changes, invested in assets for future returns and saved for future consumption in stabilisation funds. These issues require some further investigation and would be considered below.

Most of the oil exporting countries find that the size of the windfall is large compared to the national output, which puts constraints in terms of utilisation. The funds could be used gradually depending on the country's absorptive capacity. In this respect, the concept of linkages is useful. The linkage theory suggests that development does not depend on optimal choice of inputs and factors of production but on enlisting and properly utilising the resources and skills. But lack of adequate production linkages (i.e. the forward and backward linkages in the production system) in a high rent economy and higher propensity of importing goods for consumption (i.e. adverse consumption linkages) could act as hindrance to growth and development. Use of windfall through the fiscal linkage could also face difficulties:

1. Direct consumption of windfall by expansion of public services or transfer of revenues could create distortion. Moreover, expanded service becomes a pain at the time of economic downturns.
2. Investment in infrastructure in anticipation of demand to spur growth can be a deficiency-correcting measure and not a solution for long-term growth.
3. Reliance on a few decision-makers for use of windfall tends to promote large-scale, prestigious projects, making the country dependent on a few key investments instead of a diversified portfolio of investments, and thereby making the country vulnerable. As the grand projects may not be economically viable in the first place or may not contribute to the growth of the economy, there is need for appropriate prioritisation of investment projects using suitable criteria to allow the economy to diversify to move away from the boom-bust cycle of the oil dependent economy.
4. Rent-seeking could become a major concern, as bureaucrats are not remunerated for their soundness of decisions. This could promote personal enrichment instead of general development of the country.

19.5.2.1 Reallocation of Economic Activities

The oil windfall also brings with it the problems of sectoral reallocation of productive factors. This was first noted in the case of the Dutch economy and hence the term "Dutch Disease" is used to refer to this problem. There is an extensive literature on the subject (see Corden (1984) and Neary and Van Wijnbergen (1986) for details) and here a brief discussion is presented of the main issues.

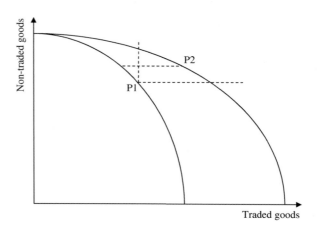

Fig. 19.12 Relocation of economic activities consequent to oil windfall

It is customary to separate the effects of oil boom on the oil-exporting economy into three major components, namely, the spending effect, the monetary effect and the resource movement effect (Auty 2002). For simplicity, assume that an economy is composed of two types of sectors: a competitive sector which sells traded goods whose prices are determined in foreign markets and a non-traded sector which is sheltered from foreign competition where prices are determined locally. If the increased wealth due to oil windfall is spent for consumption, there will be an increased demand for both tradable and non-tradable goods so long as both types of commodities are income elastic.[2] But due to competitive nature of the tradable goods, prices cannot rise above their internationally determined level (Auty 2002). Consequently, any excess demand for tradables has to be met through increased imports. This is the spending effect.

But the spending effect also leads to a resource movement effect. With the productive resources at its disposal, the country would produce and consume a certain amount of traded goods and non-traded goods, like at point P1 in Fig. 19.12. With the arrival of a sudden boom in oil, the country will have a better absorption possibility due to increased wealth. The production possibility curve shifts to the right in response to the ability to produce a much greater value of traded goods than before. But the real exchange rate has to appreciate significantly to balance the traded and non-traded sectors. This makes local products less competitive in the world market, resulting in a drop in the activities of traded sector of the economy. At the same time, it is unlikely that citizens would use all of their riches to buy traded goods. As a result, the new equilibrium will be reached at P2, where the production of traded goods is lower than the proportionate increase from P1 had the country maintained the constant share of two

---

[2] Income elasticity implies a positive relationship positive relationship between income and quantity demanded of a normal good.

19.5 Impact of Higher Oil Prices: Case of Oil Exporting Countries

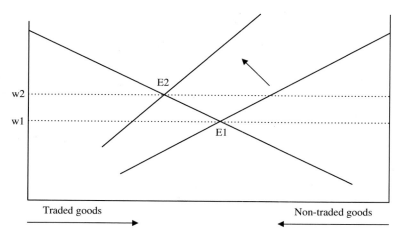

Fig. 19.13 Labour market readjustment

sectors. This in turn implies that the competitive sector shrinks as a result of the oil boom and the non-competitive sector expands.

As a consequence of the above change, the factors of production will shift from the production of non-booming tradables into the production of non-tradables. This is explained in Fig. 19.13. Assume that the total labour is allocated to producing traded and non-traded goods. Initially the economy is in equilibrium at E1 with a uniform wage rate prevailing in the economy (w1). With oil boom, the demand for goods increases due to spending effect and the resource movement effect would imply that the production of non-traded goods have to increase. In a market economy this happens through a rise in the price of the non-traded goods relative to the traded goods, providing incentives for expansion of these activities. As a result there is an increase in the labour demand and a reallocation of labour between traded and non-traded sectors. The new equilibrium would be at point E2 where the equilibrium wage will be higher. This would further accelerate the decline of the traded sector, as more firms will be out of business.

Oil windfalls increase the supply of foreign currency, lowering its price or raising the exchange rate of the domestic currency. Monetisation of these surpluses by the economy results in an increase in domestic money supply. Monetary disequilibrium (i.e. if the demand for money fails to match the supply or if there is a slow clearing money market) creates an excess supply of money which further reinforces real appreciation of the domestic currency.

Thus oil windfall in an exporting country results in a strong local currency, positive balance of payments, relocation of economic activities towards the non-tradable goods, stagnation in the industrial production, and an increase in unemployment.

19.5.2.2 Revenue Management

ESMAP (2005a) estimated the direct effects of a $10 per barrel increase in oil prices on the GDP of oil exporting countries (see Table 19.2). It can be observed that the effect is strong on low income exporting countries and the effect declines with the economy size.

The rise in export income has contributed to a significant rise in government income. For example, the revenues of Saudi Arabia have increased 86% between 2000 and 2003, while the share of revenues in GDP has increased from 23.5% in the 1990s to 68.8% in 2004 (World Bank 2005). On average, in the Middle East and North African (MENA) countries, the total revenue as a share of GDP has increased from 23.3% on average between 1990 and 2000 to 46.6% in 2004 (World Bank 2005).

Higher revenue has also prompted higher government expenditure in the MENA countries. For example, the expenditure as a share of GDP has increased from 27% on average in the 1990s to 38.7% in 2004 (World Bank 2005). Yet, the countries of the region have been successful in establishing an overall balance of 5.8% of GDP in 2004 compared to a deficit of 3.2% in 1990s.

The World Bank study also finds that the countries are saving a significant portion of their oil revenue and retiring debts. It appears that the countries have been cautious in spending their oil boom revenue this time compared to the previous oil shocks when the shocks were short lived and the rise in spending was more important. This is also supported by Saif (2009). However, a sustained increase in revenue over a period of time brings the issue of management of new found riches.

Investment in building additional supply capacity could form one option. As future oil demand growth appears to be sufficiently robust, the oil rich countries of the Middle East and North Africa are expected to face a significant share of this demand due to the relatively low production cost of MENA hydrocarbon resources and constraints on additional supply from other regions. The world market shares of MENA oil and gas could grow from the current 35% for oil and 15% for gas to around 45 and 25% respectively for oil and gas, in 2030. This could require an accelerated growth of oil production from this region: some estimates suggest that the average annual output growth has to reach 2.3% between 2004 and 2020 compared to the historical growth rate of 0.6% (between 1980 and 2003). Revenues from the oil boom could provide an easy way of funding development of hydrocarbon resources and consequent economic growth.

The oil revenue will also compete for other non-oil development activities to promote economic diversification. This would put pressures on the governments to

Table 19.2 GDP effect of $10 per barrel increase in oil price (oil-exporting countries)

Per capita income (US$)	% change in GDP
<900	+5.21
>900 and <9,000	+4.16
>9,000	+1.50

Source ESMAP (2005a)

Table 19.3 Retail prices as of Nov. 2008, US cents per litre

	Diesel	Super gasoline
Algeria	20	34
Bahrain	13	21
Egypt	20	49
Iran	3	10
Iraq	1	3
Jordan	61	61
Kuwait	20	24
Lebanon	76	76
Libya	12	14
Oman	38	31
Qatar	NA	22
Saudi Arabia	9	16
Syria	53	85
United Arab Emirates	62	45
Yemen	17	30
Rotterdam Brent price for reference	30	

Source GTZ (2009)

rethink their policies towards oil rent distribution, especially the policies related to direct energy price subsidies. Current price of energy, especially liquid fuels but also gas, are well below their opportunity costs in several MENA hydrocarbon rich countries (Table 19.3 for a rough indication of the differences between domestic prices of diesel and gasoline and an export parity price reference for crude oil), driving up internal energy demand, especially as population grows. This in turn reduces the exportable surplus and increases pressure on the local and global environment by encouraging wasteful domestic consumption in MENA.

19.5.2.3 Petroleum Funds

In order to manage the volatility of the oil revenue as well as to take care of inter-generational issues involved in the exploitation of a non-renewable resource (i.e. to ensure that the future generation are left with some resources), some countries have created a special fund which is variously called as stabilisation fund, financial fund, investment fund or simply the oil fund. There are two types of such funds—the virtual fund which appears as a special line item in the treasury account but is managed alongside all other assets of the government, and the real fund which holds the funds separately from the government assets and are managed outside the government treasury operations.

The principle of an oil fund is easy to understand using a hypothetical example (see Table 19.4).[3] In this example it is assumed that the extraction activity lasts for

---

[3] This is based on Hannesson (1998).

15 years, during which period the revenue increases slowly to reach a peak and then tapers off as the production declines. It is also assumed that the country can achieve a real rate of return on 7% on its invested capital. The present value of the revenue stream comes to 432.47 (million dollars), which yields 30.27 (million dollars per year at an interest rate of 7%. Thus, if an amount equal to 30.27 (million dollars) is used per year, the petroleum wealth will remain intact and could benefit the present and future generations. As Table 19.4 indicates, initially the revenue is not enough to meet the required flow. This leads to borrowing in the initial years but as the revenue from oil production improves, the surplus starts to build up and outgrows the amount being used every year. Finally, this leaves 432.47 (million dollars) as the accumulated fund and drawing 28.61 annually leaves the fund accumulation unchanged as long as the yield rate can be maintained. This thus leaves an asset for the future generation.

Oil funds have also been used for stabilisation purposes. As oil prices are volatile, both price increases and falls require economic adjustments. The stabilisation funds try to manage the volatility by channelling excess funds to the fund and using them during low oil prices. However, as Davis et al. (2001) indicate these funds face a number of issues including inability to stabilise or smoothing public finances

Table 19.4 A hypothetical petroleum fund (million dollars)

Net revenue	Discount factor	PV of revenue	Deposit to fund	Fund balance (beginning of year)	Fund yield	Fund balance end of year
	$r = 0.07$					
3	0.93	2.80	−27.27	0.00	0.00	−27.27
15	0.87	13.10	−15.27	−27.27	−1.91	−44.45
25	0.82	20.41	−5.27	−44.45	−3.11	−52.84
30	0.76	22.89	−0.27	−52.84	−3.70	−56.81
60	0.71	42.78	29.73	−56.81	−3.98	−31.06
80	0.67	53.31	49.73	−31.06	−2.17	16.49
100	0.62	62.27	69.73	16.49	1.15	87.38
100	0.58	58.20	69.73	87.38	6.12	163.22
100	0.54	54.39	69.73	163.22	11.43	244.37
80	0.51	40.67	49.73	244.37	17.11	311.21
60	0.48	28.51	29.73	311.21	21.78	362.72
30	0.44	13.32	−0.27	362.72	25.39	387.84
25	0.41	10.37	−5.27	387.84	27.15	409.71
15	0.39	5.82	−15.27	409.71	28.68	423.12
10	0.36	3.62	−20.27	423.12	29.62	432.47
0	0.34	0.00	−30.27	432.47	30.27	432.47
0	0.32	0.00	−30.27	432.47	30.27	432.47
0	0.30	0.00	−30.27	432.47	30.27	432.47
0	0.28	0.00	−30.27	432.47	30.27	432.47
0	0.26	0.00	−30.27	432.47	30.27	432.47
	Present value of revenue	432.47				
	Annual yield at 7%	30.27				

Source Hannesson (1998)

(because this requires fiscal policy decisions), inability of automatic saving mechanism to check government borrowing, poor control over the funds operation and spending, governance and other issues related to fund utilisation, etc.

19.6 Conclusions

Energy price volatility has remained an important issue in the energy policy debate and returns to the agenda as oil prices soar. As these episodes involve revenue transfers from importing countries to exporters, the overall global economic effects tend to be complex. This chapter has attempted to analyse various elements of economic impacts of oil price shocks and showed that simple economic tools could be used to gain important insights. It has presented how the effects of oil price shock are transmitted through the economy, how the effects vary between exporting countries and oil importing countries and possible mitigation options. The net energy importers have gained significant knowledge in this regard from the past experiences and have worked on their fossil fuel dependence. Many of them are in a better position now to deal with such issues. Similarly, the transitory phase of oil booms followed by prolonged periods of relatively low (or moderate) fuel prices cause revenue management issues for resource exporting countries. But they have also learnt from the past experiences and are dealing with such situations more prudently.

References

Auty RM (2002) Resource abundance and economic development. A study for the World Institute of Development Economics Research of the United Nations University. Oxford University Press, London

Babusiaux D, Gnansounou E, Percebois J (2007) Energy vulnerability: the right indicators. Session 31—Geopolitics II, 9th IAEE European conference, Florence, Italy

Bacon R, Kojima M (2008) Oil price risks: Measuring the vulnerability of oil importers. Viewpoint 320, World Bank, Washington, DC

Bhattacharryya SC, Blake A (2009) Analysis of oil export dependency of MENA countries: drivers, trends and prospects. Energy Policy 38:1098–1107

Bhattacharyya SC (2009) Fossil-fuel dependence and vulnerability of electricity generation: case of selected European countries. Energy Policy 37:2411–2420

Codoni R, Park HC, Ramani KV (1985) Integrated energy planning: a manual. Asian and Pacific Development Centre, Kuala Lumpur

Corden WM (1984) Booming sector and the Dutch disease economics: survey and consolidation. Oxf Econ Pap 36(3):359–380

Davis J, Ossowski R, Daniel J, Barnett S (2001) Oil funds: problems posing as solutions? Finance Dev 38(4). See http://www.imf.org/external/pubs/ft/fandd/2001/12/davis.htm

ESMAP (2005a) The impact of higher oil prices on low income countries and on the poor. Energy Sector Management Assistance Programme, World Bank, Washington, DC

ESMAP (2005b) The vulnerability of African countries to oil price shocks: major factors and policy options–the case of oil importing countries. Energy Sector Management Assistance Program, World Bank, Washington, DC

Fattouh B (2005) The causes of crude oil price volatility. Middle East Econ Surv 58(13). See (http://www.mees.com/postedarticles/oped/v48n13-5OD01.htm)

Gelb AH (1988) Oil Windfalls: blessings or curse. World Bank Publication, Oxford University Press, London

GTZ (2009) International fuel prices, 2009 edn. GTZ, Germany. See http://www.gtz.de/de/dokumente/gtz2009-en-ifp-full-version.pdf

Hannesson R (1998) Petroleum economics: issues and strategies of oil and natural gas production. Quorum Books, London

IEA (2004) Key world energy statistics 2004. International Energy Agency, Paris

IMF (2000) The impact of higher oil prices on the global economy. International Monetary Fund, Washington, DC. See http://www.imf.org/external/pubs/ft/oil/2000/

IMF (2005) World economic outlook 2005. International Monetary Fund, Washington, DC

Kaya Y (1990) Impact of carbon dioxide emission control: Interpretation of proposed scenarios. IPCC, Energy and Industry subgroup, Paris

Koyama K (2005) The recent high oil prices: its background and future prospects. Executive summary. Institute of Energy Economics, Japan. See http://eneken.ieej.or.jp/en/data/pdf/306.pdf

Krichene N (2006) Recent dynamics of crude oil prices. IMF working paper, International Monetary Fund, Washington, DC

Munasinghe M, Meier P (1993) Energy policy analysis and modeling. Cambridge University Press, London

Nakawiro T, Bhattacharyya SC (2007) High gas dependence for power generation in Thailand: the vulnerability analysis. Energy Policy 35(2007):3335–3346

Neary JP, Van Wijnbergen S (eds) (1986) Natural resources and macro-economy. The MIT Press, Cambridge

Percebois J (2007) Energy vulnerability and its management. Int J Energy Sector Manage 1(1):51–62

RBS (2004) The economic impact of high oil prices. Group Economics, Royal Bank of Scotland, Edinburgh. See http://www.rbs.com/content/media_centre/rbs_and_the_economy/downloads/world/oil_prices.pdf

Saif I (2009) The Oil boom in GCC countries, 2002–2008: old challenges, changing dynamics. Carnegie Endowment for International Peace, Washington, DC

USGAO (1996) Energy security: evaluating U.S. vulnerability to oil supply disruptions and options for mitigating the effects. Report to the Chairman, Committee on the Budget, House of Representatives, The U.S. General Accounting Office (USGAO), December 1996

WEC (2008) Europe's vulnerability to energy crisis. World Energy Council, London. See http://www.worldenergy.org/documents/finalvulnerabilityofeurope2008.pdf

World Bank (2005) Middle East and North Africa: 2005 economic developments and prospects—oil booms and revenue management. World Bank, Washington, DC

Chapter 20
Energy Security Issues

20.1 Introduction

Given the paramount importance of energy for all economic activities around the world, issues related to energy security have gained importance in the wake of recent high oil prices and the fear of supply shortages for natural gas and electricity in many countries. Energy security concerns first emerged in the aftermath of the first oil shock in the 1970s, when oil importing countries were caught unguarded and had to struggle to cope with the adverse effects of oil price rise. Since then countries have followed diverse policies to mitigate the problem. Low oil prices since mid-1980s and the shift of focus in the 1990s to market reform and restructuring meant little attention to the issue of security of supply. It was believed that markets would be able to solve the problems of the energy sector. However, concerns about peaking of oil supply and supply capacity to match the demand have brought back an era of sustained high oil prices. Once again the issue of energy security has become a major policy concern.

This chapter intends to provide an understanding of the concept, its economic dimension and an analysis of various alternative options to deal with it.

20.2 Energy Security: The Concept

"Energy security is commonly defined as reliable and adequate supply of energy at reasonable prices" (Bielecki 2002). Reliable and adequate supply implies uninterrupted supply of energy to meet the demand of the global community. This segment of the definition establishes the link between adequate supply and energy demand at any given time. Supply adequacy and reliability is not a matter of external dependency alone. In many countries (developing and developed) the internal sources of supply could equally be problematic. However, of the literature on energy security focuses on external supply alone as the control over external supply can be limited in most cases.

Reasonable price on the other hand is a more difficult term as there is no universally accepted benchmark. Economically it would mean market-clearing price in a competitive market where supply and demand balances. But as we shall see below energy security involves externality and therefore internalisation of costs would be essential for efficient resource allocation.

The term is used by different people to mean different things and accordingly, energy security has geopolitical, military, technical and economic dimensions (Bielecki 2002). There is a time dimension of it as well: in the short-term, the main concern relates to the risks of disruption to existing supplies essentially due to act of god, technical or political problem; in the long-term, the risks related to future energy supply also arise.

Like any other concept, this concept is evolving as well. For example, initially, the focus was only on oil and oil products. Now it covers all energies and various types of risks to reliable and adequate supplies (including accidents, terrorist activities, and under investment). The geopolitical, internal and temporal aspects of the issue require a multi-dimensional policy approach to deal with the problem.

The literature has focused on the oil supply security in particular and identifies a number of components of the energy security problem (Toman 2002): (a) exercise of market power by suppliers to raise prices, (b) macroeconomic disruption due to energy price volatility, (c) threats to infrastructure, (d) localised reliability problems, and (e) environmental security. But the problem is not limited to oil supply alone and recent studies focus on the entire gamut of the problem.

20.2.1 Simple Indicators of Energy Security

Two types of indicators are commonly used in the supply security literature: an indicator that expresses the level of exposure in terms of dependence level and an indicator of vulnerability. The level of import dependence of a fuel provides an idea about the price and quantity risks associated with importing the fuel and accordingly, a higher level of imports is generally considered to be a riskier option. Similarly, in the case of an electricity system, high dependence on a single fuel is considered to be a riskier option. But as the risk of supply disruption is associated with the concentration of supply sources and the probability of disruption of supply from each source, a highly import dependent system that is well diversified need not necessarily be a risky one.

20.2.1.1 Indicators of Dependence

Indicators that are relevant for energy diversity and energy security are (IAEA 2005):

(1) Import dependence—this indicator can be used for the overall supply position of a country or a region or for a particular fuel. For example, the ratio of net

energy imports to the primary energy supply in a particular year would provide how reliant the country is on imported supply. If a country consumes 100 Mtoe of primary energy and 90 Mtoe is imported, its import dependence is 90%. High import reliance normally tends to increase the price risk and volume risk related to supply interruption.

Import dependence at the fuel level shows the degree of exposure for each fuel. Often, the import dependence of different fuels varies significantly and a country could have a high import dependence for one fuel but highly self-sufficient in another.

At a more disaggregated level, the import dependence by origin of supply could provide a more accurate picture about the risk. If a country depends on a single country for its imports, the risk is particularly high. On the other hand, a diversified source of imports could reduce the risk of supply disruption.

High import dependence of a fuel does not necessarily mean high risk for a country. It depends on a number of factors: the importance of the fuel in the overall demand; how diversified is the source of supply; and the amount of market power of the suppliers. If all of these factors tend to be adverse for a country, the risk will be high.

The evolution of import dependence of a country can be viewed from a plot of the ratio over a period of time. Similarly, using supply forecasts, the expected changes in the future can be captured.

(2) Fuel Mix—this indicator basically shows the share of a particular fuel in the energy demand of a country or its importance in the energy supply. Depending on the focus of the analysis, this ratio can be determined at different levels:

(a) The primary energy consumption mix tells how diversified the overall energy demand is. For example, if a country used 90% oil and oil products and 10% gas to meet its primary energy demand, it cannot be said to have a diversified fuel mix.
(b) The final energy consumption mix gives an indication of fuel diversity at the end-user level.
(c) The sector level fuel mix provides a similar picture at the end-use sector level. The extension of the analysis at the sector level provides a clearer picture of vulnerability of different sectors. For example, if the industry relies only on electricity and natural gas for its energy needs, and if electricity is dependent on natural gas supply, then the industry is highly exposed to changes in the natural gas supply.
(d) Electricity generation mix tells which fuels (and technologies) a country uses for its electricity supply.

An analysis of the fuel mix trend can be used to identify any possible adverse changes in the fuel diversity. Corrective policies can then be considered. Similarly, forecasts of future fuel mix can suggest if the country is moving in the right direction or not. For example, the expected closure of coal and nuclear power plants in the UK by 2025 is expected to increase the share of gas in the electricity generation mix. With domestic gas supply declining, such reliance of gas-based power would necessitate gas imports, making the country vulnerable.

(3) Stocks of critical fuels—this indicates the availability of national stocks of a fuel and the length of time that the fuel could be used if supply disruption takes place, assuming current level of consumption. For example, IEA member countries maintain a 90-day stock of critical fuels.

20.2.1.2 Indicators of Concentration and Diversity of Supply

The following indicators are commonly used:

(a) Herfindahl–Hirschman index: The Herfindahl–Hirschman Index (HHI for short) is generally used for market concentration analysis. This is measured by the sum of the squares of the individual market share of each firm in the industry. The HHI ranges from 0 to 10,000, with the lower range obtained when very large number of firms exist in the industry and the higher range reached with a single producer.

The Herfindahl–Hirschman Index is represented as:

$$HHI = \sum_i x_i^2 \qquad (20.1)$$

where x_i is the market share.

The level of concentration is high with HHI above 1800. For energy security purposes, the HHI Index can be used to measure the level of concentration of imports from different sources. Thus, by considering x_i to represent the proportion of imports from supply origins, the level of import concentration can be measured.

The HHI has its own shortcomings as it fails to take into account domestic production. It cannot take the political risk into consideration. Percebois (2007) indicated that the HHI of French oil import in 2004 was 2538 and it was 2469 for natural gas. In 2005, the European Union of 25 had the HHI of 2544 for oil imports and 3538 for gas imports. These indices show high levels of import concentrations.

(b) Shannon–Wiener index: The Shannon–Wiener-Index (SWI) is a diversity index. The SW index for the share of imports from different sources is given by:

$$SW = -\sum_i x_i \ln(x_i) \qquad (20.2)$$

where x_i represents the import share from each country (or source). The negative sign at the front of the equation makes sure that the outcome of the SW index is always positive. When all imports come from a single source, the minimum value is reached (which is zero). As the number of countries supplying the fuel increases, the SW index also increases. Therefore, a higher value of the calculated SWI means good situation as regards imports diversification and supply security while a lower value means a worse situation. The main limitations of the HHI remain here also: it cannot take domestic production separately from the imports and the political risk cannot be incorporated.

The UK Energy Digest provides the SW index for the power generation diversity in the country.

(c) Adjusted Shannon–Wiener–Neumann index (SWN index): The adjusted Shannon Wiener Neumann Index (SWNI) removes the limitations of the Shannon–Wiener-Index (SWI). If the political stability factor is included alone, the index takes the form

$$\text{SWN1} = -\sum_{i} b_i x_i \ln(x_i) \qquad (20.3)$$

where b_i is the political stability factor of the country from where imports are coming. The World Bank Report on Governance Matters can be used for the political stability factor. Imports from unstable regions of the world tend to reduce the original Shannon-Wiener-Index and vice versa.

To include the share of indigenous production, the SWN index can be modified as follows:

$$\text{SWN2} = -\sum (b_i x_i \ln(x_i)(1 + g_i)) \qquad (20.4)$$

where g_i represents the indigenous production for the country in question.

20.2.2 Diversity of Electricity Generation in Selected European Countries

The diversity of fuel-mix of electricity generation in some European countries is considered below. The analysis is presented using two indices: SWI and HHI.

Table 20.1 presents the fuel mix of electricity generation in 5 European countries retained in this study for 1995 and 2005. As can be seen, coal was displaced by natural gas in the UK to a large extent and in Spain and Netherlands to a lesser extent. In Italy, fuel–oil based generation which was the dominant form of power in the mid-1990s was replaced by natural gas. Natural gas consolidated its position as the leader in the Netherlands during this period. Dependence on fossil fuels in electricity generation remained very high in the Netherlands (88%), Italy (79%) and the UK (above 70%). Spain was moderately dependent on fossil fuels in the mid-1990s but its exposure has increased in 2005 to around 60%.

Table 20.1 Fuel-mix of electricity generation in five European countries

		Coal (%)	Natural gas (%)	Oi (%)	Nuclear (%)	Hydro (%)	Others (%)
UK	1995	57.40	15.50		25.20		1.90
	2005	40.85	36.65		19.75		2.75
Germany	1995	54.02	8.05	1.67	28.73	4.51	4.02
	2005	43.46	11.03	1.70	26.29	4.31	13.21
Italy	1995	9.93	19.46	50.03	0.00	17.36	3.22
	2005	14.36	49.15	15.52	0.00	14.13	6.84
Spain	1995	34.95	2.25	8.74	33.14	14.68	6.24
	2005	25.04	26.87	8.30	19.57	7.83	12.39
Netherlands	1995	32.16	51.85	4.77	4.96	0.10	6.16
	2005	23.45	57.73	2.26	3.99	0.08	12.49

Source Bhattacharyya (2009)

Figure 20.1 presents the level of fuel-mix concentration of generation for the period between 1995 and 2005 using HHI. As can be seen, all the countries chosen in the study have HHI above 2000, indicating that the electricity supply in these countries is highly concentrated. The level of concentration has declined in the UK in the early 1990s and then stabilized. Similarly, Spain and Italy have also recorded some improvement in terms concentration in the later half of the 1990s but the improvement in these two cases were over a longer period compared to the UK. Germany did not show any change in the level of concentration of generation fuel mix over the past decade while the situation has deteriorated in the Netherlands. Of the five countries considered here, Spain had the lowest HHI since 1996 while the Netherlands, with an HHI of above 4000, had the highest over the same period. The dominant position of natural gas with a share of above 50% in the fuel mix of electricity generation has adversely affected the concentration in the Netherlands while a well distributed fuel mix of Spain has clearly improved its level of concentration.

Figure 20.2, which provides the trend of SWI of fuel mix for electricity generation in the above five countries between 1995 and 2005, also leads to the same observations as above. In all the five cases, the SWI ranged between 1 and 2, implying that these countries are not dependent on one or two fuels for their

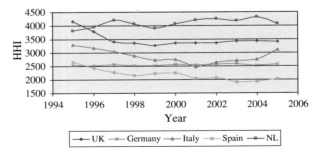

Fig. 20.1 HHI of electricity generation mix in selected European countries. *Source* Bhattacharyya (2009)

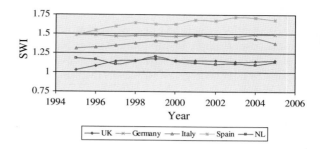

Fig. 20.2 SWI of electricity generation mix in selected European countries. *Source* Bhattacharyya (2009)

electricity generation but their fuel diversity is not highly commendable either. Spain has the most diversified generating system in the sample and the level of diversity has improved during the past decade. Germany and Italy occupy an intermediate position, where the diversity level in the German system has not changed appreciably while the Italian system has recorded an improvement until 2001 followed by a somewhat reduction in the diversity. The liberalised markets of UK and the Netherlands have the least diversified generating systems in the sample and their level of diversity did not change in the past decade.

It is clear that the above five countries rely on fossil fuels to a great extent for their electricity generation. Although their systems are not highly concentrated in terms of fuel mix, they cannot be considered to be in a highly desirable situation either. As the fossil fuel prices have risen in recent times, their electricity system is likely to be vulnerable. It is to this aspect that I now turn to.

20.3 Economics of Energy Security

Energy supply disruptions consider interruptions of supply due to a variety of factors: act of sabotage, failure of a supply technology, breakdown of supply infrastructure, etc. The level of insecurity is reflected by the risk of a physical, real or imaginary supply disruption (Owen 2004). Normally, a high level of insecurity would result in high and unstable prices over a prolonged period.

In order to understand the economics of energy security, it is important to categorise the sources of insecurity. Two types of supply disruption risks could be considered (Markandya and Hunt 2004): strategic and random. A strategic risk would arise due to political instability, market power or even inadequate investments in supply facilities. OPEC deliberately manipulating the supply and prices comes under this category. Random shocks such as terrorist acts on the other hand are more speculative in nature and may not follow any set pattern. Although these risks could affect both domestic and the international markets, the strategic risk has less relevance for the domestic systems. The domestic systems on the other hand could face supply disruption due to insufficient infrastructure, technical failures, social unrest, or due to acts of terrorism (Owen 2004).

This section focuses on the economic aspects of two main components of the energy security issue: the effect of market power on the cost of imported energy and the cost of supply disruption. Oil is used as an example as it is the most traded commodity in the world market and oil imports account for a significant share of imports in many countries. However, the same logic applies to other energies to a great extent.

First, the cost of oil imports is presented. This is followed by a discussion of the cost of supply disruption and analysis of measures to mitigate the risks.

20.3.1 External Costs of Oil Imports[1]

Although oil is a commodity, it has a certain special characteristics:

(a) oil is concentrated in a relatively small area in the Persian Gulf, which allows for monopolistic behaviour in the oil market;
(b) oil has limited (if at all) substitutes in its main uses, which removes the flexibility of users to move away from use of oil;
(c) oil supply shocks may leave nations to serious adjustment problems; and
(d) all stages of oil fuel cycle impose unintended and damaging environmental effects.

Consequently, the market failure argument applies here and the market price of delivered oil to the consumers departs from the full social cost of oil. The social costs may include costs due to non-competitive markets, costs due to environmental damages, and economic losses due to price shocks. Oil consumers do not pay for these costs in the price but the society as a whole pays for them.

One commonly identified externality related to oil import arises due to the monopsony power of certain importers that affect the price of oil in the world market. For a price taker in the international oil market, the price paid by the consumers is equal to the cost of the extra oil to the economy and hence there is no externality here. But if a consumer has a large market share in consumption (say the US), then any extra demand for imports by this consumer would adversely affect the global demand and consequently, the world oil price would increase. This raises the country's total oil import bill—for marginal and infra-marginal imports. While the private cost to consumers is the marginal cost of imports, the society bears the cost higher payments for the infra-marginal quantities, making the social cost higher than the private cost. The difference between the social and private costs is called the monopsony wedge.

The logic of externality would suggest that the market does not convey the correct signal to the consumers and accordingly, the consumption decision would

---

[1] This section relies on Leiby et al. (1997), Toman (1993), Markandya and Hunt (2004) and Huntington (2009).

Fig. 20.3 Social cost of oil imports

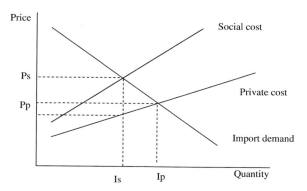

Fig. 20.4 Monopsony premium plot. *Source* Based on Parry and Darmstader (2003)

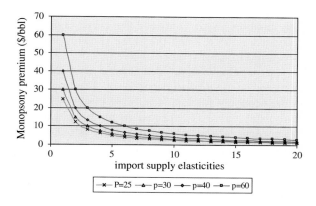

be based on private costs and not on the social costs. This is shown in Fig. 20.3. While the import based on private cost is Ip, the efficient level of import would be Is based on the social costs.

The effect monopsony power depends on two factors (Parry and Darmstader 2003): the level of import dependence and the effect of monopsony demand on the world oil market. If the country does not depend on import (i.e. import dependence is zero), there is no externality due to monopsony power. With higher level of import dependence, the monopsony wedge increases. Similarly, if the world oil market was perfectly elastic and competitive, the extra import demand from a major consumer would not have any effect on the world oil price and the externality would not exist. But the presence of the OPEC makes the supply non-elastic and the world price is affected by the supply from non-OPEC producers as well.

Parry and Darmstader (2003) suggest a simple relation to capture the monopsony premium or wedge. Generally, if P is the world price of oil and e is the elasticity of import supply, then the monopsony wedge (or premium) is given by P/e. If e is infinite (i.e. the import supply is perfectly elastic), the premium is zero. For various oil prices and import supply elasticities, the premium would vary as shown in Fig. 20.4.

As can be seen from the above plot, the premium could be high for inelastic import supply; otherwise, the premium fall quite sharply and could be low. The literature provides a wide range of estimates for the US, ranging from $0 to $14 per barrel, while Parry and Darmstader (2003) prefer to use $5 per barrel as the premium. However, most of these estimates were based on low oil prices and may not be valid in a high oil price regime. For example, Leiby (2007) estimated the monopsony premium for the US at $8.9 per barrel (at $2004 constant prices) considering the conditions prevailing in the new millennium.

20.4 Optimal Level of Energy Independence

Here the marginal cost approach is used to get some idea of optimal dependence. This requires us to construct the curve depicting marginal cost of its import dependence (MDC) and the curve showing marginal cost of security (MSC) as shown in Fig. 20.5 (Percebois 1989).[2]

The marginal import dependence cost (MDC) curve captures the costs of increased energy import dependency. This would include direct and indirect costs to the economy (including military costs, economic disruption costs, etc.). Normally, this curve is expected to be downward sloping with respect to import independence. When a country is fully self-sufficient, the marginal cost of import dependence is zero and it could be very high for 100% import dependence. It is not easy to develop such a curve as the cost depends on many factors such as import diversity, ease of energy substitution, importance given by the society on energy import, etc.

The marginal cost of security curve (MSC) on the other hand is the cost the society is willing to bear for increasing the national energy independence. A country could reduce its import dependence through energy stocks, energy rationing, promoting national supply, etc. The incremental cost of increasing independence would be captured here. It is generally assumed that the marginal cost of security is zero for domestic energy supply (although this need not be true). Costs start to increase at a faster rate with higher levels of independence. So the curve does not start at the origin (there is an offset) and has a steep slope.

The optimal rate of energy independence is given by the intersection of the two marginal curves as shown in Fig. 20.5. The graph suggests that: for an optimal level of energy independence; it is important to consider the costs of ensuring security of supply and the cost of the damage. It is not economically efficient to improve energy independence beyond the optimal level; this is so because the cost of providing the security of supply would be much higher compared to the marginal dependence cost. There is a price (P*) that the society is willing to pay to

---

[2] This part is based on Percebois (1989).

20.4 Optimal Level of Energy Independence

Fig. 20.5 Optimal rate of energy independence. *Source* Percebois (1989)

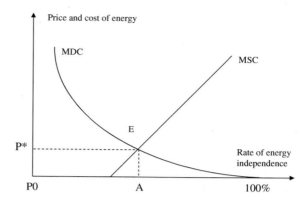

ensure the optimal level of security of supply—this is the premium that has to be paid to ensure security of energy supply.

20.5 Policy Options Relating to Import Dependence

If oil import imposes external costs to the society, what are the options available to mitigate them? The literature on energy security has considered a number of options and we discuss a few of them in the following paragraphs.

20.5.1 Restraints on Imports

Such a policy aims at imposing import restrictions through tariffs or quotas to mitigate the costs related to import dependence. Alternative policies that would eventually limit energy imports (such as tax on fuels, promotion of domestic supply, fuel substitution, promotion of alternative sources of energies, etc.) could also be considered under this category. We analyse the economic logic of using import quota and import taxes.

20.5.1.1 Effect of Import Tax and Import Restriction

Let us consider an energy importing country whose energy demand and domestic supply are given by schedules D and S respectively in Fig. 20.6. If the country does not participate in international trade, the domestic price would be the market clearing price p1. Assume that the international price p2 and is lower than p1. In an open economy, the supply would be met by a combination of local production and import. The country will produce q3 and import q2–q3. This volume of import would involve a significant foreign exchange outflow for the country.

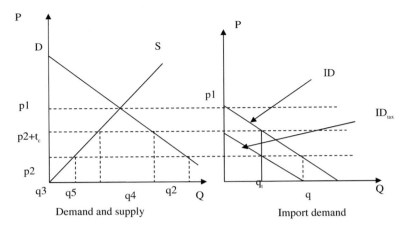

Fig. 20.6 Effect of import tax and import restriction

Consider now that the government is concerned about the energy security and that it imposes an import tax equal to t_c per unit of import. With this tax, importing energy would be costlier which makes import to shrink. The domestic supply would be encouraged at this higher level of price, as more domestic suppliers would be willing to produce. Import volume reduces to q4–q5.

The import demand function for the country is shown in the right hand panel. In absence of any import tax, the import demand is given by ID. At price p1, the import demand is zero but it increases to q when the price is p2. When the tax is imposed, the demand curve shifts to ID_{tax}. At price (p1 − t_c), the demand is zero while with tax t_c, the import volume reduces to q_t. Thus, the import schedule shifts leftwards by (q − q_t).

Now consider the effect of imposing an import quota system. Assume that the government imposes a quota at level q_t (i.e. the imports should not exceed this level). This is shown in the right hand panel. As the imported supply cannot exceed the quota, the price rises to p2 + t_c level, thereby reducing the demand as before. The domestic supply receives encouragement at this price and import remains restricted. In a quota system, the import demand function is represented by p1Aq_t. At prices below p2 + t_c, the quota is a binding constraint and the level of import remains fixed at q_t.

The tax system is a price-based mechanism. The import demand varies depending on the oil price and the level of tax. The import demand curve is shown by ID_{tax}. The effectiveness of the instrument could be less. The tax revenue accrues to the government. It does not require any additional administrative system. In a quota system, there is no ambiguity about the import level (hence a certain instrument). It requires additional administrative machinery to implement the quota system. It could also lead to corrupt practices (through grant of exemptions) or illegal smuggling of the products. More importantly, the higher revenue goes to the suppliers and not to the government. In fact, as quota is price insensitive beyond a threshold, the exporters have incentives to adjust prices to the higher level.

Thus the two policy options have different economic consequences. In the case of a quota, revenue transfer to the exporting countries would take place, if they are in a position to exploit the situation. It may also involve a higher transaction cost. While in the case of import taxes, the government could earn revenues by reducing import demand.

Therefore, for an importing country it may be beneficial to use an import tax system as long as such a system is compatible with the international trade regimes.

20.5.2 *Import Diversification*

The logic is simple: do not put all the eggs in one basket. This is because the risk of supply disruption is high when a country relies on a single source for its energy supply (i.e. becomes a captive consumer).

This risk can be mitigated through diversification of the source of supply. From an economic point of view, this implies finding the least-cost supply solution taking supply risks into consideration. However, for oil and to a lesser extent for gas, the global reliance on the Middle East is expected to increase where most of the reserves are located. This coupled with political instability of the region and increasing demand from the developing economies raise concerns for future oil supply security.

Two new developments in the area of import diversification perhaps are worth mentioning.

- The first relates to an increased level of activities and investments in production facilities by importing countries in foreign oil producing regions. Chinese oil companies are now forerunners of this trend and are investing massively around the world. Japan also relied on such a strategy in the 1970s and 1980s although may be less aggressively.
- A second trend appears to be emerging in the form of seeking cooperative solutions rather than relying on competitive outcomes. This trend is noticed in various areas:
  - *Importer-importer co-operation*: China which was engaged in competition with India through rival bidding for acquisition of energy assets elsewhere have now joined hands to jointly develop and acquire such assets. The cooperative strategy is expected to reduce the cost of procurement (and hence the supply) and better use of other resources.
  - *Importer-exporter cooperation*: Joint development by importing and exporting countries would ensure flow of required investments for the development of facilities and could reduce transactional risks.

The framework of cost-benefit analysis plays a vital role in such decisions. A nationalised company can employ a different threshold for decisions compared to a private company (regarding discount rates, profitability ratio, future market conditions, etc.). The long-term nature of these investments and uncertainties about the future as well as risk-averseness of the investors would influence the decisions. However, wrong investment decisions may lead to outgo of significant financial resources and costly supply in the future.

20.5.3 Diversification of Fuel Mix

Diversification of fuel mix in an economy tries to reduce dependence on a particular fuel and to achieve a diversified portfolio of energy supply options. For example, Salameh (2003) indicates that the US has been diversifying its fuel mix for ages to replace oil and coal by natural gas and nuclear. In the future, renewable and other technologies on which it is investing heavily could add more diversity.

The choice is often limited by: the availability of resources, available technological options to exploit such resources, costs and investment requirements, and other considerations including environmental and social concerns.

It is difficult to generalise but a few trends could be indicated.

(a) **Effects of restructuring on fuel diversity in electricity**: Reliance on market forces upon restructuring and reform of the energy industries in the 1990s led to promotion of competitive solutions in the electricity markets. This has resulted in a shift in technology choice for supply as the private investors are now looking for quick recovery of investments. Consequently, low cost options are being preferred compared to capital intensive solutions, reducing supply diversity.
(b) **Come-back fuels**: Coal and nuclear are re-emerging as preferred alternative options for power generation. Stability of coal prices, availability of technological options and higher availability of coal in the demand areas has created a positive mood, although environmental considerations act as a hindrance. Security of supply is forcing many countries to rethink about the nuclear option.
(c) **More renewable energies**: Renewable energies are being promoted for various uses to replace or reduce reliance on fossil fuels, thereby adding diversity and improving security. Various policies such as renewable energy targets or obligations, fixed feed-in tariffs, quicker depreciation and recovery of capital, and fiscal incentives are being used to promote renewable energies.

20.5.4 Energy Efficiency Improvements

Efficient use of energy reduces energy demand, which in turn reduces import requirement. This also reduces environmental damages and resource depletion.

Although significant efforts have gone into energy efficiency improvements and demand-side management programmes, availability of cheap energy has reduced their appeal in the past. With higher energy prices, it could again become easier to pursue some of these objectives.

In this respect, the importance of rational energy pricing needs to be emphasised. If domestic retail prices are maintained at inefficient levels, consumers remain insulated from the price movements and do not appreciate the need for efficient use of energies. Removal of energy subsidies could provide the necessary incentive to consumers, although efforts so far in this direction have yielded little result. The efforts are hindered by non-availability of information, need for sophisticated decision-making, use of non-standard procedures, etc.

20.6 Costs of Energy Supply Disruption

Any supply disruption imposes some costs on the economy due to loss of economic activities, price effects and costs of alternative supply arrangements. For oil, it is considered that the supply interruption will lead to higher import prices, given the dependence of the economy on the imported energy source. This then results in economic loss directly through loss of outputs, unused factors of production, cost of stand-by generation capacities, etc., and indirectly, through increased cost of business due to inefficiencies, misallocation of resources, etc.

The estimation of disruption cost involves the following steps (Razavi 1997):

- formulation of supply interruption scenarios providing information on the volume of supply unavailability over expected disruption periods; The level of insecurity is reflected by the risk of a physical, real or imaginary supply disruption (Owen 2004). Normally, a high level of insecurity would result in high and unstable prices over a prolonged period.
- assessment of how prices would be affected due to such supply interruptions.
- an estimation of GDP loss due to price increases.

Leiby (2007) suggested that the above can be represented as follows:

$$E_{\{\Delta Q\}}[C_d] = \sum \phi_j [C_{Id}(\Delta P(\Delta Q_j)) + C_{GNPd}(\Delta P(\Delta Q_j))] \quad (20.5)$$

where C_d = cost of disruption
C_{Id} = cost due to import disruption
C_{GNPd} = cost of losses due to economic dislocation
ϕ_j = annual probability of supply losses
ΔP = price change
ΔQ = quantity change
$E(C_d)$ = Expected cost of disruption

The disruption premium is obtained by considering the marginal change of the above expected cost with respect to import quantity. Leiby (2007) estimated the disruption premium for the US at $4.68 per barrel of oil at ($2004 constant prices). However, as can be imagined, the estimation of such a premium is not easy and involves a large number of assumptions and forecasts about future events. Thus the estimates vary depending quite significantly depending on the choices made.

An understanding of the disruption cost is important for deciding the mitigation strategies. If the cost is high, higher levels of supply reliability could be justified and vice versa.

20.6.1 Strategic Oil Reserves for Mitigating Supply Disruption

The Strategic Petroleum Reserve was a response of the developed countries to the oil price shocks of the 1970s. The objective was to provide a deterrent to deliberate, politically motivated reduction in supplies. This initiative was engineered by the International Energy Agency (IEA) in 1974 under the auspices of the Agreement on an International Energy Program.

Under this agreement, IEA member countries hold a stock of oil equivalent to 90 days of net imports in the previous year. Supply can be released in emergency conditions when the supply disruption exceeds 7% of IEA or any member country supply. Similarly, the EU also has adopted a comprehensive set of measures including the obligation to maintain stocks of three types of petroleum products (namely motor spirit, middle distillates and fuel oil) for at least 90 days of average daily consumption in the preceding calendar year. Although the IEA program and EU measures have some minor variations, the two serve similar purposes and member countries tend to use same stocks for complying with both the obligations (Bielecki 2002).

There are several advantages of such strategic reserves: (a) stock releases pacify markets and dampen price rises; (b) allow time for economies to adjust to the changes, (c) although a few countries are members to the plan, consumers globally benefit from the stock due to market reaction, and (d) they allow room for expanded co-operation among countries. The stockpile can be viewed 'as a publicly provided insurance policy against petroleum market shocks' (Taylor and Van Doren 2005). But what justifies public provision of this service?

Public provision of the stock may be required for a number of reasons (Taylor and Van Doren 2005; Toman 1993):

(a) *non-optimal stockpiling by the private sector*: privately owned inventory may be held at a smaller level than the economically efficient level because:

- the market price may not provide effective signals to investors about the total benefits and costs.
- the presence of externality would create a divergence between the private and social costs and benefits, requiring such an intervention.

20.6 Costs of Energy Supply Disruption

- Moreover, the private stockholder may not be able to capture the entire benefit of holding stock when there are significant macroeconomic benefits (Toman 1993; Taylor and Van Doren 2005).
- Finally, changes in the regulatory or fiscal environments could deprive the stockholder some or most of the benefits of holding the stock and thereby discourage non-optimal private stockholding (Taylor and Van Doren 2005).

(b) *Behavioural problem*: private entities guided by profit-maximising behaviour may hold stock rather than releasing it at the time of high prices in the hope of higher profits.

(c) *Cost consideration*: private stockpiling may be costly compared to publicly-owned stockpiling because of technology choice, storage location and size.

However, for any such strategic reserve, a number of issues arise (Toman 1993):

(a) *Reserve sizing*: the sizing of the stock and its use are influenced by the cost of economic disturbance to be mitigated, its probability, size and duration, and the interaction of private and public stocks could also influence the sizing decision.

(b) *Timing and method of stock utilisation*: often the literature on stockpile release profiles provide little help as the models rely on simplified assumptions.

(c) *Arrangements for stock use*: the question of institutional arrangements for using such reserves has been analysed as well. Often it is assumed that the stocks would be sold in the spot market periodically using sealed-bid auctions. Forward sales and sale of options to purchase oil from the reserve at predetermined strike prices are also possible (Toman 1993).

But such reserves also add to the cost (of building and carrying the stock among others) and hence the optimal stock size depends on the costs and benefits derived from the stockpile. Following Razavi (1997) the desired level of stock of strategic reserve (S*) could be determined using a simple framework by comparing the cost of maintaining the reserve and the benefits of avoiding a sudden supply shock (see Fig. 20.7).

Fig. 20.7 Desired level of strategic stock. *Source* Razavi (1997)

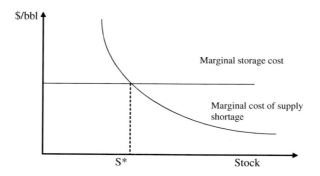

Although strategic reserves are used as a policy option, its costs are not often reflected in the pricing of energy. Taylor and Van Doren (2005) question the economic rationale for maintaining stocks as well for the following reasons:

(a) The cost for maintaining the reserve in the USA was found to be quite high compared to the oil price. They estimate that each barrel of strategic reserve costs the taxpayer between $65 and $80 and maintaining such high cost oil for shortage mitigation does not make economic sense.
(b) The amount of oil stocked is just a fraction of the global oil demand and would not be able to influence the international oil price to any significant level.
(c) The reserves have been used only three times so far in the US history and the timing and volume of stock release did not provide much comfort to the affected population.

20.6.2 International Policy Co-ordination

Security of energy supply has an international public good dimension. This is because measures taken by any country independently would also benefit (or impose costs on) others.

International policy coordination helps avoid free-riding and limit opportunistic behaviour of countries. The crisis-response provisions of the IEA form the essential mechanism for such co-ordination in industrial countries. At a regional level, ASEAN has adopted an Emergency Petroleum Sharing Scheme during shortage and oversupply to assist both importers and exporters of the region (Bielecki 2002).

Any such international mechanism would have to ensure provision of the public good in a fair, cost-sharing programme. Normally larger benefits are expected to accrue to bigger economies. This requires some sort of 'common but differentiated' responsibility approach [adopted for the Climate Change policy coordination] (APERC 2002). Similarly, it may not make sense for smaller countries to go for own strategic reserves due to adverse cost-benefit characteristics and a cooperative solution would be preferable. The possibility of economic and political policy coordination as a group could also be considered.

20.7 Trade-Off between Energy Security and Climate Change Protection

Concerns about the climate change in recent times have imposed an additional consideration in the energy security debate. The diversification of energy supply system to enhance energy security could have a bearing on the climate protection.

20.7 Trade-Off between Energy Security and Climate Change Protection

For example, if coal is locally available and to reduce dependence of imported oil or natural gas, if coal use is promoted, the carbon emission is going to increase. On the other hand, if nuclear power option is chosen, both energy security and the protection of the climate will be ensured. Given that a large number of technical options are available for abating greenhouse gas emissions with different potentials for enhancing energy security (See Fig. 20.8), there is room for a trade-off.

In Fig. 20.8, the origin represents the reference scenario based on the business-as-usual assumptions. Two policy objectives are considered from this point: enhancing energy security (along the vertical axis) and abating greenhouse gas emission for climate protection (along the horizontal axis). The figure indicates a number of alternative options—some of which predominantly offer the security benefits (such as the Strategic Petroleum Reserves (SPR) or corn-based ethanol) while some others offer predominantly climate benefits (nuclear or renewable options). There are other options in between.

To determine the optimal policy combination, Brown and Huntington (2008) suggest the following simple optimization:

Assume that there are n technologies (including energy conservation options) x_i (for $x_1, x_2..., x_n$) for protecting the climate and enhancing the security of supply, each costing c_i. Assume that the security enhancement obtained from each technology is s_i, and that the GHG abatement obtained is q_i. Then the total provision of energy security, S, is the sum of the contribution of each technology.

$$S = \sum_{i=1}^{n} s_i(x_i) \tag{20.6}$$

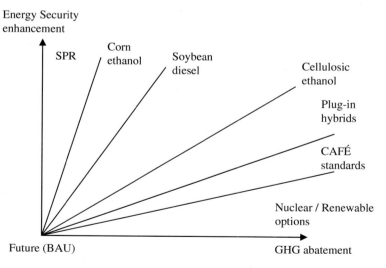

Fig. 20.8 Technology options for protection the climate and enhancing energy security. *Source* Brown and Huntington (2008)

Similarly, the total reduction in emission, Q, is given by sum of individual GHG reduction.

$$Q = \sum_{i=1}^{n} q_i(x_i) \qquad (20.7)$$

The total cost of the programme, C, is given by

$$C = \sum_{i=1}^{n} c_i(x_i) \qquad (20.8)$$

The problem then is to minimize the cost of the programme subject to the constraints of achieving a given level of supply security and GHG abatement. The Lagrangian can be written as

$$\lambda = C - \lambda_s \sum s_i(x_i) - \lambda_q \sum q_i(x_i) \qquad (20.9)$$

By setting the first derivatives with respect to x_i to zero the first order condition for optimality is obtained.

$$\frac{\partial C}{\partial x_i} = \lambda_s \frac{\partial s_i}{\partial x_i} + \lambda_q \frac{\partial q_i}{\partial x_i} \quad \text{for each } i \qquad (20.10)$$

λ_s and λ_q represent the incremental value of security enhancement and greenhouse gas abatement respectively.

The optimality condition suggests that each technology is used to the point where the marginal cost of the technology is equal to the value of additional energy security and GHG abatement it provides. Given that the right hand side contains two factors, a cost effective solution could still be obtained if one of the factors outweighs the other factor working in an opposite direction. For example, a technology that produces a large, positive $(\delta s_i/\delta x_i)$ but a small negative $(\delta q_i/\delta x_i)$ could still be part of the optimal policy solution.

If it is assumed that the policymaker is not interested in climate protection, the second term of the right hand side of Eq. 20.10 can be assigned a zero value. This leads to

$$\frac{\partial C}{\partial s_i} = \lambda_s \quad \text{for each } i \qquad (20.11)$$

Similarly, if the policymaker is not interested in energy security, the first term of the right hand side of Eq. 20.10 can be ignored. This leads to the other condition

$$\frac{\partial C}{\partial q_i} = \lambda_q \quad \text{for each } i \qquad (20.12)$$

These two equations indicate that when only one attribute is considered, each technology has to be used so that the marginal cost of additional benefit under consideration (security or climate protection as the case may be) is equal across all

20.7 Trade-Off between Energy Security and Climate Change Protection

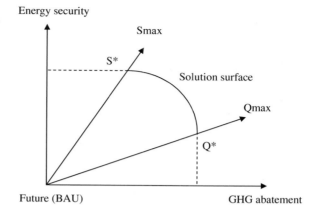

Fig. 20.9 Trade-off between climate protection and energy security objectives. *Source* Brown and Huntington (2008)

technologies. Technologies offering the largest gains at the lowest cost will be preferred in such cases. These two outcomes effectively set the upper limits of the solution surface from the perspective of each objective (see Fig. 20.9). Thus for a given cost a trade-off arises in choosing the combination of technologies through which the two objectives will be pursued.

20.8 Conclusions

This chapter has provided an overview of the energy security problem and analysed various aspects of it using simple economic principles. The chapter has presented the concepts related to the external cost of fuel imports, and the cost of supply disruption. Various options for mitigating the energy security issue are also discussed. The main message here is that a variety of options could be used but ultimately the cost of the policy and the benefits derived from it remain important. Any policy that imposes disproportionate burden on consumers is unlikely to find favour in the end.

References

APERC (2002) Energy security initiative: Emergency oil stocks as an option to respond to oil supply disruption, an APERC background report, Institute of Energy Economics, Japan (see http://www.ieej.or.jp/aperc/2002pdf/OilStocks2002.pdf)
Bhattacharyya SC (2009) Fossil-fuel dependence and vulnerability of electricity generation: case of selected European countries. Energy Policy 37:2411–2420
Bielecki J (2002) Energy security: is the wolf at the door? Q Rev Econ Finance 42:235–250
Brown SPA, Huntington H (2008) Energy security and climate change protection: complementarity or trade-off. Energy Policy 36(9):3510–3513

Huntington HG (2009) The oil security problem, chapter 16. In: Evans J, Hunt LC (eds) International handbook on the economics of energy. Edward Elgar, Cheltenham

IAEA (2005) Energy indicators for sustainable development: guidelines and methodologies. International Atomic Energy Agency, Vienna (see http://www-pub.iaea.org/MTCD/publications/PDF/Pub1222_web.pdf)

IEA (2002) Energy security. International Energy Agency, Paris (see http://www.iea.org/textbase/papers/2002/energy.pdf)

Leiby PN (2007) Estimating the energy security benefits of reduced US oil imports, ORNL/TM-2007/028. Oak Ridge National Laboratory, Tennessee

Leiby PN, Jones DW, Curlee TR, Lee R (1997) Oil imports: an assessment of benefits and costs. Oak Ridge National Laboratory, Oak Ridge (see http://pzl1.ed.ornl.gov/ORNL6851.pdf)

Markandya A, Hunt A (2004) The externalities of energy security, final report, Externe-Pol, Bath University, UK (see http://www.externe.info/expolwp3.pdf)

Owen AD (2004) Oil supply insecurity: control versus damage costs. Energy Policy 32(16):1879–1882

Parry IWH, Darmstader J (2003) The costs of US oil dependency, RFF discussion paper no. 03–59, Resources for the Future, Washington, DC (See http://www.rff.org/rff/Documents/RFF-DP-03-59.pdf)

Percebois J (1989) Economie de l'energie, Economica, Paris

Percebois J (2007) Energy vulnerability and its management. Int J Energy Sector Manage 1(1):51–62

Razavi H (1997) Economic, security and environmental aspects of energy supply: a conceptual framework for strategic analysis of fossil fuels, for Pacific Asia Regional Energy Security Project (see http://www.nautilus.org/archives/papers/energy/RazaviPARES.pdf)

Salameh M (2003) The new frontiers for the United States energy security in the 21st century. Appl Energy 76:135–144

Taylor J, Van Doren P (2005) The case against the strategic petroleum reserve, policy analysis, no. 555, Cato Institute (see http://www.cato.org/pubs/pas/pa555.pdf)

Toman MA (1993) The economics of energy security: theory, evidence and policy, chapter 25. In: Kneese AV, Aweeney JL (eds) Handbook of natural resources and energy economics, vol. 3. Elsevier Publishers

Toman MA (2002) International oil security: problems and policies, RFF issue briefs, resources for the future, Washington, DC, see (http://www.rff.org/rff/Documents/RFF-IB-02-04.pdf)

Chapter 21
Investment Issues in the Energy Sector

21.1 Problem Dimension

21.1.1 Global Investment Needs

The energy sector is capital intensive and according to WEO (2008),[1] around $26 trillion (in $2007) would be required to develop energy supply infrastructure between 2007 and 2030 to meet the global energy demand adequately and satisfactorily. On an annual average, the investment requirement turns out to be more than $1.1 trillion. Incidentally, the annual average investment requirement indicated above is almost double the requirement indicated by IEA in its 2003 report (See also Birol 2004). Clearly the energy sector needs huge investments to ensure reliable energy supply. The huge difference in the two estimates indicated above can be explained by a number of factors. First, this period saw very high oil and gas prices. Better remuneration for investment also increased the demand for supplies and the cost of inputs rose significantly between 2004 and 2008. This resulted in an escalation in investment costs. Second, the composition of the future energy supply mix changed somewhat—which also affected the cost estimates.

WEO (2003) suggested that the investment demand is expected to increase over time. Until 2020, the investment requirement would be growing slowly, but the need would increase in the next period (2020–2030) due to retirement of existing plants and infrastructure and growing demand from the developing countries. WEO (2003) estimated that in 2000 around $410 billion were invested globally. Although the need for investment estimated by that report was somewhat higher, the difference was not that alarming.

---

[1] There are other studies on this topic such as ESCAP (2008) and APERC (2003) but they have a regional focus.

However, the distribution of the demand for investment is uneven across sectors of the supply chain. For example, WEO (2008) indicates that

- the electricity supply industry would require about 52% of the investments (or around $14 trillion);
- coal appears to require the least amount of investment (3%) on a global scale for any fossil fuel—less than a trillion dollar of investments;
- the global oil industry would require close to $6.3 trillion of investment (24% share of global investment needs) while the gas industry would need an investment of $5.5 trillion (21% share).

Clearly, the cost escalation in the oil and gas industry activities has affected the overall investment estimation to a large extent. In the 2003 report (WEO (2003), IEA indicated that the electricity sector would claim about 60% of overall investment needs while oil and gas sectors would claim a share of 19% each.

The electricity industry requires such a large share of the investments for the following reasons:

- Electricity being convenient and high intensity energy, globally the demand for electricity is increasing faster than most of the other fuels. Therefore, new production capacity needs to be developed. This is true for developing countries in general.
- In many developed countries, the existing plants were set up in the 1970s just after the first oil shock. These plants have already completed their working lives and need to be replaced. WEO (2003) estimates suggest that about 51% of the above investment would be required to replace old infrastructure while the remaining 49% will cater to demand growth.

Evidently, the upstream and downstream segments of different energy industries require different levels of investments.

- For example, in the electricity sector, power generation and network development would require almost similar levels of investment. Thus upstream and downstream electricity sector investments are individually the most significant investment demands globally.
- For fossil fuels, extraction and production of energy would remain the most capital intensive activities but the shares are expected to differ by fuels:
  - mining of coal would absorb around 91% of the investment requirement;
  - for oil upstream activities absorb around 80% of the investments;
  - for gas 61% of the investment would go to production, while transmission and distribution network would need another 31% of the investment (see Table 21.1).

If the output is expressed in toe for different forms of energies, it turns out that natural gas is the most capital intensive fuel requiring $28 per toe. The capital intensity of oil follows gas closely ($22 per toe) but coal is the least capital intensive fuel requiring just $5 per toe (WEO 2003).

21.1 Problem Dimension

Table 21.1 Fuel-wise break-up of investment requirements

Fuel/industry	Investment need (trillion $2007)	Segment	Investment share (%)
Coal	0.7	Mining	91
		Shipping	9
Oil	6.3	Upstream	80
		Refining	16
		Shipping	4
Gas	5.5	Upstream	61
		LNG	8
		Transmission and distribution	31
Electricity	13.6	Generation	50
		Distribution and transmission	50

Source: WEO (2008)

21.1.2 Regional Distribution of Energy Investment Needs

The regional break down of investment needs as reported in WEO (2008) is shown in Fig. 21.1. Apparently, OECD North America has the largest demand—close to $5.5 trillion. China comes next with a $4 trillion investment need, followed by OECD Europe, Rest of Asia and Eastern Europe, each with investment needs of around $3 trillion each. The rest of the regions require an investment of $2 trillion each for the period up to 2030.

A careful look suggests that the developing countries and transition economies account for about 60% of the investment needs—close to $10 trillion. The OECD share would remain important at 40% but would be surpassed by that of the developing economies.

Figure 21.1 also gives the regional investment needs by fuel type. The domination of electricity in the investment is clearly evident. Developing countries would also spend in production capacity and other infrastructure development. In

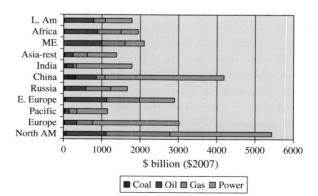

Fig. 21.1 Regional break down of investment needs for up to 2030. *Source*: WEO (2008)

fact, these countries would account for about 50% of global investment in production and other infrastructure development.

21.1.3 Uncertainty About the Estimates

Clearly, any estimates have some inherent uncertainties. The underlying basis for the above estimates was the demand–supply forecasts used made by the IEA and the market conditions prevailing at that time. The purpose of these long-term forecasts is to portray a possible outcome based on the forecaster's visualization of the future. Reviewing the past forecasts for the US, Craig et al. (2002) found that many forecasts were inaccurate and that the forecasts often overestimated the demand by 100% due to failure to anticipate structural changes, technological changes and "break-points". IEA forecasts, like any other studies, also inevitably suffer from such issues and therefore, are a source of uncertainty.

As a direct consequence, the estimates change from 1 year to another, as new information becomes available or market conditions change. IEA has been revising its estimates almost on a yearly basis. This is because of significant price rises in the sector projects noticed recently due to supply shortages and other constraints. Consequently, the estimation of the investment needs changes as well. Clearly, the effect of economic crisis of 2008/2009 will be taken into account in the future and the estimates will be further revised.

The estimation of the investment need in the IEA estimates was based on the estimation of the new capacity needs in the upstream and downstream segments of the energy industry and the unit cost of new capacity addition. This is done at the industry level and by region/country. Unit costs vary from one project to another and when investors flock together to reap the benefits offered by a particular segment, costs rise due to overheating of the market as capacity constraints in the supply-side start to bite in. Such changes are difficult to estimate and often the unit cost estimates can be averages of a range available at the time of the report. That leaves the possibility of under/over estimating the costs.

In addition, there is now a belief that the future energy demand may be significantly different in many countries. For example, WEO (2008) indicates that the transformation of the energy sector required to achieve climate stabilization at the desired 450 ppm of CO_2 equivalent level would involve an additional investment of over \$9 trillion relative to the estimates for the reference scenario (\$26 trillion). A large part of the investment will be required in non-OECD countries.

Moreover, the conservative, historical demand growth in China and India may not be representative of the future needs. For China it was assumed that coal-driven energy supply (which is less capital intensive) would continue. If however the emphasis on coal reduces in favour of other fuels, the actual demand would change. The example below from the Indian electricity sector will further clarify this point (see Box 21.1).

Box 21.1: India's Electricity Supply Expansion Plan: Is It a Tall Order?[2]

The Indian electricity industry is already plagued with significant capacity and energy shortages, poor quality of supply, limited access to population as well as poor financial performance of the utilities. High economic growth over the past few years has spurred a higher demand for electricity, which in turn increases the demand for new capacities and causes concerns for maintaining economic performance of the country facing the possibility of increased capacity shortages. With a GDP elasticity of electricity demand of about 1, Indian electricity demand increases at the same rate as that of GDP growth unless policy measures are taken to influence demand growth. In addition, the government has undertaken an ambitious plan for providing electricity supply to all households by 2012 and is investing significantly in this plan. Yet, despite a consolidation of the legal framework through the enactment of the Electricity Act 2003 and continued emphasis on sector reform, the progress, in terms of developing a viable business model, has been rather slow and a deadlock appears to have been reached in terms of reform and regulatory improvements. It is in this context we need to ask whether the capacity expansion plan is of tall order or not.

Effect of High Growth on Capacity

A few recent studies have analysed the effect of high economic growth on the electricity demand. The Expert Committee on Integrated Energy Policy (Government of India 2006) constituted by the Planning Commission of Government of India explicitly considered the high economic growth scenario and estimated energy demand, including electricity demand. The Committee has employed a simple GDP-elasticity approach[3] to forecasting and the average generation growth is forecast at about 8% for the period up to 2012, about 7% for the period 2012 and 2022 and close to 6% for the next 10 years. The generation capacity would be 233 GW by 2012, somewhere close to 450 GW by 2022 and between 800 and 950 GW by 2032.

Similarly, the Working Group on Power for 11th Plan[4] has also reviewed the demand for the 11th (2007–2012) and 12th Plan (2012–2017) periods

---

[2] Based on Bhattacharyya (2008).

[3] The GDP elasticity of electricity demand measures the change in demand for every per cent change in the gross domestic product (GDP). Based on an expert judgement estimate of the future elasticity of demand and using an assumption about future growth of GDP, the growth rate of demand is estimated.

[4] India's economic development programme is planned under five year plans developed by the Planning Commission. The 10th Plan ended in March 2007. The 11th Plan started from April 1, 2007 and will continue until 31st March 2012.

and retained a GDP-elasticity based approach to forecasting. The Group believes that the generation in the country would have to grow at 9.5% for the period up to 2012 and between 7 and 9% for the period to 2017. The Group estimated that the capacity required would increase to about 210 GW by 2012 and close to 300 GW by 2017.

However, a study by the International Energy Agency (IEA) in its 2007 World Energy Outlook (WEO 2007), indicates that the electricity generation is expected to grow at 5.2% per year on average in the 2005–2030 period but a higher growth rate of 5.7% per year is expected in the first 10 years (2005–2015). In the high growth scenario, power generation grows at 6.1% per year on average for the period up to 2030. The study suggests that in the reference scenario, the capacity requirement would increase to 255 GW by 2015 and to 522 GW by 2030. But in the high growth scenario, the overall capacity requirement could increase by another 100 GW.

The estimates for new generating capacity requirement from the three studies come out as follows (see Fig. 21.2): The conservative estimate of the International Energy Agency suggests an annual average capacity addition of 16 GW for 25 years. This average is quite similar to the Plan recommended by the Working Group but is just one half of the average capacity addition suggested for the high growth scenario indicated in the Integrated Energy Planning study. The high growth scenario of the WEO study suggests about 20 GW of capacity addition per year. To put these numbers in perspective, 20 GW capacity addition per year for 20–25 years requires more than 1.5 GW a month; which works out at an average of one decent sized plant a week, every week for 20–25 years. The challenge is surely huge.

How does this capacity expansion need compare with the past experience of capacity addition in the country? Against a target of about 7.5 GW per year over the past 15 years, just 50% did materialise and consequently only 57 GW could be installed against a target of 111 GW. Achieving a capacity addition target of 69 GW before 2012 (or 78 GW as per Integrated Energy Planning) would appear to be a Herculean task, as this would require a doubling of annual capacity addition achieved so far. Any further acceleration of capacity addition would surely look more challenging.

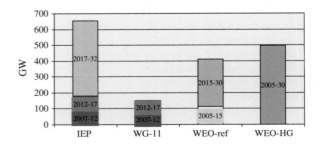

Fig. 21.2 India's generating capacity addition requirements. *Source*: Bhattacharyya (2008)

As all the above studies are based on the assumption of India's continued economic prosperity, the electricity demand and capacity expansion needs would be different if the economy slows down or in a worst case scenario faces recession. This further reinforces the need for a careful estimation of the demand and capacity addition targets.

Investment Needs for Capacity Expansion

Based on the generation and network capacity expansion forecasts, the following range of investment requirements emerges:

- IEA has estimated that India requires $956 billion in 2005 prices for the period up to 2030 in the reference scenario. This corresponds to an average annual investment of $38 billion for 25 years but increases to $50 billion per year in the high growth case.
- The Integrated Energy Policy suggests an amount of $58.5 billion per year over 25 years.
- The Working Group for the 11th Plan suggests an investment of $50 billion per year, which corresponds to the high growth scenario of the WEO-07.

However, the Consultation Paper on infrastructure investment during the 11th Plan by the Planning Commission suggests a lower level of investment programme considering the constraints in mobilising financial resources, absorptive capacity and financial positions of the utilities. The paper suggests an average investment of $30 billion per year (Planning Commission 2007).

How does the above requirement compare with the actual power sector investments in the past? It appears that only $71 billion was invested between 2002 and 2007, representing an annual average of just over $14 billion (Planning Commission 2007). Even a $30 billion investment per year would imply more than doubling the size of the investment, which requires an annual growth of more than 16% in investment outlays on a year to year basis. Targeting higher levels of investment would be more challenging. Achieving an annual investment of $50 billion as required in the high growth scenario of WEO or as estimated by the Working Group on Power for the 11th Plan would require an investment growth of around 28% per year compared to that realised during 2002–2007. This is surely a daunting task.

If the economy grows at the rate of 9%, the size of the power sector investment as a share of GDP would not be disproportionately high. The Planning Commission estimate comes to just an average of 2.3% of the GDP forecast for the 11th Plan period while the Working Group estimate comes to 3.8% of the GDP, compared to an investment in the previous Plan period of about 1.5% of the GDP (Planning Commission 2007). Although the absolute value of the investment grows substantially, a rapid growth of the economy dwarfs the investment shares to some extent. But with recent melt-down of the global financial market, it is not clear whether the growth will be sustainable or not.

21.2 Issues Related to Investments in the Energy Sector

Despite the uncertainty about the estimates of investment needs of the energy sector, there is no doubt the future growth and transformation of the sector requires huge investments. A number of issues can be easily identified in this regard, including the following: resource availability and mobilisation, the developing country perspectives, investment and pricing links, and the influence of energy sector reforms on the investment.

21.2.1 Resource Availability and Mobilisation

The first question that arises is whether such huge amounts of financial resources would be available and whether they could be mobilized as and when required. Is the resource availability a concern?

In overall terms, the investment requirement in the energy sector would be close 1% of the global GDP, on average. This is a relatively low figure. Moreover, WEO (2003) reported that $413 billion was invested in the energy sector in 2000. The 2008 estimate of the average investment requirement is more than twice that of the actual investments in 2000.[5] This implies that the future needs are increasing becoming more challenging.

Moreover, different regions and countries would have to mobilise different levels of investments compared to their GDP (Fig. 21.3). Generally, smaller economies tend to have a higher share of energy investment relative to their economic output. Most of the OECD countries would find the energy investment needs to be a very small fraction of their economic output (less than 0.5% in general). Fast growing developing countries such as China, India, Malaysia or Thailand will have a modest demand compared to their growing economies (between 2 and 3% of their GDP). Russia would need to invest around 5% of its GDP in the energy sector. African countries (not shown in the figure) will also face a high burden.

---

[5] Adjusting for the difference in the base year.

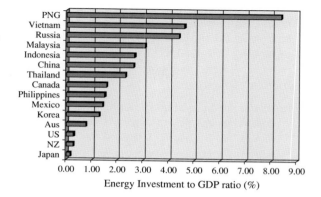

Fig. 21.3 Future energy investment needs as a share of GDP (2000–2020). *Source*: APERC (2003)

For many countries, this level of investment is often a small share of the overall savings of a country and as such resource availability may not itself be a problem. Normally countries with high levels of domestic savings (such as China with a 40% saving rate or Middle Eastern states receiving windfall gains during high oil prices) and limited capital outflows would have a positive savings-investment balance. Such countries would be in a favourable situation. But energy investments have to compete with other competing investments and demand for resources. As the state-controlled financial institutions dominate the domestic debt market, they are often rather ineffective in channelling savings to productive assets. Consequently, a part of the household savings is locked in housing while another significant share is locked in other immovable assets (such as gold), thereby reducing the funds available for productive investments. This makes domestic financing insufficient for investments in the energy sector. Further, the rate of savings in many developing countries is insufficient to support the required capital needs. For this category of countries, domestic capital will not be sufficient to meet the investment needs of the energy sector. See Box 21.2 for a rappel of the sources of funding.

In many countries, the state control over the sector is quite strong and may not allow others to enter or participate. In such cases the state-owned utilities are mandated to perform the functions. But often they are not in a position to take up the investment from their own resources due to poor performance, poor tariff/pricing policies, state appropriation of the surplus, etc. The state may not able to support the sector as before due to its budgetary constraints. This can cause a major resource mobilisation problem.

Box 21.2: Traditional and Emerging Mechanisms for Financing Energy Projects

Capital intensiveness of energy projects has always caused concerns about financing such ventures. Large energy companies, especially oil majors, often relied on financing from internal sources of the company but small and medium sized companies as well as other who are unable to mobilise the required finances on their own rely on a combination of internal sources and borrowing from various sources. The traditional sources of funds are:

(a) *Equity finance*—Equity can be raised from project partners (in the form of initial capital or through retained profits), equity funds, venture capital funds and by issuing shares to the public. As the owners of the project the equity holders bear the risks of the project and accordingly, expect a better remuneration than the debt. They also have the decision-making powers for the project.
(b) *Debt finance*—Loan funds can come from a number of sources: commercial banks, bilateral or multi-lateral funding agencies (such as regional development banks, World Bank group, etc.). In the case of state-owned companies this can involve borrowing from the government or enlisting budgetary support for capacity expansion. In addition, leasing is also used as a funding method. Lenders have the right on the interest and the capital they provide. The interest on debt is a deductible expense for tax purposes, which makes debt finance up to a certain level attractive for the project developers.
(c) *Project finance*—Here the cash flow generated by the project is used as the security to finance a project by the lenders. Normally a web of contracts is created to share the project risks and ensure project viability, recovery and repayment of capital, and investment protection. It is quite common to agree on "take or pay" type of contracts or capacity payments based on the availability of the capacity irrespective of its utilisation.

Besides these traditional mechanisms some new avenues of funding energy projects have emerged in recent times. These are more commonly used for renewable energies or carbon emission mitigation projects. Two project-based mechanisms, namely Joint-Implementation and the Clean Development Mechanism, are promoting cleaner technologies for emission reduction. Similarly, the Global Environment Facility is another such funding mechanism. In addition, some countries have also instituted specific funds to support renewable or clean technologies which provide either subsidies or specific supports.

21.2.2 Foreign Direct Investments

If domestic capital availability is not enough in some countries, can they attract foreign capital? This possibility exists given that surplus funds are available in other parts of the world. Figure 21.4 shows the trend of private investment in the energy sector since 1990. It reveals that private investment projects in the energy sector peaked up between 1990s— in 1997. Since then, there was a steady decline until 2004 when things started to work in favour of the sector. But since the 2008 financial crisis, the interest in the energy sector of developing countries has reduced. It is also important to note that a significant part of the private investment has been used to acquire existing assets—thus investments in new projects only accounts for about one half of the total private investment in the sector.

The regional distribution of private investment flow in the energy sector is more revealing (see Fig. 21.5). Most of the private investment was concentrated basically in two regions: Latin America and East Asia. These two regions have attracted 63% of the private capital inflow. Europe and Central Asia (Eurasia in the figure) and South Asia have attracted about 14–15% each of the total flow for the energy sector. The Middle East, North Africa and Sub-Saharan Africa did not really benefit from this.

The data in Fig. 21.5 clearly shows that funds tend to flow to those areas where the returns on the investments are higher and are commensurate with the risks

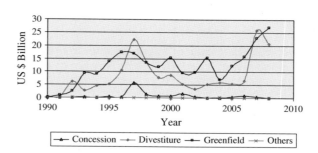

Fig. 21.4 Private investment projects in the energy sector. *Source*: PPIAF database (World Bank)

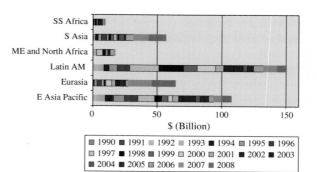

Fig. 21.5 Regional distribution of private investment in energy. *Data*: PPIAF database

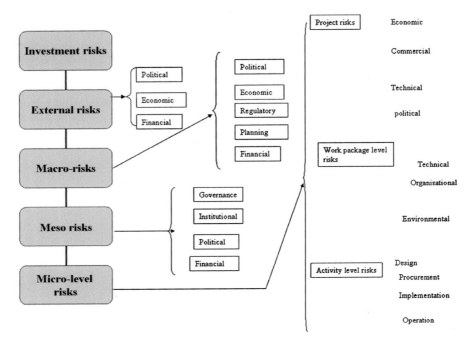

Fig. 21.6 Hierarchical presentation of investment risks. *Source*: Based on Bhattacharyya and Dey (2007)

involved. The countries facing resource constraints are likely to be more risky places to invest. As a result, funds may not flow where they are needed or flow in only limited quantities.

21.2.3 Risks in Energy Investments

Energy investments normally face a large number of risks working at different levels, which can be presented in a hierarchical manner[6] at four levels: external, macro, meso and micro levels (see Fig. 21.6). At the external level, changes in the global market conditions or financial markets can affect the investment conditions. There are political and regulatory influences as well that could affect a specific country (through embargoes or sanctions) or changes in the international trade, environment or other laws. One of the major factors affecting energy investment is the price of oil but as discussed in Chap. 14, the volatility of oil prices makes investment decision difficult. Although high oil prices make energy investments profitable, periods of

---

[6] This is based on the idea presented in Bhattacharyya and Dey (2007).

low prices leave investments vulnerable. As such episodes leave long-term impressions on the investors, conservative decisions generally rule supreme.

At the macro level, country-specific risks arise due to possible changes in the political condition, regulatory environment or economic and financial conditions of the country. A large number of factors affect economic viability of any investment including the possibility of economic downturn, level of inflation, changes in currency valuation, and possibilities of labour unrest. An economic slow down will affect resource mobilisation and shift economic priorities. The level of inflation introduces the possibility of changes in the monetary policies and thereby affects the business environment. Changes in currency valuation affect the foreign exchange positions and the balance of payment position, as some components of the project may require import of equipment, spare parts and materials. Labour unrest due to industrial actions of organised unions disrupts industrial activities, thereby creating a negative environment for investment in the industry. Similarly, a number of legal and regulatory risks surround investment, implementation and management of energy investments. Non-transparent regulatory processes, adverse changes to regulatory processes, and ambiguities and discretionary powers in favour of consumer protection, prevention of misuse of monopoly power and protection of investment of the supplier affect the investment conditions.

At the meso and micro levels, similar risks arise. Governance-related issues can be a major concern in some areas at the meso level. Three factors could be highlighted at the meso level: (1) law and order situation in some areas may not be conducive for undertaking such investment projects. Terrorist activities, kidnapping and other anti-social activities may prevent implementation of projects in some areas. (2) Similarly, high levels of corruption could adversely influence investment decisions. (3) Finally, politicisation of projects could lead to delays and law-order problems and other conflicts as well, affecting progress of the project. Similarly, at the micro-level, economic risks, namely volume risk, rate risk and consumer risk, commercial risks as well as project implementation risks can influence investment decisions. Table 21.2 provides a list of risks related to energy investments.

Normally, oil companies tend to invest in the oil business rather than investing elsewhere because of the high returns. However, in recent times many companies have distributed the profits amongst their shareholders. This could be for several reasons including: lack of profitable areas to invest due to limited acreage available to them; conservative norms used for profitability assessment for investment projects, reducing the investment prospects; excess finance available compared to the need for investment in profitable ventures; and a gesture of appreciation and wealth sharing during difficult economic times.

21.2.4 Energy Pricing-Investment Link

Investment in the energy sector is closely related to the pricing policies followed for the energy services. Any investor is interested in the cost-benefit of her

Table 21.2 Risks in energy investments

Risk type	Risk sub-type	Example
Economic risk	Market risks	Price risk, demand risk
	Macro-economic risks	Currency risk, inflation, economic recession, etc.
	Construction risks	Cost and time overruns
	Operation risks	Unsatisfactory performance, inadequate reserves, inadequate capacity utilisation, etc.,
Political risk	Nationalisation/ expropriation of assets	Change of ownership rules due to political changes in the host country.
	Embargoes/investment bans	Restriction on resource movements imposed by individual countries or by a group of countries for political reasons.
Legal risks	Regulatory	Legal changes affecting the business environment
	Contractual risks	Possibility of unilateral changes in the contract terms
	Judicial risks	Risks related to the quality of judiciary and the level of judicial activism
Force majeure risks	Natural disasters	
	War/civil unrest	
	Strikes	

Source: Birol (2005) and WEO (2003)

investment and energy prices influence this analysis significantly. Prices provide the signal to both consumers and the suppliers. The price charged to the consumers should be such that the investments are profitable and investors are attracted to the sector. Unless the investors are adequately remunerated for the risks taken by them, they would not invest in the country. At the same time, inappropriate prices tend to distort the demand, thereby requiring costly investments.

As long as the prices are providing correct signals to both consumers and the investors, the balance is maintained. But any breaks in the loop create mismatches in investment needs and resources available for investment, causing problems for the sector (see Fig. 21.7). Concurrently, investment in energy influences the cost of supply and given the lumpy nature of the investment, the prices tend to be volatile with investments. Thus, there is a two-way relationship between investment and pricing.

21.3 Developing Country Perspectives on Investment

The failure of many developing countries in attracting investments in the energy sector stems from the factors discussed above. Poor, non-remunerative prices are not sufficient to generate surpluses for the sector and create poor financial positions for the energy companies. Distorted pricing is also fuelling high demand for

21.3 Developing Country Perspectives on Investment

Fig. 21.7 Investment–price relationship

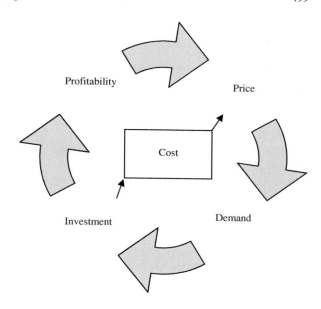

energy and leading to uneconomic use of resources. The supply–demand relation is not in equilibrium and the investment-pricing loop is broken. In addition, time and cost over-runs of projects, inadequate capacity utilisation, poor maintenance of the system and inefficient operation add to the cost of supply. Because many developing countries also lack technical and managerial capabilities in running the sector, cost increases further.

In addition, lack of transparent regulation, discretionary use of regulatory powers, regulatory capture and manipulation, and lack of other institutional capabilities (credible judiciary, credible law enforcement, etc.) also increase the risk of investors and this is reflected in the cost of capital through a higher premium. Investors thus cannot investment unless there is a good prospect of recovering adequate remuneration. But the poor pricing policies and limited paying capacity of the consumers do not offer such remuneration, leading to capacity shortages and unreliable supply.

Some resource-rich countries on the other hand are in a completely different situation during the periods of high energy prices. Their worry is to economically use windfall gains and ensure economic growth. As many producers are worried about the decline in resource prices as a consequence of excessive investment and may not be able to sustain a higher level of economic activities should prices fall, there is a tendency to delay investments for resource capacity expansion. Such strategic behaviour either at a country level or at a group level acts as a barrier to ensure future energy supply.

21.4 Reform and Investment

The general economic prescription from the international organisations is:

- to correct price distortions to establish cost-based pricing that would provide correct signals to the consumers and the investors. The drive for correct energy pricing has been sustained by such organisations for more than two decades now, but in general, energy prices in most places remain inappropriate. This shows the difficulty of implementing this prescription.
- to reform the industry so that the private sector is allowed to enter and participate with an objective of developing a more competitive supply industry. This idea has been pursued since 1990s but the scorecard of reformed industries remains unimpressive. Even many developed markets have not been reformed and are not competitive.

Further, the reformed energy sector may also face additional problems. Generally the private sector would be interested in less capital intensive, quickly recoverable projects. This may not be always in the best interests of the host country given the question of energy security and the existence of strategic resources or fuels. Competitive markets may not provide sufficient incentives for developing costly infrastructure unless some sort of support is provided. The support mechanism may go against the competitive structure of the industry as well. Poorer countries also face the question of access, affordability and viability of supply in rural areas. Private investors would be unlikely to invest in uneconomic ventures or in areas that would require a long time to break-even. In such a situation, the state would remain as the option of last resort. If the government has to carry the uneconomic activities, does it make sense for it to sell the profitable segments of the business?

Moreover, investments cannot wait for reforms to take place but unless some improvement occurs, the sustainability of investments will be in question. There lies the dilemma.

21.5 Global Economic Crisis and the Energy Sector Investments

The deepening of the financial crisis in the Western banking and financial sector after the collapse of one of the most prestigious Wall Street player, Lehman Brothers, when it filed for bankruptcy protection in September 2008 and subsequent loss of face of the financial sector has triggered the worst global economic crisis since the Great Depression. Although the problems of investing in toxic financial instruments have started to emerge since the sub-prime crisis hit the US economy in late 2007, the crisis deepened in late 2008.

The global bail-out for the financial crisis has cost trillions of tax-payers' money. This has in turn resulted in de-facto (and de-jure in some case) nationalisation of many Western banks—at least temporarily. The stock market indices around the world have reached record low levels and trillions of dollars of market capitalisation has vanished in the process. The crisis has come at a time when the global economy was already showing signs of distress. The developed economies have slowed down significantly but the developing economies are expected to continue with their moderate to high growth, which in turn improves the overall effect to some extent.

Consequently, the energy and commodity prices have fallen sharply since the crisis. Crude oil prices and along with it other energy prices have seen a sharp fall. Low energy prices in turn made many investment projects unviable. Concurrently, the liquidity problem in the financial system has created credit crunch. Although governments have injected fresh capital to the banking system, the support is not coming free of cost. The banks will be charged for taking the government guarantee support, which will be passed on to consumers. The cost of borrowing has increased and is expected to remain high, which in turn will put off the consumers and industries from heavy borrowing. The creditworthiness of banks and the credibility of sovereign guarantees came under pressure and many countries became prone to crisis subsequently.

The effect of the financial and economic crisis on the energy sector is already evident. Many projects are being delayed and some will be abandoned. In such a situation, the investment programmes for future energy supply will face greater challenges. Although the energy demand has fallen in the short-term, the long-term effects are more likely to emerge in due course and lack of investment could implant the seed for future crises in the sector (see IEA (2009) for further details).

21.6 Conclusions

This chapter has highlighted the enormous investment needs of the energy sector and has presented a broad overview of the investment needs by energy industries and by region. It has also highlighted the issues related to the mobilisation of resources as well as the effects of recent economic slow-down. Investments in the energy sector would remain a major challenge in the coming decades and the size of investment need only shows the huge future potential of the industry.

References

APERC (2003) Energy investment outlook in the APEC region. Asia Pacific Energy Research Centre, Institute of Energy Economics, Tokyo

Bhattacharyya SC (2008) Investments to promote electricity supply in India: regulatory and governance challenges and options. J World Energy Law Bus 1(3):201–223

Bhattacharyya SC, Dey PK (2007) Managing risks in a large rural electrification programme in India. Impact Assess Proj Appraisal 25(1):15–26

Birol F (2004) World energy prospects and investment challenges. Oil and Gas Review 18–22. http://www.touchbriefings.com/pdf/951/birol.pdf

Birol F (2005) The investment implications of global energy trends. Oxford Rev Econ Policy 21(1):145–153

Craig PP, Gadgil A, Koomey JG (2002) What can history teach us? A retrospective examination of long-term energy forecasts for the United States. Annu Rev Energy Environ 27:83–118

ESCAP (2008) Energy security and sustainable development in Asia and the Pacific. Economic and Social Commission for Asia and the Pacific, Bangkok

Government of India (2006) Integrated energy policy—report of the expert committee. Planning Commission, New Delhi

IEA (2009) The impact of financial and economic crisis on global energy investment. In: IEA background paper for the G8 Energy Minister's meeting, 24–25 May, 2009, International Energy Agency, Paris

Planning Commission (2007) Projections of investment in infrastructure during the eleventh plan, consultation paper. The Secretariat for the Committee on Infrastructure, Government of India, New Delhi. http://infrastructure.gov.in/pdf/Inv_Projection2.pdf

WEO (2003) World energy outlook 2003 insights: global energy investment outlook. WEO, Paris

WEO (2007) World energy outlook 2007. International Energy Agency, Paris

WEO (2008) World energy outlook 2008. International Energy Agency, Paris

Further Reading

Blackman A, Wu X (1999) Foreign direct investment in China's power sector: trends, benefits and barriers. Energy Policy 27:695–711

Hamilton M (2003) Energy investment in the Arab world: financing options. OPEC Rev 27(3):283–303

Islas J, Jeronimo U (2001) The financing of the Mexican electricity sector. Energy Policy 29:965–973

Martinot E (2001) Renewable energy investment by the World Bank. Energy Policy 29:689–699

Chapter 22
Energy Access

22.1 Problem Dimension

This section presents the gravity of the energy access issue by looking at the present situation and expected future outlook considering the business as usual scenario. Most of the information below is based on IEA reports on the subject.

22.1.1 Current Situation

The most commonly cited figures on the lack of access to energy indicate that there are about 2 billion people without adequate access to clean cooking energy and about 1.7 billion people are without access to electricity (WEA 2000).[1] The origin and genesis of these figures are not easy to find. WEA (2000) does not elaborate on the source of the estimation or the estimation procedure. The World Bank report on Energy Services for the Poor (World Bank 2000) indicates that the estimate of 2 billion people is perhaps outdated. Estimation is difficult due to imprecise definition of the term "access" and lack of good quality data arising from poor understanding of the traditional energy use due to dispersed and distributed nature of this energy and focus on supply of commercial energies in the national energy balances and less focus on where it is used and by whom. Although traditional energies play an important role in many developing countries, the statistics is not reliable and household surveys are not common in all developing countries.

Information on access to electricity is somewhat better. According to WEO (2002), which provided detailed country-wise electricity access information, about 1.64 billion or 27% of the world's population did not have access to electricity in 2000. Since then, IEA has been updating the information on electrification on a

---

[1] Similar figures are quoted in DfID (2002).

regular basis and the most recent information suggests that about 1.4 billion population do not have access to electricity. The regional distribution is given in Table 22.1. It shows that two regions have large concentrations of people without access to electricity—South Asia (614 million or 42% of those lacking electricity access globally) and Sub-Saharan Africa (587 million or 40% of those lacking access to electricity).

A closer look at the data shows that about 70% of those lacking access to electricity reside in just 12 countries while the rest 30% is dispersed in all other countries (see Table 22.2). The rural population in most of these countries is lacking access, although in a few countries the urban population also lacks access. While the total number of people without access to electricity is high in South Asian countries, Sub-Saharan Africa fares worse in terms of rate of electricity access. In fact, out of 10 least electrified countries in the world, nine are from sub-Saharan Africa and Myanmar is the only country from Asia (see Table 22.3).

It can also be noted that most of these countries:

(a) have low per capita GDP compared with the world average. Except Indonesia, all countries in Table 22.1 have national average per capita GDP less than 10% of the world average.
(b) have low per capita primary energy consumption, ranging from 8% to 42% of the world average.
(c) Have very low per capita electricity consumption -the national average per capita electricity consumption in these countries ranges between 1% and 15% of the world average.

WEO (2002) provided some details about biomass use in the developing countries and estimated that about 2.39 billion people use biomass for cooking and heating purposes in these countries. This information is available at an aggregated level, which indicates inadequate knowledge about this important source of energy and points to poor quality of information. Subsequently, in 2006, IEA revised the

Table 22.1 Level of electrification in various regions

Region	Population without electricity (Millions)	Electrification rate (%)		
		Overall	Urban	Rural
North Africa	2	98.9	99.6	98.2
Sub-Saharan Africa	587	28.5	57.5	11.9
Africa	589	40.0	66.8	22.7
China and East Asia	195	90.2	96.2	85.5
South Asia	614	60.2	88.4	48.4
Developing Asia	809	77.2	93.5	67.2
Middle East	21	89.1	98.5	70.6
Developing Countries	1453	72.0	90.0	58.4
Transition economies and OECD	3	99.8	100.0	99.5
Global total	1456	78.2	93.4	63.2

Source: WEO (2009)

22.1 Problem Dimension

Table 22.2 Major concentration of population with access to electricity

Country	Rank in terms of population	Population without electricity access (Million)	Share of population without access (%)		
			Urban	Rural	Total
India	2	404.5	6.9	47.5	35.5
Bangladesh	7	94.9	24	72	59
Indonesia	4	81.1	6	48	35.5
Nigeria	8	80.6	31	74	53.2
Pakistan	6	70.4	22	54	42.4
Ethiopia	15	68.7	20	98	84.7
DR Congo	19	57	75	96	88.9
Myanmar	24	42.8	81	90	87
Tanzania	30	36.6	61	98	88.5
Kenya	32	32.8	48.7	95	85
Uganda	37	29.1	57.5	96	91
Afghanistan	44	23.3	78	88	85.6

Source: WEO (2009)

Table 22.3 Reliance on biomass for cooking energy needs in 2004

Region	Total population		Rural		Urban	
	%	Million	%	Million	%	Million
Sub-Saharan Africa	76	575	93	413	58	162
North Africa	3	4	6	4	0.2	0.2
India	69	740	87	663	25	77
China	37	480	55	428	10	52
Indonesia	72	156	95	110	45	46
Rest of Asia	65	489	93	455	35	34
Brazil	13	23	53	16	5	8
Rest of Latin America	23	60	62	59	9	3
Total	52	2528	83	2147	23	461

Source: WEO (2006)

estimate upward to 2.5 billion. This remains the most recent estimate on the use of biomass for cooking purposes. Table 22.3 presents some details about traditional energy consumption in developing countries.

Clearly, such a heavy reliance on traditional energies imposes economic cost on the society. Combustion of household fuels leads to air pollution. As biomass is often used in inefficient stoves, one-fifth of the fuel may be diverted as products of incomplete combustion, thereby creating health hazards. Air pollution is also a concern where coal is used as household energy. Coal smoke contains particulate matters as well as emission of health damaging contaminants. The local level pollution arising from liquid and gas based petroleum products is relatively less due to higher efficiency of cook stoves and better fuel quality.

Combustion of biomass energy indoor is a major source of indoor air pollution. The timing of such pollution (when most of members of the family are present) and the level of exposure due to poorly ventilated houses make poor households vulnerable to serious health effects. Four main health effects are attributed to household use of solid fuels (WEC 2000):

(a) infectious respiratory diseases;
(b) chronic respiratory diseases;
(c) premature deaths
(d) blindness, asthma, heart diseases etc.

As a consequence, 1.5 million premature deaths occur that is directly attributable to high indoor air pollution (WEO 2006), which represents a major heath risk in the developing countries. The regional distribution of these pre-mature deaths follows the biomass use patterns and South Asia and Sub-Saharan Africa suffer the maximum loss in this respect. Many millions also suffer from other lung and respiratory diseases as a result of pollution from burning traditional energies. As women and children are more exposed to such conditions, they are more vulnerable.

22.1.2 Future Outlook

But more importantly, forecasts by IEA suggest that unless policies are implemented to address the access issue, the number of people without access will not decline in the 2030 horizon. Although 75 million is expected to gain access to electricity every year until 2030 (WEO 2002), increases in the population in developing countries of South Asia and Sub-Saharan Africa will mean that electricity access will remain a problem. According to WEO (2002) 680 million in South Asia and 650 million in Sub-Saharan Africa will still live without electricity access. Significant improvements in the rest of the world are expected by this study (see Table 22.4).

The situation will be quite similar in the case of traditional energy use for cooking purposes (see Table 22.5). WEO (2006) suggested that the number of

Table 22.4 Expected future electrification rates (%). *Source* NSSO (2001b)

Region	2002	2015	2030
Sub-Saharan Africa	24	34	51
North Africa	94	98	99
South Asia	43	55	66
China and East Asia	88	94	96
Latin America	89	95	96
Middle East	92	96	99
Total	66	72	78

Source: WEO (2004)

22.1 Problem Dimension

Table 22.5 Outlook for biomass use for cooking in 2015 and 2030 (million)

Region	2004	2015	2030
Sub-Saharan Africa	575	627	620
North Africa	4	5	5
India	740	777	782
China	480	453	394
Indonesia	156	171	180
Rest of Asia	489	521	561
Brazil	23	26	27
Rest of Latin America	60	60	58
Total	2528	2640	2727

Source: WEO (2006)

people using biomass will increase in the 2030 horizon. Most of the population relying on biomass for cooking will live in Asia and Sub-Saharan Africa.

22.2 Indicators of Energy Poverty

As energy is an essential input for economic development of any country, consequently low access to clean energy hinders economic growth and therefore, requires special attention. However, the empirical evidence of energy-poverty link is often presented in simple graphs showing that energy consumption increases with income, or the human development index (HDI) improves with income and higher energy use. Pachauri et al. (2004) indicate that there are three types of measures normally found in the literature to indicate existence of energy poverty: 1) economic measures such as energy poverty line, 2) engineering measures of minimum energy needs, and 3) measures based on access to energy services.

The economic approach tries to find out how much consumers lying below the national poverty line spend on energy and how this expenditure compares with the overall household expenditure. If for example, a consumer spends more than 10% of her expenditure on energy, the consumer may be regarded as lying below the fuel poverty line. Such a definition is used in the U.K. However, the expenditure depends on the fuel mix, level of efficiency of the appliances, size of the household and prices in the market. Therefore, while a large budget share could indicate fuel poverty, it may give wrong signals as well.

The engineering approach uses an estimation of the energy needs to satisfy the basic requirements of any household. These are normative levels often used by government authorities to plan for energy needs of a community or a country. They are based on some assumptions about the types of activities generally performed by households and the energy requirement using available technologies. Clearly, such a norm will vary from one country to another and can vary over time. However, an understanding of the basic needs can help analyse various implications of non-availability of such supplies to the target groups.

Finally, the approach based on access to services departs from the above two in the sense that it tries to find out whether consumers have physical access to the

Table 22.6 Factor goalposts for EDI in 2002

Factor	Maximum	Minimum
Per capita commercial energy consumption	9.4 toe (Bahrain)	0.01toe (Togo)
Share of commercial energy in total final energy	100%	8% (Ethiopia)
Electrification rate	100%	2.6% (Ethiopia)

Source: WEO (2004)

Table 22.7 Example of EDI for India

Factor	Formula	Indicator
Per capita commercial energy consumption	(0.33−0.01)/(9.4−0.10)	0.034
Share of commercial energy in total final energy	(56−8)/(100−8)	0.519
Electrification rate	(46−2.6)/(100−2.6)	0.445
Average index	(0.034+0.519+0.445)/3	0.332

Source: WEO (2004)

supply of energy, and access to markets for equipment. Generally, the poor will have limited choice in terms of access to fuels and equipment choices compared to the well-off consumers. However, this is more data intensive and it may be difficult to compare two situations quantitatively using this approach.

WEO (2004) has presented an index, Energy Development Index (EDI) along the line of HDI. EDI is composed of the following three factors:

- per capita commercial energy consumption,
- share of commercial energy in total final energy use,
- share of population with access to electricity.

An index is created for each factor by considering the maximum value and minimum values observed in the developing world and determining how a particular country has performed. The following formula is used for this index

$$\text{Factor index} = \frac{(\text{Actual value} - \text{minimum value})}{(\text{maximum value} - \text{minimum value})} \quad (22.1)$$

The goalposts (maximum and minimum values) are taken from the observed values within the sample of developing countries considered. For example, for calculating the factor goalposts for EDI in 2002, WEO (2004) used the values shown in Table 22.6.

The simple average of three indicators gives the overall EDI.

For any country, e.g. India, EDI can be calculated using Eq. 22.1 and noting the goalposts as well as actual data for the country. For 2002, India's per capital commercial energy consumption was 0.33 toe, share of commercial energy in the final energy was 55.75% and the rate of electrification was 46%. The individual indicators are shown in Table 22.7.

Although this indicator provides a numerical value, it is not devoid of problems. It perpetuates the idea that higher level of energy consumption is synonymous to

22.2 Indicators of Energy Poverty

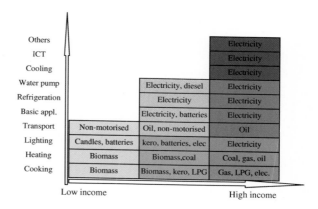

Fig. 22.1 Energy ladder example. *Source* WEO (2002)

economic development. Accordingly, countries in the Middle East with high per capita energy use rank better in this index. It also assumes that biomass energy use represents a symbol of under-development, which need not be the case, depending on how it is used. Finally, it also assumes that the grid-based electrification is essential for development, which is not true either.

22.3 Energy Ladder and Energy Use

It is normally noticed that the energy mix varies significantly among the poor and the rich. Normally, people in the lower income group tend to use more traditional energies to meet their needs. But with higher income people tend to move up the energy ladder and tend to use more commercial energies and less traditional energies. The general idea is presented in Fig. 22.1. As energy is a derived demand, the ability to use any modern fuel is dependent on the affordability of energy-using appliance and the ability to pay for the fuel on a regular basis. This can be an issue with the poor and hence they tend to rely on cheap technology and fuels.

The issue is not restricted to rural areas alone—often the poor in the urban setting are also using traditional energies, but there are urban–rural differences in energy consumption patterns.

An example from India is shown in Fig. 22.2, which shows that 76% of the rural households in 1999–2000 relied on firewood and chips, while only 22% of the urban households used this fuel. Urban households relied more on commercial fuels (LPG, kerosene) and the situation changed quite significantly between 1993–94 and 1999–2000. Indian Census 2001 reported that more than 139 million households in India (72% of all households)[2] rely on traditional energies for their

---

[2] According to Census 2001, there were around 192 million households in India, of which around 72% reside in rural areas and the rest in urban areas. The average household size was 5.3 persons in 2001.

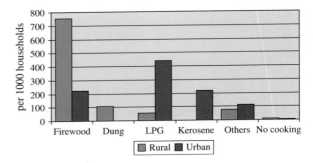

Fig. 22.2 Urban-rural differences in cooking energy use in India. *Source* NSSO (2001a)

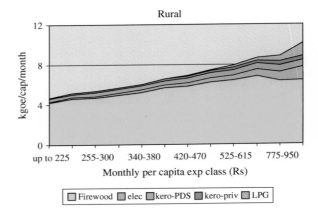

Fig. 22.3 Energy consumption pattern in rural India by expenditure class. *Source* NSSO (2001b)

cooking needs. Out of this, more than 124 million households reside in rural areas, while the remaining 15 million live in urban areas.

Energy consumption pattern by different income groups is more difficult to obtain. Sample data from NSSO (2001b) was used to generate energy consumption[3] by different expenditure classes separately for rural and urban areas (see Figs. 22.3 and 22.4). Figure 22.3 suggests that firewood is the main cooking energy in rural India irrespective of income level, although its share falls from around 90% for the lowest expenditure class to around 64% in the highest expenditure class. Yet, as the higher expenditure classes consume more cooking energy per capita, firewood consumption in absolute terms is more for the higher expenditure classes and the highest expenditure class consumes almost 50% more wood fuel compared to the lowest class. This clearly indicates that the issue of

---

[3] The data for Figs. 22.3 and 22.4 covers all household energy consumption and does not differentiate between cooking and lighting. However, it is reasonable to assume that electricity is mainly used for lighting while firewood and LPG are used for cooking. Kerosene may be used for both lighting and cooking. NSSO (2001b) provides data in physical units (kg or litres). The following conversion factors were used to arrive at ton of oil equivalent figures: firewood—0.32 kgoe/kg, electricity—0.086 kgoe/kWh, kerosene—0.836 kgoe/l, LPG—1.13 kgoe/kg and coal—0.441 kgoe/kg.

22.3 Energy Ladder and Energy Use

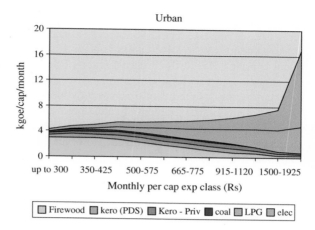

Fig. 22.4 Energy consumption pattern in urban India by income class. *Source* NSSO (2001b)

access to clean cooking energy in rural areas has a much bigger dimension and is not limited to the poor households alone.

The picture changes significantly in urban areas. The use of firewood diminishes quite appreciably with higher expenditure class while the use of cleaner fuels such as LPG or electricity increases. Even at the lowest expenditure class firewood plays a significantly lower role compared to the rural areas (around 70% share compared to 90% in rural areas). High levels of electricity and LPG use by higher expenditure classes suggest that they are unlikely to have affordability problems. Therefore, the problem of access to clean energies in urban areas is a problem faced by the poor households to a large extent. Figures 22.3 and 22.4 also suggest that there is not much difference in the per capita energy consumption in the lower expenditure classes between urban and rural areas. But the highest expenditure class in urban area has a much higher per capita consumption compared to the rest of the households in the country.

This brings us to the drivers influencing the choice decision. This is discussed below.

22.4 Diagnostic Analysis of Energy Demand by the Poor

Energy demand in poor households normally arises from two major end-uses: lighting and cooking (including preparation of hot water).[4] Cooking energy demand is predominant in most cases and often accounts for about 90% of the energy demand by the poor. Such a high share of cooking energy demand arises

---

[4] In some climatic conditions space heating may also be an important source of energy demand. However, for this discussion space heating demand is not considered.

partly from the low energy efficiency and partly due to limited scope of other end-uses.

Any energy use involves costs and resource allocation problems. Both traditional energies (TE)[5] which play a crucial role in the energy profile of the poor, and modern energies impose private and social costs. The private cost may be in monetary terms or in terms of time spent by the family members to collect the TEs. For collected TEs, the problem of valuation of the cost arises and the collected fuel is considered as free fuel by many, even perhaps by the poor themselves, as no monetary transactions are involved. However, depending on the quantity of collected fuel, its source and the type of labour used in the collection process, the private cost and social cost can be substantial. The social cost arises due to externalities arising from pollution and other socio-economic problems related to particular forms of energy use.

The entire decision-making process for use of any modern energy form (electricity, kerosene or LPG, or renewable energies) as opposed to any other form of traditional energies revolves around monetary transactions. Any commercial energy requires monetary exchanges and the decision to switch to commercial energies can be considered as a three-stage decision-making process. First, the household has to decide whether to switch or not (i.e. switching decision). Second, it decides about the types of appliances to be used (i.e. appliance selection decision). In the third stage, consumption decision is made by deciding the usage pattern of each appliance (i.e. consumption decision).

While the costs do not always lend themselves to monetary-based accounting, the switching decision is largely determined by monetary factors: the amount and regularity of money income, alternative uses of money and willingness to spend part of the income to consume commercial energies as opposed to allocating the money to other competing needs. Appliance selection is affected by similar factors: cost of appliance, the monetary income variables described above and the availability of financing for appliance purchases through formal and informal credit markets. Finally, the consumption decision depends on, among others, family size, activities of the family members, availability of appliances and family income.

This framework of three-stage decision-making (presented in Chap. 3) helps in analysing the problem in a logical manner. The poor normally lack regular money income flows due to unemployment or part-employment, both of which sometimes produce in-kind payments as compensation. Moreover, they often participate in informal sector activities, where barter rather than monetised transactions prevail. It is rational for any household or individual to focus on private monetary costs rather than social and/or non-monetised costs due to the inherent subjectivity and complexity of the valuation problem. Moreover, any modern energy has to compete with other goods and services (including saving for the future) procured by

---

[5] I have preferred to use the term traditional energies to non-commercial energies to avoid any confusion arising out of monetisation or commercialisation of some of such fuels. .

22.4 Diagnostic Analysis of Energy Demand by the Poor

the household for an allocation of monetary resources. Given above characteristics and constraints, it is quite logical for the poor to have a natural preference for the fuel that involves no or minimum money transactions. Reliance on firewood and other traditional energies used for cooking, which constitute the major source of energy demand by the poor, can be explained using this logic.

For any commercial energy to successfully penetrate the energy demand of the poor would then require satisfaction of the following economic factors:

(a) The energy should be suitable and perhaps versatile for satisfying the needs;
(b) It should have a competitive advantage that would place no or little demand for money transactions (in other words, the low cost supplies) in the present circumstances, and/or
(c) the use of modern energy should result in supply of adequate money flows to the poor so that they become willing to spend some part of the money on purchasing commercial energies.

Other supply- and demand-related issues and social factors (such as availability of fuel, social acceptance, ease of use, pollution, etc.) will also affect fuel choice and its use, but they are secondary to economic factors.

The second stage (i.e. appliance selection decision) has a deciding influence on energy demand. Often energy appliance has a relatively long life (5–10 years) and its initial costs are high relative to the income level of the poor. In order to, in a sense, amortise the costs the appliances will likely have to be used for sometime, thereby introducing strong path dependence in energy demand. Strong path dependence affects fuel switching possibility and responsiveness of the consumers to external changes. Fuel switching option will be limited by the appliance choice decision and will involve potentially sizeable capital expenditure. The rigidity or strong path dependence leaves limited options to consumers in the event of sudden changes in prices or supply conditions in the short run, who have to depend on their existing stock of appliances in any case.

The appliance selection decision has important bearings for the poor as well. First, high initial cost of appliances for using modern energy is a major deterrent. Consumers naturally prefer low cost appliances, although they are often energy inefficient. This also results from the difficulty of mental calculations for an economic appliance selection that involve factors such as operating costs, discount rates and appliance life. Second, appliances which the poor consider as essential and affordable will be selected, thereby restricting the choice to a bare minimum. Third, the poor are inherently adverse to experimentation and are unlikely to commit themselves to uncertain and unproven technologies on their own. Fourth, strong path dependence of modern energies is likely to add to the reluctance of the poor to invest in modern energies.

Once a decision is made to switch to a modern fuel and the appliance is purchased, the only variable left in the hand of the user is its utilisation. The short term response of consumers to demand arises from this factor, which is quite limited.

22.5 Evaluation of Existing Mechanisms for Enhancing Access

Although a wide range of options are adopted to enhance access to energy, the existing policies rely on the state to provide access by subsidising supply to consumers. A number of energies come under this purview: kerosene for lighting and cooking purposes, LPG for cooking purposes and electricity. Subsidies for such energies could be supported from social considerations: as some minimum amount of energy is required for sustaining livelihood, those who are unable to procure such energies could be supported to procure them. This is essentially the argument behind using lifeline rates for electricity. This is explained in Fig. 22.5 below. If the price is p_e, then consumers with low income will not be able to enter the market as they cannot afford the service. If the consumer surplus of low income consumers multiplied by an appropriate social weight is greater than the social cost of supply, adoption of a lifeline rate could be justified. This does not affect the overall efficiency of the pricing scheme as those having demand above the minimum level of demand Qmin would face the rate at p_e.

The externality argument could also be used to support subsidies: as the use of traditional energies imposes considerable health effects on the population, by switching to clean energies the social cost of health damage could be reduced. As long as the benefits of fuel switching are greater than the social cost, such a subsidy scheme could be followed.

But subsidised energy supply in developing countries has come under scrutiny and the following criticism can be identified:

(a) the subsidy is not targeted, implying that the benefits do not reach the desired group. In many cases all consumers have been given the benefits of subsidy for administrative simplicity, which allows the rich to benefit more as their per capita consumption is higher than the poor. Where the benefit is restricted to

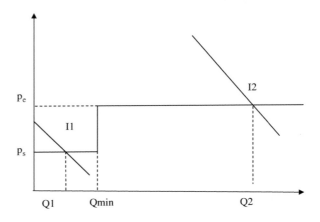

Fig. 22.5 Lifeline rates

the poor, lack of administrative verification and monitoring allows considerable leakage of subsidy, allowing others to benefit.
(b) As energy cannot be used without owning appliances and as subsidies are granted for energy consumption (not for appliance ownership), subsidised supply helps those who can afford appliances. Thus subsidies for LPG and electricity often accrue to the rich.
(c) Continued use of subsidised supply has given rise to a sense of right to this privilege, making subsidy removal politically difficult.
(d) Subsidised supply distorts price signals and increases demand, which in turn requires more investment for supply systems. This can be seen from Fig. 22.6 (also see Chap. 13). As most of the residential consumers contribute to the peak demand, higher consumption requires extra peaking capacity, which is costly but at the same time may not be remunerative for the supplier. Capacity shortage results in absence of new capacities, imposing social costs due to non-availability of supply.
(e) Subsidy imposes revenue burden on the supplier and the state, and when the subsidy is not timely provided, the financial performance of the supplier gets affected.
(f) Inefficient energy use through subsidies adds to pollution and contributes to the climate change problem.

Getting energy prices right essentially means rebalancing the prices by removing subsidies and cross-subsidies. There are two issues involved here: correct prices would make energy supply a commercially attractive proposition but at the same time, commercial energies will become less the competitive as compared to the traditional energies. However, as observed earlier, the subsidy system for petroleum products is not targeted to the poor and such improperly targeted

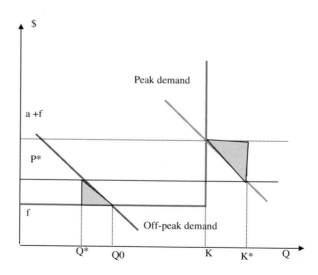

Fig. 22.6 Inefficiency of inappropriate pricing

subsidies could be removed without much effect on the poor. But strategic subsidies would remain a key policy tool of the state to promote commercial energies amongst the poor.

22.6 Effectiveness of Electrification Programmes for Providing Access

To resolve the energy access problem, rural electrification initiatives need to be analysed considering the factors presented in the diagnostic analysis. A number of observations/inferences can be made:

- Electricity is mainly used for lighting purposes and accounts for a minor share of households' energy needs. In order to resolve the energy access problem through electrification, electricity use has to meet the cooking energy requirements of the poor. A number of issues arise in this respect:
  - *Competitiveness*: electricity is unlikely to be competitive when compared with traditional energies used for cooking purposes. Subsidized supply to household belonging to lower income groups normally will allow them to use electricity for lighting. Promotion of electricity supply is unlikely to reduce reliance on traditional energies for cooking per se.
  - *Quality of supply*: As the power supply to rural areas gets low priority, even when access is available, actual supply may be limited, especially during peak demand periods due to prevailing capacity shortage conditions. Lack of adequate supply acts as a hindrance to expansion of electricity use in productive and other activities.
  - *Initial investment*: Use of electricity for cooking entails significant initial investment when compared with traditional energy use. Cash-strapped poor households are unlikely to switch to electric cooking even if quality electricity is available at an affordable rate.

Thus, electricity has a less chance of succeeding in the cost competition with other fuels. This in turn implies that demand for lighting cannot justify the investment in electrification of an area. Consequently, rural electrification alone cannot resolve the problem of energy access in rural areas, as other fuels would be used by the poor to meet cooking demand. It appears that policy makers tend to ignore or forget this simple truth, may be because of better prestige and visibility of electrification projects (and hence for better political mileage).

For economic and financial viability of rural electrification projects, expansion of productive use of electricity is essential. Integrating other rural development programmes with rural electrification could create a synergy for promoting agro-based industrial activities and productive use of electricity in rural areas.

Additionally, countries with a poor record of credible subsidy management system due to resource constraints, sustainability of subsidized schemes is highly doubtful. A credible alternative has to rely on development mechanisms that ensure adequate money supply to the poor on a regular basis. This makes it necessary for rural energy supply issues to be set in a broader canvass of overall development. But experience so far hardly supports the catalytic role of electrification programmes. Rural industry or commerce has not developed as a thriving business proposition so far in many rural areas. Thus, sustainability of subsidized rural electrification system may remain a thorny issue for a long time to come.

Energy sector reform has not been a great success in countries where most of the poor are concentrated and is progressing quite slowly. Electricity reform has not produced the desired results so far and even the progress has been dismal in most areas. Simultaneously, the state funding for electricity has been drastically reduced, without any concomitant participation from the private sector. Private participation in power distribution does not seem to be gaining momentum and it is quite likely that the privately-owned distribution companies will be least interested in undertaking a loss-making activity. Depending on reforms for solving the energy access problem will be synonymous to inaction. This is not suggest that reforms are not required or should not be followed. Energy sector reforms are essential but being a politically sensitive process, making it a pre-condition for providing access to the poor is not a logical approach.

22.7 Renewable Energies and the Poor

Many place great hopes on new technologies for solving the problem (WEC 2000; DfID 2002; World Bank 1996). New technologies that are suitable for distributed energy supplies are now available and can be cost-effective compared to grid-based supplies. Such technologies often have the added advantage of being environment friendly and hence their promotion would be beneficial for the world as a whole. However, despite extensive research and commercialisation efforts over past three decades, these energies are not competitive yet, without subsidies of some sort or other. Using subsidies for creating a market for new technologies has the disadvantage that subsidy removal becomes difficult, as the LPG case demonstrates. The technical fix of the problem does not appear to be an answer.

Consider now the case of renewable energies to analyse whether they meet the above requirements indicated earlier in the diagnostic analysis. As cooking and lighting constitute two major energy demands of poor households (excluding space heating), we consider these two separately. As there are different types of renewable energies (solar, wind, hydro and even sustainable biomass), we focus on solar energy here. Similar arguments can perhaps be advanced for other energies as well.[6]

---

[6] The specific arguments may have to be adjusted in some cases but the generic argument remains valid.

Solar energy is available abundantly and from time immemorial, mankind has been using this form of energy for various purposes. The poor economies are major direct users of such energy, especially for drying purposes, although such information does not enter the official statistics. So long as the use of this energy remains traditional, the poor find its use helpful. However, modern ways of using this energy move it away from the poor. The more sophisticated the conversion process and higher the degree of usability, the further removed it becomes from the poor.

Consider solar home systems (SHS) - a technology on which great hope rests. SHS is available for one or two lights and a few small electrical or electronic appliances like TV or radio. This allows multiple use of solar energy and could make poor homes a better living place. The system size is typically between 10 to 100 watts and involves an initial cost of a few hundred dollars and the replacement of batteries and lamps also involves some costs at regular intervals. Does this technology satisfy the factors we identified in the previous section?

First, let us verify the versatility of the SHS. It essentially competes with the lighting energy demand of the poor and is not suitable for cooking purposes. Therefore, it is not capable of meeting the most important energy need of the poor. Moreover, lack of alternative lighting fuels forces the poor to use some form of commercial energies in any way, mostly in the form of kerosene or electricity. Environmental benefits of SHS are thus limited compared to a solution that satisfies cooking energy needs of the poor. Second, the cost of SHS is significantly higher than the income level of the poor (IEA 2002). Therefore, it does not offer any competitive advantage in terms of costs. Third, the system cannot be used in any direct production process, although households may utilise manual labour over an extended period of time with these facilities, thereby indirectly benefiting from extra income. Given this indirect nature of income support, the impact of SHS in poverty alleviation is limited.[7] Hence, SHS does not appear to satisfy any of the criteria mentioned above in general and is not an attractive option for the poor.

As the versatility of and the income generation prospect from SHS can hardly be changed, only its cost aspect can be reworked to promote such a technology amongst the poor. From the perspective of the poor, a gift would be an ideal or most preferable financing option, as it imposes no monetary cost on the poor for using a renewable energy in a modern way. However, the cost of such a 'solar gift' could easily run into a few hundred billion dollars if SHS has to be provided to the two billion poor.[8] Had SHS been a substitute for cooking energy demand, the environmental benefit could have compensated a significant part, if not all, of the cost of the gift. In the absence of a sizeable environmental benefit, SHS appears to be a costly gift.

---

[7] See IEA (2002).

[8] Assuming 500 dollars on average for a SHS and a family of 5 members sharing a SHS (i.e. 400 million poor households, according to IEA (2002)), the cost comes to $200 billion.

22.7 Renewable Energies and the Poor

Admittedly, innovative financing schemes are being used to promote SHS successfully around the world. However, according to IEA (2002), all such projects, promoted nationally or internationally, are supported by direct and indirect subsidies. Subsidies go against the principle of getting prices right and the experience of liquid petroleum gas (LPG) suggests that the removal of subsidies becomes a politically challenging task. If a technology or energy has to depend on subsidies and favourable incentives for its financial viability, we are going to make the same mistakes made in promoting some other energy forms (such as LPG subsidies, kerosene subsidies, etc.). The culture of subsidies would not lead to a sustainable development.

Let us now turn to the application of solar energy for cooking purposes. Solar cooking is a safe and simple way of cooking food using solar energy. Solar cookers of various types are available at relatively cheaper costs (as low as 25 dollars).[9] Solar cooking is smoke-free and fire-risk free. It also eliminates the need for fuel gathering, an activity that consumes a significant amount of time of the women in poor families. By addressing cooking demand—the major energy demand of the poor—solar cooking has the potential of satisfying the first criteria set out in the previous section. Moreover, in countries where the poor rely mainly on biomass, solar energy is available abundantly and can be easily harnessed. It is comparatively cheap, much cheaper than the SHS and thus imposes a relatively low monetary burden on the poor. It comes close to satisfying the second criteria as well. Finally, the time saved by eliminating fuel collection could be used productively to generate income for the family. The cooking system can also be used for food preservation, which may save some expenses as well. Although it does not lead to direct income generation, it provides indirect avenues of income generation and hence satisfies to some extent the third criteria as well. Despite such advantages solar cooking has not yet taken off and is perhaps least popular, except for drying food grains in a traditional way. It has not succeeded in enticing the poor households.

This failure stems mainly from two factors.[10] The first relates to our very poor knowledge about how to manage small energy systems and how to organise the activities around them. Managing such decentralised systems is different from the traditional approach of providing energy through centralised supply systems. The skill set required for such new types of activities may not be available locally and there is a need to focus on such capacity development. Second, there appears to be a market failure here as low marginal cost of supply does not attract private investors to provide the good. The commercial profitability of such ventures being low, promotion of solar cooking is unlikely to be taken up by private investors

---

[9] Solar Cookers International website http://solarcookers.org/order/cookers.html.

[10] As expected, there is a large volume of literature on this issue, which identifies hundreds of barriers. While many of those barriers may be valid, the factors mentioned here are in my opinion most important.

having profit motives. Some alternative form of supply is required, which has not yet emerged.

Should governments intervene and undertake supply of solar cooking appliances? Traditional economics would suggest yes but as mentioned earlier, governments do not have any special skills for managing the activities of decentralised energy supply systems. Most governments work in a centralised manner and tend to impose its decision at all levels. There can however be a case for governments providing capacity building facilities and opportunities for undertaking such activities, at least initially.

Financing options can also improve acceptance of solar cooking. Providing solar cooking facilities as a gift to 2 billion poor would cost 20 times less than providing SHS.[11] It could be vastly effective in reducing global climate change problems and environmental benefits of such a system could even justify the gift.

From above we can conclude that although the standard recommended strategies for enhancing energy access to the poor often stress on three R's: energy market Reforms, Renewable energy technologies and getting prices Right (World Bank 1996, 2000), they are unlikely to resolve the access issue.

22.8 Alternative Solutions

Having established the need for an alternative strategy above, it is logical to think about the outline of a policy or strategy that could solve the problem. It is important to mention at the outset that we neither have any ready-made strategy for solving the energy access problem nor we have undertaken any experimental/pilot study in this regard. Instead, our idea is to suggest a holistic approach that places the issue of energy access in its proper context of income generation, monetisation of income and provision of affordable supply. We also indicate areas where further work is necessary.

As discussed earlier, sustainable, long-term solutions for energy access problem cannot rely on subsidized supply of clean energies. Similarly, piece-meal solutions that address only a part of the problem will not help either. What is required in the long term is to ensure adequate supply of monetary resources to households to sustain a life style that relies on clean energies and other monetized inputs. Thus the energy access issue joins here the problem of ensuring economic development, which in turn calls for an integrated approach of combining various development efforts at a decentralised level (DfID 2002 and WEC 1999) as opposed to treating electrification or energy supply issues in an isolated manner.

---

[11] Based on a $25 kit for solar cooker, the cost for 400 million households comes to $10 billion, which is 20 times less than $200 billion required for providing SHS to the same number of households.

22.8 Alternative Solutions

Given the diversity of a country in terms of energy use, resource availability and other conditions, it is desirable to search for appropriate local solutions instead of universal or global solutions to the problem. The policy objective should be to promote innovative solutions rather than prescribing templates for adoption. Thus each decentralised unit (which may be at the block level for rural areas and municipal level for urban areas) will have to search for its own solutions

(a) Focusing on the creation of opportunities for higher income generation in monetary terms, as opposed to in-kind income. Unless money flow increases to the poor, commercial energies stand little chance of competing with traditional energies.
(b) by developing local energy markets taking into account the specificities of local energy situation (resources, needs, capacities, strengths and constraints) and adopting appropriate supply mechanisms and organizational structures to cater to the local needs.
(c) by selective and judicious use of market interventions to make energy supply affordable but ensuring financial viability of energy supply. Unless the supply is financially viable, it cannot be sustained.
(d) by ensuring local community participation in the decision-making and policy implementation process.

Such a bottom-up policy is inherently multi-dimensional and necessarily complex. It comes as a contrast to the existing policies which are top-down in nature and essentially imposed on the population. Implementation of such a policy would require development of a common framework that can be adapted to each situation, creation of an organizational set up to carry out the policy, building organizational capacity, adequate funding arrangements, and above all a complete review and perhaps an overhaul of the mode of functioning of the government, existing organizations and the economic activities to facilitate decentralized mode of functioning of the economy.

For example, various barriers affect creation of opportunities for higher income generation. They include lack of infrastructure (roads, storage facilities, telecommunication facilities, access to markets), restrictions on movement/sale of agricultural products, market distortions (government interventions in the market imposing price/quantity restrictions), trade restrictions and lack of information. Removal of such barriers can lead to better remuneration for the agricultural output which being a traded good will command international prices. Opening of the agricultural sector can also bring a structural change in the form of a shift towards agro-industries (rather than trading primary agricultural outputs). A comprehensive review of policies would be required to ensure compatibility with the decentralized mode of working and removal of barriers for creating better opportunities for rural development as such as well as for rural income generation.

Similarly, several barriers impede with rural energy market development and lead to market failure. Better income opportunities would improve affordability and thus create new demand. But lack of effective delivery mechanisms could require intervention and development of a conducive environment. While the local development could be energy driven and energy-centric, through use of local resources for energy production, (especially using locally available biomass or mini hydro or similar resources), which in turn helps in local income generation through employment and payments for resources, the direct employment opportunity in energy sector may be limited in most cases. However, indirect or induced employment and income generation potential remains which in conjunction with removal of barriers for the agricultural sector can bring about a significant change. Given initial low demand, innovative ideas such as one-stop shop concept offering multiple energy services or products (TERI 2002) would need to be explored.

Finally, it may be easier to implement a targeted subsidy mechanism at the local level than through a centralized system. Instead of using uniform price subsidies all over the country, it would be possible to adopt need-based subsidies or support programmes.

Surely, the strategy outlined above is essentially a long-term one and crucially depends on the decentralization of the political system in the country. Such a decentralized system may lack effectiveness due to a number of barriers involving policy, administrative and structural aspects and requires strengthening. This remains a challenging task by any standard. In the shorter term, the solutions cannot aim for curing the problem but should try to contain it and provide temporary relief. State support for employment creation for the poor, ensuring minimum amount of commercial energy supply at affordable rates for the poor and subsidy for appliances and connections will remain important.

22.9 Conclusion

This chapter has presented an overview of issues related to energy access in developing countries. The chapter has indicated the current and future status and presented the commonly used indicators of access, including the energy development index suggested by IEA. It then analysed the issue by considering the three-stage decision making process and highlighted the inadequacy of existing solutions. The chapter suggested a bottom-up approach to resolve the problem, although such a policy is inherently difficult to implement. Further research is required in this area to identify practical, viable solutions.

References

DfID (2002) Energy for the poor: underpinning the millennium development goals. Department for International Development, UK

IEA (2002) Financing mechanisms for solar home systems in developing countries: the role of financing in the dissemination process, IEA Report PVPS T9-01:2002, Paris

NSSO (2001a) Energy used by Indian households, 1999–2000, NSS 55th round, National Sample Survey Organisation, Government of India

References

NSSO (2001b) Consumption of some important commodities in India 1999–2000, National Sample Survey Organisation, Government of India

Pachauri S, Mueller A, Kemmler K, Spreng D (2004) On measuring energy poverty in Indian households. World Dev 32(12):2083–2104

TERI (2002) Defining an integrated energy strategy for India: a document for discussion, debate and action. The Energy and Resources Institute, New Delhi

WEA (2000) Energy and the challenge of sustainability, World Energy Assessment 2000, UNDP, New York

WEC (1999) The challenge of rural energy poverty in developing countries. World Energy Council, London

WEC (2000) Energy for tomorrow's world—acting now. World Energy Council, London

WEO (2002) World energy outlook 2002. International Energy Agency, Paris

WEO (2004) World energy outlook 2004. International Energy Agency, Paris

WEO (2006) World energy outlook 2004. International Energy Agency, Paris

WEO (2009) World energy outlook 2009. International Energy Agency, Paris

World Bank (1996) Rural energy and development: improving energy supplies for two billion people. World Bank

World Bank (2000) Energy services for the world's poor. World Bank

Part V
Economics of Energy–Environment Interactions

Chapter 23
The Economics of Environment Protection

23.1 Introduction

So far, we have considered various aspects of energy supply, demand, markets and issues related to energy. But we did not consider one important aspect—the effect of energy sector on the environment. This is addressed in this chapter. This chapter presents an overview of the energy–environment interactions and introduces the economic instruments for dealing with such problems.

23.2 Energy–Environment Interactions

Economic activities make use natural resources and other inputs (labour, capital, etc.) to transform them into usable outputs. The production and consumption process generates wastes, which are rejected into the environment. Similarly, the living beings on the earth use air and water provided by the environment for sustaining life. The environment thus provides a number of services to facilitate economic activities. These include:

- A resource base, which can be put to extractive or amenity uses. These are similar to non-renewable energies that are consumed once and for all.
- A life support system through provision of air and water;
- A waste sink where the wastes of the production process are rejected. These wastes can be in solid, liquid or gaseous forms.

These resources and services are scarce in nature and increased use of these facilities risk the danger of deteriorating the quality of the environment.

When the environment is used as a receptor of wastes, there is a limit up to which it can absorb the wastes and assimilate in its system. This capacity is known as the absorptive capacity or assimilative capacity. Beyond this capacity, the wastes accumulate in the environment and the concentration of waste level increases with addition of new wastes. The environmental problem starts when the waste rejection crosses the assimilative capacity of the environment. As the users who do create the problem are not often responsible for bearing the cost of the damage, it creates an externality as well.

Environmental damages can be viewed as a multi-dimensional problem. The following are important considerations in this regard:

(a) Environmental media: The pollutants can be released in different media, causing damages to them. Air, water, land and natural ecological systems are most commonly considered media.
(b) Spatial dimension: The damages have a spatial dimension in the sense that they affect the households at the local level and local communities (rural areas, urban areas) where emissions are released. A part of the pollutants is transported through the media (air or water), causing regional damages (e.g. acid rain). Some pollutants affect the entire world through their effects (e.g. global warming).
(c) Origin of pollutant: Pollutants may come from static sources or mobile sources.
(d) Time pattern of pollution generation: Pollutants may occur in a continuous or discontinuous manner. Emission from a power plant chimney is an example of continuous pollution. Pollution from forest fire is an example of intermittent pollution.
(e) Quality of pollutants: Some pollutants can be hazardous or poisonous in nature, mere presence of which can cause damage. There are others which when exposed to above threshold levels cause damage. There can be others which do not affect the health as such (neutral) but can be harmful for the overall environment (e.g. CO_2).
(f) Life of pollutants: Different pollutants have different lives. Some decay after periods (i.e. short-lived) while others take longer time to decay.

Environmental pollution can be caused by natural phenomena (known as biogenic sources of pollution) or from human activities (known as anthropogenic sources). The extent of disruption caused by human activities is indicated in Table 23.1 below in descending order. Human disruption index (i.e. the ratio of human-generated flow of pollutants to the natural baseline) is dangerously high for toxic materials such as lead and cadmium. Oil added to oceans is 10 times the natural level of emission of 2,000,000 tonnes per year. Sulphur and methane emissions are 3–4 times the natural level of emissions. On the other hand, CO_2 emission is only 5% of the natural baseline. In all these cases, energy sector plays an important role in the emission of pollutants.

Different activities in the energy system (production, conversion and utilisation) lead to various environmental impacts. Table 23.2 provides a summary of salient impacts of coal and oil and gas cycles. Detailed information on damages from energy activities can be obtained from Externe UK report, Vukina (1992), and elsewhere.

23.2 Energy–Environment Interactions

Table 23.1 Environmental damages due to human activities

Insult	Natural baseline	Human disruption index	Share of human disruption caused by			
			Commercial energy supply (%)	Traditional energy (%)	Agriculture (%)	Manufacturing, others (%)
Lead	12,000 t/year	18	41			59
Oil added to oceans	200,000 t/year	10	44			56
Cadmium	1,400 t/year	5.40	13	5	12	70
Sulphur	31 Mt S/year	2.70	85	0.50	1	13
Methane	160 Mt/year	3.75	18	5	65	12
Nitrogen fixation	140 Mt N/year	1.50	30	2	67	1
Mercury emissions	2,500 t/year	1.40	20	1	2	77
Nitrous oxide	33 Mt/year	0.49	12	8	80	
Particulates	3,100e Mt/year	0.12	35	10	40	15
Hydrocarbon emissions	1,000 Mt/year	0.12	35	5	40	20
CO_2	150 BtC/year	0.05	75	3	15	7

Source UNDP (2004)

Table 23.2 Environmental effects of coal and oil fuel cycles

Burdens		Solid wastes	Liquid wastes	Gaseous	Noise	Accidents
Production	Coal	Damaged landscape, earth deposits	Mine drainage	Methane, dust, CO_2, SO_x, NO_x	Blasting, heavy equipment	Flooding, implosion, fire
	Oil and gas	Mud	Oil discharge, drain water, rain water	Methane, CO_2, NO_x, SO_x, flaring	Blasting	Fire, explosion
Conversion to electricity	Coal	Ash, sludge	Hot water, ash slurry, plant waste water	Particulates, CO_2, SO_x, NO_x		Fire, explosion
	Oil and Gas		Hot water, oil discharge, plant waste water	Particulates, CO_2, SO_x, NO_x, gas leaks		Fire, explosion
Final use	Coal	Ash, sludge	Hot water, plant waste, ash slurry	Particulates, CO_2, SO_x, NO_x		Fire
	Oil and Gas		Hot water, oil discharge	Particulates, CO_2, SO_x, NO_x, gas leaks		Fire, leaks

23.2 Energy–Environment Interactions

Table 23.3 Effects of gaseous pollutants

Effects	CO	CO_2	CH_4	NOx	SOx	Ozone
Green house effect		+	+		−	+
Ozone depletion		±	±	±		
Acid deposition				+	+	
Smog				+		+
Corrosion					+	
Decreased visibility				+	+	
Decreased self cleansing of atmosphere	+		±	−		−

+ contribution to the effect, − amelioration
Source Bhattacharyya (1995)

Combustion of fossil fuels produces a number of gaseous pollutants, including sulphur oxides, nitrogen oxides, carbon mono and di-oxides, and particulate matters. These gaseous pollutants cause a number of effects such as smog, acid rain, ozone depletion and global warming. SO_2 released into the atmosphere when combines with water forms sulphuric acid, and reaches the earth as acid precipitation. The acid rain causes damages to the forests, aquatic life and corrodes buildings and materials. Similarly, nitrogen oxides when combine with water produce nitric acid, which has similar effects. Table 23.3 summarises the effects of some gaseous pollutants.

23.2.1 Energy–Environment Interaction at the Household Level

At the household level, a variety of energy services are used ranging from cooking to lighting and heating/cooling. At one end of the spectrum, biomass-based fuel is used while at the other end, modern energies such as natural gas and electricity are used to satisfy the needs. Historically it is found that people tend to move up the ladder of higher fuel quality with higher income (fuel ladder concept discussed in Chap. 22). Yet, more than 2.5 billion population is still dependent on poor quality, traditional energies for household needs (WEO 2006).

Fuel harvesting and combustion are two major sources of environmental impacts at the household level of energy use. Fuel-wood harvesting has contributed to depletion of forests in many countries (although this need not be the only or most important reason for deforestation). Post-harvest burning of farmland is a major source of local level, seasonal pollution as well. Similarly, these activities impose significant occupational health hazards, particularly on women and children who are involved in collection and use of household energies. Harvesting of crop residues and dung may also lead to deprivation of soil nutrients and other conditioners (WEC 2000).

Combustion of household fuels leads to air pollution. Incomplete combustion of fuels used in inefficient stoves (mainly for biomass and coal in poor households) creates health hazards. The local level pollution arising from liquid and gas based

Table 23.4 Emission factors for energy used by households (kg/TJ of energy input)

	Methane
Residual fuel oil	1.4
Diesel oil combustors	0.7
LPG furnace	1.1
LPG stoves	0.9–23
Kerosene stoves (wick)	2.2–23
Bituminous coal stoves	267–2,650
Wood stove conventional	932
Charcoal stoves	275–386
Stoves for agricultural wastes	230–4,190
Dung stoves	230–4,190

Source IPCC (2006)

petroleum products is relatively less due to higher efficiency of cook stoves and better fuel quality. The timing of such pollution (when most of members of the family are present) and high level of exposure due to poor ventilation in houses make poor households vulnerable to serious health effects, leading to pre-mature deaths or chronic respiratory or infectious diseases.

Similarly the household use of energy also has global warming implications as well. IPCC (2006) provides some information on the GHG emissions from some important household fuels (see Table 23.4). Low efficiency of many residential energy use technologies results in higher environmental impacts.

23.2.2 Community Level Impacts

At the community level, the major environmental impact of energy use is the level of urban pollution, although industrial and other activities also contribute to this to a lesser extent. According to UN (2009), about 3.2 billion people lived in urban areas in 2005 but about 72% of them are from the developing world. Rapid urbanization of the developing countries in Asia and Africa is expected to incnrease the urban population to close to 5 billion by 2030, when 79% of the urban population will be in the developing world.

There has been a dramatic change in the geographical location of urban centres in the world. In 1900, 9 out of 10 largest cities were located in North America and Europe (UNEP 2002). By 1950, 7 out of 10 were located in the developed world. In 2005, only three (Tokyo, New York and Mexico City) are located in the developed world (UN 2009).

Combustion of fuels (either in power generating stations or in industry or in residential or commercial activities) and use of energy in transportation are the major sources of urban air pollution. Particulate matters, Nitrogen oxides (NOx), Sulphur oxides (SOx), ozone formed at the ground level through interaction of NOx and volatile organic compounds, carbon mono-oxide and lead are the major pollutants which are normally monitored for health purposes.

The urban air pollution problem is not a new phenomenon. Coal use was prohibited in London in the thirteenth century due to its health effects. Smog was a problem during the Industrial Revolution.[1] In the early twentieth century, the problem was noticed in the cities of developed countries (London Smog for example). Through a combination of measures (e.g. use of strict environmental regulation, ban on use of coal in city limits, switch over to natural gas and nuclear power), the situation has been reversed in most developed countries. The air pollutant emission has stabilised or declined in developed countries (UNEP 2002) despite higher levels of power generation and transport use.

Although the situation has improved in the developed countries, during winter smog conditions, the air quality is normally below the WHO recommended guidelines. In some cases, high concentrations of ozone are found during summer smog due to special meteorological conditions or site characteristics (Schwela and Gopalan 2002).

The same however cannot be said about the cities of the developing countries. Most of these urban areas suffer from poor air quality. These so called megacities[2] impose heavy energy-related environmental burdens due to:

(a) Reliance on solid fuels and biomass by a large section of the poor living in these cities.
(b) Use of inefficient fleet of transport: motorcycles and 3 wheelers form the backbone of transport in many places. Old vehicles continue to ply and the traffic management is poor, resulting in congestion and consequent high pollution.
(c) The urban areas were often established for industrial activities, many of which were set up without much regard or respect to environmental conditions. Along with major industries, small industrial activities also grow up and due to their choice of cheap fuel, they tend to be polluting in nature.

23.2.3 Impacts at the Regional Level

This can be viewed as an intermediate level of impacts—lying in between local level problems and global level problems (such as climate change). The regional problems include acid rain issues, tropospheric ozone and suspended fine particles (WEC 2000). The regional impact occurs due to transportation of pollutants over long distances due to wind and other climatic conditions. During this transportation, the pollutants undergo chemical reaction with water vapour and other pollutants and transform into acids or ozone.

---

[1] Encyclopedia of the Atmospheric Environment, http://www.ace.mmu.ac.uk/eae/english.html.
[2] Megacities are cities with a population exceeding 10 million.

SOx and NOx are two acid forming pollutants. SOx emitted from sulphur containing fuel when comes in contact with water vapour forms sulphuric acid. Similarly, NOx emitted from vehicular traffic or in other combustion processes forms nitric acid in contact with water. These acids when precipitate in liquid or solid forms on the earth (forests, land) or water bodies can affect the soil and vegetation; water in the lakes and rivers; aquatic life; forests, and human health.

Energy activities play an important role in anthropogenic sources of SOx and NOx emissions and contribute significantly towards CO, methane and volatile organic compounds emissions. These pollutants are mainly responsible for acid formation. Emission of these pollutants from fossil-fuel based power plants and from industrial sources is mainly responsible for the problem. Menz and Seip (2004) indicate that acid rain has mostly affected Europe, North America and South-East Asia, especially central and southern China. While the acid rain problem has been noticed since the 1970s in many parts of the developed world, the intensity of the problem is declining there due to drastic reduction in pollutant emissions. For example, Menz and Seip (2004) report that suplhur emissions in Europe and North America have fallen in the past two decades by 65 and 40%, respectively. On the other hand, acid forming pollutants are on the rise in Asia and can pose a major threat. GEO 3 indicated that around 0.28 million ha of forest land was reported to have been damaged due to acid rain in China (UNEP 2002).

23.2.4 *Global Level Problems: Climate Change*

The link between the climate and the GHG is complex and is complicated by a large number of natural processes. While the knowledge in this respect is still incomplete, there is now an agreement among researchers that the anthropogenic sources of pollution affect the climate. The Intergovernmental Panel on Climate Change (IPCC) set up by the United Nations has collected a large set of data and evidence and suggests that the global temperature has been increasing in an unprecedented manner during the past two centuries (see IPCC (2007)).

Continued reliance on fossil fuels has increased CO_2 emissions globally (see Fig. 23.1). It shows the increasing trend of CO_2 emission from fossil fuels but unlike in the past, CO_2 emissions from the developing countries are increasing at a faster rate than that of the developed countries. Although CO_2 emission per person is much lower in the non-OECD economies due to higher population base, the issue of limiting emission becomes important when most of the incremental emission is expected to come from a few large developing economies. However, given the low per capita emission level in these countries and their need for economic growth to sustain a reasonable quality of life, it would be quite difficult to impose any binding commitment on them. This makes the climate issue even more challenging.

Research by the IPCC suggests that the global temperature can increase between 1.8 and 4.0 degrees C by 2100 relative to that in 1980–1999. This is

23.2 Energy–Environment Interactions

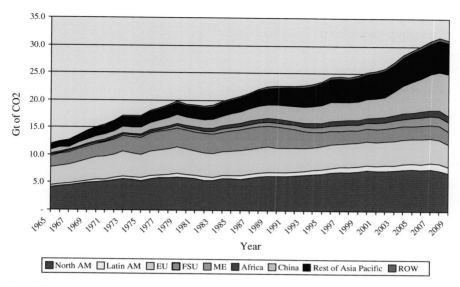

Fig. 23.1 CO_2 emission from fossil fuels. *Data source* BP statistical review of energy, 2010

unprecedented during the past 10,000 years and such an increase can have major effects, including (IPCC 2007):

- Sea level changes: may change between 0.18 to 0.59 metres between 1990 and 2100.
- Changes in precipitation levels: larger year to year variations in precipitation is projected and an increase in precipitation is expected.
- Changes in climatic zones and
- Changes in severity of frequency of extreme events.

As a consequence, the human society faces new pressures and risks. The impact of these changes can be significant (including loss of low lying countries, changes in living conditions, economic losses, loss of biodiversity, displacement of large number of people, etc.), estimation of which is fraught with theoretical and practical difficulties. We will look at these challenges in Chap. 26.

23.3 Environmental Kuznets Curve

As developing countries would demand more energy to drive economic growth, an important policy issue arises whether they should follow the industrial countries' policy of polluting first and cleaning later or leap-frog towards cleaner technologies and avoid the mess in the first place. The economic logic behind the first comes from the observation known as the Environmental Kuznets Curve which suggests an inverted U relationship between per capita pollution and per capita

Fig. 23.2 Hypothetical environmental Kuznets curve

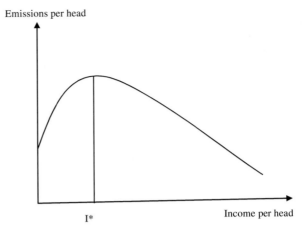

income (see Fig. 23.2 for a hypothetical curve).³ Such a relationship suggests that in the initial phase of increasing per capita income, the citizens may be willing to accept a poor environmental quality but as the income improves, a turning point will be reached and the demand for better environment in any case will arise. Increases in wealth will lead to further reduction in the environmental pollution. If such a relationship holds, economic growth will cure the environmental problem and the damage may be a transitional one.

The supporters of this idea argue that generally, if there is no change in the economic structure or technology use, economic growth leads to proportional increases in the environmental effect, known as the scale effect. However, economic development normally happens in stages—initially agricultural activities dominate, which is followed by industrialisation of the economy and finally a shift towards the information and service oriented activities takes place. This structural change along with better environmental awareness, technological changes, and better environment management initiatives improves the environment.

However, there are several criticisms against this idea:

(a) There is no standard shape or relationship that is valid for all pollutants or for all regions.
(b) The result is highly dependent on the econometric method used and the data analysed. Different researchers have reported different results for the same pollutant, indicating the lack of robustness of the results.
(c) The research that finds such relationships are unable to explain the process through which the cure takes place.

---

³ The idea was first indicated in the World Development Report 1992 of the World Bank. A large volume of literature in this area is now available. See Webber and Allen (2010), Stern (2003), Grossman and Krueger (1995) for further details.

(d) The estimation often ignores the feedback from environmental damage to economic activities. If the environmental damage is sufficiently strong, it is likely that the economic activity will be affected and therefore, ignoring this does not provide a correct picture.

Stern (2003) concludes that there is no strong evidence that countries follow a common inverted U shaped pathway for environmental damage as their income rises. Webber and Allen (2010) conclude that there is no single relationship between environmental quality and income. Moreover, it may take decades before reaching the trend inversion and accordingly, waiting for such times will not make sense. Proactive policies and measures would be required to mitigate the problem.

23.4 Economics of the Environment Protection

In Chap. 12, the basic market model was presented. The existence of such a market is contingent upon a number of strict conditions, including the following:

- Efficient property rights structure that ensures exclusivity, transferability and enforceability. Exclusivity implies that the ownership of the good is exclusive to the person buying it. This excludes the possibility of consumption of the good by others. Transferability implies the ability of transferring the ownership to others while enforceability implies that transactions and contractual obligations could be enforced. As we will see, some of these elements are often violated in the case of environmental goods.
- Completeness of markets for all goods and services;
- Perfect divisibility of goods and services: assumes that it is possible to divide goods and services so that there is no joint supply. As we will see below, many environmental goods violate this requirement.
- Information is available freely, perfectly and all agents are informed.
- There are no entry or exit barriers.

Violation of these requirements leads to market failures. This is quite common in the case of energy industry and for environmental goods. In economic treatment of the environmental effects of energy use relies on the concept of externality. This is considered in this section.

23.4.1 Externalities

Although the term externality is widely used, it is difficult to offer a precise definition. Many authors have used the term in different ways: some have used it so broadly that all market failures are covered, while others tend to give a restrictive meaning to the term. The definition used here is adapted from Baumol and Oates (1988). An externality occurs when:

- an action of an economic agent affects the utility or production possibilities of another agent;
- the welfare of the agent depends on the activities of some other agent;
- the agent does not bear all consequences of his/her action (costs or benefits not fully compensated).

If there is compensation or payment in return for damages, then the externality does not exist to the extent of compensation. It is considered to be internalised. In other words, an externality is an unpriced, unintentional and uncompensated side effect of one agent's action affecting another agent's welfare (Sundqvist 2004) and Sundqvist and Soderholm (2002).

Externalities can be of different types:

a) *Positive/negative*: when an externality provides net benefits to others, it is considered as a positive externality. Scenic beauty of a private garden provides an external economy to all who pass. Discomfort from smoking cigarettes to non-smokers is an example of negative externality.
b) *Private (depletable)/public (undepletable)*: This distinction relies on the idea of jointness of supply of the externality. If an external effect is such that consumption by one leaves no external effect for others, then the effect is considered as private or depletable. For example, if I dump trash in my neighbour's backyard, it is unlikely to be dumped to somewhere else. Therefore, it is depleted. This kind of private externality is limited in nature.

Most of the environmental externalities are of public variety. This means that consumption by one does not reduce the availability to others. For example, my smoking of polluted air does not reduce the amount of polluted air available to others. In this case, victims are jointly affected.

c) *Pecuniary/Technological*: This is a particular type of externality that arises from a change in the prices of inputs or outputs in the economy. For example, an increase in oil prices in the international market leads to an increase in electricity prices. The consumers of electricity are affected and they suffer negative external effects. But unlike a true externality, this oil price effect does not change the technical relationship of electricity production. Once the oil price effect is withdrawn, the old position could be reached. A technological externality on the other affects the input–output relationships. All commonly considered externalities are of this variety.

23.4.2 Spectrum of goods

Goods and services could be classified using two characteristics: jointness of supply and exclusivity. Jointness of supply is related to the question whether the consumption is rival or not. Consumption of a good is rival when its consumption by one does not leave others to consume the same. The good is depleted. For

23.4 Economics of the Environment Protection

Table 23.5 Classification of goods

Jointness exclusivity→↓	Divisible	Indivisible
Exclusive	Pure private	Quasi-public
Non-exclusive	Quasi-private	Pure public

example, my consumption of a cake does not allow others to enjoy the same. If on the other hand, consumption of a good by one does not diminish (or deplete) the good, the good is called non-rival in consumption. For example, my viewing of natural scenery does not reduce its availability to others.

Jointness of supply is related to the cost of supply of a good. Once a good is made available, if an additional unit can be supplied at no additional cost (i.e. the marginal cost of supply is zero), the good is said to be non-rival or joint. This characteristic of the good makes it difficult for the market to provide it privately. As discussed earlier, in a competitive market, goods are priced at the marginal cost. If the marginal cost is zero, the good is supplied free of charge and there can be no private supplier who will be willing to provide the good free, thereby incurring losses.

Exclusivity on the other hand relates to the possibility of excluding potential consumers, often through pricing mechanism. The source of this problem is often technological in the sense that with the available technology it may be extremely costly to exclude potential consumers. But there are cases where it may be impossible to do so. For example, it is impossible to exclude someone from breathing air. Non-exclusion leads to legal access to multiple users and sets in motion a race for capturing the benefits before others. As a result, the resource is overexploited.

From a practical point of view, non-exclusivity and jointness of supply give rise to the problem of free-riding. A free-rider is one who benefits from a good without contributing to its provision or supply. Free-riding leads to a situation where each consumer will expect others to pay for the supply and in the process nobody or very few will pay. This will reduce the supply and creates supply shortages or problems.

Based on the above two characteristics of goods, four types of goods and services can be identified (see Table 23.5). Examples of various types of goods include:

a) Pure private goods: oil, gas, electricity, etc.
b) Pure public goods: air, national defence, police;
c) Quasi private: fish stocks, ground water, oil reservoirs,
d) Quasi public: private beach, etc.

The characteristics of a pure public good (i.e. non-rival consumption and non-exclusive) give rise to practical problems of ensuring supply and pricing through the market mechanism. As the marginal cost of supply is zero, there is no incentive for the private sector to engage in the supply activity. As a non-exclusive good, it is difficult to ensure the good is used by those paying for the good (i.e. it is difficult to check free-riding). This also deters private investors. In such cases, state

23.4.3 Private Versus Social Costs

The existence of externalities creates a wedge between the private cost and the cost to the society. For any decision-making, economic agents rely on the private costs (i.e. costs borne by them). But when an economic activity generates an externality, the society bears the cost of its effects. As a result, the social costs are more than the private costs. This is shown in Fig. 23.3. As a competitive market will consider the private costs of supplying a good, the equilibrium output will be Q0, whereas the output considering the social cost of supply will be Q*. Correspondingly, the price in a competitive market will be P0 whereas the price increases to P* when externality is considered. Thus markets tend to use scarce resources in an inefficient way.

The process of reaching the optimal or efficient solution in presence of an externality is called internalisation of externalities. This essentially requires correcting the price signal in the market so that the price is raised to P*. There are two basic economic approaches that could be followed to do this directly (see Fig. 23.4):

a) One is to impose a tax on per unit of emission and the other is to restrict the quantity of output to Q*. These options are shown in Fig. 23.4. The tax option is known as the Pigouvian tax, where a tax equal to the marginal external cost has to be imposed to correct the problem of negative externality. Similarly, a subsidy equal to the marginal external benefit is required to resolve the positive externality problem.

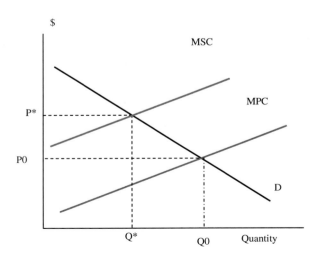

Fig. 23.3 Non-optimal allocation of resources in presence of an externality

Fig. 23.4 Internalisation of externalities

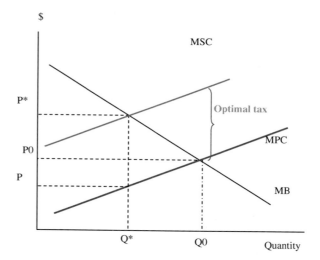

b) The quantity restriction can be imposed by imposing a cap on the output level. While both the tax and quantity options will yield the same results, they have different practical implications. In the case of tax, the tax revenue goes to the government while in the case of quantity restrictions, the revenues from higher prices will accrue to the producer.

23.5 Options to Address Energy-Related Environmental Problems

There are alternative ways of classifying the instruments available for environmental management (or pollution control) but the following categorisation based on Levinson and Shetty (1992) identifies three elements of the control system:

- Behaviour control mechanism: Two common alternatives rely on regulation or incentives,
- Level of control: whether the instrument directly or indirectly influences the emissions; and
- Control variables: Three important parameters here are price, quantity and technology. Quantity-based policies fix the level of environmental damage but allow prices to change. On the other hand, price-based policies fix the cost of controlling environmental damage but allow the pollution level to change in response to economic conditions. Technical solutions focus on technology-driven fixes to generate an acceptable level of damage.

Of the twelve options identified in Table 23.6, two are not feasible (non-incentive price-based options) and are left blank. The rest ten provide alternative ways of controlling the externality problem.

Table 23.6 Alternative policies to reduce pollution

		Price	Quantity	Technology
Incentive (economic)	Direct	Emission tax	Tradable emission permits	Technology tax on presumed emission
	Indirect	Fuel tax	Tradable production permits	Subsidise R&D, fuel efficiency
Non-incentive (regulation)	Direct	–	Emission standards	Technical standards
	Indirect	–	Product standards, bans, quotas	Efficiency standards

Source Levinson and Shetty (1992)

The level of control can be better understood looking at Eq. 23.1 below:

$$\text{Total emissions} = (\text{emission/input}) \times (\text{Input/Output}) \times (\text{Output}) \quad (23.1)$$

For example, assume 1 kWh of electricity requires 0.7 kg of coal of a given variety. If 1 kg of coal produces 2 kilograms of CO_2, and if the power plant produces 1 GWh of electricity, total CO_2 emission from the plant is

$$= 2 \times 0.7 \times 10^6 \, \text{kg of } CO_2 = 1,400 \, \text{tons of } CO_2$$

Any policy that influences emissions is a direct policy. Examples of direct policies include emission taxes, tradable permits and emission standards. As these policies directly target the pollution, pollution measurement, monitoring and compliance verification are essential.

Normally the transaction cost is low for large, point sources of pollution. Direct policies appear to be most justified in environmental problems that involve large, publicly noticeable companies such as power utilities, large industries, big mining companies, etc. Indirect policies have to be relied on where direct instruments are not feasible. They influence either any of the components of the right hand side of Eq. 23.1 or a combination of them: waste generation process (emission per unit of input), efficiency of the conversion process (input per output), or the demand for the pollution-intensive product (output).

Examples include tax on inputs, tax on related goods, product standards, etc. These policies are normally less efficient because they work only on a component of the problem. Only in exceptional cases, an indirect policy can be as efficient as a direct policy.

Below we take a look at a number of alternative mitigation options available for environmental problems of the energy sector.

23.5.1 Regulatory Approach to Environment Management

This approach uses the coercive power of the state to influence polluter behaviour and limit emissions below a prescribed quality target. The mechanism is simple:

1) Statutes and rules prescribe an acceptable behaviour (specifying the quantity of pollutants that can be emitted, obligations to reduce a given amount of pollutants, stipulations on technology or types of inputs to be used, restrictions or bans on specific product use or production, restrictions on firm sites, etc.).
2) Compliance is checked against the set behaviour.
3) Violations lead to penalties.

Because it relies on state power, this approach is also known as a command-and-control approach.

Normally a state agency is empowered to prescribe the acceptable behaviour, which often takes the form of a standard. A number of standards are used for this purpose, including ambient air quality standards, process standards, industrial emissions standards, fuel quality standards, etc. See Box 23.1 for a brief discussion on these standards.

Box 23.1: Brief Description of Various Standards

Ambient air quality standards: Ambient air is the air breathed by the general public. The ambient air quality standards specify the upper limit of concentrations of pollutants. These standards may be based on specific national studies or based on recommendations made by international organisations such as the World Health Organisation (WHO). The standards may be uniform for a country or may be differentiated by area. In some cases, the ambient standards may be differentiated by effects. For example, US EPA indicates two sets of standards: primary and secondary. Primary standards apply to everyone while the secondary standards are set to prevent unacceptable effects on public welfare (such as unacceptable damage to crops, ecosystem etc.). These ambient standards are set to protect the population and cost consideration does not normally play any role here.

Industrial emission standards: These standards specify the maximum limits of pollution from industrial activities and are specified for each type of activity and pollutant. They also normally vary by type of technology, vintage of the plant, and even by plant location. Normally older plants are allowed to pollute at a higher level, considering the difficulty in meeting stricter targets by them. These standards also vary across countries.

Process/performance standards: These specify the minimum performance requirements of processes or technologies that are used in controlling pollution. Such standards include electrostatic precipitator (ESP) performance, flue gas desulphurisation (FGD) performance, etc.

Fuel quality standards: The minimum quality of fuels that could be used in production or energy generation purposes may be specified.

Normally, the regulatory approach specifies: A set of emission rules that apply to all polluters of a particular pollutant. Same regulations often apply all over the country. These features can lead to inefficiencies, because regulations are not based on the marginal cost of pollution control of each polluter and force all polluters to control pollution to a particular level. This can be a costly option. Marginal cost of pollution control can vary between locations (say rural and urban areas). Consequently, forcing same level of control to all areas can lead to inefficient use of resources. This will then justify a rural–urban dichotomy in setting standards, which may however be difficult to monitor and implement.

This is the most commonly used approach for environmental problems. The regulatory approach has the following advantages: 1) there is long-standing experience in using regulatory instruments and therefore, from a governance point of view, this is a familiar instrument; 2) setting such a regulation is often guided by health concerns and does not require information on individual firms and their characteristics; 3) Once a target is set, it can be achieved if the regulation is effectively implemented. 4) The users also have confidence in this measure and the instrument has a high level of acceptance. 5) It may be the only option for dealing with toxic and hazardous pollutants. 6) It can be applied at short notice and can be effective in the short term. 7) As a non-revenue generating instrument, it is easy to garner political support. 8) It is often considered as fair, as all users face the same targets.

The weaknesses of the regulatory approach are: 1) effective enforcement is difficult; 2) bargaining and negotiations weaken regulations. Various pressure groups often influence the regulators to get waivers, concessions or relaxed standards. The regulation may thus be subjected to bargains and regulatory captures. 3) It is often a costly option, as the cost aspect is not considered in setting the standard. Moreover, when the abatement or damage costs vary, uniform standards impose burdens on polluters whose marginal abatement or damage costs differ from that implicitly used in setting the standard. 4) A standard only requires meeting the target and hence there is no incentive for out-performing the standards.

The domination of the regulatory approach over economic instruments can be explained by the following factors: It may be the only option for controlling emission of hazardous pollutants. It works better than the economic instruments when the abatement cost is low but the monitoring capabilities are deficient (i.e. there is no cheap option for ensuring measuring and monitoring emissions) (Cole and Grossman 1999); In non-market oriented but law abiding economies, where the state plays an important role and where the profit-motive is not decisive, regulations can be effective. Firms are comfortable with the approach because they can lobby for exemptions and concessions and consequently can obtain a cheaper solution. Environmental pressure groups were hostile to use of economic instruments for pollution control, which they considered as a licence to polluters to pollute. Concerns of missing the targets, consequent health damages, and local hot spots, were also raised. Moreover, EIs may be more difficult to tighten. Similarly, legislators also prefer this approach because of their familiarity with this system, hidden costs, reliability of the instrument and opportunities for symbolic politics in the sense that the standard may be very stringent but the enforcement may be lax. It also gives them opportunities to satisfy their constituencies by offering exemptions and concession.

23.5.2 Economic Instruments for Pollution Control

A number of alternative economic instruments are available for controlling environmental pollution. These include fiscal instruments (taxes), charge systems (fees), financial instruments (subsidies), property rights, market creation (trading), bonds and deposit-refund systems and liability rules. We discuss taxes, property rights and market creation.

23.5.2.1 Taxes and Charges

The theoretical underpinning of taxes was discussed earlier. A tax equal to the marginal environmental damage can be used to internalise externalities. Such a direct tax on per unit of pollution (known as Pigouvian tax) is an economically efficient mechanism of internalisation. A tax on CO_2 emission or SOx or NOx emission is a direct tax on pollution. A few countries have started such environmental taxes but often the tax rate does not have much relation with the external cost.

Imposition of a tax requires the generator of externality to compare the cost of pollution reduction with the tax payment obligation. This arbitrage leads to a decision about whether it is worthwhile to abate pollution and up to what level (or to pay tax). The polluter will choose an option depending on its cost-benefit.

So far, it was assumed that the pollutants mix uniformly once discharged and there is no distinction between the polluting sources. In this case, the tax rate remains same for all the polluters. However, in reality, all sources are not similar and the effects vary depending on the location and characteristics of the source. In such a situation, each source has to be considered separately. The tax rule remains the same (i.e. the unit tax shall be equal to the marginal social cost) but as the marginal social cost varies from source to source of pollution, tax rates will be different for different sources.[4]

As a Pigouvian tax can raise significant revenues for the government, the issue of budget balancing arises. To keep the budget balanced, the government must offset by an equal increase in its distributions. This is similar to marginal cost pricing subject to a budget constraint. The solution to the problem is known as Ramsey-Boiteux pricing principle. This suggests that pricing based on marginal

---

[4] See Chap. 4, Baumol and Oates (1988) for details.

costs has to be adjusted in accordance with the price elasticity of demand and supply. Marginal cost based pricing is not efficient in such a situation.

The issue of victims of externalities often appears in the literature: what should be done about them? There are suggestions that they should be compensated for the damages they suffer while others suggest that the victims should be taxed in some circumstances. According to Baumol and Oates (1988), 'so long as the number of victims is large, the efficient treatment of victims prohibits compensation...Moreover, taxation of victims is equally inappropriate...'.[5] Compensation will offer economic incentive to accept the damaging effects and the victims will not undertake sufficient defensive actions. Compensation will also encourage more people to enter the affected area or activity. Therefore, it is not appropriate to provide compensation.

There are a few issues related to Pigouvian taxes: (1) it requires direct measurement of pollution, which is difficult for small sources; (2) setting the tax at the appropriate level (equal to marginal damage cost) requires knowledge about the damages caused by pollution and their monetary evaluation. This is often very difficult. (3) Administration of the tax system may be difficult as well.

When direct taxes are not practical, taxes on outputs and inputs are also used but as they are indirect measures, they are less efficient compared to direct measures. Indirect taxes are more commonly used and for energy, oil products used in transport are often targeted (although not necessarily for environmental reasons).

Given that tax and standards are two common instruments, it is logical to ask how a tax instrument compares with standards. Figure 23.5 suggests the difference in economic terms. A tax allows the polluter to control pollution according to its marginal abatement costs, while a standard prescribes a quantity limit irrespective of the cost. The figure considers two cases:

(a) If the cost of control is uniform across the industry and is represented by MC1, then both tax and standard yield the same result.
(b) When the cost of control varies, then a tax will allow a low cost firm (MC2) to reduce its pollution to a greater extent (i.e. up to Q2) than the high cost firm (MC1). A tax thus provides the flexibility to the polluters to decide about their level of control, while the standard imposes the decision on the polluters.

23.5.2.2 Who Bears the Tax Burden?

The tax imposed on the generator of external effect may appear to affect the generators at the first instance but as the goods are sold in the market, consumers pay for the taxed good. This might suggest that consumers ultimately bear the burden of tax. In fact, both the generators and the consumers share the burden

---

[5] Chapter 3, Sect. 4 of Baumol and Oates (1988).

Fig. 23.5 Tax versus standards

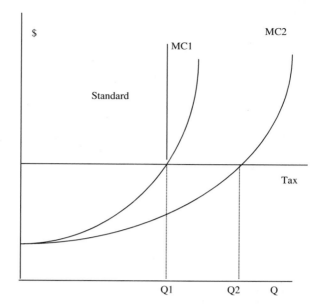

Fig. 23.6 Who pays for a pollution tax?

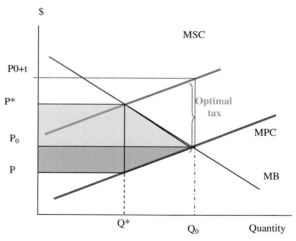

but the level or degree of burden sharing depends on the elasticity of demand.[6] Figure 23.6 explains.

Before the tax is imposed, producers were receiving P_0 price and selling Q_0 amount of goods. After imposition of a Pigouvian tax, the price increases to P^* and the quantity of goods sold reduces to Q^*. But the producer now receives a price P (which is equal to P^*-t) that is less than P_0. Similarly, the sales volume reduces to Q^* from Q). Thus the producer receives fewer surpluses under the tax.

---

[6] See Pearce et al. (1994) for more details.

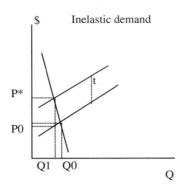 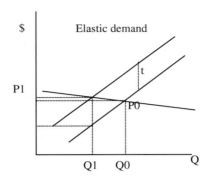

Fig. 23.7 Tax burden sharing

Similarly, the consumer pays a higher price and consumers fewer goods than she was doing under the competitive condition. The government now removes a part of the surplus through taxation. Thus both consumers and producer pay for the pollution tax.

The sharing of the burden depends on the elasticity of demand. When the demand is inelastic, changes in prices will not affect the demand for the good. This results in transferring the burden to the consumers. On the other hand, when the demand is elastic, consumers will switch to other products and the demand will be affected substantially. In this case, the producers will bear a larger share of the burden. This is shown in Fig. 23.7.

23.5.2.3 Property Rights Approach to the Externality Problem

One of the sources of externality is the lack of well-defined property rights. From the characteristics of the goods indicated earlier, we noted that exclusivity is an essential requirement for any transaction of a good and for the private sector to be interested in supplying it. This requirement is frequently violated in practice—in the case of public goods or quasi-private goods.

Types of Property Rights[7]

There are, in general four types of property rights systems:

- Privately owned.
- State-property regimes: where the state has the ownership rights of the property instead of any individuals. This is the case of sub-soil resources in many countries. The state is presumed to protect the collective interest of the society. However, as the state acts or exercises its ownership rights through

---

[7] For further details consult, Tietenberg (2001).

Fig. 23.8 Tragedy of the open-access resources

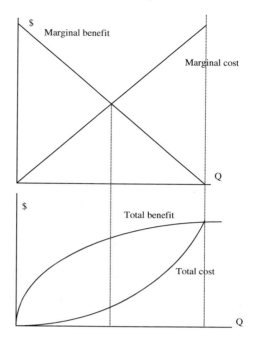

agents (bureaucrats or officials), the motives of the agents can be different from that of the collective interests. This leads to the principal-agent problem and can affect management of the state-owned properties for environmental purposes.
- Common-property resources: there are some resources which are not owned privately but collectively by a number of users. This is a form of institutional arrangement for managing resources and excluding others to use the benefits. Often formal or informal rules are devised to govern the resource use. Due to multiplicity of ownership, cheating and free-riding issues appear.
- Open-access resources: no group or individual has ownership rights of the property. This implies that none has legal power to restrict access. In such a case, each user has the incentive to look at his/her total costs and benefits and not at the marginal costs and benefits. Consequently, the resource will be utilised until the total cost equals the total benefits, as shown in Fig. 23.8. In a cost-benefit framework, the objective is to maximise the net gain, which occurs when marginal cost equals marginal benefit. Overexploitation of the resource results in inefficient allocation.

The property rights approach to the externality problem attempts to work through proper redefinition and enforcement of property rights where possible. In case of pure public goods, there is limited possibility of developing or attributing property rights. But in the case of quasi public or private goods, redefinition of property rights might be a solution. For example, if an open-access resource is converted to a private ownership by changing the legal status of ownership, the overexploitation problem can be resolved, as the private owner will maximise the net benefit and produce at an optimal level.

The Coase Theorem

The fundamental theorem in the area of externality and property rights was developed by Ronald Coase. The theorem states that "where there is costless bargaining between the generator and the victim of an externality, the optimal outcome will emerge so long as either party holds the pertinent property right—it does not matter which one".[8] Coase considered the issue of cattle grazing. Consider that Farm A raises cattle but the cattle trespasses into Farm B and damages the crop. Thus the cattle imposes an external cost on Farm B. Coase suggested that irrespective of which party was assigned the right (either to Farm A to allow cattle straying or to Farm B), the result will be the same. This is explained through a simple example below.[9]

Assume that a plant rejects effluents to a nearby river. The citizens living nearby use the water from the river and the effluents from the plant would inflict a damage of £500. It costs £100 for the plant to set up an effluent treatment plant whereas it will cost £300 for the citizens to arrange for water purification systems.

Consider that the plant has the right to pollute. The citizens will be willing to offer £300 as the upper limit of payment to the plant to purify the effluent. The plant is willing to accept a minimum of £100 (which is its cost of installation). The outcome is that the citizens will pay the plant £100 to set up an effluent plant.

Consider now the opposite case: the victims have the property right. The plant is willing to offer £100 (which is the cost of setting up the effluent treatment plan) as the maximum amount of compensation to the victims but the victims will accept nothing less than £300 (which is the cost of its purification system). The outcome is that the plant sets up the effluent plant at a cost of £100. There is no monetary transaction in this case.

In both the cases, the monetary cost is £100 but in the first case the victims subsidised the plant whereas in the latter case the plant spends £100 to set up the effluent plant. The two outcomes have different equity implications.

Although the theorem is interesting, it suffers from a number of practical limitations. These include:

- *Small number case*: Coase theorem applies to the small number case where the number of victims and externality sources are limited and where negotiation is possible. In most of the environmental externality concerns (SO_2, NO_X, particulate emissions, GHG emissions, etc.), large number of victims and externality sources are involved. Where a large number of participants are involved, the transaction cost of negotiation can be prohibitive.

---

[8] Extracted from Baumol and Oates (1988).

[9] This example is taken from Viscusi et al. (2000).

- *Negotiation difficulties*: Negotiation in such a large number case may be practically impossible, as each participant may have her own stand point and no solution may emerge. Similarly the polluters may have more bargaining power than the victims, which can affect the outcome.
- *Establishing property rights* is not easy in most cases of practical relevance.
- *Long-term efficiency*: While in the short-run the results of both the alternative property right assignments had equity implications, it can turn out to be a long-term efficiency concern as well. If an industry is subsidised by the citizens, more plants will enter the market to take the benefits, thereby providing wrong signals to the economy.
- As indicated in the example, alternative assignment of rights might have significant equity implications. This may involve significant wealth transfers, which may not politically or socially acceptable.

23.5.2.4 Tradable Permits

This is a form of market creation that has been used to a limited extent compared to taxes and charges. The basic idea here is simple:

(1) The total quantity of emissions should not exceed the tolerable limit;
(2) Within this limit, each polluter should be free to decide how to restrict her emission. Each polluter decides whether to reduce pollution and sell the permits, or pollute and buy permits or pollute and pay penalties.
(3) Based on cost and price signals, polluters can benefit from exchanging the credits of extra emission reduction with others.
(4) Non-compliance would lead to penalties.

Box 23.2: Basics of Emissions Trading

Consider the example below to understand the mechanics of emissions trading (Fig. 23.9). Assume we have two installations in the system and the overall cap is set at 800 tons of SO_2. Each unit is given 400 permits allowing them to pollute up to 400 tons of SO_2. But one unit finds the cost of pollution control too much and decides to emit 440 tons, which exceeds the limit by 40 tons. The other unit was able to reduce its pollution to 360 tons and had 40 extra units of permits. In a regulatory system, the compliant unit would not get any credit for reducing emissions below the required level while the non-compliant unit would have to pay a penalty. In the trading system, another option arises: the non-compliant unit can buy unused 40 permits from the other unit and overall they can meet the target. This buying and selling of permits create a market for credits, which can mitigate the problem at a cheaper cost to the society.

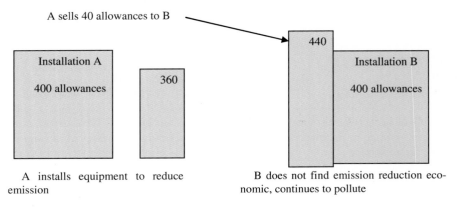

Fig. 23.9 Emissions trading example

Such a system allows trading of emission permits. The mechanism normally consists of the following (see Box 23.2 for a simple example):

(1) *Overall cap*: The total quantity of pollution load is decided based on a historical analysis or based on studies indicating the loading required to achieve a given level of environmental quality. This cap should be lower than the expected level of emissions that would arise in business as usual case.
(2) *Permit distribution*: The permissible emission quantity is then distributed among existing polluters either free of cost or through auction. Each polluter receives permits proportional to its historical emission level or on some other basis. Holding allowances permits a polluter to pollute up to that level.
(3) *Trading mechanism*: If the cost of pollution control is less, a polluter may reduce its emission and trade the balance amount of permits. A market is created for selling and buying permits and the interaction of supply and demand decides the market clearing price.
(4) *Monitoring and recording*: Each participating unit is monitored and proper record of transactions, emissions, penalties is required to be kept.

This approach (also known as 'Cap-and-Trade') has been employed in the USA for controlling emissions of acid rain producing gases (SOx and NOx). This is discussed in Chap. 24. More recently, the Kyoto Protocol has included the Emissions Trading as one of its flexible mechanisms. A grand experiment on emissions trading has started in the EU to reduce greenhouse gas emissions. This is presented in Chap. 25.

This approach is appealing because: by fixing the total quantity of pollution, it limits pollution and thus achieves the environmental target. Being a direct control method this is an efficient option. The mechanism provides flexibility to polluters and obtains the least cost solution. New entrants can be accommodated and accordingly, this mechanism ensures environmental protection even at higher

23.5 Options to Address Energy-Related Environmental Problems

levels of economic activities. There is no need for inflation correction either. As this approach has important regulatory features, the regulators feel comfortable with it.

Benkovic and Kruger (2001) suggest the following criteria to be considered before applying this approach to any particular circumstances:

- Is a flexibility mechanism suitable from damage or health point of view?
- Can emissions be monitored and measured consistently?
- Do abatement costs vary across sources?
- Is there a sufficient number of participants?
- Does a system of property rights exist?
- Are there credible regulatory agencies and regulatory capabilities?

Clearly, the approach would not be appropriate for all conditions.

23.5.2.5 Technological Options

Technical options related to pollution control can be grouped under different categories ranging from control technologies to fuel switching to renewable technologies. Demand-side management options to reduce demand are also considered. These options are discussed in detail in Chap. 24 in relation to pollution control from stationary sources.

23.5.3 Assessment and Selection of Instruments

Choosing an instrument for environmental damage control is not easy as stakeholders have differing views and conflicts may arise. It is then essential to follow a systematic approach (see Box 23.3 for an example) but it needs to be kept in mind that no single instrument would satisfy all criteria. Compromises would be required to find an acceptable solution.

Box 23.3: Selection Criteria Suggested by Panayotou (1994)

Panayotou (1994) suggests the following nine questions to assess and select appropriate instruments for a particular country considering its specific characteristics and conditions:

Environmental effectiveness: Will the instrument achieve the environmental objective within the specified time span and what degree of certainty can be expected? Normally for irreversible losses (of species or biodiversity) the acceptable margin of error will be lower compared to losses that can be reversed or remedied.

Cost effectiveness: Will the instrument achieve the environmental objective at the minimum possible cost to the society? While considering costs to the society, a holistic view has to be taken. Costs include compliance cost, administrative costs related to monitoring and enforcement, and any cost induced by distortions.

Flexibility: Is the instrument flexible enough to changes in technology, the resource scarcity and market conditions? For example, if an instrument is dependent on a particular technological choice, its effectiveness will be reduced when new technologies arrive and change monitoring, cost of compliance, other factors. The instrument should reflect the level of scarcity and at higher scarcity level, the value should rise. Similarly, with changes in the inflation level, the instrument should adjust in order to remain effective.

Dynamic efficiency: Does the instrument provide incentive to technological innovation? Does it promote environmentally sound infrastructure and economic structure? A dynamically efficient instrument encourages flow of resources to the areas where the country has comparative advantages, promotes technological innovation and environmentally sound infrastructure.

Equity: Will the costs and benefits of the instruments be equitably distributed? Normally the poor will have lower willingness to pay for environmental benefits but being vulnerable, they may gain more from the environmental protection. The distributional impact of an instrument is an important consideration.

Ease of introduction: Is the instrument consistent with the country's legal framework? Does it require new legislation? If so, is it feasible? Does the regulatory body have the requisite administrative capacity to administer new instruments? Normally instruments that do not require new legislation and new organisations are preferred.

Ease of monitoring and enforcement: How difficult or costly will monitoring and enforcement be? A country with limited monitoring and enforcement capability will choose instruments requiring limited monitoring and enforcement efforts.

Predictability: Does the instrument combine predictability and flexibility? An instrument becomes effective if it remains in force in the long-run and thus imposes predictable costs on the users. This provides better signals for investment decisions and brings desired changes. Uncertainty and unpredictable instruments reduce effectiveness.

Acceptability: Is the instrument understandable by the public, acceptable to the industry and politically saleable? Finally, if the users are not able to understand an instrument, its effectiveness is expected to be lower. Similarly, acceptability to industries and politicians also affect the instrument selection.

Fig. 23.10 Externality and monopoly markets. *Source* Baumol and Oates (1988)

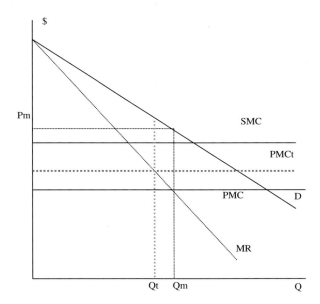

23.6 Effects of Market Imperfection[10]

So far we have considered the externality problem in a competitive market environment. In this section, we will consider how externality affects a monopoly market structure and how to correct it.

A monopolist decides the price based by equating marginal revenue to marginal cost (the principle is same in the competitive market). But as the monopoly faces a negatively sloped demand curve, market price exceeds the marginal revenue and the marginal cost at the optimal output level. As Fig. 23.10 shows, the monopolist restricts output to Qm and charges higher prices (Pm) compared to the competitive markets (PMC). Thus a monopolist is less harmful in terms of environmental pollution as the output is restricted (and conserves more resources).

In the case of a polluting monopolist, the private marginal cost (PMC) is less than the social cost (SMC). But imposing a tax on the monopolist may be too much of a good thing as the tax will further reduce the output, which reduces the consumer surplus further. In Fig. 23.10, PMCt indicates the private marginal cost plus a tax t and Qt is the output after imposition of tax. Note that Qt is less than Qm.

The tax on pollution encourages the monopolist to change his production process in such a way that lower emission will result. This reduces the social marginal cost. On the other hand the tax raises the private marginal cost to PMCt. At the optimal level, the new SMC and PMCt should be equal where all external

---

[10] This is based on Baumol and Oates (1988).

Fig. 23.11 Internalisation of monopoly externality. *Source* Baumol and Oates (1988)

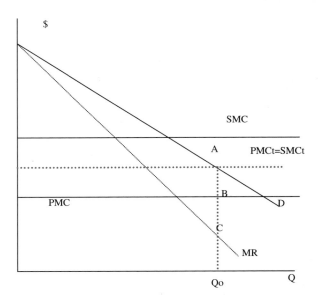

effects are internalised. This is shown in Fig. 23.11. The output at the least social cost should be equal to Qo.

In order to correct the problem of externality of a monopolist and achieve a competitive market-like solution, two instruments are required:

- A tax on pollution and
- A subsidy per unit of output (equal to AC) to encourage the monopolist to produce at a level similar to the competitive market conditions.

However there are practical difficulties in applying such a solution. Providing subsidies to a monopolist will be difficult for any regulatory agency. The alternative to the two-instrument policy is to adjust the tax in such a way that the welfare loss associated with monopolistic restrictions can be minimised. Such a solution requires a tax which is less than that prevailing in the competitive market. Such a fee is given by

$$t^* = t_c - \left| (P - MC) \frac{dy}{ds} \right| \tag{23.2}$$

The second term on the right hand side of the above equation depends on the price elasticity of demand and thus depending on the elasticity of demand, the fee should vary by source.

The determination of such a policy requires an enormous amount of information. It is also unlikely that such a policy of two instruments will be favoured by politicians and legal systems. The solution for other types of market imperfections will be more complicated.

23.7 Valuation of Externalities

Two basic approaches are used for the valuation of external costs in the energy sector: the abatement cost approach and the damage cost approach.

The abatement cost approach uses the cost of mitigating damage or meeting regulations. It considers that the regulations are set based on the willingness to pay of the public. It assumes that the decision makers make optimal decisions and that the abatement strategy is based on least cost control. As the policy and regulations change, the estimates have to be revised. The damage cost approach on the other hand aims at measuring the net economic damage arising from negative externalities.

Two categories of methods are commonly used: top-down and bottom-up. The top-down method focuses at the national or regional level and use aggregate studies with aggregated data. Typically, such a study uses total quantities of pollutants at the national level and considers the share of damage caused by an activity to arrive at an estimate of damage per unit of pollutant from an activity. The valuation is based on existing estimates and approximations and it cannot take site specificity into account.

The bottom-up approach on the other hand uses a disaggregated method of identifying the impacts caused by an activity and estimating appropriate monetary values of such impacts. Impact-pathways method is such a bottom up approach, which typically involves five steps (see Externe project website, and Sundqvist 2004):

- Identification and quantification of burdens considering fuel cycle, technology, and location of activities.
- Analysis of dispersion and changes in pollution levels. This uses modelling techniques for evaluating pollution transport under different atmospheric conditions and the consequent changes in ambient pollution levels.
- Use of dose-response functions to estimate the physical impacts of changed concentrations. This step uses relationships that relate changes in concentrations of pollutions to effects such as changes in mortality rates, changes in other health conditions, effects of buildings and structures, etc.
- Translation of physical impacts into economic damages and benefits through economic valuation functions.
- Finally to distinguish between how much of the identified damages and benefits have been internalised already and how much remains external.

Figure 23.12 presents the steps involved in the impact pathways method.

Clearly, this is an involved task and consequently, no valuation can perhaps capture the entire range of damages. Moreover, the valuation of the damages when there is no market for the good can be challenging and debatable. A lot of progress has been made in this area (and it is beyond the scope of this book to cover this aspect) but to say the least, the debate in this area remains unsettled.

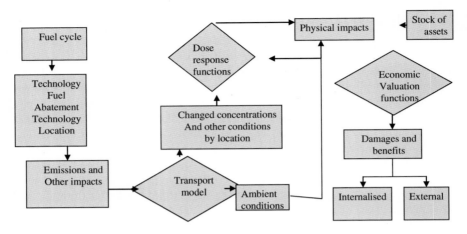

Fig. 23.12 Steps in impact pathways method

Consequently, a number of concerns arise in relation to external cost estimations. These are:

(a) *Coverage*—Ideally, all possible types of environmental impacts should be evaluated for a correct and comprehensive analysis of damage cost. However, this is impossible in practice due to limitations in data, knowledge, resources and even technical capabilities. So compromises have to be made and the general practice is to consider only major types of impacts, though the process of selection of priorities is an iterative process. Similarly, the entire fuel cycle should be considered but often only a part of the cycle is considered in more detail while the rest is either ignored or analysed to a lesser extent.

(b) *Areas of uncertainty*—A large number of uncertainties and information constraints affect any evaluation process. As the study is specific to a particular location, the characterisation of the ambient environment plays an important role. However, this requires information on local ecological conditions, polluting activities, and an inventory of buildings and materials as well as information on population. For any practical modelling purposes, it is not possible to capture all the relevant details and omissions and commissions are normal. The polluting processes and activities will have different technical characteristics. Again, only limited varieties could be included in any model and this requires aggregation/generalisation, which introduces some loss of detail. Finally, analysis of dispersion of pollution requires atmospheric modelling. Results vary depending on the type of model used, the atmospheric conditions considered in the analysis and other assumptions. All these factors critically affect the results and introduce uncertainties.

(c) *Impact estimation*: One major constraint in such a study remains the appropriate dose–response functions. In many cases, relevant functions for the local conditions are not available and adaptations from other countries are

commonly used. Even for the Externe project, the dose functions based on the US studies were employed. The problem becomes more dramatic in the case of developing countries where epidemiological studies are less commonly carried out. The question about the credibility of the study arises because the damage functions, being a function of local mode of life, market structure, individual preferences, etc., are specific for a location and their transportability to conditions far removed from the original study remains questionable. An additional problem relates to the valuation of ecosystems and other non-use and option values.
(d) *Valuation issues*—Transportability of results is a main concern as the studies carried out on specific locations. As indicated earlier, putting monetary values on goods and services is not always easy. Similarly, variations in value of life depending on method used and across countries make comparisons difficult. Finally, social values change over time, making transportability of results through time difficult.

Therefore, there is no surprise that Sundqvist (2004) found that the external cost valuation varied widely for electricity generation depending on the method used, coverage, fuel choice and country considered. He concluded that "the disparity arises due to methodological reasons and due to problems in the application of these methods". This makes internalization of externalities more difficult.

23.8 Government Failure

So far we noticed that when markets fail due to the presence of externality or market imperfections, government intervention is justified. Government intervention can take various forms:

- Corrective taxation or work through other economic instruments to impose quantity restrictions;
- Use regulatory approach to deal with the problems (say use of standards or regulations), and
- Manage common property resources;
- Assignment of property rights and ensuring enforcement of rights.

Yet, government intervention is no guarantee for successful solution of the problems. Governments also fail to deliver due to different reasons:

- Rent-seeking behaviour of interest groups: various interest groups try to influence the government, legislators and the regulators to extract the best deal for them. In return they provide political support to the government. This process leads to the problem of protection from competition, price ceilings, etc. which protect interests of groups but may not be appropriate from social perspectives.
- Information problem: Most of the issues require collection and processing of large quantities of information. Availability of information may be a problem.

Moreover, the capacity to deal with such quantities of information is also limited. This results in poor decision-making.
- Agency problem: as discussed earlier, government actions often lead to conflicts between social interests and the interests of those executing/implementing the decisions.
- Lack of appreciation of complex problems: Governments often appreciate the complexity of the problems and try to fix either on a piece-meal basis or inadequately.

Therefore, relying too much on governments can be a problem as well.

23.9 Conclusion

This chapter has provided a review of the environmental issues related to energy use and the options available to remedy them from an economic perspective. Using simple economic principles, the chapter has explained the economic thinking on this important issue and considered alternative options to deal with the problem. Clearly, the economic principles provide certain solutions but whether they are applied or not depend on many other factors. In practice, a combination of approaches is commonly used to make sure that the overall objective of a better environment is obtained.

References

Baumol WJ, Oates WE (1988) The theory of environmental policy. Cambridge University Press, Cambridge

Benkovic S, Kruger J (2001) To trade or not to trade? Criteria for applying cap and trade. The Scientific World, 1. See (http://www.epa.gov/airmarkets/articles/tradingcriteria.pdf)

Bhattacharyya SC (1995) The power generation and environment: a review of the Indian case. Int J Energy Res 19(3):185–198

Cole D, Grossman PZ (1999) When in command and control efficient? Institutions, technology and the comparative efficiency of alternative regulatory regimes for environmental protection. Wis L Rev 1999:887–938

Grossman GM, Krueger AB (1995) Economic growth and the environment. Q J Econ 110(2):353–377

IPCC (2006) Guidelines for National Greenhouse gas inventories, IPCC National Greenhouse Gas Inventories Program, Intergovernmental Panel of Climate Change. Institute for Global Environment Strategies, Japan

IPCC (2007) Climate change 2007: synthesis report, intergovernmental panel on climate change, Geneva

Levinson A, Shetty S (1992) Efficient environmental regulation: case studies of urban air pollution, Los Angeles, Mexico City, Cubatao and Ankara, policy research working papers, WPS -942, World Bank, Washington, DC

Menz FC, Seip HM (2004) Acid rain in Europe and the United States: an update. Environ Sci Policy 7(4):253–265

References

Panayotou T (1994) Economic instruments for environmental management and sustainable development, UNEP paper 16, UNEP, Nairobi

Pearce D, Turner K, Bateman I (1994) Environmental economics: an elementary introduction. John Hopkins University Press, Baltimore

Schwela D, Gopalan H (2002) Ambient air pollution and emerging issues in megacities. In: Haq G, Han W, Kim C (eds) Urban air pollution management and practice in major and mega cities of asia. Korea Environment Institute, Korea

Stern DI (2003) The environmental Kuznets curve, Encyclopedia of Ecological Economics, http://www.ecoeco.org/pdf/stern.pdf

Sundqvist T (2004) What causes the disparity of electricity externality estimates? Energy Policy 32(15):1753–1766

Sundqvist T, Soderholm P (2002) Valuing the environmental impacts of electricity generation: a critical survey. J Energy Lit 8(2):3–41

Tietenberg T (2001) Environmental Economics and Policy. Addison Wesley, Boston

UN (2009) World urbanisation prospects: the 2009 revision, Department of Economic and Social Affairs, United Nations, New York

UNDP (2004) World energy assessment: overview 2004 update. United National Development Programme, New York

UNEP (2002) Global environment outlook 3: past, present and future outlook. United Nations Environment Programme, Nairobi

Viscusi WK, Vernon JH, Harrington JE Jr (2000) Economics of regulation and antitrust. MIT Press, London

Vukina T (1992) Energy and the environment—some key issues. EDI working papers, World Bank. Washington, DC

Webber D, Allen D (2010) Environmental Kuznets curves: mess or meaning? Int J Sustain Dev World Ecol 17(3):198–207

WEC (2000) World energy assessment 2000, Chapter 3: energy, the environment and health. World Energy Council, London

WEO (2006) World energy outlook 2006. International Energy Agency, Paris

World Bank (1992) World development report, World Bank

Chapter 24
Pollution Control from Stationary Sources

24.1 Introduction

Stationary sources of pollution are those which are non-moving, fixed sources. They are of two types: point sources and area sources. Point sources are large sources which emit significant levels of pollutants. Area sources on the other hand are small sources of pollution which are distributed over a large area and individually, each source is not a significant emitter but combined together, they can be a significant source of pollution.

Energy plays an important role in both point and area source pollutions. For example, power stations form a major point source of pollution. Around 65% of world's electricity is generated from fossil-fuels and coal plays an important role in this regard. Around 60% of world coal consumption is used in power generation. Industrial use of energy is the other main point source of pollution. Similarly, wood fuel used for cooking and heating in houses is a major source of area pollution as well as indoor pollution.

Pollution from stationary sources can take various forms: air pollution, water pollution, land degradation through solid wastes and noise. Both air and water pollution from stationary sources can have local as well as regional and international dimensions. As Table 24.1 indicates, air pollution can cause health and welfare effects. Trans-boundary transportation of air pollutants has caused acid rains and other problems in many parts of the world (including Asian Brown Cloud). Energy producing facilities, being major water users, are also responsible for significant water pollution in the form of thermal shock, discharge of uncontrolled contaminants, and even discharge of hazardous pollutants (such as heavy metals and trace metals).

24.2 Direct Pollution Control Strategies

Following the distinction presented in Chap. 23, both direct and indirect policies could be used to control pollution from stationary sources. Three commonly used

Table 24.1 Health and welfare effects of air pollution

Pollutant	CO	SO$_2$	NO$_2$	Ozone	Lead	Particulate matter (PM)
Description	Colorless, odorless gas	Colorless, forms acid with water, and interacts with others in the air	Reddish brown, highly reactive gas	Gaseous pollutant when it is formed in the troposphere	Metallic element	Very small particles of soot, dust, or other matter
Health effects	Headaches, reduced mental alertness, heart attack, cardiovascular diseases, impaired fetal development, death	Eye irritation, wheezing, chest tightness, shortness of breath, lung damage	Susceptibility to respiratory infections, irritation of the lung and respiratory symptoms (e.g., cough, chest pain, difficulty breathing)	Eye and throat irritation, coughing, respiratory tract problems, asthma, lung damage	Anemia, high blood pressure, brain and kidney damage, neurological disorders, cancer, lowered IQ	Eye irritation, asthma, bronchitis, lung damage, cancer, heavy metal poisoning, cardiovascular effects
Welfare effects	Contribute to the formation of smog	Contribute to the formation of acid rain, visibility impairment, plant and water damage, aesthetic damage	Contribute to the formation of smog, acid rain, water quality deterioration, global warming, and visibility impairment	Plant and ecosystem damage	Affects animals and plants, affects aquatic ecosystems	Visibility impairment, atmospheric deposition, aesthetic damage

Source EPA website

24.2 Direct Pollution Control Strategies

direct policies in this regard are: pollution standards, emission taxes and emission trading.

24.2.1 Pollution Standards

As discussed in Chap. 23, pollution standards can be of different types. Any point source has to comply with the requirements of the ambient quality standards (for example see US National Ambient Air Quality Standards or EU directive 1999/30/EC). Thus the pollutant concentration in the ambient atmosphere as a result of pollution form a source should not exceed the limits specified in the standards. As the pollution in the ambient atmosphere results from the aggregated effect of all polluting sources, it is not easy to identify the responsibility of a particular source without conducting detailed modelling exercises.

One such problem arose in India when the yellowing of the Taj Mahal in Agra was popularly attributed to the pollution from a refinery in a nearby area. However, upon thorough investigation it was found that the SOx emission from the refinery was minimal but the emission from local foundries and other small-scale industrial activities was responsible for the damage because of their reliance on coal, inefficient technologies and intensive usage pattern.

Industries also face specific pollution standards as well. For example, the European directive on air pollution from large combustion plants (2001/80/EC) stipulates the limits for air pollution from plants having a capacity in excess of 50 MW irrespective of fuel used.

- A review of the directive suggests that stricter standards apply to plants licensed on or after 1st July 1987, whereas older plants were either given exemption from the compliance requirements subject to satisfaction of certain conditions or allowed additional time to reduce emissions following a national emission reduction plan.
- Similarly, specific derogations were given to certain plants or member countries. For example, plants of 400 MW or higher sizes which operate less than 2,000 h per year (on a 5-year rolling average) up to the end of 2015 are allowed to emit 800 mg/Nm^3 of SO_2, which is almost twice the standard for new plants.
- New plants face emission standards based on the best demonstrated technology. Thus differential standards to take care of country specificities, vintage consideration and technical differences are quite common even in developed countries. They often result from the bargaining process through which these standards are developed.

Process standards are design standards or equipment standards that the process of equipment has to satisfy. These can apply to important pollution control devices such as flue gas de-sulphurisation (FGD), electrostatic precipitators (ESP), waste water treatment plants, etc.

Finally, fuel standards are also used to control the quality of fuels to be used in a process or system. Although this is more relevant for vehicles, fuel quality control in stationary sources can also generate environmental benefits. The fuel standard is often linked with the appliances/processes used. Thus gas quality that could be used in a turbine has to meet certain manufacturer's specifications (which may or may not be based on any specific standard).

Standards often specify or require use of a particular technology for emission control. This reduces the flexibility of the polluter to control the pollution.

24.2.2 Emission Taxes and Charges

Taxes on emission constitute a direct policy instrument to address pollution control from stationary sources. These taxes are imposed on the emission of pollution directly and cover air pollution, water pollution or noise. As a pollution control measure these taxes do not specifically target energy industries but major stationary energy supply sources normally come under their purview. Figure 24.1 shows the level of emission taxes on some air pollutants in Europe. As can be seen, there is significant variation in terms of emission charges across countries and type of pollutants.

Yet, the overall influence of these taxes does not appear to be high. The overall revenue generated through these taxes remains miniscule compared to other taxes and charges for environmental purposes.

24.2.3 Emissions Trading

Another direct pollution control policy is the emissions trading. A number of emissions trading programmes have been used in the past to deal with stationary

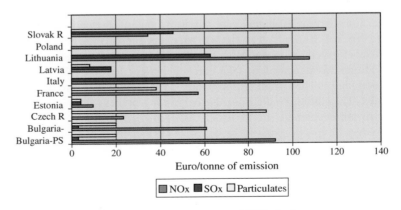

Fig. 24.1 Emission charges in some countries. *Source* OECD database

pollution control, with a high emphasis on energy-related pollution. The lead trading programme was one of the early precursors to the emissions trading. But the most important experiment was that of the US acid rain programme.[1] These are discussed below.

24.2.3.1 US Lead Phasing Out Programme

Lead was used as an additive in gasoline to improve engine performance since 1930s but lead is a toxic element that damages brain development of children and causes other health problems to adults (such as hypertension, heart attacks and premature deaths, etc.). Japan was the first country to control lead in gasoline in the 1970s responding to concerns of high level of lead in the air. It followed a rapid lead phase-out programme to eliminate lead from gasoline and since 1986 Japan has not produced any leaded gasoline (Lovei 1998).

- The US EPA started its lead phase out programme in 1973 intending to reduce lead content in gasoline to 0.6 g per gallon by 1978. Subsequently, the lead content was reduced further to 0.1 g per gallon before complete phase out.
- In 1982, the Environment Protection Agency introduced the lead trading programme which allowed refiners to trade lead credits. Refiners were given tradable credits to add specified quantities of lead to gasoline based on their production level and the lead standard at the time. Those with higher cost of mitigation of lead content could buy the credit from other refiners with low cost of control (thus complying with the regulation). Initially banking of credits were not allowed, meaning that if lead credits were neither used nor sold in the quarter in which they were created, they would expire.
- The programme ran for five years (1982–1986) and in 1985, EPA allowed banking of credits, which ran for two years (1985–1987). The industry reached the goal of 0.1 g per gallon of lead norm by the end of the banking programme.

This was a limited scale trading programme used to control lead emission in the oil refining industry.

24.2.3.2 The Acid Rain Programme[2]

This programme is a pioneering experiment of the application of the tradable permit mechanism in controlling environmental pollution. Prior to this programme, air pollution from power plants were controlled depending on their vintage and technology used. This followed the command-and-control

---

[1] The Emissions Trading Programme of the EU is considered in Chap. 26.
[2] See Burtraw and Palmer (2003), Burtraw and Szembelan (2009), EPA (2010) Environmental Defense (2000) and UPEPA website for further details.

instruments. The Clean Air Act Amendments of 1990 allowed the use of tradable permits to control SOx emissions from power plants. It involved capping the SO_2 emission at 8.95 million tons, representing a significant reduction from the 1980 levels (by about 10 Mt). Each power plant was allocated allowances free of charge and they were allowed to transfer, sell or bank their permits. Each permit allowed a plant to emit one ton of SO_2. Those who had low abatement costs, found it cheaper to reduce emissions and sell the permits. Others having higher control costs procured additional permits from those willing to sell.

The first phase of the programme started in 1995. All dirty, large power plants participated in the first phase of the programme. Other industries were not required to participate but they were allowed to join the programme voluntarily. 263 mandatory units and 183 voluntary units joined the programme in 1995. In the second phase, the programme was expanded to all plants above 25 MW. The number of participants increased to more than 2,000 by 2000. By 2009, more than 3,500 installations participated in the programme and the compliance rate was 100%.

Figure 24.2 compares the actual level of SO_2 emissions with the allowed level. It can be seen that the programme performed better than the target. The emission level in 2009 was much below the target for 2010 set at 8.95 million tons. This was mainly due to low demand for electricity in that year due to economic slow down but EPA suggests that lower demand contributed about 30% of the total Sox reduction while the rest came from fuel switching and sulphur control technologies. The trend of declining sulphur emissions clearly shows the effectiveness of the programme.

Since 2005, the Environment Protection Agency has issued the Clean Air Interstate rule to permanently cap SOx and NOx emissions in the eastern states of the US. This is clearly reducing the emissions in the country at a faster rate as can be seen from Fig. 24.2.

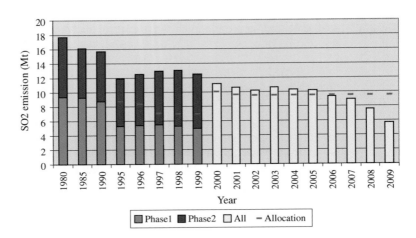

Fig. 24.2 Performance of the acid rain programme. *Source* EPA (2010)

24.2 Direct Pollution Control Strategies

With the introduction of this programme, the power plants had the options to: use low sulphur fuel; remove sulphur chemically from the exhaust; use efficient power plant technologies; buy allowances; and pay penalties. This programme helped generators achieve their targets at a low cost, as is evident from the permit prices. It was expected to vary between $300 and $1,000. In reality the prices were low: between $69 and $350. This reduced the compliance cost substantially and increased the rate of compliance.

The programme has been successful because:

- There were enough participants in the system; more than 250 compulsory participants initially and the number has increased to more than 3,000.
- The regulatory system required to monitor and implement the programme was available.
- The cost of controlling the pollution was not uniform in the industry. There were high cost and low cost polluters, which made trade possible and feasible.
- The regulatory aspect of the programme was familiar to the participants. This made acceptance of the programme easier.

A study by Chestnut and Mills (2005) found that the acid rain programme not only reduced Sox and NOx emissions but due to fuel switching, it contributed to the reduction of mercury emissions as well as reductions in fine particulate matters and ozone. Consequently, the air quality improved and generated health benefits and improvements in visibility. Chestnut and Mills (2005) estimated that the benefits for 2010 at $122 billion while the costs were estimated at £3 billion, yielding a 40:1 benefit to cost ratio. Most of the benefits ($119 billion) result from the reduction in the mortality rate attributable to the improvements in fine particle emissions. The high benefit to cost ratio clearly demonstrates the effectiveness of the acid rain programme.

24.2.3.3 US NOx Trading Programme[3]

The NOx reduction programme under the Clean Air Act was not a cap-and-trade programme but used a flexible unit-specific emission rate and in some cases an averaging plan. But due to difficulties in achieving air quality standards, a progressive shift towards the market-based trading system took place. These developments were region-specific: the south coast and the north-east were major non-attainment zones and the initiatives involved these areas.

The first large-scale urban NOx trading programme was introduced in 1994, called, RECLAIM (Regional Clean Air Initiatives Market). For the northeast region, the Ozone Transport Commission (OTC) was created to control ground level ozone formation (or smog). The OTC established a multi-state NOx emissions trading programme (called the NOx Budget Program) which sets the budget

---

[3] This section is based on OTC (2003a, b), Burtraw and Szambelan (2009) and EPA (2010).

or cap for NOx emission from power plants and other combustion sources during the ozone season (May through September). The first phase of the programme ran between 1999 and 2002 and the participating states agreed to collectively reduce NOx emission within the region much below the level required under the regulation by cutting down the emission compared to 1990 baselines. Each member state allocated the emission allowances to the participating sources (called budget sources) in accordance with the state share of the regional budget. Each allowance allowed the source to emit 1 ton of NOx during the ozone season. The source had to demonstrate that its emission during the ozone season would not exceed the allowances held by it and that it would comply with the regulatory requirements. Any saving in allowances could be traded or banked for future use.

During the first phase of the NOx Budget programme, the total NOx emission of the region was capped at 219,000 tons during the ozone season, which represented more than 50% reduction compared to the 1990 baseline of 490,000 tons. This cap remained valid until 2003, when it was reduced to 143,000 tons. The programme applied to more than 1,000 large combustion facilities, including over 900 electricity generating stations.

Although the programme was not concerned about NOx emission reductions on a short-term basis, the results showed that daily averages and peaks have declined during the ozone season between 1999 and 2002. Moreover, although there was a concern about shifting the polluting activity in other areas not covered by the programme, in reality this did not happen. In fact, an increase in nuclear generation appears to have offset the reduction in fossil fuel generation during the ozone season.

Since 2003, the NOx trading programme has been replaced by the NOx State Implementation Plan call (NOx SIP Call for short). This followed the OTC Budget programme but it gave flexibility to the states to decide the sources to be controlled to achieve the overall state NOx budget. A total of 2,570 sources were affected by this programme in 2004 but 87% of them are electricity generating units (EPA 2005). The Clean Air Interstate Rule (CAIR) introduced in 2005 and took effect for NOx in 2009 has now expanded the trading programme to 25 eastern states. However, some uncertainty has arisen due to court decisions asking the EPA to adjust the rule to take care of the court's concerns.

The trading programmes are considered as a successful implementation example of the cap-and-trade instrument. Figure 24.3 suggests that the states kept NOx emissions within their budgets collectively. Thus the programme was successful in reducing NOx emission by a factor of 3 between 1990 and 2009.

24.3 Indirect Policies

24.3.1 Pollution Control Technologies

Pollution control technologies normally come under two types:

24.3 Indirect Policies

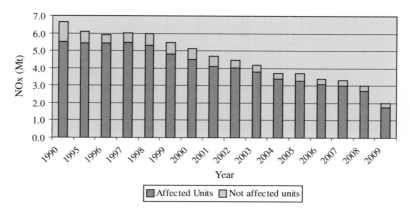

Fig. 24.3 NOx control performance. *Source* EPA (2010)

- *End-of-pipe solutions*: these treat exhausts from chimneys (or tail ends) from power stations or vehicles to limit the emission to the acceptable level.
- *Technical fixes* that reduce pollution emissions—these either clean the fuel or burn them differently to reduce pollution.

Common end-of-pipe technologies used in the power sector include:

(a) Mechanical and electrical devices for removing particulates such as filters and electrostatic precipitators, which were introduced in the industrialised countries over the past 50 years and can remove more than 99% of particulate matters.
(b) Flue gas desulphurisation technologies for removing sulphurous emissions from the waste gas are now available which can remove more than 90% of the sulphur emissions.
(c) NOx reduction technologies are also available that reduce emissions of nitrogen oxides by using catalysts and lowering combustion temperatures and avoiding excess air.

Among the technical fixes used in the power sector, two commonly used options are:

(a) *Coal-cleaning technologies*: coal-cleaning methods remove impurities from coal and reduce non-burning mineral matters that produce ash. They also typically remove some sulphur from coal.
(b) *Fluidised bed combustion*: fluidised bed boilers allow burning of coal or other solid forms of energy in a strong rising current of air. The combustion takes place in presence of other chemicals such as limestone, which absorbs sulphur emission, thereby eliminating the need for flue gas desulphurisation. A large part of the ash is collected in the bed as bottom ash as opposed to fly ash in a pulverised fuel boiler.

Most of the technical options are implemented through standards, which specify a particular level of emission to be achieved.

24.3.2 Options Related to Fuels and Conversion Processes

These indirect policies options related to conversion systems offer two possibilities: fuel switching and improved conversion processes. Switching to low carbon fuels or better quality fuels emerges as an option when alternative fuels are economically available.

24.3.2.1 Switching to Natural Gas

Switching from coal or oil to natural gas was a common example in the 1990s, often spurred by the deregulation of the gas and electricity markets complimented by the availability of better, efficient conversion technologies. Moreover, gas offered some advantages: low gestation period gas-based electricity generation, low capital intensiveness due to modular specification and higher proven reserves of gas meant easy access to gas in many cases. Natural gas being less carbon intensive, CO_2 emission reduces. At the same time, less particulates and SOx are produced due to lesser impurities and solid residues in gas. The environmental benefits have helped gas to consolidate its position as well.

The main constraint for gas market development remains the infrastructure necessary for establishing physical links between the supplier and the consumer, which is a capital-intensive activity. The growth has also been impaired by: non-availability of gas markets close to the source of production, geo-political factors affecting development of gas infrastructure, and risks involved in the business due to lack of appropriate regulatory regimes as well as lack of adequate and financially sound customers. Linking gas prices to oil prices also affected gas use in the power sector in recent times.

24.3.2.2 Renewable Energies

Renewable energies offer another option for fuel switching. This has been considered in Chap. 11. Renewable energies have emerged as a preferred solution to reduce environmental damages and to enhance diversity of supply. But there are still considerable barriers to be overcome, including: access to grids, regulatory and other restrictions on connectivity for decentralised systems, appropriate tariff and accounting mechanisms for sale and purchase of power from such sources, high costs of some energies such as solar, intermittent nature of supply, and long-term commitment for subsidies.

24.3.2.3 Process Improvements

As indicated earlier, losses in the conversion processes constitute an area of focus to reduce environmental pollution. Technical changes have always attempted to improve efficiency of conversion. Advanced technologies for electricity generation include combined cycle gas turbine, integrated coal gasification combined cycle, and various technologies using fluidised bed boilers. Compared to the conventional technologies, the advanced technologies use energy resources more efficiently and achieve better efficiencies at a relatively low incremental cost. Table 24.2 provides some examples.

24.3.2.4 Management of Demand

Another indirect policy to reduce environmental damage lies in reducing the demand for the pollution generating output. In the case of energy-related pollution

Table 24.2 Technological solutions for pollution control in power generation

Fuel and plant type	Emission control	Percentage reduction in relation to base case			Th. effy	Added costs
		PM	SO_2	NO_x		
Base						
Coal, conventional	None	0	0	0	34	
Improvements and control						
Conventional boiler	Mechanical cleaning	90			34	<1
	Fabric filters	>99			34	2–4%
	ESP	>99			34	2–4%
	ESP/coal cleaning	>99	10–30		34	4–6%
	ESP/SO_2 controls	>99	90		34	12–15%
	ESP/SO_2 & NO_x controls	>99	90	90	33.1	17–20%
FBC	ESP	>99	90	56	33.8	<0–2%
PFBC/combined cycle	ESP	>99	93	50	38.9	<0–2%
IGCC	None	>99	99	50	38	<0–2%
Fuel and plant type	Emission control	Percentage reduction in relation to base case			Th. effy	Added costs
Residual fuel oil						
Conventional boiler	None	97	30	12	35.2	
	ESP/SO_2 control	>99.9	93	12	35.2	10–12%
Combined cycle	ESP/SO_2 & NO_x controls	>99.9	93	90	34.4	13–15%
Natural gas						
Conventional boiler	None	>99.9	>99.9	37	35.2	
	NO_x controls	>99.9	>99.9	45	35.2	
Combined cycle	None	>99.9	>99.9	62	44.7	

Source World Bank (1992)

from stationary sources, this essentially implies reduction in the demand for energy products. This has been presented in detail in Chap. 6. To recapitulate, there are alternative demand management approaches:

- *Energy efficiency improvements*: the efficiency of appliances decides the quantity of energy required to meet the energy needs. Efficient appliances reduce energy demand, and consequently reduce environmental damages. A wide range of energy saving opportunities can be identified in any country—these include demand for lighting, motors, cooling and heating in various sectors of the economy including industries, services and residential houses.
- *Efficient and adapted prices*: low prices lead to higher demands and higher levels of emission. Correcting price distortions and giving correct signals to consumers can help reduce demand.
- *Even direct control of loads*: direct control of demand through curtailment is also quite common in many countries around the world. In the case of electricity, this is done by shedding loads of required size, often through involuntary means. This arises from an inefficient management of the system in most cases and is not considered as a desirable control option.

Demand management and efficiency improvement constitute one of the low-cost options to deal with the environmental problem.

24.4 Indoor Air Pollution[4]

The problem of indoor air pollution has received attention more recently than other regional and global pollution problems. More than 2.5 billion people in the world rely on biomass fuel for their cooking and heating energy needs. If coal is added, then about one-half of the global population relies on solid fuels for cooking energy needs. Biomass and coal combustion leads to emission of a large number of pollutants including particulate matters, carbon monoxide, sulphur dioxide, nitrogen oxide and other chemicals. Often the level of pollution is much higher than the EPA guidelines (see Table 24.3) by a factor of 100 or more.

Prolonged exposure to such pollutants is believed to cause a number of diseases, including acute respiratory infection, chronic lung diseases, lung cancer and birth-related problems. There is a social dimension to the problem as well—most of the environmental burden falls on the women, children and the elderly, making these sections of the population more vulnerable. It is believed that indoor air pollution causes between 1.5 and 2 million deaths worldwide, accounting for 3–4% of the mortality worldwide (Ezzati and Kammen 2002). Warwick and Doig (2004) indicate that:

---

[4] This section relies on Ezzati and Kammen (2002), Warwick and Doig (2004)

24.4 Indoor Air Pollution

Table 24.3 Typical indoor pollution conditions in developing countries

Region	Fuel/technology	Particulate level ($\mu g/m^3$)	CO (ppm)
Africa	Wood, coal and charcoal	531–1,998	
Asia	Wood, coal, dung	330–35,000	1.6–26.2
Americas	Wood, dung	3–27,200	
Oceania	Wood	600–2,000	
US EPA guidelines		150 (24 h average)	9 (8 h average)

Source Bruce et al. (2002)

- About 36% of the lower respiratory tract infections are due to excessive exposure to indoor pollution.
- Similarly, 22% of the chronic pulmonary obstructive disease (COPD) is related to indoor pollution. A woman using biomass for cooking is 2–4 times prone to COPD than her counterpart not using such energies.
- People using wood fuel are 2.5 times more prone to active tuberculosis.

Considering the significant health impact of indoor air pollution, it becomes a high priority area in terms of policies to provide access to clean energies.

Although energy use is the main source of indoor air pollution, it is not the only source. US EPA identifies among others: fuel combustion, tobacco smoking, central heating and cooling systems, humidification systems, building materials, household cleaning products and furniture. Moreover, although the problem has reached a serious level in developing countries, the issue is not limited there alone. Developed countries also face the health effects but often from different combinations of the polluting sources.

Given the importance of the problem, it is necessary to look into the possible solutions. Technical solutions to the problem include making arrangements for better smoke removal from houses, using efficient cooking or energy using devices and improving houses so that ventilation problem is reduced. Despite continued research on better cooking stoves, little achievement is noticed in terms of acceptance of such technologies. Better ventilated houses with smokestacks received recent attention in poor houses and some improvements at a low cost may be possible. There remains a lot of scope for further work in these areas.

A set of policy options could also help reduce the social cost of indoor air pollution. These include fuel switching, providing credits for better houses and efficient appliances, raising awareness about health dangers of indoor pollution, ensuring better availability of cleaner fuels, etc. Renewable energies are often identified as the potential candidates for clean energies. Despite their environment friendly nature, not much progress has been made so far. A holistic, integrated, bottom-up approach to development and energy access issues is required, which could require departure from traditional policies and delivery mechanisms.

24.5 Conclusion

This chapter has presented the options for mitigating pollution from stationary sources. The energy sector is a major contributor to such pollution and alternative measures have been adopted to mitigate the problem. Although market-based instruments such as cap-and-trade have now gained momentum in some countries, the use of command-and-control method still dominates in most of the countries. However, there is a growing trend towards using a combination of technological solutions and market as well as non-market based approaches.

In addition, the issue of indoor pollution due to traditional energy use in the developing world escapes any regulatory intervention or market-based mitigation approaches and has not received adequate attention yet, despite its heavy health and other economic impacts. This joins the issue of clean energy access in the developing world and requires a far greater careful consideration than is being given now.

References

Bruce N, Perez-Padilla R, Albalak R (2002) The health effects of indoor air pollution exposure in developing countries. World Health Organisation, Geneva

Burtraw D, Palmer K (2003) The paparazzi take a look at a living legend: the SO_2 cap-and-trade program for power plants in the United States. Resources for the Future Discussion Paper 15 (see http://www.rff.org/rff/Documents/RFF-DP-03-15.pdf)

Burtraw D, Szembelan SJ (2009) U.S. emissions trading markets for SOx and NOx. RFF-DP-09-40, Resources for the Future, Washington

Chestnut LG, Mills DM (2005) A fresh look at the benefits and the cost of the US acid rain program. J Environ Manage 77(3):252–266

Environmental Defense (2000) From obstacle to opportunity: how acid rain emissions trading is delivering cleaner air. Environmental Defense, New York

EPA (2005) Evaluating ozone control programs in the Eastern United States. Focus on NOx Budget Trading Program, Washington (see http://www.epa.gov/airtrends/2005/ozonenbp.pdf)

EPA (2010) Acid rain and related programs: 2009 emission, compliance and market analyses. Environment Protection Agency, Washington

Ezzati M, Kammen DM (2002) Household energy, indoor air pollution and public health in developing countries. RFF issue brief 02-26. Resources for the Future, Washington

Lovei M (1998) Phasing out lead from gasoline: worldwide experiences and policy implications. World Bank, Washington (see http://siteresources.worldbank.org/INTURBANTRANSPORT/Resources/b09phasing.pdf)

OTC (2003a) NOx budget program: 1999–2002 progress report. Ozone Transportation Commission, Washington

OTC (2003b) 2002 OTC NOx budget program compliance report. OTC, Washington (see http://www.otcair.org/document.asp?fview=Report#)

Warwick H, Doig A (2004) Smoke—the killer in the kitchen. ITDG Publishing, London

World Bank (1992) World development report 1992, development and the environment. World Bank, Washington

Further Reading

EPA (2003) Tools of the trade: a guide to designing and operating a cap and trade programme for pollution control. US Energy Protection Agency, Washington

Eskeland GS, Devarajan S (1996) Taxing bads by taxing goods: pollution control with presumptive charges. World Bank, Washington

Krupnick A, Morgenstern R, Fisher C, Rolfe K, Logarta J, Rufo B (2003) Air pollution control policy options for Metro Manila, RFF discussion paper 03-30. Resources for the Future, Washington

EC (2001) Directive 2001/80/EC on the limitation of emissions of certain pollutants into the air from large combustion plants

Vehmas J, Kaivo-oja J, Luukkanen J, Malaska P (1999) Environmental taxes on fuels and electricity—some experience from the Nordic countries. Energy Policy 27(6):343–355

World Bank (1999) Pollution prevention and abatement handbook: toward cleaner production. The World Bank, Washington

Chapter 25
Pollution Control from Mobile Sources

25.1 Introduction

A mobile source of pollution is a source that changes its location over time and accordingly, the pollution affects different locations. This is the case of the transport sector, which has emerged as a major source of final demand for energy in many countries. Globally, about 2.3 billion toe of energy was used in 2007 by the transport sector, which represented about 20% of the global primary energy demand. But the transport sector is heavily dependent on fossil fuels—about 97.5% of the total demand was met at the global average level, although some regions have tried to reduce their dependence by moving to renewable transport fuels. According to WEO (2008), energy demand for the transport sector will increase to 2.6 billion toe in 2015 and 3.2 billion toe by 2030, and 92% of this demand will be met by using oil (Fig. 25.1). Excessive dependence on fossil fuels and a rapid growth of the sector, especially in the developing world is the main source of concern for mobile pollution. This chapter presents a brief introduction to mobile pollution and discusses the mitigation options, considering an economic perspective.

The vehicle population has seen an average growth of 4.6% per year between 1960 and 2002 (Dargay et al. 2007). The fleet increased from 122 million in 1960 to 812 million in 2002. According to Dargay et al. (2007) the vehicle population is likely to grow at a slightly slower rate to reach the 2 billion mark by 2030. Six countries, namely China, USA, India, Japan, Brazil and Mexico, will account for more than 50% of the global vehicle population by 2030. About two-thirds of the increase in vehicle population will come from the non-OECD region. Rapid urbanisation and rising income explains the growth in transport fleet in these countries, which in turn contribute to the urban air problem and increasing carbon emissions.

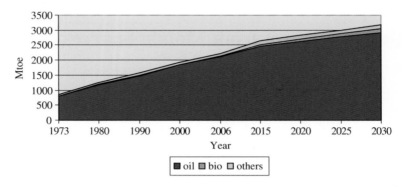

Fig. 25.1 Growth in global transport energy demand. *Data source*: various WEO reports

25.2 Special Characteristics of Mobile Pollution

Mobile sources of pollution have certain special characteristics as indicated below:

- There are numerous small sources which make monitoring difficult. Controlling pollution in each individual case may lead to poor performance as well. As the number of potential regulated entities is large, the number of possible interventions increases as well.
- As the mobile source changes its location, the emission of pollution also moves with it. The environmental problem arises only when the source is in the wrong place at the wrong time. This makes it difficult to tailor emission control for different locations.
- The source of pollution is durable in nature and the changes in the technology mix take time. The performance of the equipment deteriorates over time as well. Because of long-life and capital intensiveness of the energy-using equipment, the holding pattern and the average age of the vehicle in operation varies. The rate of pollution also varies by make, model and vintage of vehicles and depending on the usage patterns, road and traffic conditions (Harrington and McConnell, 2003).
- The timing of emission is important, as emissions during certain environmental conditions may contribute to air quality deteriorations. The effect can be seasonal and over short spans, which in turn requires mitigation options to be time-specific.

As a result, a somewhat different treatment is required to mitigate pollution from transportation compared to the stationary sources of pollution. Moreover, the external costs of using the transport services are not only related to pollution alone (as discussed below) which further complicates the issue. The external cost varies by mode of transport as well but often the road transport received greater attention because of its domination in the transport modes. This chapter focuses on the road transport mode only.

Table 25.1 Accounting of full cost of road transport

User costs	Government costs	Societal costs
1. Vehicle purchase and debt	1. Capital investment (land, structures, vehicles)	1. Parking—free private
2. Gas, oil, tires	2. Operations and maintenance	2. Pollution—health care, cost of control, productivity loss, environmental harm
3. Repairs, parts	3. Driver education and motor vehicle administration	3. Infrastructure repair, vibration damage, etc.
4. Auto rentals	4. Police, justice, fire	4. Accidents—health insurance, productivity loss, pain and suffering
5. Auto insurance	5. Parking—public, tax breaks	5. Energy—trade effects
6. Tolls	6. Energy—security	6. Noise
7. Transit fares	7. Accidents—public assistance	7. Land loss (urban, crop value, wetlands)
8. Registration, licensing, annual taxes	8. Pollution—public assistance	8. Property values, aesthetics
9. Parking—paid		9. Induced land use patterns
10. Parking—housing cost		
11. Accidents—private expenses		
12. Travel time		

Source: Delucchi (1998) and Anderson and McCullough (2000)

25.3 Social Costs of Transport Use[1]

Any transportation activity by any mode imposes a certain amount of costs but some of which are borne by the user, some by the government and the society. For example, Table 25.1 provides a list of costs borne by various agents in respect of road transport.

Although a part of the cost is paid by the users, the full cost is generally not borne by them. This leaves some gap between the full- and the private cost of transportation service, resulting in external costs to the users. In general, three types of external costs can be identified for the transport sector. The first are those experienced primarily by concurrent road-users, but decisions on activity levels do not take this cost into account. Road accidents, road damages and congestion fall into this category (infrastructure usage related-costs). The second category of cost arises from environmental pollution and this can have local, regional and even global effects at present and in the future. The third category concerns the

---

[1] Delft (2008), Newbery (1988, 1995), Silberston (1995), Bhattacharyya (1996) and RFF (2003) for further details.

interaction between transport investment, transport demand and land-use (infrastructure planning and investment-related costs).

25.3.1 Infrastructure Usage Related Costs

Let us consider the three elements separately

- Road damages inflict two types of costs: increased cost of road repairing after the damage and increased vehicle operating costs to subsequent vehicles. The damage to the road infrastructure is caused by two factors: weathering and the passage of heavy vehicles. It is normal for any road to wear out due to weather effects (heavy snow, heavy rain, etc.) and therefore, only a part of the road damage cost can be attributed to the traffic. On the other hand, the damage inflicted by the vehicles depends on the vehicle characteristics and the type of road (paved, unpaved, etc.). For any given road, the extent of damage is proportional to the fourth degree of the axle power. Trucks normally inflict the damage, while the impact of personal cars is relatively insignificant. However, Newbery (1988, 1995) argued that the road damage externality turns out to be insignificant compared to the weathering damage. In such a case, the marginal cost of road infrastructure use is equal to the cost of repair and maintenance of the present network (Newbery 1988, 1995).
- Marginal accident costs are somewhat difficult to define.[2] Any vehicular accident imposes a number of social costs:
    - costs due to an increased risk of accident—the costs faced by any person meeting with an accident consist of costs to himself, family and friends due to loss of utility, medical costs, police costs and output loss;
    - costs due to loss to other concurrent users: this arises as the increased accident risk affects the risk of accident of other users as well;
    - costs due to damages to the infrastructure; and
    - the avoidance cost of changed traffic behaviour due to changed traffic condition.

Not all of these costs are private, neither they are external always. It depends on the insurance pricing and compensation and liability rules in operation. The marginal external accident costs can be estimated as follows (Delft 2008):

$$\text{Marginal external accident cost} = \text{traffic volume} \times \text{risk elasticity} \times \text{unit cost per accident} \times \text{external part}. \quad (25.1)$$

The risk elasticity measures the risk of an additional accident at the actual level of traffic volume. Delft (2008) provides some examples of accident-related externalities for European countries (Table 25.2). Clearly, the cost varies by location, by type of network and vehicle type.

---

[2] Mayeres (2002) for details.

25.3 Social Costs of Transport Use

Table 25.2 Examples of external accident costs (€cents/vehicle kilometer)

Vehicle and road type	Germany	France	Italy	Sweden	UK
Cars in urban roads	4.12	6.69	4.78	2.68	2.61
Cars in motorways	0.29	0.48	0.34	0.19	0.19
Motorcycles in urban roads	30.29	49.25	35.17	19.72	19.19
Motorcycles in motorways	0.2	0.32	0.23	0.13	0.12
Heavy vehicles in urban roads	10.49	17.04	12.18	6.83	6.64
Heavy vehicles in motorways	0.29	0.48	0.34	0.19	0.19

Note: Central estimates
Source: Delft (2008)

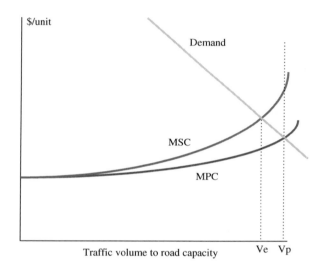

Fig. 25.2 Congestion inefficiency

- Similarly, during congestion, the traffic volume reaches road capacity, which in turn leads to reduction in traffic flow and reduction in the speed. Lower speed affects the operating costs of other road users and increases time costs due to cost of extra time and loss of opportunity to work. As the user does not consider these costs imposed on others, congestion also imposes external costs because by neglecting the social cost of congestion, over consumption of road space takes place (Fig. 25.2).

To determine the externality and identify ways of internalizing it, the marginal social cost of congestion has to be found. It is common to use speed–flow characteristics to determine the incremental travel cost due to congestion. The following basic relationship captures the external cost:

$$\text{External congestion cost} = \text{increased journey time} \times \text{value of time} \times \text{traffic volume} \quad (25.2)$$

Table 25.3 Benchmark external congestion costs (€/vehicle-kilometers, 2000)

Area and road type	Passenger car	Goods vehicle
Urban motorways in large urban areas (>2 million)	0.50	1.75
Urban collectors in large urban areas (>2 million)	0.50	1.25
Local street centres in large urban areas (>2 million)	2.0	4.0
Urban motorways in large urban areas (<2 million)	0.25	0.88
Urban collectors in large urban areas (<2 million)	0.3	0.75
Local street cordons in large urban areas (<2 million)	0.3	0.60
Rural motorways	0.1	0.35
Rural trunk roads	0.05	0.13

Note: These are central estimates for morning peaks
Source: Delft (2008)

As can be imagined, the input requirement for the above equation is demanding because the increase in journey time is obtained through specific speed–flow relationships, which are site-specific and time dependent. Similarly, there is no consensus on the value of time as this depends on the purpose of the travel, earning power of the travellers and the distribution of the travellers at different points of time. Accordingly, the estimation can vary widely spatially, temporally and by network. Delft (2008) provides some estimates of benchmark external congestion costs in Europe for different locations and vehicles (Table 25.3).

25.3.2 Environmental Pollution Costs

Environmental pollution due to vehicle use imposes external costs due to effects on health, buildings and materials, agricultural crops or biodiversity. Motor vehicles are major contributors of particulate matters (PM), lead, sulphur oxides, and NOx. The emission patterns are different for gasoline and diesel vehicles. Gasoline vehicles normally emit carbon monoxide (CO), volatile organic compounds (VOC), NOx and lead while diesel vehicles are major sources of PM and NOx. Accordingly, the cost of damage varies as well. A number of factors affect vehicular emissions. These include: type of engine, age of vehicle, axle power, time of running, operating speed, driving characteristics and emission control device. Generally, the health effect of air pollution is the most important element of cost in this category.

The external costs of environmental pollution due to vehicle use can be estimated using the impact-pathways method. The steps involved for this are (as discussed in Chap. 23): estimation of the emission; determination of the dispersion of emission and its effect on ambient air quality; use of dose–response relations to determine the physical impacts and finally monetary valuation of the impacts. As before, the estimates are site-specific and vary by type of vehicle and local environmental conditions. An example for Germany is provided in Table 25.4.

The climate change-related costs are also added in some studies but the valuation is more controversial. This is because the identification and estimation of

25.3 Social Costs of Transport Use

Table 25.4 External air pollution costs (€cents 2000/vehicle-kilometer) in Germany

Vehicle	Size	Euro class	Average cost
Petrol car	<1.4 l	Euro 0	2.0
		Euro 1	0.9
		Euro 2	0.4
		Euro 3, 4, 5	0.1
Diesel car	<1.4 l	Euro 2	1.1
		Euro 3	1.1
		Euro 4	0.6
		Euro 5	0.4
Trucks	<7.5 t	Euro 0	9.1
		Euro 1	5.4
		Euro 2	5.0
		Euro 3	4.0
		Euro 4	2.3
		Euro 5	1.4

Source Delft (2008)

Table 25.5 Transport-related external costs of climate change (€/ton of CO_2)

Year of application	Low estimate	Central estimate	High estimate
2010	7	25	45
2020	17	40	70
2030	22	55	100
2040	22	70	135
2050	20	85	180

Source: Delft (2008)

damages as well as their valuation faces more uncertainty. Accordingly, depending on the scenario used, the estimation of climate-related costs can vary widely (Table 25.5). The table shows that the cost varies by a factor of 6–9 between the low case and the high case.

The climate-related external costs when converted to cost per vehicle kilometers become comparable to other external costs those shown in Table 25.4. These are presented in Table 25.6. It is clear that the climate-related externality can be quite high compared to other externalities and in the high estimate the cost could be significantly high. According to Timilsina and Dulal (2010) the cost of transport-related air pollution can cost between 1 and 3% of the GDP of a country.

25.3.3 Infrastructure-Related Costs

Infrastructure-related costs are generally borne by the government but this provides benefits to all road users. The investment brings positive benefits to the economy and therefore, it creates positive externality. But at the same time, while

Table 25.6 Climate-related externals costs for road transport in Europe (€cents/vehicle kilometer)

Vehicle	Size	Euro class	Climate-related cost
Petrol car	<1.4 l	Euro 0	0.5
		Euro 1	0.5
		Euro 2	0.4
		Euro 3, 4, 5	0.4
Diesel car	<1.4 l	Euro 2	0.3
		Euro 3	0.3
		Euro 4	0.3
		Euro 5	0.3
Trucks	<7.5 t	Euro 0	1.2
		Euro 1	1.0
		Euro 2	1.0
		Euro 3	1.1
		Euro 4	1.0
		Euro 5	1.0

Note: these are central estimates
Source: Delft (2008)

some users bear the cost of using the road, some may not be paying for this. It is also possible that the users pay only a part of the cost of developing or replacing the infrastructure. Consequently, some externality arises.

25.3.4 Internalisation of Externalities

The relative importance of different components may vary within a country and across countries according to the infrastructure available, the locality, the technology and fuel used, atmospheric conditions and so on. This is shown in Tables 25.2 and 25.6.

The question that arises immediately is whether fuel taxes constitute a good instrument for internalising all three types of externality of transport use. Problems arise here. The efficacy of taxing transport fuels to cover external costs depends on two factors: the possibility of substitution between fuel and other inputs, and the correlation between the vehicle use and the externalities it generates. For example, an additional fuel tax induces economy of fuel consumption and improves the technical efficiency of combustion, although it does not necessarily reduce other external costs. Since the possibility of technical improvement is rather limited for diesel vehicles and the cost of pollution and road damage can be attributed largely to heavy vehicles, a tax on diesel constitutes a good method of internalising certain externalities. The case is quite different for gasoline. The price elasticity of demand for gasoline is more than that of diesel, and the possibility of substitution and improvement of technical efficiency is not negligible. There is no direct case between gasoline-driven vehicles and the externalities they generate. In such a case, a tax on gasoline is not a good way of internalising externalities (Newbery 1989).

Moreover, problems arise with congestion costs, as they are not related directly to energy use. The transport service, rather than transport fuels, is to be subjected to tax, in order to correct this market failure. A fuel tax does not reflect the time- and location-dependent nature of congestion cost and penalises those users who are not contributing to congestion. This is especially true of rural users in both developing and developed countries and of users in a large number of small towns and cities in developing countries. Taxing fuels has economic implications, as it impedes economic growth of a country and leads to the drainage of resources from rural to urban areas to protect urban externalities.

It would be better to take recourse to other direct instruments than to use a fiscal instrument ineffectively. Yet a fuel tax is preferred to other instruments in a real-life situation for the following reasons: its administrative simplicity, the flexibility offered to users in choosing a response (technology, type of fuel), the possibility of distinguishing fuels by pollution damages, the influence on the supply-side options by encouraging efficient vehicles, and so on (Silberston 1995).

25.4 Mitigation Options

Emission from transport can be expressed as follows (IEA 2000):

$$G = A.S.I.F \tag{25.3}$$

where G = emission from transport, A = transport activity (expressed in passenger-kilometre travelled or ton-km of goods transported), S = modal share (i.e. share by mode of transport, e.g. air, water, road and rail this can also be split by type of vehicle and purpose of travel), I = modal energy intensity (i.e. energy consumption per unit of travel activity), F = emission factor per unit of energy use.

Any direct mitigation policy would focus on total emission control from transport while indirect policies would focus on one or more components on the right side of the above expression. Moreover, a decomposition analysis of transport-related pollution can provide interesting insights about the influence of each component on the overall pollution. This insight could be used to develop alternative policies as well.

As pollution from mobile sources is a local issue, depending on the relative importance of the causal factors the mitigation options have to be designed. This in other words implies that specific solutions have to be developed rather than generic solutions. A comprehensive control strategy would focus on the following four elements (ADB 2003):

- improvements in emissions standards and technology,
- cleaner fuels,
- improved transport planning and traffic demand management and
- improved inspection and management (ADB 2003).

These are considered below.

25.4.1 Vehicle Emission Standards and Technologies

Vehicle emission is controlled through two types of standards: standards for new vehicles and standards for in-use vehicles. Normally, stricter standards are applied to new vehicles in a progressive manner. This is done to ensure bringing changes to manufacturing processes and building supply capabilities. For example, Euro standards, which have emerged as the industry-wide standards in many parts of the world, have been progressively tightened (Figs. 25.3, 25.4).[3] Similarly, in the case of US, the Corporate Average Fuel Economy (CAFÉ) programme remains the main policy tool for fuel economy in the transport sector. However, the European policy was supplemented with high fuel taxes, which resulted in a long-term change in terms of vehicle size and efficiency. Low fuel tax rates in the US did not yield similar results as in Europe but with tighter standards now in operations, improvements are visible.

As standards come under national jurisdictions, often developing countries maintain a time lag in introducing tighter standards. This allows them to learn from the experiences of the developed countries and take advantage of better technologies. Type approval tests on new vehicles are carried out to determine whether the emission standards are satisfied by a type of vehicle. In addition, roadworthiness tests are also conducted. But as the modal share of new vehicles remains low initially, the impact on pollution control is felt over a longer time.

In-use vehicles are required to attain a certain level of pollution (which is normally less strict than new vehicle standards) and failure to do so makes them road unworthy. In-use standards often allow exemptions or special considerations depending on the vintage of vehicles. This is done to reduce negative impacts on users of such vehicles. But in developing countries where vehicles live longer, such a policy can make implementation difficult and cumbersome. This also reduces the effectiveness of stricter control for new vehicles. Accordingly, the in-use standards should also be tightened progressively, which could lead to phasing-out of old, polluting vehicles. In some countries, incentives are provided to retire old vehicles on a regular basis.

However, to ensure compliance with standards for in-use vehicles, an effective, routine inspection and maintenance programme is required. Mandatory inspection at a regular interval, supplemented by road-side apprehension programmes for verifying compliance is required. It is also important to use appropriate testing methods. Many developing countries, where facilities for proper testing are lacking, rely on simpler tests using cheaper equipment which cannot determine whether sophisticated pollution controls are operating properly or not.

Vehicle technologies are also important as well. As vehicles live long, the stock changes slowly. Similarly, the composition of the stock by technology type is

---

[3] Euro 1 was introduced in EU from 1 July 1992. Euro 2 was introduced from 1 January 1996. Euro 3 was introduced from 1 January 2000 and Euro 4 from 1 January 2005. Euro 5 was introduced in September 2009 and Euro 6 will be introduced in September 2014.

25.4 Mitigation Options

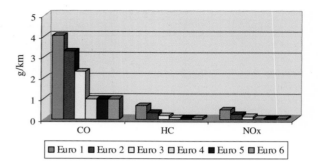

Fig. 25.3 Evolution of Euro standards for gasoline passenger cars. *Source*: European Automobile Manufacturer's Association

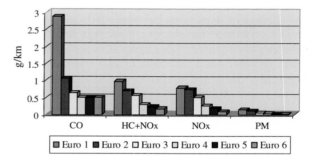

Fig. 25.4 Evolution of Euro standards for diesel cars. *Source*: European Automobile Manufacturer's Association

important for overall emission. For example, high reliance on two-wheelers in some developing countries and high growth of stock of this category of vehicles significantly influence vehicular emission. Generally two-stroke engines used in two wheelers create more environmental pollution due to incomplete combustion of fuel than four-stroke engines. As older vehicles are less fuel efficient, in addition to pollution they lead to higher fuel demand.

In addition, advanced vehicle technologies can play some role in reducing urban air pollution and damage. Some such options include electric vehicles, hybrid vehicles and fuel cell vehicles. Electric vehicles use energy stored in batteries but the replenishment of batteries and their costs are major concerns for viability. Hybrid cars use dual fuels—traditional oil and another fuel (electricity or others). Fuel cells generate hydrogen to propel the car. As more users adopt these technologies, the unit cost is expected to fall.

25.4.2 Cleaner Fuels

A related element in the vehicular emission control is the quality of fuel that is used or allowed to use. Better the quality of fuel, less pollution it generates. The drive for better fuel use has long been driven by public health concerns. For

Table 25.7 Better fuel quality for improved emission standards for gasoline vehicles

Fuel quality	Unit	Euro 2	Euro 3	Euro 4
Lead			Nil	Nil
Sulphur	ppm	500	150	50
Benzene	% (v/v)	5	1	1
Aromatics	%	–	42	35
Olefins	%	–	18	18
Oxygen	% (m/m)	–	2.7	2.7
Reid vapour	kPa	35–100	45–100	45–100

Source: European Automobile Manufacturer's Association

example, lead in gasoline has been used since 1930s but its health effects became a major concern in developed countries in the 1970s, leading to lead-free fuel use. Developing countries were slow in switching to lead-free regime but a considerable progress has been made in this respect worldwide. Similarly, concerns for visibility, smog and ozone formation at lower level led to the drive for lower sulphur use in fuel.

Moreover, these standards are often directly related to the vehicle emission standards. For example, unleaded gasoline is a prerequisite for introducing catalytic converters in vehicles. Similarly, low sulphur content in fuel is required to ensure compliance with advanced emission standards (see Table 25.7 for a comparison of fuel specification for Euro standards).

In addition to use of better quality fuels, use of alternative fuels can also be considered for transport purposes to reduce environmental damages and dependence on oil as well as improve diversity and use of local resources. Such alternative fuels include a wide range of possibilities including use of compressed natural gas, LPG, biodiesel, methanol, ethanol, vegetable oils, synthetic oils derived from coal, electricity and hydrogen (ADB 2003). IEA (1999) provides a comparative analysis of the characteristics of different alternative fuels (Table 25.8), while Table 25.9 provides an overview of supply costs of alternative fuels based on Concawe (2007).[4]

Some of the technologies are quite mature while others are emerging. Alternatives like CNG and LPG have quite matured—CNG for public transport purposes, while LPG for personal use or in taxis. Depending on the availability of gas and LPG, these alternatives could be promoted. As gaseous fuels, adulteration is difficult and they are in particulate emission. But conversion of diesel engines for gas use is difficult and leads to higher NOx emission (ADB 2003). Replacement may be a better option.

Similarly, methanol has attractive fuel properties (e.g. lean combustion capability, low temperature combustion and low photochemical reactivity) and can be

---

[4] Torchio and Santerelli (2010).

25.4 Mitigation Options

Table 25.8 Alternative fuel characteristics

Fuel	Energy content (MJ/l)	On board storage		Safety				Air pollution	Ground water pollution
		Mass, % of gasoline	Vol, % of gasoline	In open air	In closed areas	In tank	Toxicity		
Gasoline	31.2	100	100	*	–	–	**	**	***
Diesel	35.7	100	88	–	–	–	*	**	***
LPG	24.2	180	154	**	**	–	*	*	*
NG	23.3	240–490	360	–	**	–	–	–	*
Methanol	15.6	191	176	–	*	**	***	***	**
Ethanol	21.2	165	151	–	*	**	*	*	**
Biodiesel	32.8	117	100	–	–	–	*	–	–
Hydrogen	8.9	154–269	376–515	***	***	***	–	–	–
DME	18.2–19.3	147	166	**	**	–	*	*	*

Note: More * indicates higher influence of a factor, – indicates negligible influence
Source: IEA (1999)

Table 25.9 Supply cost of alternative transport fuels

Fuel	Technology	Cost of substitution (€/100 km)	Cost of CO_2 avoided (€/tCO_2 eq)
CNG	PISI-Bifuel	1.32	579
LPG	PISI-Bifuel	1.35	672
Ethanol (pulp to heat)	PISI	2.10	198
Biodiesel (RME, glycerine as chemical)	CIDI + DPF	1.80	217
Synthetic diesel ex-natural gas (NG)	CIDI + DPF	0.21	–
Hydrogen from thermal processes ex-NG	ICE PISI	5.03	–
Hydrogen from thermal processes ex-coal	ICE PISI	5.68	–
Hydrogen from electrolysis (electricity from NG)	ICE PISI	6.80	
Hydrogen from electrolysis (electricity from nuclear)	ICE PISI	8.12	566
Hydrogen from electrolysis (electricity from wind)	ICE PISI	8.05	568

Note: Data corresponds to crude oil price scenario of $25/bbl
ICE internal combustion engine, PISI port injection spark injection technology, CIDI compression ignition direct injection technology, DPF diesel particulate filter
Source Concawe (2007)

produced from natural gas found in remote places, far away from markets. But cost is the main barrier in the case of alternative fuel use. It can be seen that many of the alternative fuels were not cost effective at moderate level of oil prices prevailing at the time of the study but at their viability is believed to have improved with higher oil prices prevailing now. Ethanol based alternative technologies have been widely used and experimented in Brazil during the periods of oil shocks. But fall in oil prices in the 1990s resulted in a reverse switching to oil-based vehicles, resulting in economic loss due to unused economic capacities. The interest in biofuels has re-emerged with oil prices reaching high levels. The EU has now decided to achieve a 5.75% biofuel target by 2010 and 10% by 2020. But the cost and the wider impacts of biofuels, especially its effect on food security, require a more careful consideration of this alternative source.[5]

Use of electricity remained confined mostly in tramways and trains but with electric vehicles, urban commuting may become feasible as well. Although the environmental effect of electricity vehicle depends on the source of electricity, with higher emphasis on renewable energies and nuclear, the environmental effects could be mitigated at least in urban areas. As indicated in Table 25.9, the cost per ton of CO_2 avoided is quite high for alternative transport fuels, thereby implying

---

[5] For example Eide (2008) and FAO (2008) for a detailed analysis on this issue.

that it makes better economic sense to adopt other low cost options before turning to carbon abatement through alternative transport fuels.

In addition to environmental benefits, these alternative fuels enhance energy security by reducing oil dependence. In fact, this effect can be considered as an external benefit and accounted for in deciding their use. As the transport sector remains the only captive market for oil, emergence of alternatives could change the oil industry dynamics and bring new challenges.

25.4.3 Traffic Management and Planning

This is part of an overall management of emission from the transport sector. The focus is how to meet the travel needs efficiently, cost-effectively and without imposing environmental damages. This goes beyond the environmental aspect of energy use and considers land use, efficient infrastructure development and travel demand management. In this section, we just focus on travel demand management through energy pricing.

Efficient energy pricing could influence demand for travel and provide signals for modal shifts and technology choices. Often the transport sector is conveniently chosen for tax purposes. Wide varieties of taxes are used in transport apart from taxing transport fuels. These include: registration taxes, annual circulation tax, and user charge. Some countries use registration charges to promote certain types of vehicles and discourage others. For example, Singapore and some Nordic countries use prohibitive registration charges to discourage personal vehicle use. Circulation tax is generally imposed for revenue generation, although a part of the revenue is used for road maintenance. Typical user charges are tolls and congestion charges. These are less commonly used than the other two.

The role of public transport in this regard needs to be highlighted. Subsidised public transport in many countries is provided for social reasons but can also be justified for the positive environmental externality it is expected to generate. This is because the emission per passenger-kilometer is much lower compared to equivalent emission from individual vehicles. However, in many cases public transport remains unused while commuters still rely on personal modes of transport, thereby reducing the effectiveness of the system.

Further, transport fuels are often taxed at much higher rates than any other fuel. While these taxes are imposed for revenue generation purposes, some countries try to justify them on environmental grounds as well. However, there is often little correlation between the environmental damage and the tax imposed. The wide variation in tax rates even within similarly developed countries cannot be explained using environmental logic. Similarly, subsidised fuel prices in many countries also distort the market and encourage wasteful consumption as well as environmental degradation. This in turn creates traffic management issues as well.

25.5 Conclusion

This chapter has briefly introduced the concepts of external costs related to the transport sector and presented some information on various elements of the external costs. The growing demand for transport services in the developing countries and the domination of fossil fuels in the sector contribute largely to for urban air quality issue in large urban areas of these countries. Mitigation of the problem requires a comprehensive strategy that considers emissions standards, clean fuels, transport planning and management and monitoring and inspection mechanisms. A lot of coordination and effort is required to address this complex problem and there is no easy way out.

References

ADB (2003) Reducing vehicle emissions in Asia. Asian Development Bank, Manila. http://www.adb.org/documents/guidelines/Vehicle_Emissions/reducing_vehicle_emissions.pdf

Anderson D, McCullough G (2000) The full cost of transportation in the twin cities region. Centre for Transportation Studies, University of Minnesota, Minneapolis

Bhattacharyya SC (1996) A preliminary estimate of the external costs of energy use in the transport sector: a case study of India. Opec Review, September 1996, pp 247–260

Concawe (2007) Well-to-wheels analysis of future automotive fuels and powertrains in the European context: well-to-wheels report

Dargay J, Gately D, Sommer M (2007) Vehicle ownership and income growth: 1960–2030, New York University. http://www.econ.nyu.edu/dept/courses/gately/DGS_Vehicle%20Ownership_2007.pdf

Delft (2008) Handbook on estimation of external costs in the transport sector: internalisation measures and policies for all external costs of transport, Delft, the Netherlands

Delucchi MA (1998) The annualized social cost of motor vehicle use in the US, 1990–91, summary of theory, data, methods and results. Institute of Transportation Studies, University of California, Davis, California

Eide A (2008) The right to food and the impact of liquid biofuels (agrofuels). Food and Agricultural Organisation of the United Nations, Rome

FAO (2008) The state of food and agriculture 2008. Biofuels: prospects, risks and opportunities. Food and Agricultural Organisation of the United Nations, Rome

Harrington W, McConnell V (2003) Motor vehicles and the environment, RFF Report, Resources for the Future, Washington, DC

IEA (1999) Automotive fuels for the future: the search for alternatives. International Energy Agency, Paris

IEA (2000) Flexing the link between transport and greenhouse gas emissions: a path for the World Bank. International Energy Agency, Paris

Mayeres I (2002) Taxes and transport externalities, working paper number 2002-011, KU Leuven. http://www.econ.kuleuven.be/ete/downloads/ETE-WP-2002-11.PDF

Newbery DGM (1989) Road transport fuel pricing policy. Annu Rev Energy 14:75–94

Newbery D (1988) Road user charges in Britain. Econ J 98(390):161–176

Newbery D (1995) Royal Commission Report on transport and the environment—economic effects of recommendations. Econ J, 105(Sept):1258–1272

RFF (2003) Motor vehicles and the environment, RFF report available at http://www.rff.org/rff/Documents/RFF-RPT-carsenviron.pdf

Silberston A (1995) In defence of the Royal Commission Report on Transport and the Environment. Econ J, 105(Sept):1273–1281

Timilsina G and Dulal HB (2010) Urban road transportation externalities: costs and choice of policy instruments. The World Bank Research Observer. doi:10.1093/wbro/lkq005

Torchio MF, Santarelli MG (2010) Energy, environmental and economic comparison of different powertrain options using well-to-wheels assessment, energy and external costs—European market analysis. Energy 35(10):4156–4171

WEO (2008) World energy outlook 2008. International Energy Agency, Paris

Further Reading

Faiz A, Weaver CS, Walsh MP (1996) Air pollution from motor vehicles. World Bank, Washington, DC

Safirova E, Gillingham K (2003) Measuring marginal congestion costs of urban transportation: do networks matter, RFF discussion paper 03-56, Resources for the Future, Washington, DC http://www.rff.org

IEA (2000b) The road from Kyoto: current CO_2 and transport policies in the IEA. International Energy Agency, Paris

Chapter 26
The Economics of Climate Change

26.1 Climate Change Background

The issue of climate change has become a contemporary subject in any discussion on energy and the environment. The scientific reason for global warming is attributed to a phenomenon called the greenhouse effect. In simple terms, this works as follows: the ultimate source of energy for the climate system is the radiation from the Sun. The wavelength of the radiation varies inversely with the temperature of the source. Most of the radiation from the Sun reaches the earth in shorter wavelengths (0.2–0.4 μm). This heats up the earth's surface and the earth emits radiation in long wavelengths (4–100 μm).[1] There are some gases which are transparent to shortwave radiation but opaque to long waves. These gases do not allow long waves to pass through, thereby entrapping the radiation. This is called the greenhouse effect.

26.1.1 The Solar Energy Balance[2]

Every square metre of the earth's surface on average receives 342 Watts of solar radiation (W/m^2).

- About 107 W/m^2 (or 31%) of the radiation is immediately reflected back by the clouds, atmosphere and the earth's surface.
- Out of the remaining 235 W/m^2, a small amount (67 W/m^2) is absorbed by the atmosphere and the rest (168 W/m^2) warms up the earth's surface (IPCC 2007a).

To maintain a stable climate, the incoming energy must remain in balance with the outgoing energy. This requires on average re-radiation of 235 W/m^2.

---

[1] Cline (1991).
[2] This section is based on IPCC (2007a) and CBO (2003).

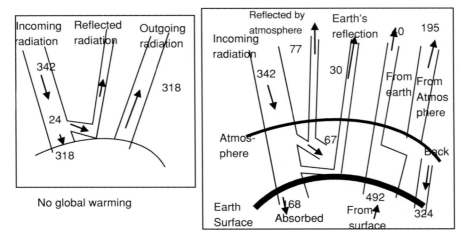

Fig. 26.1 Earth's average energy balance. *Source*: IPCC (2001a, Chap. 1, Fig. 1)

The typical temperature of the source should be $-19°C$ for the long wave radiation of this amount. This temperature is much lower than the average temperature on the earth. What happens instead is known as natural greenhouse effect. The greenhouse gases (GHG) trap heat in the atmosphere and radiates long wave radiations in all directions. A part of it reaches the earth, which warms up the overall surface temperature to 14°C. This in effect shifts the effective emission temperature of $-19°C$ to 5 km above the earth's surface, thereby facilitating a pleasant condition for the living beings. Figure 26.1 captures the above details.

In a state of equilibrium, the incoming radiation is equal to the outgoing amount at the outer atmosphere. Any changes in the solar radiation or re-radiation from the earth or atmosphere changes this balance (i.e. creates an imbalance which is called the "radiative forcing") and the climate system has to adjust to bring the balance back. The system is sensitive to external factors such as changes in solar radiation or volcanic eruptions as well as changes due to human activities. If the concentration of the GHG changes due to human activities, keeping everything else constant, the radiative forcing will change.

26.1.2 GHGs and Their Global Warming Potential

The most commonly available greenhouse gas (GHG) is water vapour. Other GHG include carbon dioxide (CO_2), chlorofluorocarbons (CFCs), methane, nitrous oxide (NO_x) and ozone. Different gases trap different quantities of energy; they have different life periods and they also differ in terms of their ability to react with others. Normally, CO_2 is assigned the global warming potential (GWP) of 1 and all others are ranked with respect to it. Table 26.1 provides the details on life span of gases and their GWP.

26.1 Climate Change Background

Table 26.1 Life and GWP of some GHGs: changing pattern

Gas	Life (years)	GWP for 100 years time horizon		
		SAR	TAR	AR4
CO_2	5–200	1	1	1
Methane	12	21	23	25
Nitrous oxide	114	310	296	298
CFC-12	100	8,100	10,600	10,900
HCFC-22	11.9	1,500	1,700	1,810

SAR second assessment report, *TAR* third assessment report, *AR4* fourth assessment report; *Source*: IPCC (2001a, Chap. 6, Table 6.7) and IPCC (2007a, Chap. 2, Table 2.14)

Table 26.2 Changes in concentration of selected GHGs

Gas	Concentration in 1750 (ppm)	Concentration in 1998 (ppm)	Concentration in 2005 (ppm)	Change in concentration (ppb) 1998
CO_2	278	365	379	14
CH_4	0.732	1.745	1.774	0.029
N_2O	0.270	0.314	0.319	0.005
SF_6	0	0.0042	0.0056	0.0014
CF_4	0.040	0.080	0.074	−0.006

Source: USEPA (2002) and IPCC (2007a)

The GWP values are used to determine the CO_2 equivalent of different emissions. This is done by multiplying the quantity of emission of a gas with its GWP value. For example, 1 ton of methane emission is equivalent to 62 tons of CO_2 for a time horizon of 20 years but over a 100 year period, the global warming potential changes to 23 tons of CO_2. Also note that the IPCC revises its GWP factors. In its 1996 report, the GWP for methane was taken as 21 for the 100 year horizon but was increased to 23 in its 2001 report. In the recent report (2007), the value has been increased to 25.

The concentrations of these gases are found to rise as a result of human activities. Table 26.2 indicates the changes in concentrations of some GHGs.

According to IPCC (2001a, 2007a), the concentration of GHG has increased steadily since the industrial revolution. The changes in concentrations of CO_2, methane and NO_x increased the radiative forcing due to these GHGs. On the other hand, due to industrial activities the emission of sulphates has increased significantly. The sulphate concentration leads to negative forcing and attenuates the global warming problem.

As can be seen from Table 26.2,

- CO_2 concentration has increased from 280 ppm in the pre-industrial age to 379 ppm in 2005, an increase of 35% (IPCC 2007a). This level of concentration has not been noticed in the past 650,000 years (IPCC 2007a).[3] Burning of fossil

---

[3] Summary for Policy makers, P.2.

fuels contributes mostly to this increase in CO_2 emission while changes in the land use contribute to some extent.
- Methane concentration has increased from 732 ppb in the pre-industrial era to 1,774 ppb in 2005, registering an increase of above 140% (IPCC 2007a). Fossil fuel use and agriculture contribute mainly to the anthropogenic sources of methane in the atmosphere but the relative shares are not well determined (IPCC 2007a).
- Nitrous oxide concentration on the other hand has increased relatively less, from 270 ppb in the pre-industrial era to 319 ppb in 2005, representing an increase of 18%. About one-third of N_2O emission is anthropogenic in nature, coming essentially from agriculture.

The combined radiative forcing due to the increase in concentration of these emissions is estimated at 2.3 W/m^2 (IPCC 2007a). Such an increase is unprecedented over the past 10,000 years. IPCC (2007a) indicates that the contribution of the anthropogenic sources of emission to this radiative forcing is 1.6 W/m^2, implying a warming effect (see Table 26.3).

Table 26.3 Components of global radiative forcings

RF components	Nature	RF values (W/m^2)	Spatial scale	Level of scientific understanding
CO_2	Long-lived gases, anthropogenic	1.66 (1.49 to 1.83)	Global	High
Methane	Long-lived gases, anthropogenic	0.48 (0.43 to 0.53)	Global	High
NO_x	Long-lived gases, anthropogenic	0.16 (0.14 to 0.18)	Global	High
Halocarbons	Long-lived gases, anthropogenic	0.34 (0.31 to 0.37)	Global	High
Ozone	Stratospheric, anthropogenic	−0.05 (−0.15 to 0.05)	Continental to global	Medium
Ozone	Tropospheric, anthropogenic	0.35 (0.25 to 0.65)	Continental to global	Medium
Stratospheric water vapour from methane	Anthropogenic	0.07 (0.02 to 0.12)	Global	Low
Surface albedo	Land use, anthropogenic	−0.02 (−0.04 to 0)	Local to continental	Medium to low
Total Aerosol	Direct effect, anthropogenic	−0.5 (−0.9 to −0.1)	Continental to global	Medium to low
Total Aerosol	Cloud albedo effect	−0.7 (−1.8 to −0.3)	Continental to global	Medium to low
Linear contrails	Anthropogenic	0.01 (0.003 to 0.03)	Continental	Low
Total net	Anthropogenic	1.6 (0.6 to 2.4)		
Solar irradiance	Natural	0.12 (0.06 to 0.30)	Global	Low

Source: IPCC (2007a)

26.1 Climate Change Background

IPCC (2007a) reports that:

- There is clear evidence of warming of the climate system.
- Over the past 100 years the temperate has increased by 0.74°C but the increase in the past 50 years was double that of the 100 year average.
- The sea level on average has increased 1.8 mm per year between 1961 and 2003, with a clear indication of higher rise between 1990 and 2003. It is however not clear whether this represents a long-term change. The overall sea level rise in the twentieth century is estimated at 0.17 m.
- The snow and glacier cover is declining in the mountains, leading to sea level rise.

Precipitation level has also changed in various parts of the world. Significantly higher precipitation was observed in the eastern parts of North and South America, northern Europe, and northern and central Asia. On the other hand, drying has been noticed in Sahel, the Mediterranean, southern Africa and parts of southern Asia. The frequency of heavy rain-days has increased. Similarly, there is a reduction in the cold events (cold nights, frosts, etc.) while the hot events have increased.

IPCC (2007a) suggests that "most of the observed increase in global average temperatures since the mid-20th Century is very likely due to the observed increase in anthropogenic greenhouse gas concentrations". It also concludes that "it is very unlikely that climate changes of at least the seven centuries prior to 1950 were due to variability generated within the climate system alone."

The report presents model simulation results to show that when the natural factors are only considered, the observed pattern of temperature rise does not match with the model results. But the best agreement results when both natural and anthropogenic factors are combined. Thus the global warming problem is attributable to human activities.

The primary source of warming CO_2 accumulation involves small changes in CO_2 flows relative to large stocks. Emissions from fossil fuel burning add 6.3 Gt C/year (compared to a stock of fossil fuels of 16,000 Gt C). Deforestation adds another 1.6 Gt C/year. Plants absorb 2.3 Gt C/year (Carbon stock of 500 Gt C while land holds 2,000 Gt C) while another 2.3 Gt C/year is reabsorbed into the ocean (the stock of Carbon here is 39,000 Gt C). This leaves 3.3 Gt C/year of annual accumulation in the atmosphere, which causes the greenhouse effect. This is shown in Fig. 26.2. The annual emissions are less than 1% of the stock of carbon contained in the atmosphere (750 Gt C). Different segments of the natural carbon cycle operate on different time scales. For example, the accumulation of carbon takes millions of years to form fossil fuels or carbonate rocks (RCEP 2000). On the other hand, carbon is absorbed by the growth of vegetation on a regular basis. A small change in the natural flows and stocks could have significant impact on the global carbon balance and management (Cline 1991).

The Fourth Assessment Report compared the results of the previous Assessments with the actual observed data. The First report suggested an increase of 0.3°C per decade between 1990 and 2005. The Second report revised this to

Fig. 26.2 The carbon cycle.
Source: IPCC (2001a)

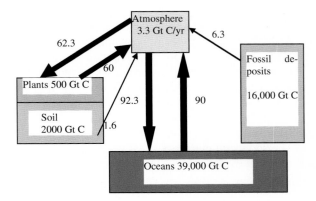

0.15°C per decade by considering the cooling effect of aerosols—a knowledge which was not available in the First report. The Third report followed the Second report in its projection. The observed data broadly confirms the above forecasts, thereby adding confidence to the process.

For the Fourth Assessment Report special scenarios were considered to analyse the effects of different levels of concentrations of GHG emissions by the end of the twenty-first century. One scenario assumed the concentrations constant at 2,000 levels while others assumed concentrations varying of 600, 700, 800, 850, 1,250 and 1,500 ppm in 2100.

Based on the analysis reported there, IPCC (2007a) suggested that the following effects could be expected:

- The earth surface temperature is expected to rise between 1.8 and 4°C. The warming expected to be highest over land and most high northern latitudes and least Southern Ocean and parts of the North Atlantic Ocean. Even if the concentrations are held at the 2,000 levels, the global warming is expected to continue in the short term due to "committed warming".
- The global mean sea level is projected to rise by 0.18–0.59 m between 1990 and 2100. The lower range is almost double the range indicated in the Third report but the upper range is significantly lower.
- Ice cover is expected to contract and sea-ice is expected to shrink.
- Heat events are likely to be more frequent in the future, with frequency of tropical cyclones increasing with higher peak wind speed and higher precipitation.
- Changes in the precipitation are likely as well. Increases in the precipitation in higher altitudes and decreases in subtropical regions are quite likely.

IPCC (2007b) reports the impacts of climate change based on models and other studies. Possible adverse effects include:

- Increases in fresh water availability are expected in high altitudes but decrease in draught-affected areas. The risk of flood increases as well.

- The global crop production potential is likely to increase with an increase in global temperature of 1–3°C but above this the crop yield reduces.
- Many millions of people are expected to be affected by flood every year due to sea-level rise by 2080. The coastal areas are likely to be prone to such disasters.
- The effects of climate change on the health will vary from one area to another. Those who are unable to adapt or have difficulties to adaptation will be adversely affected.
- Least developed countries are more vulnerable to climate change damages.

As some of the GHGs have long lives, their effect will be felt even after many years of their emission. This has two implications: the present problem is essentially due to the emissions released into the atmosphere earlier, which creates the problem of ownership responsibility of the problem. The second aspect of this is that even after the GHG emissions are stabilised, their effects will be felt for decades, albeit to a lesser extent. This suggests that lower the level of stabilisation, the smaller the temperature change (IPCC 2001a).

In order to control the greenhouse effect, actions will be necessary. This will be considered in the next section where economic issues related to climate change will be considered.

26.2 The Economics of Climate Change

26.2.1 Problem Dimension

As we discussed in Chap. 23, there are certain goods which are non-rival in consumption and are difficult to exclude. The Earth's atmosphere and the oceans show such characteristics. They are difficult to carve up into private property and their marginal cost of supply is zero. They are owned as common resources in the sense that everybody and anybody has rights to use them, the access is not restricted (i.e. they are open-access resources) and nobody has to pay for the use of these resources. As is common with such properties, as nobody owns it, nobody cares for it but everybody depends on these resources. There is overuse of the resource and degradation of the resource detriments all. Thus they are public goods in a open-access property regime and constitute a classic case of externality.

In addition, the climate has a global dimension. Everyone in the world pollutes the atmosphere through energy-use and other activities. Similarly, the effects of the atmospheric and climate change will affect everybody as well. The global nature of the problem adds complexity in managing the problem (CBO 2003):

- There are multiple actors and reaching a collective agreement is a challenging task. The sources of emissions vary across the regions of the world and there is uncertainty about the quantum, responsibility and the effects of such emissions.
- As is common with public goods, there is the free-rider problem. Each member has an incentive towards cheating by taking no action for containment of the

problem, considering that others will take the necessary action. If everybody cheats, the collective action of containment is sure to fail.
- The historical and economic contexts of the member countries are not identical. The responsibility towards the problem and the capacity to take containment action vary significantly. This makes an agreement difficult.

Finally, there is the time dimension of the problem. Being a long-term problem, it affects not only the present generation but also future generations. If actions are taken now, the future generations may have less wealth but a better environment. Similarly, if actions are not taken, the future generations may be left with somewhat more wealth but a degraded environment. This leads to the questions whether to sacrifice now for the future generations or not. In any case, the decisions have to be taken without consulting the affected parties—the future generations.

These three aspects complicate the economics of climate change.

26.2.2 Overview of GHG Emissions

According to UNFCCC (2009), the total GHG emission (without counting for the removal from LULUCF (Land Use, Land Use Change and Forestry) from Annex 1 Parties decreased from 18.8 Gt in 1990 to 18.1 Gt of CO_2 equivalent in 2007, representing a decrease of 3.9%. But as the economies in transition have seen a rapid fall in their emissions, the total emission from non-transition countries has actually increased by 11.2% between 1990 and 2007. Of all GHGs, CO_2 account for more than 80% of the emissions. Similarly, the energy sector emitted about 15 Gt of CO_2 equivalent, accounting for 80% of the total GHG emissions from Annex 1 Parties. Industrial processes and agriculture account for the majority of the remaining emissions. The regional trend of CO_2 emission from the energy use is shown in Fig. 26.3.

The information in respect of non-Annex 1 Parties is available to a lesser extent. The UNFCCC reports that in 1994, 122 countries emitted 11.7 Gt of CO_2 equivalent in 1994, of which 7.4 Gt of CO_2 equivalent in the form of CO_2.

Among Annex 1 Parties, Australia, Canada, Japan and the United States have failed to reduce their greenhouse gas (GHG) emissions below 1990 levels in 2007 while the European Union has been successful in controlling its emissions. The European Environment Agency in its recent submission to the UNFCCC (EEA 2009) has reported that in 2007 the EU-15 emitted 4.3% less of GHG relative to 1990 levels excluding LULUCF contribution, while the 27 members of the EU is have achieved an overall reduction of 9.3% compared to the 1990 level (see Fig. 26.4).

In 1990, the GHG emission from the fifteen members of the European Union was 4.2 Gt excluding LULUCF. This volume has declined to 4.06 Gt in 2007, thereby recording a fall of about 4.3% (EEA 2009). About 77% (or two-thirds) of

26.2 The Economics of Climate Change

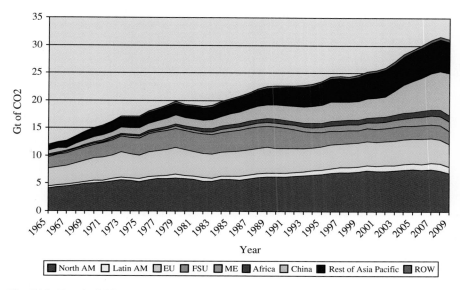

Fig. 26.3 Trend of CO_2 emission. *Data source*: BP Statistical Review of Energy (2010)

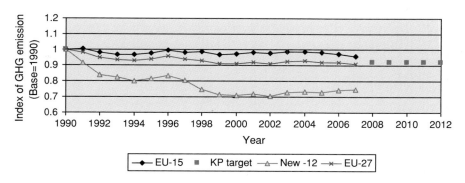

Fig. 26.4 GHG emission trend in EU. *Source*: Bhattacharyya and Matsumura (2010)

the GHG emission in 1990 came from energy-related activities while the remaining originated from other industrial processes and activities. The share of the energy-related emission has slightly improved in 2007, reaching almost 80% of all GHG emission in the region.

Clearly, different members contributed differently to the GHG emissions in EU15 (see Fig. 26.5). Germany and the United Kingdom are the dominant emitters in the region, contributing about 47% of the emissions in 1990 but in 2007, their share has fallen to 39%. France and Italy are two other large emitters in the region—each contributing about 13% each to the regional GHG emission. Spain accounts for about 10% of the emissions. Taken together, these five members account for about 75% of the GHG emission in EU15.

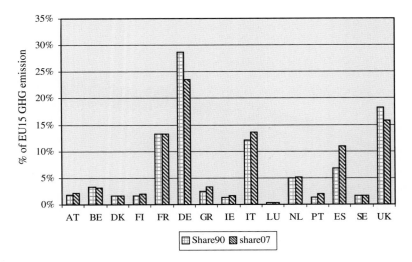

Fig. 26.5 Contribution of EU-15 in GHG emission reduction. *Source*: Bhattacharyya and Matsumura (2010)

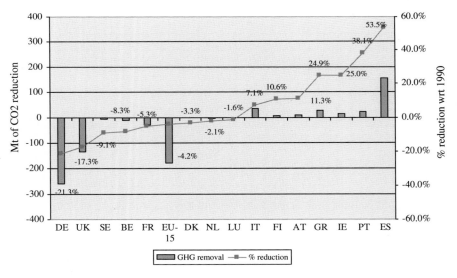

Fig. 26.6 GHG emission removal in EU15. *Source*: Bhattacharyya and Matsumura (2010)

The total GHG removal in EU15 was about 178.5 Mt between 1990 and 2007 (Fig. 26.6). Germany and the United Kingdom were most effective in mitigating the emission: Germany reduced 259 Mt of GHG while the United Kingdom mitigated about 134 Mt. On the other hand, Spain, Italy, Portugal, Ireland, and Greece (among others) contributed negatively (i.e. their emission increased within

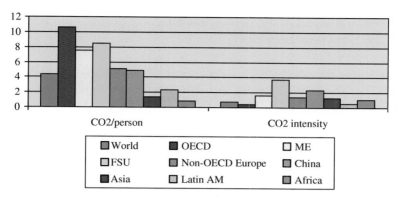

Fig. 26.7 CO_2 emission intensity. *Note*: CO_2 per person is in t CO_2eq per person, CO_2 intensity in kgCO_2eq per US$2000. *Data source*: IEA (2010)

this period): Spain added 154 Mt of extra emission while Italy added another 36 Mt.

Figure 26.6 clearly shows the disparity in the emission reduction levels between 1990 and 2007 by member states. About one half of the members were not able to reduce their emissions, which cancelled out other countries' better performance. All together, EU15 has achieved a reduction of 4.3% in the period.

However, in terms of emission per person there is a wide variation regionally and between the lowest and the highest emitters globally (see Fig. 26.7). For example, the global annual average of CO_2 emission was 4.39 ton per person in 2008 while a person in Africa, on average, emits just 0.9 ton. On the other hand, any person in the OCED countries on average emits more than 10 ton of CO_2 per year. But the picture changes completely when emission intensity in terms of economic activity is considered. Globally, the average emission per dollar of economic output (in constant 2000 prices) in 2008 was 0.73 kg whereas the OECD countries emitted just 0.41 kg for every dollar of output. The countries of the Former Soviet Union emitted much more—about 3.71 kg per dollar of output. In fact, all the regions outside OECD countries had higher CO_2 intensity than the OECD average for 2008. Figure 26.7 captures the debate and the dilemma very well: the developing countries with higher population base are showing significantly lower levels of emissions per person but their economic activities are relatively polluting, thereby providing an opportunity for emissions reduction.[4] On the other hand, the developed countries are consuming much more energy and emitting more on a per person basis but their large economic output brings the emission intensity down. The search for an acceptable solution for all parties has to recognize these differences.

---

[4] It is important to highlight that the emission in developing countries for goods exported to the developed countries appears in their account, which distorts the picture to some extent.

26.3 Economic Approach to Control the Greenhouse Effect

The basic approach to economic analysis of the climate issue relies on cost-benefit analysis. The idea is to capture the costs and benefits of the greenhouse effect and the policies through two fundamental functions: the climate change damage function and the climate change abatement cost function.[5]

The damage function captures the cost of the global warming effect to society and would include the cost of changing crop production, land loss to oceans, loss of biodiversity, forestry, fishery, loss to human settlements, buildings, energy sources, etc. On the other hand, the abatement cost function captures the costs borne by the economy in pursuing policies to slow or prevent global warming. These policies would include cost of fuel switching, CFC substitution, creating coastal structures, etc.

The objective of the analysis is to determine the efficient strategies to reduce the costs of climate change. Such a strategy will maximise the net benefit or gain. This is carried out using the marginal damage cost and marginal abatement cost functions. The marginal abatement cost function is an upward sloping curve showing the incremental cost of reducing GHG emission by one unit. Zero market price on GHG emissions makes the first units of GHG reduction virtually free. Therefore, the curve starts at zero. But as the GHG reduction increases, the abatement cost increases as well. This gives rise to the upward sloping curve. The marginal damage function on the other hand measures the incremental cost to the economy of an extra unit of GHG emission. This function is less understood but it is generally considered that higher levels of GHG will hurt the global economy. The curve has been drawn as wavy curve to reflect the uncertainty about its shape. The efficient level of control is at point E in the Fig. 26.8.

Similar to any policies, climate policies involve trade offs between:

- Investing now to reduce damages in the future; as there are alternative options possible (adaptation, mitigation or no action), the decision involves which one or which combination to follow and to what extent.
- Climate policies with non-climate policies that compete with the funds for other benefits such as education, health, etc.

As in any investment decision, the choice of discount rate plays an important role. In the climate policies, the importance of discount rate is even higher due to the long time horizon involved. Normally people place higher preference to present than to the future. This practice is known as discounting. Those who place high value to present are said to have a high discount rate compared to those who prefer to save for the future. Normally the market interest rates reflect the arbitrage between present spending and future saving. As public investments compete with

---

[5] There is now a well-developed body of literature on this subject. Nordhaus (1991), Stern (2007), CBO (2003), and Goulder and Pizer (2006) for further details and simple analysis of the issue.

Fig. 26.8 Efficient level of GHG emission control.
Source: Nordhaus (1991)

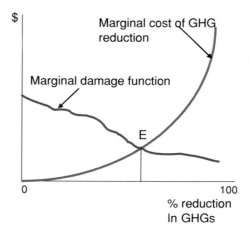

private investment for social welfare, normally economists apply market interest rates for investment appraisals. The rate of return on investment is expected to be at least equal to the interest rate (or cost of capital).

However, for long term investments, the choice of the discount rate becomes a matter of controversy. This is because the cash flow at a distant future appears to be worth little at present even at moderate discounting. There is no consensus on the discount rate to be used but a low discount rate can be argued for the following reasons (CBO 2003):

- uncertain investment opportunities in the long term;
- the possibility of risk free investment for long term does not exist;
- historical market interest rates may not capture people's attitude toward distant future.

A lower discount rate will then justify increasing investment in providing benign environment and other stock expansion thereby reducing the current consumption in favour of wealth creation for the future generation.

26.3.1 Integrated Assessment

Numerous studies have attempted to analyse the benefits and costs of mitigating climate change problem (see IPCC (2007c) and Stern (2007) for example). Many such exercises attempt to integrate economic development and the climate problem in an integrated manner (and hence the name integrated assessment). Essentially, these models attempt to see the impacts of alternative socio-economic development paths on the emissions, climate and the effects on the economy and the society. These analyses try to capture:

- impacts of climate on human beings and other natural systems;
- costs and benefits of alternative mitigation options;

- emission reductions due to these options;
- their impact on economic growth; and
- the distributional impacts.

As can be imagined, such assessments face a large number of uncertainties. The nature of the uncertainty is both scientific and economic. Scientific uncertainties include (CBO 2003):

- Given GHG emission, how much will accumulate in the atmosphere?
- How will a given change in the concentration affect the global climate?
- What will be the distribution of that global climate change and when is that going to happen (i.e. how rapidly?)
- How the regional climate change affect the other natural and economic systems (sea levels, agriculture, forestry, fishing, etc.)?
- Will global warming lead to sudden changes in climatic conditions?

The economic uncertainties include (CBO 2003):

- How the world population and economic grow during the time horizon of assessment?
- What will be the composition of economic activities and how much energy will come from fossil fuels?
- How will the mitigation policies affect the accumulation of GHGs?
- How much will these policies cost?
- How much future generations will value these policies?

As a complex global problem, the climate change issue raises a large number of fundamental issues. These include:

- What should be the overall reduction in GHG emission? At what price?
- Who should reduce the emission, how much and when?
- Should there be income transfers from high to low income groups?
- How to prevent free-riding and cheating?
- Can an international co-operative solution be achieved?

No general or consensus answers exist to the above questions. In what follows, a brief review of alternative options to mitigate the problem is presented.

26.4 Alternative Options to Cope with Global Warming

26.4.1 Generic Options

The question that arises is what actions can be taken to cope with the problem of global warming. Three generic mitigation options can be considered (see e.g. Cline 1991):

1. *Slow or prevent global warming*: This has received the most attention in public debate. The focus has been on reducing emissions from fossil fuels by fuel

switching or by reducing emission intensity (i.e. emissions per unit of economic output). This could be done by switching to low-carbon fuels or by reducing energy consumption. Preventing CO_2 from entering the atmosphere or by increasing removal of emission from the atmosphere can also provide effective solutions. RECP (2000) indicates that removing CO_2 from the exhaust using "end-of-pipe" technology used for treatment of pollutants is technically feasible. CO_2 so removed could then be disposed of deep into the geological strata such as depleted oil and gas field, deep underground formations containing saline water, deep coal formations, etc. The report suggests that 200 Gt C could be stored in Europe, which is equivalent to 770 years of emission from European power plants.[6] Similarly, increased tree plantation can be used to remove atmospheric CO_2 emissions. RCEP (2000) indicates that for significant impact, large areas have to be covered with plants or large deforested areas have to be reforested. The cost for such options may vary depending on the sites and their social, political and environmental impacts have to be ascertained before launching such large programmes.
2. *Offset climate effects*: A second option stems from the possibility of climatic engineering whereby carbon is removed from the atmosphere through technological solutions. Proposals in this category include: shooting iron particles in the atmosphere and fertilising the ocean with trace iron. RCEP (2000) suggests that the biological activity of the oceans could be increased by sprinkling iron on the ocean surface. However this may require a large quantity of iron to be sprinkled over the ocean surface and may be a costly option. Moreover, the technology has not been applied on a large scale.
3. *Adapt to warmer weather*: A third option is to accept the warm weather and adapt. This can be the result of automatic response of the society to the gradual change in the weather. Governments could also take pre-emptive measures to limit harmful climatic impacts. These include building dykes, R&D on heat resistant crops, etc.

In most cases, the first category of options are considered and evaluated, as preventive steps have to be taken well in advance. Within the broad category, the options related to the energy supply sector are considered here. Other important sectors include energy using sectors like industry and transportation, agriculture, forestry, etc. Table 26.4 provides a list of options in various sectors.

26.4.2 National Policy Options

A large set of policy options is available to any individual country to choose from. These are normally categorised as market-based instruments, regulatory

---

[6] See RCEP (2000, Chap. 3) for details.

Table 26.4 Mitigation options at the sector level

Sector	Option
Energy	Efficiency improvements for end-use appliances and energy supply technologies
	Improvements in transmission and distribution systems
	Switching to low carbon fuels and renewable energies
Transportation	Fuel efficiency improvements
	Fuel quality improvements
	System planning and demand management
	Switching to non-energy intensive modes
	Modal shifts and fuel switching (including bio-fuels and other alternatives)
Forestry	Better forest management and control of forest clearing
	Growing forests and re-forestation
	Re-use products and reduce demand
Agriculture/land use	Better fertiliser use practices and control systems
	Crops of large carbon sequestration capabilities
	Livestock management
	Use agro products as a fuel

instruments and voluntary agreements. Similarly, a group of countries could decide to act collectively and employ one or more policy instruments.

The national policy options could be divided into two categories: those which are not specifically designed for climate change issue but have impacts of GHG emissions and those which are related to climate and other environmental policies. The first category of options include: structural reform policies, price and subsidy reform policies and liberalisation of energy industries. During the 1990s, many countries embarked on market oriented structural reform of their economies. Such reforms included trade liberalisation, financial deregulation, privatisation of state-owned enterprises, tax reform, etc. (IPCC 2001b, 2007c). These reforms have affected the GHG emissions. For example, IPCC (2001b) notes that energy use in China has increased by 40% since 1978 but its energy consumption per unit of output fell by 55% between 1978 and 1995. Such improvements in energy intensity result in lower emissions of GHG, even without adopting specific mitigation measures.

Distorted energy prices are quite common in many countries. Many fuels used by domestic and industrial users are provided at a subsidised rate. IEA (2010) suggests that in 2008, the global energy subsidy amounted to $557 billion, of which oil products received $312 billion and natural gas received $204 billion. According to the same source, the removal of subsidies could reduce global energy demand by 5.8% by 2020 and could reduce CO_2 emissions by 6.9% by 2020. This will amount to a reduction of 2.4 Gt of CO_2 emission by 2020. However, generally such studies do not capture the possibility of returning to the traditional energies when subsidies are removed, especially by the poorer sections of the population. If each household needs a certain amount of energy to meet their basic needs, it is quite likely that the market-based prices will drive some consumers out of the

26.4 Alternative Options to Cope with Global Warming

modern energy path to the traditional energies. The health and environmental consequences of such damages can be significant (see Bhattacharyya 1996 for such a study).

Liberalisation of energy markets provides better choice to suppliers and consumers of energy. The energy market is many countries is characterised by state-owned, vertically integrated monopolies. The opening of the market to increase competition and choice can have significant environmental effects. The impact of liberalisation can be positive or negative in environmental terms. For example, the British experience resulted in a dash for gas and closing of coal-based plants. This has resulted in an environmental dividend. On the other hand, in Japan, post-liberalisation, independent power producers entered the market but relied on coal and fuel oil which are dirty fuels. This is expected to have a negative impact in terms of GHG emissions (IPCC 2001b).

Specific policies for climate and environmental purposes come under different forms and include:

- *Regulatory approaches*: As discussed in Chap. 23, there are different types of standards which are used for environmental purposes. Energy efficiency standards are one such option for controlling GHG emissions. These standards are widely used and are growing in number.
- *Carbon tax and other charges*: The basic principle here is similar to the environmental taxes and charges. As climate change is a free, global public good, it generates externality. A tax can be used to internalise the externality. As each polluter faces a uniform tax on emissions per ton of CO_2eq, the tax would result in a least cost solution in a first–best world. However, in a real world, the first–best conditions are hardly met and therefore, carbon taxes may not result in least cost solutions (IPCC 2001b). Some advantages and disadvantages of carbon tax as a policy tool are indicated in Table 26.5. In practice, some countries have

Table 26.5 Advantages and disadvantages of a carbon tax

Advantages	Disadvantages
Corrects externality	Target reduction may not be achieved unless relevant elasticities are known
Raises revenue Can be used to offset other distorting taxes	Impose deadweight losses Can be significant for some countries
Low compliance cost for industry Those with lower abatement cost will abate	Expected to be regressive in nature
Incentive to adopt cleaner technology and energy conservation	A new tax may be a politically sensitive issue
Can be modified relatively easily	A new tax may be a politically sensitive issue Needs to be adjusted to inflation, technical progress and increases in emissions

Source: Nordhaus (2001) and Bohringer (2003)

adopted carbon tax as a mitigation option. Nordic countries such as Sweden, Norway, Denmark and Finland have introduced energy taxes partly based on carbon content. UK, Germany, Switzerland and Italy have also introduced carbon taxes to achieve their climate change commitments (IPCC 2001b). However, the effectiveness of these taxes does not appear to be significant because of unilateral nature of the taxes that tend to affect the performance of the industries and the economy.

- *Tradeable permits*: The principles are same as discussed earlier for other environmental pollutions. Each participating polluter receives an allocation of permits (gratis or by auction) initially. If a polluter pollutes less than the permits it holds, it can sell the excess permits to those who exceeds their quota. The market price of the permits provides the signal for corrective action: whether to reduce the pollution below the allowable limit and sell the permits or buy permits to meet the target. For GHG emission, the European Union experiment is discussed below.
- *Voluntary agreements*: There is no internationally agreed definition of voluntary agreements (VA). However, IPCC 2001b) defines VA as "an agreement between a government authority and one or more private parties, as well as a unilateral commitment that is recognised by the public authority, to achieve environmental objectives or to improve environmental performance beyond compliance." Voluntary agreements take various forms: they can be between the government and firms; between government and the industry association; VAs can relate to general issues such as R&D, energy efficiency but they can also specify quantified targets in some cases. Most of the VAs are not legally binding.

This is a relatively new instrument. Since 1996, more than 300 VAs have been singed in the EU. This instrument is being used in Japan, Germany and the Netherlands. The advantages of this system are: the transactions costs are low. It relies on a consensus approach rather than imposing a target or policy or instrument on the parties. The parties are free to choose their method of achieving the targets assigned to them. It appears that this option works when low cost options, often technical solutions to the problems, are available.

- *Informational instruments*: Although in standard economic models it is assumed that information is available freely and fully, and all the agents are fully informed, in reality information is costly to obtain. Information gaps result in poor decisions, increased uncertainties and higher risks (IPCC 2001b). Policy instruments are used at three levels to improve information: to raise awareness about climate change; to stimulate research on mitigation options and climate issues; and to help implementation of measures (IPCC 2001b). Options for promoting information include: educational programmes and labelling.
- *Subsidies and other incentives*: Subsidies are similar to taxes but instead of payments made by the polluters, they receive payments for every unit of GHG emission reduction. In theory, both tax and subsidies produce the same results in the short run but subsidies encourage entry into the polluting activity and

therefore lead to inefficient use of resources in the long run. Besides subsidies, other incentive policies are possible to promote GHG mitigation and include demand side management, promotion of green power and research and development policies.

26.4.3 Emissions Trading System (ETS) of the EU

From 2005, the European Union (EU) has embarked on a novel project of emissions trading involving GHGs. It is considered a new "Grand Policy Experiment" because of its scope, size, complexity and international nature (Kruger and Pizer 2004). The program started on January 1, 2005 and applied to member countries. The first phase (or the warm up phase) ran up to 2007. The second phase will coincide with the first commitment period of the Kyoto Protocol (KP), i.e. 2008–2012. Thereafter, the program will run in 5 year phases.

Initially, the program will cover CO_2 emissions from four broad sectors: production and processing of iron and steel, minerals (such as cement, glass or ceramic production), energy and paper and pulp (see Fig. 26.9). Any installation with an emission exceeding a certain limit will be included in the program. Around 12,000 installations (see Fig. 26.10 for the distribution by country) are participating in the first phase of the program, covering 46% of the EU CO_2 emission.

The initial allocation of permits was free of charge. Each member state allocated a part of its GHG reduction target under the burden sharing arrangement of the EU associated with the KP for use in the ETS. Then the state decided how much of this target will be allocated to each participating sector and how the sectoral target will be assigned to each installation. This was done through a National Allocation Plan (NAP). The Plan was reviewed by the European

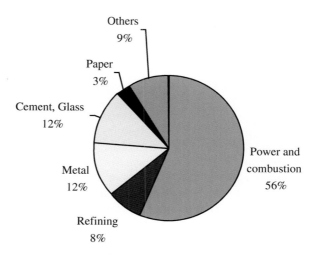

Fig. 26.9 ETS coverage by industry sector. *Source*: CITL website

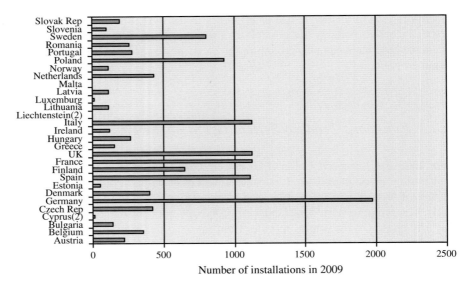

Fig. 26.10 Number of installations covered by ETS by country. *Source*: CITL website

Commission. Although the allocation scheme was supposed to be ready by October 2004, in some cases the approval from the Commission came in May 2005. The Commission revised the allocations in a number of cases.

Each member state maintains its own registry of transactions, although two or more states may join together. The Commission operates an independent transaction log to verify transfers between national registries. Continuous emission monitors are optional; it is expected that installations will rely on emission factors, production and fuel use to estimate emissions.

Banking and some amount of borrowing are allowed within any compliance period. Banking between first two periods is also allowed. Installations face a penalty for emissions in excess of surrendered permits. In the first period, the penalty is €40/ton of CO_2 but increases to €100/ton of CO_2 in the second period.

The results of the first year of warming up shows that most countries issued excess permits than required to meet the target. This resulted in a collapse of the permit prices. There were some other teething troubles as well: recording and monitoring was not ready in some countries. For the current phase, the EU reviewed the NAP and tried to ensure that the excess allocation does not happen in the future. However, the actual phase started with an economic crisis of 2008 and consequently, the demand for energy fell, thereby requiring even lesser permits for compliance. Thus the problem of oversupply continued in the recession-hit economy. The compliance rate has improved and only 2% of the installations did not surrender their allowances by due date. The market appears to work as such but the debate over the appropriate economic tool for emission control has resurfaced again.

26.4.4 International Policy Options

The Kyoto Protocol (KP) recognised three flexibility mechanisms, namely, international emissions trading (ETS), and two project-based mechanisms, i.e. Joint Implementation and the Clean Development Mechanism. Other instruments that have been discussed in the literature include international carbon tax, direct financial transfers, etc. We briefly discuss these policy options below (see IPCC 2001b for further details).

International Emissions Trading: The KP allows trading of emission quotas agreed by Annex I Parties amongst themselves during the first commitment period of 2008–2012. But Sect. 17 of the KP specifies that "any such trading shall be supplemental to domestic actions for the purpose of meeting quantified emission limitation and reduction commitments". The issue of "hot air" (where the initial allocation to some countries could be more than their expected emissions) becomes important here. If the "hot air" is allowed to trade, it may be cheaper for other countries to meet their KP obligations, while restrictions on hot air trade would effectively require more reduction of GHGs. The EU ETS comes under this system and will integrate with the instruments of other systems.

Joint-Implementations (JI)[7]: The Kyoto Protocol allows an Annex I country to participate in a project in another Annex I country and to receive emission reduction units (ERUs) from the project, which could then be used to meet the investor country's KP obligations. The JI has been implemented in Annex I Parties of the economies in transition in Eastern Europe.

Any JI project has to be approved by the Parties to the Annex I of the KP and the ERUs can only be transferred between Parties. JI projects have to satisfy an additionality criterion which requires that the reduction in emissions should be additional to reductions without the project. Nuclear facilities cannot be used to generate ERUs. Projects starting from 2000 and meeting the requirements of JI may be listed as JI but ERUs will be applicable to the period from 2008 onwards.

Clean Development Mechanism[8] : This mechanism allows Annex I Parties to implement GHG mitigation projects in non-Annex I countries and to receive certified emissions reductions (CERs), which is expected to be a tradable commodity. This is the only mechanism in which non-Annex I parties could participate. The CDM aims to help developing countries achieve sustainable development and provide the developed countries with the flexibility of achieving their emission reduction targets by taking credit from emission reductions in developing countries. We will discuss the CDM separately in Chap. 27.

Direct International transfers: It is now recognised that developing countries need "new and additional" financial resources as well as technology transfers to achieve sustainable development (SD) and to implement mitigation options limiting climate change. This funding should be in addition to the Overseas

---

[7] See http://unfccc.int/kyoto_mechanisms/ji/items/1674.php for further details.
[8] See http://unfccc.int/kyoto_mechanisms/cdm/items/2718.php for further details.

Table 26.6 Price or quantity instruments for climate change

Price	Quantity
Performs better under uncertainty	It is efficient if properly designed and quantity set correctly
Revenue neutral taxes avoid additional deadweight loss	High transaction costs for monitoring and enforcement
Monitoring and enforcement difficult	More linked to technologies
Cheating through compensatory taxes highly likely	Allowance price may be volatile
Asymmetric impact on countries	Works better when costs are better known
No guarantee that targets will be met	

Development Assistance (ODA) received by developing countries. However, in practice, it is noticed that the ODA has shown a declining trend in the late 1990s and lack of financial resources in developing countries is a major concern. Similarly, transfer of environment-friendly technologies to developing countries is a requirement for achieving SD and mitigating climate change. Such technology transfers should be on a "grant or concessional basis on non-commercial terms" (IPCC 2001b). However, there exist a number of barriers affecting technology transfers, including high transaction costs, lack of understanding of local demand and lack of vision about emerging technologies (IPCC 2001b).

International carbon tax: This is an extension of the national tax logic. The carbon tax will be harmonised at the international level which implies that carbon emissions in each country will be associated with a uniform price. This allows comparison of options at the international level and those policies/options which costing less than the tax payments will be implemented. There are two main concerns about this approach (IPCC 2001b): first, taxes do not ensure whether a particular level of emission control will be achieved unless the taxes are correctly set. Setting the tax correctly is difficult due to lack of knowledge on damage costs. Second, it may be difficult to negotiate an agreement on international taxes. A comparison of price (tax) and quantity (emission trading) options for climate change purposes is presented in Table 26.6.

26.5 Climate Change Agreements

26.5.1 UNFCCC[9]

After a decade of climate diplomacy, in 1992 at the United Nations Conference on Environment and Development (UNCED) in Rio de Janeiro, Brazil (popularly

---

[9] See http://unfccc.int/essential_background/feeling_the_heat/items/2914.php; RCEP (2000) for further details.

26.5 Climate Change Agreements

known as the Rio Summit), an overwhelming majority of the world leaders agreed to a Framework Convention on Climate Change (known as UNFCCC). The Convention was ratified by more than 180 nations and came into force 21 March 1994. The Convention recognised that there is a problem and it set up a process for agreeing to specific actions later. It agreed to stabilise GHG emissions to prevent climate change in such a way to allow the eco-system to adapt naturally. The Convention did not specify the concentration at which the climate will be stabilised; it only specified that the concentration should not reach a dangerous level.

Recognising that the past and current emissions have to a large extent originated in the developed countries, the Convention accepts a "common but differentiated" responsibility. It places the heaviest burden for combating the problem on the developed (rich) countries according to the "capabilities and social and economic conditions". The Convention accepts that the poor nations have a right to economic development and that they are more vulnerable to the effects of climate change. It called for transfer of environmentally sound technologies and additional financial resource to the developing countries to ensure their development.

The Convention placed an obligation on developed countries to reduce their GHG emissions to the 1990 levels by 2000. Although many of them have not met this requirement individually, it appears that they have collectively satisfied this requirement.[10]

26.5.2 The Kyoto Protocol

Although the Convention provided the basic framework, it did not specify detailed mechanisms to control the climate problem. The Conference of Parties (COP) in their first meeting in Berlin in 1995 decided to launch a new round of negotiations on stricter and more detailed targets. After intense negotiations for more than 2 years, the Kyoto Protocol (KP) was adopted in 1997 in Kyoto, Japan. Similar to the Convention, the KP did not flesh out all the operational rules of the Protocol. In its seventh meeting (COP 7) in Marrakesh, the rule book was adopted.

The developed countries agreed to reduce their collective emissions by 5.2% below 1990 levels during the first commitment period: 2008–2012. Table 26.7 provides the list of individual targets. The KP covers six main GHGs: three major gases—carbon dioxide, methane and Nitrous oxide and three long-lived gases—hydrofluorocarbons (HFCs), perfluorocarbons (PFCs) and sulphur hexafluoride (SF6). The Protocol become legally binding when at least 55 countries including developed countries accounting for least 55% of developed countries' 1990 CO_2 emissions have ratified it.

Upon receiving the required support in terms of ratification by states and emission coverage, the Protocol entered into force on 16 February 2005. However,

---

[10] UNFCCC website claims this.

Table 26.7 KP targets

Country	Target (1990[a] to 2008/2012) (%)
EU-15[b], Bulgaria, Czech Republic, Estonia, Latvia, Liechtenstein, Lithuania, Monaco, Romania, Slovakia, Slovenia, Switzerland	−8
US[c]	−7
Canada, Hungary, Japan, Poland	−6
Croatia	−5
New Zealand, Russian Federation, Ukraine	0
Norway	+1
Australia	+8
Iceland	+10

[a] Some EITs have a baseline other than 1990
[b] The EU's 15 member States will redistribute their targets among themselves, taking advantage of a scheme under the Protocol known as a "bubble". The EU has already reached agreement on how its targets will be redistributed
[c] The US has indicated its intention not to ratify the Kyoto Protocol
Note: Although they are listed in the Convention's Annex I, *Belarus* and *Turkey* are not included in the Protocol's Annex B as they were not Parties to the Convention when the Protocol was adopted
Source: UNFCCC website

the USA under the Bush administration decided not to participate in the Protocol. Absence of the USA reduces the effectiveness of the mechanism to a great extent. The American logic for abstaining from the Protocol is that it places an undue burden on them and that the Protocol does not include those developing countries who are important polluters.

As indicated earlier, the KP envisages three flexibility mechanisms: emissions trading, Joint Implementation and the CDM. These will not be repeated here. Although the Protocol employs economic instruments, it puts in place a regulatory system for verification and monitoring of the emission reduction. This adds to the cost of mitigation.

The Protocol has decided about the targets for the first commitment period. No long term commitment has been set. Negotiations at each stage are time consuming, costly and often difficult. Similarly, the penal provisions appear to be weak. Given that the compliance in the first commitment period may be low, strong penal provisions are required. Finally, it remains to be seen whether the Protocol is able to promote sustainable development in developing countries through better technologies and fund transfers.

26.6 Conclusion

This chapter has provided an overview of the climate challenge and presented the scientific reason behind the problem. It then discussed the problem dimension by looking at the emissions from different sources and regions. The basic economic framework was then discussed to find out alternative options, with a special reference to the energy sector. The chapter has then introduced the national and international initiatives in dealing with the problem. The evidence so far suggests that although the members of the European Union have been successful in reducing their emissions, other Annex 1 Parties are struggling in normal economic times. At the same time, the fast growing developing countries are emitting significantly higher volumes of GHG, which will de-stabilize the climate unless coordinated efforts are made.

Obviously, the Herculean challenge requires an out-of-the-box approach. There is no ready-made template as no country has resolved the problem yet. Therefore, each country would have to find its own solution through trial and experimentation. This calls for a multi-faceted strategy, which I have called elsewhere as MAGIC (Bhattacharyya 2010). The elements of this strategy are: effective demand Management, Adoption of internationally best-practice technologies, Good governance, effective use of Indigenous resources and Clean energy for all. It is important to recognise the need for a transition to a sustainable path and initiate actions so that in the long-term a better future can be ensured for all.

But managing such a change faces a number of major challenges—managing supply, involving better project management approaches, resource management in a broader sense, and environment and social responsibility management. The task is challenging and requires well co-ordinated efforts at all levels. Each country has to find its own road to energy sustainability and it requires an inclusive, open, co-operative solution that will benefit each country in particular and the world in general.

References

Bhattacharyya SC (1996) Impacts sectoriels de la suppression des subventions énergétiques: le cas de l'Inde'. Revue de l'Energie 479(July–August):376–390

Bhattacharyya SC (2010) Shaping a sustainable energy future for India: management challenges. Energy Policy 38(8):4173–4185

Bhattacharyya SC, Matsumura W (2010) Changes in the GHG emission intensity in Eu-15: lessons from a decomposition analysis. Energy 35(8):3315–3322

Bohringer C (2003) The Kyoto protocol: a review and perspectives. Oxford Rev Econ Policy 19(3):451–466

BP Statistical Review of Energy (2010) Downloadable data from www.bp.com

CBO (2003) The economics of climate change: a primer, a CBO study. In: The congress of the United States, congressional budget office. http://www.cbo.gov/showdoc.cfm?index=4171&sequence=0

Cline WR (1991) Scientific basis for the greenhouse effect. Econ J 101:904–919

EEA (2009) Annual European Community greenhouse gas inventory 1990–2005 and inventory report 2007. European Environment Agency, Copenhagen

Goulder LH, Pizer WA (2006) The economics of climate change, RFF DP 06–06. Resources for the Future, Washington

IEA (2010) World energy outlook, looking at energy subsidies: getting prices right. International Energy Agency, Paris 479 (July–August)

IPCC (2001a) The climate change 2001: scientific basis, chapter 1: the climate system: an overview. Cambridge University Press, London. http://www.grida.no/climate/ipcc_tar/
IPCC (2001b) Climate change 2001: mitigation. Cambridge University Press, London
IPCC (2007a) Working group I report, the physical basis. Cambridge University Press, Cambridge. http://www.ipcc.ch/ipccreports/ar4-wg1.htm
IPCC (2007b) Working group II report, impacts, adaptation and vulnerability. Cambridge University Press, London. http://www.ipcc.ch/ipccreports/ar4-wg2.htm
IPCC (2007c) Working group III report, mitigation of climate change. Cambridge University Press, London
Kruger J, Pizer WA (2004) The EU emissions trading directive: opportunities and potential pitfalls, Resources for the Future, Discussion Paper 04-24, Washington
Nordhaus WD (1991) To slow or not to slow: the economics of the greenhouse effect. Econ J 1991:920–937
Nordhaus WD (2001) After Kyoto: alternative mechanisms to control global warming. Paper for a joint session of the American Economic Association and the Association of Environmental and Resource Economists, Atlanta
RCEP (2000) Energy and the climate change. Royal Commission on Environmental Pollution, London. http://www.rcep.org.uk/newenergy.htm
Stern N (2007) The economics of climate change, the stern report. Cambridge University Press, Cambridge
UNFCCC (2009) National greenhouse gas inventory data for the period 1990–2007, FCCC/SBI/2009/12. UNFCCC Secretariat, Bonn. http://unfccc.int/resource/docs/2009/sbi/eng/12.pdf
USEPA (2002) Greenhouse gases and global warming potential values, EPA, US greenhouse gas inventory program. http://yosemite.epa.gov/oar/globalwarming.nsf/UniqueKeyLookup/SHSU5BUM9T/$File/ghg_gwp.pdf

Chapter 27
The Clean Development Mechanism

27.1 Basics of the Clean Development Mechanism

The Clean Development Mechanism (CDM) is one of the flexibility mechanisms of the Kyoto Protocol (KP), the other two being emissions trading and Joint Implementation. The Kyoto Protocol was adopted in the third Conference of Parties to the United Nations Framework Convention on Climate Change (UNFCCC) at Kyoto, Japan in 1997. This is a project-based mechanism of emissions trading and is the only mechanism involving non-Annex 1 parties[1] (or developing countries without any obligation to reduce the GHG emission).

The idea behind this cooperative mechanism is that one ton of greenhouse gas (GHG) reduced anywhere in the world has the same effect on the climate. Hence, economic principles would suggest that the least-cost options should be exploited first, wherever they may be located. Annex 1 countries can take credit of emissions reduced in a CDM project in a developing country to meet their targets. This adds more choice and flexibility to comply with the targets and offers economic solutions. The non-Annex 1 countries in turn receive capital for investments in projects and clean technologies to reduce their emissions and enhance socio-economic well-being.

Thus, the CDM has two key goals[2]:

- to promote sustainable development (SD) objectives in the host country (i.e. non- Annex 1 countries);
- to assist Annex 1 Parties to meet their GHG reduction targets.

---

[1] UNFCCC lists 38 counties in its Annex 1 who have a GHG reduction obligation. Any country not included in this list is called non-Annex 1 Parties. The KP puts the Annex 1 countries in its Annex B. Annex 1 parties are referred to as Annex B parties in the KP.

[2] Article 12.2 of the KP states: "The purpose of the Clean Development Mechanism shall be to assist Parties not included in Annex 1 in achieving sustainable development and in contributing to the ultimate objective of the Convention, and to assist Parties included in Annex 1 in achieving compliance with their quantified emission limitation and reduction commitments under Article 3." See KP text at http://unfccc.int/resource/docs/convkp/kpeng.pdf (last visited on December 02, 2004).

A CDM project activity in a non-Annex 1 country produces certified emission reductions (CERs). The Annex 1 Parties can use these CERs towards partial compliance of their emission reduction targets.

27.1.1 CDM Criteria

According to Sect. 12.5 of the KP, a CDM project has to satisfy the following criteria:

- parties involved in the project activity do so voluntarily and both approve the project;
- the project must produce real, measurable and long-term benefits to the mitigation of climate change; and
- the emission reductions should be additional to any that would occur without the project activity (commonly known as the "additionality" criterion).

In addition, article 12.2 of the KP states that the purpose of the CDM is to assist non-Annex 1 Parties in achieving sustainable development. This is interpreted to suggest that the project activities should be compatible with the SD requirements of the host country. However, neither the KP nor the subsequent COPs have attempted to provide any guideline on the issue of sustainability, leaving the decision to the host countries.

Considerable attention has been paid to the requirement of "additionality" criterion. We will look into it in more detail.

27.1.2 Participation Requirement

The Conference of the Parties (COP)[3] at its seventh meeting in Marrakech in 2001 has decided the modalities and procedures for the CDM.[4] The participation requirements for the CDM are:

- the participation in the CDM activity is voluntary;
- participating parties must designate a national authority for the CDM;
- participating parties must have ratified the KP;

---

[3] This is the highest decision-making body of the Convention and KP. It is an association of all the countries that are Parties to the Convention.

[4] See http://cdm.unfccc.int/Reference/COPMOP/decisions_15_17_CP.7.pdf for decisions taken at Marrakesh by COP-7.

27.1 Basics of the Clean Development Mechanism

In addition, the eligibility criteria for any Annex 1 Party to be able to use CERs are as follows:

- the assigned amount[5] has been calculated based on the modalities approved by the COP[6];
- it has put in place a system for the estimation of anthropogenic emissions by sources and removals by sinks;
- it has put in place a national registry;
- it has submitted annually the most recent required inventory pertaining to emissions of GHGs;
- it submits supplementary information on assigned amounts.

Both private and public entities can participate in the CDM activity with permission from the Party. They can acquire and transact the CERs if the Party is eligible for doing so. In any event, the Party remains the responsible for meeting its GHG reduction obligations under the KP.

27.1.3 Eligible Projects

Annex A to the KP provides a list of sources/sectors of GHG emissions which are covered under the Protocol. Table 27.1 reproduces the list.

Section 2.1 of the KP suggests that to achieve GHG reduction and sustainable development, the Annex 1 parties shall implement and/or further elaborate policies and measures suitable for the national context. Such actions can include:

- energy efficiency improvement in various sectors of the economy;
- sustainable forest management, afforestation and reforestation;
- promotion of sustainable agriculture;
- protection and enhancement of sinks and reservoirs of GHGs;
- Research, promotion, development, and increased use of renewable energy technologies, environmentally sound technologies and carbon sequestration technologies;
- Limitation of methane emissions from waste management and energy production and supply;
- Measures to limit GHG emissions in the transport sector;
- Phasing out of market imperfections, fiscal incentives, tax/duty exemptions and subsidies in all GHG emitting sectors;
- Encouraging reforms in all relevant sectors to reduce GHG emissions.

---

[5] Assigned amount is the total amount of GHGs that the Annex 1 parties have decided not to exceed in the first commitment period.

[6] Decision 19 of COP 7 prescribes the modality. See footnote 4 for the document.

Table 27.1 Sector/Sources of GHG emissions covered under the KP

Sector	Source category
Energy	Fuel combustion: energy industries, manufacturing industries and construction, transport, other sectors, other
	Fugitive emissions from fuels: Solid fuels, oil and gas, other
Industrial processes	Mineral products, chemical industry, metal production, other production, production of halocarbons and sulphur hexa-fluorides, consumption of halocarbons and sulphur hexa-fluorides, other
Solvent and other product use	
Agriculture	Enteric fermentation, Manure management, rice cultivation, agricultural soils, prescribed burning of savannas, field burning of agricultural residues, other
Waste	Solid waste disposal on land, wastewater handling, waste incineration, other

Source Annex A, Kyoto Protocol

While the KP does not specify any specific actions or projects for the CDM nor does it specify any appropriate actions for the Non-Annex 1 parties (because of absence of any GHG reduction targets for them), it can be considered that projects or actions eligible for Annex 1 parties shall apply to non-Annex 1 parties as well. Accordingly, the following projects are considered to be eligible for implementation under the CDM:

- end-use energy efficiency improvement projects;
- supply-side efficiency improvements;
- renewable energy projects;
- fuel switching;
- agriculture (reduction of methane and nitrous oxide);
- Industrial processes
- Sink projects (afforestation and reforestation)

27.1.4 CDM Entities/Institutional Arrangement

27.1.4.1 Conference of the Parties

This is the ultimate decision-making body under the Convention and KP. It has the authority over the CDM and provides guidance to the mechanism and to the Executive Board (EB). Based on the recommendations of the EB, it takes decisions on rules of procedure, designation of the operational entities approved by the EB, and other recommendations made by the EB. It also reviews the annual report of the EB, the distribution of operational entities and takes decisions to promote entities from developing countries, the distribution of CDM activities and assists in arranging funding of CDM project activities.

27.1.4.2 Executive Board (EB)

The Marrakech meeting of the COP decided on the organisational structure and functions of the Executive Board. It is the executing authority with all powers and duties to manage the programme. It is responsible and accountable to the COP. The EB is a 10 member body with five members elected from five UN regions, two members each from non-Annex 1 and Annex 1 parties and one member from the small island nations. Each member is elected for two years and can serve two consecutive terms. A chairperson and a vice-chairperson are selected from amongst the members. The chairperson and the vice-chairperson shall be alternately from the Annex 1 and non-Annex 1 parties and shall alternate annually. The EB is expected to take decisions by consensus but where a consensus is remote, the decision has to be taken by a 3/4th majority of the members present. The EB is required to meet at least three times a year.

The functions of the EB include, inter alia:

- accreditation of the designated operating entities;
- operationalisation of accreditation procedures and standards;
- manage and make publicly available a repository of rules, procedures and standards;
- manage a CDM registry and CDM project database;
- report to COP on EB activities on an annual basis, etc.
- approve baseline methodologies, project boundaries, monitoring plans, etc.
- review provisions regarding small scale projects.

27.1.4.3 Host Country Responsibilities

According to Marrakech Accords, the host country of the CDM activity shall designate a national authority (DNA) for the CDM. The DNA of each party involved shall be required to give written approval of voluntary project participation by the participants. The host country government shall also certify that the project conforms to its objectives of SD and that it helps in achieving SD. For many host countries, the CDM is a novel and intricate experience involving international, national and local level players. Managing and directing these diverse interest groups is a complex process, which requires project management skills, knowledge of CDM rules and procedures, coordination skills, etc. (UNDP 2003). It is suggested that host countries seeking active involvement in the CDM will also be engaged the following (UNDP 2003):

- national policy and regulatory framework for promoting CDM transactions;
- setting SD priorities and policies related to SD;
- CDM project approval and registration mechanisms;
- Management of retained CERs;

- Participation in ongoing KP negotiations;
- Capacity development for the CDM.

27.1.4.4 Designated National Authority

Establishing a designated national authority (DNA) is a requirement for participation in the CDM. DNA is a key organisation of the host country for controlling CDM activities in the country. It is responsible for confirming that the project assists in achieving sustainable development of the country and that the participants are involved voluntarily. As a key national entity, it can play an important role in the following areas:

- co-ordination of agencies responsible for setting sustainable policies, environmental and investment regulations and organizations involved in CDM project development;
- provide guidelines on national project approval criteria, guidelines on priority projects, selection, consultation and monitoring processes;
- decide procedures for processing project applications, authorizing verification organizations;
- support in managing the risks associated with CDM projects; and
- facilitate projects by providing information
- marketing and promoting CDM projects.

27.1.4.5 Designated Operational Entities (DOE)

As CDM aims at real, measurable and long-term benefits from projects, entities are required to validate and monitor the claims of project proponents. To make the process transparent and independent, CDM has created entities called designated operational entities (DOE). They are accredited by the EB subject to meeting the accreditation standards and are designated by the COP. The DOEs are accountable to the COP through the EB. A DOE performs the following functions:

- validate CDM project activities;
- verify and certify the emission reductions by source of GHGs;
- maintain a list of CDM projects validated or verified and certified by it;
- submit an annual report to the EB;
- make public information obtained from CDM project participants.

As will be clear in the CDM project cycle, DOEs are involved at two stages of the project: one for project validation and the other at the time of certification of emission reduction. Normally, two separate DOEs will be involved at these stages, although the EB may allow a single DOE to perform both the activities for a particular project.

27.1.4.6 Project Participants

Project participants include project developers and investors. Project developers are those who are interested in developing and perhaps operating a project in a host country. They can be any legal entities (government bodies, private sector, financial institutions, NGOs, etc.). A project developer prepares a Project Design Document (PDD) in a standard, prescribed format, which is the starting point of the CDM project cycle. They are also responsible for project monitoring. Investors on the other hand are entities that purchase the emission credits from the project. They are Annex 1 country players (government bodies, private sector or NGOs).

27.1.5 CDM Project Cycle

Figure 27.1 presents a schematic diagram indicating the steps involved in a CDM project. The stages of the project cycle include[7]:

Project identification: The CDM project activity starts with the identification of a project idea. This will normally be done by the host country participants. The project idea has to be verified for its eligibility under the CDM. These can be ascertained by asking the following questions (UNDP 2003, Chap. 2):

- Does the host country meet the participation requirements? (see the requirements above)
- Does the project fall into one of the eligible project categories? (see eligibility requirements above)
- Does it involve a proven/established and commercially feasible/replicable technology?
- Does it satisfy the SD criteria of the host country?
- Does it produce unacceptable environmental/social effects?

Project Idea Note (PIN): This is not a requirement of the CDM process but preparing such a note may be helpful. This note provides basic information about the project idea (project type, location, anticipated GHG reduction, crediting life, expected CER price, environmental and other benefits, financing structure, etc.). The PIN may be circulated to interested/potential credit buyers to gauge their interest in the project.

Project Design Document (PDD): This is a required step in the CDM process. The EB has developed a standard format of the PDD[8] and each project developer

---

[7] This applies to standard projects under the CDM. Small-scale projects follow a slightly different cycle while the Programme-based activities introduced in 2006 are governed by different regulations. New regulations were introduced in 2009 to address concerns raised by stakeholders.

[8] The current version of the PDD is available at UNFCCC website.

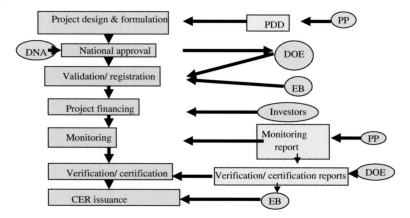

Fig. 27.1 CDM project cycle. *Source* based on UNEP (2002)

has to prepare such a document for the project. The PDD must contain the following elements[9]:

- general description of project activity
- application of a baseline methodology
- duration of the project activity/crediting period
- application of a monitoring methodology/plan
- estimation of GHG emissions by sources
- environmental impacts
- stakeholders' comments.

This is a complex and crucial document as it will be used for the host country approval, for DOE validation and by the investors. It is essential to establish that the project will help achieve sustainability criteria of the host country and will meet additionality requirement based on an appropriate baseline methodology. Preparation of the PDD is a costly affair and hence proper liaison with national authorities may help (UNDP 2003, Chap. 3).

National approval: The project participants submit the PDD to the designated national authority (DNA) for national approval. The DNA verifies whether it satisfies the SD criteria of the host country. As the DNA is responsible for screening projects in a transparent and expeditious manner, standardisation of the approval procedures and checks could facilitate the process. At the end of this phase, the DNA issues statements confirming the voluntary participation of the project participants and certifying that the project enhances sustainable development of the host country.

Project validation/registration: The project participants submit the PDD to any designated operational entity (DNA) of its choice. The DOE initiates the validation

---

[9] For detailed guidance see UNEP (2004) and UNDP (2003).

27.1 Basics of the Clean Development Mechanism

process, which entails an independent evaluation of the project activity against the requirements of the CDM as defined by various COP decisions.

According to the Annex to decision 17 of COP 7, the DOE reviews the following requirements:

- the participation requirement;
- comments are invited from local stakeholders and a summary of the comments are forwarded to the DOE. The project participants should also indicate how they are taking care of the comments.
- Documentation on environmental impacts has been submitted and if the impacts are substantial, an environmental impact assessment has been conducted following the host country procedures;
- The project satisfies the "additionality" criteria in terms of emissions reduction;
- The baseline and monitoring methodologies comply with the governing requirements; this in other words implies that the DOE verifies whether the methodology for baseline is an approved methodology or a new methodology. If it is a new methodology, the DOE submits the methodology and the project documentation to the EB for review. If the EB approves the methodology, it becomes an approved one and can be used by other PPs. The DOE continues with completes the validation process. If the EB disapproves the method, it cannot be used by any other project developer. The project participants will then be required to revise the baseline methodology.
- The project complies with the monitoring, verification and reporting requirements of the CDM;
- The project satisfies with all other requirements for CDM project activities.

If DOE finds that the project meets the above requirements, it submits a validation report to the EB as well as to the project participants. The DOE also requests the EB to register the project. However, before submitting the validation report, the DOE has to satisfy the following additional requirements:

- it must receive a written confirmation from the DNA about the voluntary participation of the participants and the compatibility with the host country sustainability criteria;
- it must make public the PDD and allow 30 days to submit comments from Parties, stakeholders and UNFCCC accredited NGOs and make the comments public;
- determine whether the project can still be validated in view of the comments received.

If the DOE rejects the project activity, it shall provide reasons for doing so to the project participants.

Once the EB receives a request for registration, the project activity shall be registered within 8 weeks (4 weeks for small scale projects), unless a Party involved in the project activity or at least 3 members of the EB request for a review. The EB has to conclude the review within two meetings of the EB and shall inform the project participants and the public about the review outcome.

Project financing: Once the project is registered, it can be implemented. However, for the initial few years it was decided that projected commenced from 2000 will qualify for the first commitment period. During this period, it may so happen that the project was implemented even before its registration. In a normal case, upon registration the project will be implemented. Investors will invest at this stage.

Monitoring: As a CDM project has to produce real, measurable and long-term emission reductions, for the purpose of determination of the CER, the project needs to be monitored. The CDM procedure requires that the PDD shall include a monitoring plan providing the following details:

- collection and archiving of all relevant data for estimation or measurement of anthropogenic GHG emissions from the project boundary;
- collection and archiving of all relevant data for estimation or measurement of baseline within the project boundary;
- identification of sources and collection and archiving of all relevant data on GHG emissions from sources outside the boundary of the project but which could be reasonably attributed to the project;
- collection and archiving of information on environmental impact of the project;
- quality assurance and control procedures for the monitoring process;
- procedures for periodic calculation of emission reduction by the project activity and for leakage effects; and
- documentation of all steps involved for the calculation of environmental impacts.

The project participants can follow an approved methodology for monitoring or propose a new methodology. In case of a new methodology, the DOE shall refer the method to the EB for decision. The procedure is similar to that of baseline methodology. The DOE determines whether the methodology for monitoring is appropriate for the type of project activity and whether it has been applied successfully elsewhere.

The project participants have to implement the registered monitoring plan and if they wish to modify the monitoring plan to improve monitoring, the revision has to be submitted to the DOE for validation. The implementation of the monitoring plan is a pre-condition for the issuance of CER.

Verification and certification: These two activities are defined as follows: "Verification is the periodic independent review and ex post determination by the designated operational entity of the monitored reductions in anthropogenic emissions by sources of greenhouse gases that have occurred as a result of a registered CDM project activity during the verification period. Certification is the written assurance by the designated operational entity that, during a specified time period, a project activity achieved the reductions in anthropogenic emissions by sources of greenhouse gases as verified."[10]

---

[10] Annex to Decision 17 of COP 7.

27.1 Basics of the Clean Development Mechanism

Project participants are required to contact a DOE for verification and certification of the project activity. The DOE shall:

(a) establish whether the monitoring documentation project documentation is in line with the requirements of the registered PDD;
(b) carry out on-site inspections, as appropriate;
(c) utilise additional data from other sources if required;
(d) reassess monitoring results and check whether the monitoring methodologies have been applied correctly and whether adequate and transparent documentation is maintained;
(e) advise on appropriate changes to the monitoring methodology for any future crediting period, if necessary;
(f) Determine the reductions in anthropogenic emissions by sources of greenhouse gases which is additional;
(g) Spot any divergence between the actual project activity and the PDD and inform the project participants, who in turn are required to address the concerns and supply relevant additional information; Produce a verification report and make it available to the project participants, the Parties involved and the executive board. The report shall be made publicly available.

Subsequently, the DOE shall certify in writing the amount of additional emission reduction achieved by the project activity and the submission of the certification report to the EB is considered to be a request for issuance of CER equal to the amount of additional reduction certified by the DOE. The EB shall issue the CER within 15 days of receipt of the request, unless a party to the project or at least three members of the EB requests for a review. Such a review is limited to the fraud, malfeasance or incompetence of the DOE and the EB is required to complete the review within 30 days. If the EB approves issuance of the CER, the CDM registry administrator shall issue the CER and deduct the quantity of CER required to meet the administrative expenses and to meet the cost of adaptation. The remaining CER shall be forwarded to the registry accounts of the participants.

This completes the CDM project cycle. It is a complex system with a significant amount of regulatory control compared to standard project activities. This makes a CDM project different and unless a country has the requisite capacity to deal with such regulatory aspects, it will be difficult to attract CDM projects.

27.1.6 Additionality and Baseline

Baselines and additionality are two key concepts used in any CDM project. The Annex to decision 17 of COP 7 defines the terms as follows[11]:

---

[11] See Annex to decision 17/CP.7.

Fig. 27.2 Baseline explanation

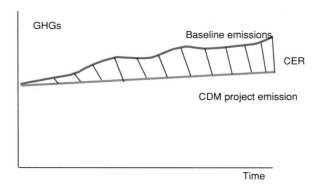

"A CDM project activity is additional if anthropogenic emissions of greenhouse gases by sources are reduced below those that would have occurred in the absence of the registered CDM project activity.

The baseline for a CDM project activity is the scenario that reasonably represents the anthropogenic emissions by sources of greenhouse gases that would occur in the absence of the proposed project activity. A baseline shall cover emissions from all gases, sectors and source categories listed in Annex A within the project boundary."

These two concepts are used to determine whether the project activity shall lead to real, measurable and long-term emission reductions. The baseline paints the picture of what would have happened to emissions in the absence of the project activity. In common parlance this is also referred to as the 'business-as-usual' scenario (BAU). The idea essentially comes from the project evaluation methodology where the benefits of a project are measured by comparing two situations: with project and without project. In the case of a CDM project, the benefits of the project in terms of GHG emissions are measured by comparing with a situation where the project does not materialise. This is explained in Fig. 27.2.

Establishing the baseline is a crucial element of the project design document. According to the Marrakech Accord, the baseline shall satisfy the following conditions:

- It must based on either the approved methodologies or a new methodology subject to the EB approval;
- It should be specific to the project activity;
- It should follow a conservative and transparent approach;
- It should reflect the specific conditions, policies and circumstances of the host country;
- It should be such that no CER could be earned due to decreases in activity levels outside the project boundary or due to force majeure.

The Marrakech Accord approved the following baseline methodologies:

- Existing actual or historical emissions, as applicable; or

- emissions from a technology that can be considered as an economically attractive option, given the barriers to investment; or
- "The average emissions of similar project activities undertaken in the previous five years, in similar social, economic, environmental and technological circumstances, and whose performance is among the top 20 per cent of their category."[12]

To develop and define a credible, transparent baseline, the following steps have to be followed (UNDP 2003, Chap. 3):

- choosing a baseline approach;
- adopting or creating a baseline methodology;
- defining the project boundaries: the project boundary shall be such that all anthropogenic emissions by sources of GHG that are under the control of the project participants and are significant, and reasonably attributable to the project activity shall be covered.
- Forecasting emissions under the business-as-usual scenario;
- Forecasting the future emissions from the project;
- Assessing leakage: According to the Marrakech Accord, "Leakage is defined as the net change of anthropogenic emissions by sources of greenhouse gases which occurs outside the project boundary, and which is measurable and attributable to the CDM project activity." Emission reductions have to be adjusted for leakage. According to UNDP (2003), two effects that should be considered to assess leakage are activity shifting and outsourcing. Activity shifting implies that emissions are not permanently reduced but shifted from one place to another. Outsourcing on the other hand means that purchasing or contracting out products or services that were produced or provided on-site earlier.
- Estimation of emission reductions from the project for the purpose of verification.

27.1.7 Crediting Period

According to the Marrakech Accord, two options are available for the crediting period (for a standard project):

- a maximum period of seven years renewable two times (i.e. a total crediting period of 21 years) but here the DOE will ascertain before each renewal whether the original baseline is still valid or it needs to be updated based on new information;
- a maximum period of 10 years without any renewal option.

---

[12] Annex to decision 17/CP.7.

Clearly, the crediting period can affect the project viability significantly. The 10-year period provides more certainty in the sense that there is no possibility of revision to the baseline. But as the CERs can be used for a maximum period of 10 years, the financial gain from selling the CERs will be limited. On the other hand, the renewable option allows for taking credit of the CER revenue for a maximum of 21 years but here the project participants bear the risk of CER volume changes due to revision of the baseline. A choice has to be made depending on the type of the project and by comparing the net benefits from the alternative options.

27.2 Economics of CDM Projects

27.2.1 Role of CDM in KP Target of GHG Reduction

As mentioned earlier, the CDM is one of the flexibility mechanisms adopted under the KP. In order to meet the KP requirement of GHG reduction, Annex 1 Parties have the following options available to them:

- Domestic policies and measures to reduce GHG emission. These include the options discussed in the lecture on climate change and include non-market policies (such as structural reform, energy sector reform) and specific climate change policies (such as regulatory standards, taxes and charges, tradable permits, voluntary agreements, etc.). It is a requirement under the KP that each Annex 1 party has to take certain domestic measures and cannot rely solely on the credits from projects outside its territory.
- Constrain operation: This implies that by constraining the activity, a country can achieve its target.
- Buy CERs from the CDM;
- Buy Emission Reductions (ER) from Jointly Implemented projects;
- Buy credits from Emission Trading System; or
- Pay penalty for non compliance.

Therefore, in deciding the mix of options, the cost and benefits of alternative options will play an important role. The Annex 1 Parties will compare the marginal cost of abatement from each option and shall opt for the least-cost options, if regulation allowed so. Considering the benefits are same from all the options, the factor that becomes important is the cost of reducing per ton of GHG emission. Globally CDM is expected to play a minor role at least in the first commitment period. Therefore, the projects that offer cheaper and large volume CERs will be preferred to those with high cost and low volume CERs. As the competition builds up, all the developing countries can potentially participate in the CDM, which makes the competition real tough. This is verified in the EU member states. In 2009, only 4.1% of the total allowances used in the EU-ETS came from the CDM.

52% of these CERs were supplied by China, 21% from India, 14% from South Korea and 9% from Brazil. Only 4% of the CERs used by EU member states came from other countries.[13] This confirms the limited role of the CDM in the current compliance period.

27.2.2 Difference Between a CDM Project and an Investment Project

A CDM project is like an investment project but it has certain special characteristics. In financial terms an investment project produces a stream of financial return during its project life. A CDM project also yields financial returns but in addition it generates carbon credits and other sustainable development benefits. The carbon credits have a monetary value and will affect the net financial returns. On the input side, it uses normal investment and a special kind of investment called carbon investment. Carbon investments will enter because of potential gains from carbon credits obtainable from the project (Spalding-Fecher 2002). This is shown in Fig. 27.3.

27.2.3 CDM Transaction Costs

Table 27.2 provides a list of transaction costs identified by Michaelowa and Stronzik (2002). More important of these are:

- As discussed in the CDM project cycle, for the development of a project, the project participants have to prepare the PDD (or project document). The PDD is more than the pre-feasibility and feasibility studies normally carried out for an investment project. Because of its critical importance in the project cycle and a complicated document, its preparation involved significant upfront costs.
- Contracting a DOE for the purpose of validation and registration is another important cost.
- Contracting a DOE for the purpose of verification and certification of CER also constitutes an important cost.
- Registration fees and EB administrative charges: The EB has decided the fees for registration of CDM projects depending on their size (see Table 27.3).
- Other charges: According to the Marrakech Accord, each CDM project will pay 2% of its project proceeds to an adaptation fund. This is compulsory for all projects except projects in least developed countries.
- Host countries may also impose tax or levies on the CDM project proceeds.

---

[13] Europa Press release IP/10/576 of 18th May 2010.

Fig. 27.3 CDM and conventional investment projects. *Source* Spalding-Fecher (2002)

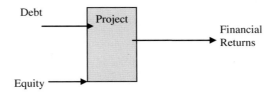
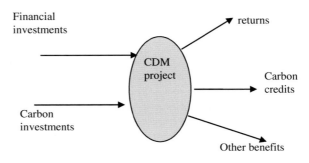

Table 27.2 CDM project transaction costs

Cost components	Description
Search costs	Costs incurred by investors for seeking mutually advantageous projects
Negotiation costs	Costs for project formulation, preparing PDD and arranging public consultation with stakeholders
Baseline determination costs	This can be part of the PDD preparation or can be outsourced. This cost arises when outsourced
Approval costs	Cost of obtaining national approval
Validation costs	Costs for engaging a DOE for reviewing the PDD. Cost for revising the PDD can also apply
Review costs	If a review is requested, additional costs will be involved
Registration costs	Cost of registering the project with the EB
Monitoring costs	Cost of collecting, archiving and reporting data/information in accordance with the Monitoring Plan
Verification costs	Cost of engaging a DOE for undertaking verification of the monitoring report
Review costs	Cost of review of monitoring report if requested
Certification costs	Cost of issuance of the CER by the EB
Enforcement costs	Cost of administrative and legal measures to ensure enforcement of the project activities
Transfer costs	Brokerage costs
Registration costs	Cost to hold an account in national registry

Source Michaelowa and Stronzik (2002)

27.2 Economics of CDM Projects

Table 27.3 Examples of registration fees

Project size (GHG reduction)	Registration fee
Less than 15,000 tons per year	$5,000
Between 15,000 and 50,000 tons per year	$10,000
Between 50,000 and 100,000 tons/year	$15,000
Between 100,000 and 200,000 tons/year	$20,000
More than 200,000 tons/year	$30,000

Source UNFCCC-CDM website

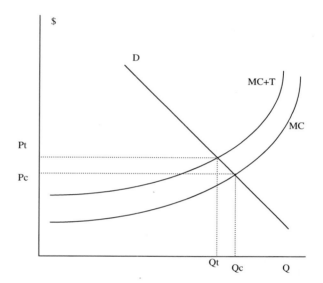

Fig. 27.4 Effect of transaction costs

Transaction costs are expenses incurred in carrying out a transaction. These are costs above the marginal costs and they raise the price of the output. This is shown in Fig. 27.4. Transaction costs by increasing the price reduce the demand for the product. Ultimately, they affect the project viability: in the case of a CDM project, the break-even CER price will rise or for a given range of market CER price, the price of the product has to increase perhaps to such a level that the product becomes uncompetitive. Thus, the project loses its viability. The range of some CDM-related transaction costs are given in Table 27.4.

Transaction costs can be a serious problem for small scale projects with low volume of outputs and high share of transaction costs due to fixed nature of some costs. Table 27.5 suggests the relation between project size and these costs.

Krey (2004) surveyed 15 unilateral potential CDM projects in India and found that the average transaction costs range between $0.06 and $0.47 per ton of CO_2 eq. For projects with high emission reduction, the costs accruing from the

Table 27.4 Transaction cost estimates

Cost element	Frequency	Cost (US '000$)
Feasibility assessment	One time	5–20
PDD preparation	One time	25–40
Registration	Per year	10 (can vary between 5 and 30)
Validation	One time	10–15
Legal work	One time	20–25
Monitoring and verification	Per year	3–15

Source based on UNDP (2003)

Table 27.5 Project size and transaction cost relationship

Size	Type	Reduction (t CO_2)	Euro/ t CO_2
Very large	Large hydro, gas power plants, large CHP, geothermal, landfill/pipeline methane capture, cement plant efficiency, large scale afforestation	>200,000	0.1
Large	Wind power, solar thermal, energy efficiency in large industry	20,000–200,000	0.3–1
Small	Boiler conversion, DSM, small hydro	2,000–20,000	10
Mini	Energy efficiency in housing and SME, mini hydro	200–2,000	100
Micro	PV	<200	1,000

Source Michaelowa et al. (2003)

adaptation fee is the most important component, while for projects with low emission reduction, the PDD, the search cost, the adaptation fee and validation costs were the major contributing factors.

27.2.4 CER Supply and Demand

As mentioned earlier, the Kyoto targets of Annex 1 countries will be met by a combination of different measures. It is difficult to clearly forecast the demand for CERs from the CDM activities due to changes in the Parties' position, economic conditions and therefore permit needs and changes in the supply conditions.

The carbon market has grown very significantly during this decade.[14] The total volume of carbon transactions in 2003 was 78 Mt of CO_2 eq. This has increased to 4.8 Gt CO_2 eq by 2008 and 8.7 GtCO_2 eq by 2009. The market growth has therefore been spectacular. The composition of the market has also equally

---

[14] This is based on Point Carbon (2003a, b), Point Carbon (2007), Capoor and Ambrosi (2007), and various issues of State of the Carbon Markets.

27.2 Economics of CDM Projects

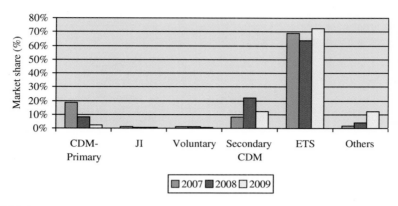

Fig. 27.5 Structure of the carbon market. *Source* state of the carbon markets (various)

changed. In 2003, most of the market was dominated by project-based pre-Kyoto compliance instruments while in 2009 the allowance-based instruments have gained a market share of 85%. This complete reversal of the situation has been due to the introduction of EU ETS and other allowance trading markets. The trend of the carbon market structure is shown in Fig. 27.5.

The primary CDM market has shrunk over the years: in 2007 the CDM volume was 552 Mt CO_2 eq but in 2009, it has reduced to 211 Mt CO_2 eq. The economic recession in Annex 1 parties has reduced the need for Kyoto instruments for compliance and accordingly, the over supply situation has eroded the CDM market share. The composition of the sellers and buyers has changed as well quite significantly over the decade. For example, in 2002–2003, Latin American countries were the major suppliers with a 40% market share. Asia and Transitional economies came next. More than 50% of the CER supply in that year originated from Brazil alone. In 2009, the Latin American share is limited to 7% only, while China dominated with a 72% market share (see Fig. 27.6). The Chinese domination in the supply started in 2005 with its large-scale industrial GHG reduction projects and continues. This has completely changed the supply market and has influenced the overall market growth.

On the other hand, the composition of the buyers has also undergone a sea change (see Fig. 27.7). In 2002–2003, the Netherlands and Japan were the major buyers. In fact, the project-based market was highly influenced by the Dutch government's CER procurement programmes. In 2009, the main buyer was the United Kingdom with a 37% share. However, this is due to the fact that most of the financial players involved in the market are located in the UK and not because these units are used in the UK (Kossoy and Ambrosi 2010). Germany, Sweden and other Baltic Sea countries have bought 20% of the CDM CERs while Spain, Portugal and Italy have procured another 7% of the supply. The Netherlands and other European states now have a share of 22%.

Fig. 27.6 Sellers of CDM CERs in 2009. *Source* Kossoy and Ambrosi (2010)

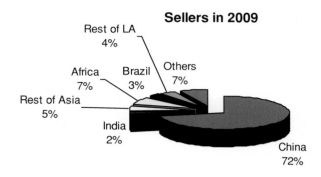

Fig. 27.7 Buyers of CDM CERs in 2009. *Source* Kossoy and Ambrosi (2010)

The pricing mechanism has also undergone a vast change in line with the market structure. The market was fragmented in the early days (2002–2004) and the prices vary by type of project and technology used. At that time, the renewable energy projects were commanding a higher CER price, while fuel switching and methane capture attracted a low price (see Table 27.6).

In the early days, the government programme schemes dictated the buyer's price. This was ended by the Chinese domination in the supply market which led to a price influenced by the Chinese suppliers. This also led to a more remunerative price for the suppliers. The idea of a floor price was emerging and viability of the projects was improving. But oversupply in the market is likely to increase the price risk for the investor and affect the investment decisions. Clearly, the excess supply of Kyoto Protocol units could affect the market price of CERs.

The outlook for CDM remains clouded now with the reduced requirement for the first compliance period due to economic downturn. Kossi and Ambrosi (2010) indicate that the total Kyoto asset demand in the 2008–2012 period is expected to be 1222 $MtCO_2$ eq but the supply could be well over 3000 $MtCO_2$ eq, with Green Investment Scheme from Russia and other East European countries potentially supplying more than 1800 $MTCO_2$ eq. Thus the market is likely to end up with oversupply. Haites (2007) also reported a similar concern - that the supply may increase in the first period of KP commitment due to projects in the pipeline and

Table 27.6 CER prices in the early days

Item	Price range (per tCO$_2$e)	Comments
Prototype carbon fund (World Bank)	USD 3–3.5	US$ 0.5 per tCO$_2$e extra for developmental projects
CERUPT (Netherlands)	€ 5.50	Renewable projects
	€ 4.40	Biomass
	€ 4.40	Energy efficiency
	€ 3.30	Fuel switch and methane
Finnish Gov.	€ 2.47–3.2	

Source Pacudan (2004)

the possibility of flooding the market by Russian and East European Assigned Allocation Units (AAUs) remains.

27.2.5 Risks in a CDM Project

Any investment project faces a number of risks, including country risk, construction/implementation risk, technical risks, environmental risks, financial, legal and operational risks. These are also present in a CDM activity. Moreover, a CDM project faces a number of additional risks. They are:

- baseline risks (such as: Are the baseline assumptions robust? Will they remain valid over the crediting period? Will the project perform as assumed in the project design document?),
- validation and registration risks (whether the project will be approved by the host country, whether the project will meet the additionality requirement, will the stakeholders perceive the project as acceptable, etc.),
- monitoring and verification risks (i.e. whether the records are being kept as proposed in the PDD, whether the emission reduction verifies with the claim, etc.).
- market risk (i.e. how the CER price will be? Whether there is sufficient demand for CER);
- Transaction risks (delivery risk, timing risk, credit risks, etc.)

CDM projects are now taking on average 572 days for the validation and registration and another 607 days before CER issue. This long gestation period introduces risks to the project and affects investor interest (Kossoy and Ambrosi 2010). However, for a new mechanism, the interest in CDM has been quite encouraging: According to the CDM/JI pipeline database of UNEP Riso, at present 5443 projects are in the pipeline, of which 2344 projects have been registered while another 169 projects are in the process of registration. 2930 projects are at various stages of validation. For a new system, this represents a very high level of activity and interest.

Normally, a project with higher risk should yield higher returns and unless this happens, investors may not show interest in taking up projects. The slow growth of the CDM market verifies these concerns. Moreover, the investment has flown to places where more viable projects are available. This has in fact defeated the basic idea of the programme to promote clean technologies in the developing world. The concentration of projects in a few countries and the limited involvement of the poorer countries have reduced the effectiveness of the programme as a development tool.

27.3 Conclusions

This chapter has provided an overview of the project-based flexibility mechanism that was included in the Kyoto Protocol to integrate developing countries in the effort to reduce greenhouse gas emissions and promote sustainable development in the developing world. The chapter has presented the regulatory process and the project cycle in detail. It has provided a brief review of the changes that have taken place in the market and discussed some issues like the influence of transaction costs on the project viability.

The CDM was an innovative idea but the complexity of making it operational and practical has proved to be quite difficult. Kossoy and Ambrosi (2010) highlight that over-regulation, regulatory inefficiencies and capacity bottlenecks have somewhat tarnished the CDM's reputation. Limited financial benefits have reduced the investor interest in the programme and can erode it further with the prevailing market conditions of oversupply and low prices. Although attempts have been made to simplify rules and allow programme-based activities, the regulatory and practical challenges remain. The long-term prospects of the programme look uncertain and it will require a thorough re-appraisal of the developments to unleash its potentials.

References

Capoor K, Ambrosi P (2007) State and trends of the carbon market 2007. World Bank, Washington DC
Haites E (2007) Carbon markets, a report for the UNFCCC. http://unfccc.int/files/cooperation_and_support/financial_mechanism/application/pdf/haites.pdf
Kossoy A, Ambrosi P (2010) State and trends of the carbon market, carbon finance. The World Bank, Washington DC
Krey M (2004) Transaction costs of CDM projects in India—an empirical survey. Hamburg Institute of International Economics, HWWA, Hamburg

Michaelowa A, Stronzik M (2002) Transaction costs of the Kyoto mechanisms. HWWA, Hamburg
Michaelowa A, Stronzik M, Eckermann F, Hunt A (2003) Transaction costs of the Kyoto mechanisms. Climate Policy 3:261–278
Pacudan R (2004) CDM market overview, URC-IGES workshop on capacity development for CDM in Asia. Asian Institute of Technology, Bangkok, September 29-October 1, 2004, Thailand
Point Carbon (2003a) Annex 1 parties current and potential CER demand, Report prepared for the Asian Development Bank and International Emissions Trading Association
Point Carbon (2003b) CDM market overview, Presentation made to SE Asia Forum on GHG Mitigation, Manila, Philippines. http://www.pointcarbon.com/wimages/SE_Asia_Forum_100903_Jorund_Buen6.ppt
Point Carbon (2007) Carbon 2007: a new climate for carbon trading, Point Carbon
Spalding-Fecher R (2002) Financial and economic analysis of CDM projects. In: Davidson O, Sparks D (eds) Developing energy solutions for climate change: South African research at EDRC, energy and development research centre. University of Cape Town, South Africa
UNDP (2003) The clean development mechanism: a user's guide. UNDP, New York
UNEP (2002) Introduction to the CDM, UCCEE. Riso, Denmark
UNEP (2004) CDM Information and guidebook. Riso, Denmark

Part VI
Regulation and Governance of the Energy Sector

Chapter 28
Regulation of Energy Industries

28.1 Introduction

This chapter introduces the concepts of regulatory economics relevant for the energy industry. The energy industry in general and the network-based energy sector in particular show the problems of natural monopoly and monopoly features. Two solutions have been commonly used to resolve the problem: either to allow monopolies to operate under a system of regulation or to use public ownership to capture monopoly rents and redistribute the proceeds through state policies. Regulation is an essential component of the first solution and is discussed here.

In the world of perfect competitive markets, little regulatory effort is perhaps required. But when such ideal markets do not or cannot exist, regulations then aim at controlling the behaviour of the market agents to control the market failures and to produce competitive market-like results through the use of coercive power of the government or its agencies that restrict or constraint the decisions of the economic agents. Such regulatory interventions are quite pervasive in reality and can be categorised under three main groups: economic regulation, regulation of anti-competitive behaviour and social regulation. The first type is concerned with the issue of economic efficiency (and equity) in a non-competitive market environment. Economic regulations refer to government imposed restrictions on firms' decisions through the control of price, or quantity or control of entry or exit or some combinations of these. The second is concerned with anti-competitive behaviour and protection against anti-competitive practices. The third is a relatively new but more pervasive type of regulation—dealing with the protection against socially undesirable behaviour and promotion of socially desirable goods.[1]

---

[1] See for example Viscusi et al. (2005), Shleifer (2004) and Newbery (1999, Chap. 4) for further details. See also the Body of Knowledge on Utility Regulation, an annotated bibliography and for further references. See (http://www.regulationbodyofknowledge.org/).

In this chapter we consider only economic regulation. Within the category of economic regulation, a number of alternatives exist: including the traditional cost of service regulation, or alternative modern variants like the incentive regulation, conduct regulation and regulation by contract. We focus on two most commonly used forms of economic regulation, namely the traditional regulation and the incentive regulation. First, we present the traditional regulation, then the incentive regulation and finally we discuss some issues and experiences related to regulation, especially in developing countries.

28.2 Traditional Regulation

The traditional approach to regulation is known as the rate of return regulation. This approach attempts to balance the total costs and the total revenue of a utility by considering all relevant components. These components can be expressed as follows Eq. 28.1:

$$\sum_{i=1}^{n} p_i q_i = \text{Expenses} + s.RB \qquad (28.1)$$

Where P_i = price of ith good,
Q_i = quantity of ith good,
n = number of goods,
s = rate of return on investment;
RB = rate base, which indicates the total worth of the investment made by the utility.

This type of regulation can be viewed as a two-part regulation: rate level regulation and rate structure regulation. The rate level regulation looks at the right hand side of Eq. 28.1 and decides about the appropriateness of the expenses, the allowable rate of return and the rate base. The rate level regulation attempts to regulate the costs of the utility. The rate structure on the other hand deals with the structure of rates to be paid by consumers. The rate structure regulation aims at regulating the revenue earned by a utility. We consider the two parts separately.

28.2.1 Rate Level Regulation

In order to perform the rate level regulation, the regulatory bodies try to build up, item by item, the total permissible expenditure for any given time period, say one year. This is usually based on a test year—usually a past year adjusted for known and measurable changes so as to reflect conditions expected to prevail during the time the proposed tariff will be in effect. This permissible expenditure is known as revenue requirement of the utility, which it will then be allowed to recover through

28.2 Traditional Regulation

charges and income. As indicated in Eq. 28.1 above, this requires a detailed scrutiny of the expenses and decisions about whether these expenses are permissible for inclusion in the cost-of-service computations, and whether these permissible expenses will be treated as operating expenses[2] or as capitalised expenses. This is a demanding task for any regulatory body following the traditional approach.

The choice of test year is an area of concern. A test year can be a historical test year. In this case, the latest available actual performance data is used for determining the revenue requirement. The historical basis allows easy verification but the disadvantage with this method is that the tariffs are unlikely to match with the future condition when the tariff will be applied. A common variation of this method is to allow for adjustments for "known and measurable" changes. Variations in costs due to new power purchase agreements or due to inclusion of new major industrial customers could be considered as "known and measurable changes" for an electric utility. The utility must provide documentary evidence to the regulators to justify all changes in costs and sales.

In some cases a "forward-looking test year" is used as the basis for revenue requirement and allowed revenue determination. The "forward-looking test year" costs and sales are estimated based upon a forecast of future costs and future load. While the forecast may rely heavily on past experience, all expected changes are incorporated, not just "known and measurable ones." The costs here will be based on projected performance, which is difficult to verify ex-ante.

Why is the scrutiny of expenses required? Kahn (1989) suggests five reasons as follows:

1) *Exaggeration of costs*: regulated utilities may exaggerate costs and fool the regulators to allow more than required revenues. The exaggeration may take the form of padded expense figures and inflated capital investments.
2) *Depreciation allowance*: Depreciation is allowed for using up of capital assets. The rate of depreciation is a matter of judgement and there is room for dispute over the appropriate level. The utility may take advantage by exaggerating the cost.
3) *Imprudent expenses*: Regulated utilities may be involved in imprudent expenses that are not really for the benefit of their consumers but if they can get such expenses included in the revenue requirement, they will be encouraged to follow such lavish styles.
4) *Business with affiliates*: It is also possible that the regulated monopoly is involved in transactions with other affiliates and uses these as a conduit for higher profits by offering higher than normal rates for supply of goods and services.

---

[2] Operating expenses are directly included in the revenue requirement and the entire amount will be recoverable through tariffs over the test year. Capitalised expenses on the other hand will form part of the investment and depreciations will be allowed for use of the capital and the undepreciated investments will receive a rate of return.

5) *Lack of competitive pressures*: The regulated industries do not face competition and hence there is no incentive or pressure on them to keep costs down. As regulators act as surrogates for competitive markets, they are required to exercise some control on the costs presented to them by the utilities.

While the scrutiny of expenses is required, the regulatory body cannot substitute for the management of the utility and can only allow or disallow some costs ex-post. Hence, although operating costs typically account for a majority of the costs (80–85% according to Viscusi et al. 2005), regulators often have limited their focus to regulating profits and not detailed scrutiny of expenses.

It needs to be mentioned that inclusion of fixed costs such as depreciation and return on investment marks the departure from the marginal cost pricing because marginal cost is a measure of changes in variable costs alone. Therefore, regulatory agencies require the revenue requirement to be in line with the full-cost or average cost pricing (Kahn 1989).

28.2.1.1 Expenses

Expenses included in Eq. 28.1 could be rewritten as
$E + D + T$ where
E = operating expenses, including cost of inputs (such as cost of power in an electric utility), remuneration for labour and other administrative costs;
D = depreciation expenses; and
T = taxes on income and other taxes.
Thus Eq. 28.1 can be rewritten as

$$\sum_{i=1}^{n} p_i q_i = E + D + T + s.RB \qquad (28.2)$$

Often cost of inputs dominate the operating expenses. For example in a power utility, the cost of power purchase and cost of power generation constitute the major operating cost. Employees cost and administration and maintenance (A&M) costs are the other two components of the operating costs. Cost of power being the most important element of cost of service, regulators tend to exercise some control on the price and quantity of purchase from various sources by establishing some economic rules. The scope for scrutiny of employee costs and A&M charges is quite limited.

Although depreciation is included as an expense, it has a different character. Unlike other expenses, it does not constitute a cash outflow for the utility. The purpose of depreciation is to account for the wear and tear and obsolescence of the capital assets. The value of the assets declines over the economic life and this has to be recuperated from the consumers. The depreciation in a sense belongs to the owners and forms part of the return on investment. Accordingly, Kahn (1989) suggests that return on capital has two components: "return *of* the money capital invested over the estimated economic life of the investment and the return (interest and net profit) *on* the portion of the investment that remains outstanding." These two components are inter-linked and hence, it is better to consider depreciation along with return on investment. Depreciation can be used to pay down the debt as the asset for which the capital was provided depreciates or, if the investment was made from equity, increase retained earnings or buy back equity.

Taxes paid by the utilities are considered as part of their allowed expenses. Taxes paid in any year are influenced by the depreciation expenses because the depreciation rates for tax purposes are often different from that for the regulatory purposes. Two practices are normally followed in this respect:

- flow-through accounting which involves the taxes actually paid in a year, and
- normalised accounting which includes tax paid on average over the period of plant life

Normalised accounting removes the fluctuation of taxes due to depreciation rules and is commonly used.

28.2.1.2 Rate Base

The rate base essentially indicates the aggregate investment made by a regulated utility on which a return has to be allowed. The rate base largely includes net plant in service plus adjustments that vary with regulatory commission. Net plant is the gross value of the plant less accumulated depreciation already charged. In addition, the rate base may include cost of the inventory, work in progress and working capital allowance. The rate base is composed of physical assets and is different from the capital structure of a utility (Voll et al. 1998).

In order to determine the rate base, the American regulators use two concepts: whether the assets were used and useful in delivering the goods and services offered by the utility; and whether the investment was prudent. A prudence test asks whether the investment or expense was prudent (least cost) given what the decision-maker knew, *or should have known*, when the investment was made. The used and useful test asks whether the investment or expense was (and is) necessary for the provision of supply. A commission may disallow any expense or asset that fails *either* test (Voll et al. 1998). The utility has the burden of proving to the regulatory body's satisfaction that the plant in question has been used and was useful, and the investment was made prudently.

The method of valuation of the rate base has been a major issue of debate. The assets are often valued at original cost less depreciation. On the other hand, economists often suggest valuation based on reproduction or replacement costs. The first method considers the value originally paid by the utility for its plant and equipment less accumulated depreciation. This information is easy to obtain and there is little debate over this data. The problem with this approach is that the cost of old plant might have been cheap but it will cost the utility much more to reproduce the same assets now. Hence, as economic principles suggest that prices should reflect current marginal costs, setting tariff based on historical costs may lead to too low prices. The replacement cost method on the other hand uses the cost to replace the capacity at today's prices. This better reflects the cost of the plant and equipment but as no such accounting data is available in the records of the utility, the value is less amenable to regulatory verification. This method may also be illegal in some jurisdictions.

It may be worthwhile to mention here that the relevant economic pricing principle is short run marginal cost, where the marginal cost is the variable cost. Hence the fixed cost does not enter into picture. When the regulators consider depreciation or rate of return, they essentially consider full cost or average cost pricing and not the marginal cost pricing (Kahn 1989).

28.2.1.3 Rate of Return

Equation 28.10 can be rearranged as follows:

$$(\text{Revenue-Expenses}) = \text{Net income} = s.RB \qquad (28.3)$$

or, s = Net income/Rate base = Net Income/Investment, where investment is equal to rate base.

Viewed from this angle, the rate of return is the income which investors are allowed to earn per unit of investment. The regulatory agency sets a rate of return (unless it is already set by the government in a law) and allows the utility an opportunity to earn on its investment. Depending on the actual performance (i.e. revenues earned, expenses incurred and investment made), the rate of return actually earned may deviate from the allowed rate of return. The utility is said to be "under-earning" when "s" actually earned is below that authorized and vice versa. The Commission may ask the utility to adjust the tariff in both the cases.

In trying to fix a rate of return the regulatory agencies attempt to strike a balance between the interests of the investors and that of the consumers. According to Kahn (1989) "there is no single, scientifically correct rate of return, but a 'zone of reasonableness' within which judgment must be exercised." The lower limit of this range is the requirement of the utility to be able to attract capital for investment. The upper limit could be the most profitable return that could be obtained from investments of similar type elsewhere (perhaps in a free-from competition environment). As can be imagined, this zone of reasonableness is quite broad.

A typical approach towards determination of the rate of return begins with estimation of the cost of capital.[3] This relies on the idea that the rate of return is the return *on* the investors' and lenders' capital. The return on the Rate Base should be sufficient to service the capital that was raised to provide the physical assets of the utility. This in turn requires an estimation of the costs based on the utility's capital structure (i.e. its sources of funds: debt and equity capital).

---

[3] Cost of capital forms part of an extensive literature in finance. We will not enter into the details of this here.

28.2 Traditional Regulation

The estimation of cost of debt [4] is fairly straightforward as the interest to be paid on debt is known in advance. This will be available in the debt agreement entered between the lenders and the utility. The annual interest payment to be made on the outstanding amount of the debt along with any other legitimate cost of borrowing provides the cost of debt. The calculation is normally done for each individual debt instrument, as the terms and conditions vary. The sum of interest payable divided by the outstanding amount of debt gives the weighted average cost of debt.

The cost of equity is on the other hand more difficult to estimate. Utilities compete with other businesses for various inputs of the production process including capital. As the utility has to go to the capital market to sell its stocks, it should be in a position to pay the price at which investors are willing to supply the capital. The problem is then to determine what level of return do investors require in order to provide the capital for the utility's needs. A number of methods are available for this purpose. The most common methods are: discounted cash flow (DCF), risk premium method and the capital asset pricing model (CAPM). The DCF is most common, and the risk premium method is used as a "sanity check". The capital asset pricing model is less commonly used. Implementation of the methodologies requires different types and amounts of data, but all require considerable judgment on the part of the analyst.

Discounted Cash Flow Method[5]

Assume that the stock of a utility is currently traded at a price P. This price can be viewed as equivalent to the present value of the dividends expected by the investors. This can be written as:

$$P = \frac{D_1}{1+k} + \frac{D_2}{(1+k)^2} + \cdots + \frac{D_i}{(1+k)^i} + \cdots \qquad (28.4)$$

where P = price of stock,
D_i = expected dividend in year i
k = cost of equity capital

From Eq. 28.4, the cost of equity capital is simply the discount rate used by the investors to reflect their time preference. This is the return on their next best opportunity at the same degree of riskiness. If investors expect their dividends to grow at a constant rate g, then

$$D_2 = D_1(1+g); D_3 = D_2(1+g) = D_1(1+g)^2 \qquad (28.5)$$

Substituting Eq. 28.5 in Eq. 28.4, we get

---

[4] Normally debt includes both short- and long-term debts unless a special provision is made to account for the cost of short-term debts.
[5] See Viscusi et al. (2005) or (Appendix A to Chap. 2 of Kahn 1989).

$$P = \frac{D_1}{1+k} + \frac{D_1(1+g)}{(1+k)^2} + \cdots + \frac{D_1(1+g)^{i-1}}{(1+k)^i} + \cdots \qquad (28.6)$$

Summation of the above series yields

$$P = \frac{D_1}{(1+k)} \left[\frac{1}{1 - \frac{(1+g)}{(1+k)}} \right] \qquad (28.7)$$

Simplifying Eq. 28.11 and rearranging we get,

$$k = \frac{D_1}{P} + g \qquad (28.8)$$

For example, if the current yield (D_1/P) is 5% and the expected growth in dividends over time is 3%, the cost of capital is 8%.

For the US, the data for dividends and stock prices can be obtained from sources such as Standard and Poor's, the Wall Street Journal and Value Line. However, measurement of yield and the growth expectation of dividend raise number of questions. Viscusi et al. (2005) indicate the following issues:

- should the yield be measured as the last year's dividend divided by the average stock price?
- Should the growth be the average over past few years? If so, what is the surety that the future dividend growth will follow past trends?
- Should the cost of capital be measured for a particular industry or for a group of industries with same degree of risk for investors?

The calculation therefore depends on the analyst's judgement, and hence could be controversial.

Risk Premium

This approach attempts to determine what premium investors would expect over the going debt rate. Given that servicing debt takes precedence to paying dividends in case of a financial difficulty, equity is considered as a riskier investment than debt. Accordingly, equity should receive a higher return commensurate with the risk undertaken. The idea here is to estimate the premium that an equity capital should command. This is done by comparing the average equity price in the capital market and cost of risk free bond rates. The premium for equity thus obtained is a general reflection of the risk premium expected by the investors. In the utility industry, the average stock price may be different from the average stock price in the capital market. Thus a utility specific rate may also be determined. The premium together with the risk free rate gives the cost of equity capital.

For example, if the risk free long-term bond rate is 5% and the risk premium is 4%, the cost of capital is 9%.

28.2 Traditional Regulation

CAPM: The capital asset pricing model (CAPM) is another method of estimating risk premiums. The model uses a standardized formula for the calculation of the rate of return on equity as shown below:

$$r_e = r_f + Beta_* (r_m - r_f) \tag{28.9}$$

where
  r_e is the rate of return on an equity
  r_f is the risk-free rate of return (e.g., government bonds)
  r_m is the market rate of return (i.e., the returns on equities in general)
  Beta is the coefficient reflecting the volatility (risk) of the utility's stock relative to the market

For example, if the risk free bond rate is 5%, market rate of return is 10% and Beta is 0.6, then the return on equity is 5% + (10−5)*0.6 = 5% + 3% = 8%.

As can be imagined, this method requires information on Beta and market rates of return. This is widely used by the financial investors but less commonly used by regulatory bodies.

Weighted Average Cost of Capital (WACC)

The overall cost of capital is obtained by averaging the cost of debt and the equity by using their shares as weights. For example if the capital structure is composed of 80% debt and 20% equity, and the weighted cost of debt is 8% and the estimate of the required return to equity is 12%, "s", the total overall cost of capital (WACC) is:

$$80 \times 8\% = 6.4\%$$
$$20 \times 12\% = 2.4\%$$
$$\text{Total overall} = 8.8\%$$

This figure would then be used in determining the return component to be allowed in the revenue requirement.

28.2.1.4 Problems in Measuring the Cost of Capital

As indicated above, a number of problems are encountered while measuring the cost of capital. Following Kahn (1989), the following problems can be identified:

(a) Whose cost of capital? In the utility regulation, the question that arises is whether the cost should be for the particular utility under consideration or whether it should be applicable to a representative group of companies. In case it is for an individual company, the issue of efficiency of the company will enter into picture. If it is well-run, it might have secured the funds at a cheaper rate but permitting lower return for this reason is penalising efficiency of the

firm. The opposite argument can be true for inefficient companies. In case of a group of companies, the composition of the group needs to be determined.
(b) Cost of capital at what time? Historical or current? This issue is similar to that of valuation. Before performing the cost of capital exercise, it needs to be decided whether we are interested in the cost of capital at the time of raising it (i.e. historical) or its present cost. In practice, the debt is estimated at the historical rate while equity is valued at current rate. Is this mixing appropriate or sensible?
(c) Measuring cost of equity capital: As indicated above, the cost of equity capital is not easy to determine, because there is no objective measure of investors' anticipated earnings. Current yields can either overestimate or underestimate the actual cost of capital.
(d) Inconsistency between rate base and cost of equity: If the rate base is valued at book value (i.e. original cost less depreciation) and the rate of return on equity is based on the ratio of earnings to market price of the stock, the two factors are not determined in a consistent manner.
(e) Effect of capital structure on cost of capital: It is generally agreed that the overall cost could be reduced by borrowing up to a certain level. This is because the cost of borrowed capital is an allowable expense for tax purposes, while the return on equity is subject to tax on profit. For example, if the cost of debt is 5% and that of equity is 10%, and if a utility wants to raise $100, borrowing adds only $5 to the cost. If the income tax on profit is 50%, the utility has to recover $= 10/(1-0.5) = \$20$ from the charges, $10 for the stockholders and another $10 for the government. As the share of the debt increases, the utility has to pledge equally larger share of its income to service the debt. The cost of debt and equity would increase in recognition of this extra risk. This aspect is not considered in the cost of capital exercise.

28.2.2 Rate Structure Regulation

The objectives of rate design is to strike a judicious balance so that the tariff set becomes economically efficient and fair and at the same time it provides the regulated utility a reasonable opportunity to recover costs including return on investments. Often regulators attempt to achieve a number of competing objectives while setting tariffs, including (Weston 2000):

- revenue related objectives such as cost recovery, predictable and stable revenues, predictable and stable rates;
- cost-related objectives such as rates to promote efficient consumption decision, fair rates without undue discrimination, etc.;
- practical considerations such as simple, understandable and acceptable rates, ease of administration, non-controversial rates, etc.

Table 28.1 Classification of functionalised costs for an electric utility

Cost items	Dem	Ener	Customer
Power cost	Y	Y	
Transmission	Y	Y	
Distribution-related			
Wages	Y		Y
Depreciation	Y		Y
A & M	Y		Y
Return	Y		
Total	A	B	C

Note: Y: relevant. A, B and C are totals of each column

Once all the elements on the right-hand side of Eq. 28.1 are available, the total revenue requirement is determined. The next step involves allocating the costs to various consumer categories and designing tariffs to be charged for their consumption. The cost allocation process attempts to attribute costs to various classes of consumers so as to reflect the cost of providing utility services to each class. Tariff design on the other hand involves deciding the mechanism of recovering the allocated costs. As in any design, tariff design goes beyond economic aspects and involves non-cost issues as well.

The cost allocation process involves three steps: cost functionalisation, cost classification and cost allocation to customer classes. Cost functionalisation entails separating costs according to different functions undertaken by the utility. For example, an electric utility performs generation or purchase of electricity, transmission, distribution, retail supply, customer service and administration and general functions. Costs for each function have to be identified.

Cost classification involves attributing the functionalised costs by cost causation. In an electric utility, demand related costs, energy-related costs and customer-related costs are typically identified. Costs that vary with load (kW) are considered as demand-related. Costs that vary with energy consumption or supply are considered as energy-related costs. Costs that vary with the number of consumers served are classified as customer-related costs. An example is given in Table 28.1.

Allocation of classified and functionalised costs to consumer class is the last step of the cost allocation process. Consumers of a utility are grouped under broad categories based on their characteristics (such as load, voltage of supply, whether interruptible or not, etc.). The costs are then allocated to each class of consumers using some allocation principle (for example fairness, contribution to the cost causation, etc.).

There are two commonly-used cost-based methodologies for allocation: embedded cost (or fully distributed cost) approach and marginal cost approach. The embedded cost approach uses historical accounting information while the marginal cost approach uses the economic theory of marginal costs for the allocation.

According to Parmesano et al. (2004), there is neither any universally accepted principle nor any engineering or economic theory guidelines for classifying and allocating costs using an embedded cost approach. However, the following factors are often used: 1) type of generation plant, 2) planning and operating constraints,

3) load patterns faced by the utility, 4) system load factor, 5) contributions of the classes to the total demand, 6) kilowatt-hours purchased by each class as a percent of total sales, 7) the number of customers in the class as well as many other factors and combinations thereof (Parmesano et al. 2004; Voll et al. 1998).

In the marginal cost-based approach, the marginal cost of supplying each consumer class is determined. The expected revenue from the marginal cost-based tariff is then compared with the overall revenue requirement. The gap, if found, is then closed using alternative schemes (fixed charges or Ramsey pricing).

Rate makers often confronted with the issue of fairness of rates in the sense of whether one group is subsidising another group. This equity dimension of the problem, while often ignored in theoretical literature, is of great importance for practical tariff setting. As the embedded cost approach uses cost allocation factors without any theoretical backing, they often tend to be controversial in this respect. Similarly, economic efficiency may sometimes require pricing different consumers differently, which may be in conflict with common notions of fairness.

28.3 Problems with Traditional Regulatory Approach

A number of problems related to the traditional regulatory approach have been identified in the literature. Some of these are discussed below:

(a) Cost minimising behaviour: one of the objectives of regulation is to ensure market like decision-making in a regulated industry. In a competitive market, competition ensures that the firm minimises its costs. Efficient firms receive rewards in the forms of high profits while the inefficient ones are punished by low profits or losses. In a regulatory environment, there is no incentive for the regulated utilities to minimise costs below that allowed by the regulator, as the rate of return is regulated. As indicated earlier, it is difficult for the regulatory agencies to scrutinise all expenses incurred by a utility without assuming the managerial role of the utility, which is clearly beyond the scope of the regulators. In fact, the regulatory mechanism creates strong incentives to inflate expenses at no cost to stockholders (for example offering higher salaries and expenses to management) and to engage in transactions with affiliates (Kahn 1989).

However, regulatory lags tend to mitigate the above problem to some extent. Regulatory lag refers to the time delay between two consecutive cases for tariff determination. Once a tariff is set, it remains in force until a new tariff is determined. During this intervening period, the utility may earn higher profits by reducing costs and keep the higher profits. Thus rate freeze provides some incentive to the utilities to be cost efficient. Moreover, regulators tend to allow the utility to earn profits within a 'zone of reasonableness', which also provides a similar benefit (Kahn 1989). A similar mitigation effect arises due to the threat of

28.3 Problems with Traditional Regulatory Approach

disallowed costs. Regulators can disapprove costs that are considered as unreasonable. This tends to keep a check on the costs.

(b) Lack of recognition of efficiency: Regulators while deciding the rate of return often do not pay much attention to the performance of the utility. If a utility is able to attract low cost capital, the rate of return allowed to it will be low, which does not recognise the efficiency of the management in attracting low cost capital. Return is not linked to performance in the case of regulated utilities, which does not provide correct signals to the players.

(c) Averch-Johnson effect: A traditionally regulated monopoly may engage in socially undesirable investments (i.e. social costs may be higher than the social benefits) because such investments can allow them to earn higher returns due to expanded rate base and higher cost of capital. Such investments can also allow them to charge higher rates to recover revenue deficiencies arising out of such investments. These incentives can encourage them '(1) to adopt an excessively capital-intensive technology and (2) to take on additional business, if necessary at unremunerative rates' (Kahn 1989).

Although the above claims proved to be difficult to be verified empirically, Kahn (1989) has listed a number of areas which might be influenced or might result due to the Averch-Johnson (A-J) effect. These include:

- unwillingness of utilities to adopt peak-load pricing, which reduces peak demand and thereby reduces capacity expansion needs;
- desire to maintain a large reserve margin[6];
- resistance to adopt integrated regional capacity investment planning and integrated form of power pooling.
- Resistance to the introduction of capital saving technology;
- Resistance to lease facilities to others;
- Adopt an excessively conservative reliability standards for generation, with implications for high cost of assets and reserve margins;
- Less hard bargaining tendency when purchasing equipment from others;
- A tendency to reach out to expand business, even at unremunerative rates.

(d) Problem of inter-company co-ordination: Technological progress brings new opportunities for business and scope for cost reduction. For example, electric utilities can benefit from scale economies and diversified plant mix by using interconnected transmission system and by entering into power pooling and interchange arrangements. These arrangements allow them to use larger generating plants, to reduce costs of operation and to achieve higher capacity utilisation (Kahn 1989). In a competitive market, players will seize such opportunities by entering into agreements or through mergers and acquisitions.

---

[6] Reserve margin is the excess capacity over peak demand which is maintained to tide over emergency conditions.

However, traditional regulation may not be successful in ensuring such inter-company co-operations due to lack of adequate incentives or due to passive or negative nature of regulation.

(e) Interventionist nature of regulation: RoR regulation requires time consuming detailed scrutiny of the costs and income of the utility. The regulator also tends to ask whether the decision of the utility was prudent and reasonable. Regulators also try to direct the utility what to do and what not. By nature, this form of regulation tends to be interventionist and heavy handed.

(f) Cost of regulation: RoR regulation tends to impose high costs on both sides. On the regulatory side, it requires scrutiny while the compliance cost of the utility is also high. The determination of rate base and cost of capital, regular reviews and hearings and subsequent litigations (where they arise) impose cost burdens on the participants (Intven 2000).

(g) Lack of innovation: RoR regulation does not provide the utility with a strong incentive to develop innovative products, as there is no incentive for doing so.

28.3.1 Regulatory Alternatives

A number of regulatory alternatives could be identified from the literature. They include: incentive regulation, conduct regulation and regulation by contract. We discuss these alternatives below.

28.3.1.1 Incentive Regulation

Incentive regulation process has been defined as "the use of rewards and penalties to induce the utility to achieve desired goals where the utility is afforded some discretion in achieving goals" (Berg undated). The above definition has three important components: first, the system of rewards and penalties replaces the command and control form of regulation; second, the goals are not unilaterally decided by the regulator but the utility assists in deciding the goals; and finally, the utility decides how to achieve the goals as opposed to regulator prescribing specific actions (Berg undated).

Any incentive regulation attempts to follow the following two principles (IPART 1999):

- competition is preferable to regulation, that is wherever possible, competitive solutions shall be used or promoted;
- regulation should emulate competitive outcomes. The reference standard is thus the outcomes of a competitive market and the industry is encouraged through incentives to move towards efficient frontier.

An incentive regulation can take two generic forms: individual incentive regulation or yardstick regulation. In an individual incentive regulation, the regulator would regulate the utility based on some of its own observable measures (Chong 2004). A number of alternative schemes are commonly discussed under the first category. These are: price cap regulation, revenue cap, and hybrid caps. A price-cap sets the maximum allowable change in prices (for individual services or products or a basket of services). A revenue cap on the other hand sets the maximum revenue that a regulated entity is allowed to earn and the utility then decides a set of prices that will allow it to earn the set revenue. A hybrid cap on the other hand combines the features of a revenue cap and a price cap.[7]

Yardstick competition on the other hand uses comparative statistics from other utilities to regulate the utility. A number of other variant schemes are also used; we will look into performance based regulation as well.

28.3.1.2 Regulation by Contract

This alternative form of regulation has been proposed recently notably by Bakovic et al. (2003). The privatisation of state enterprises in general and energy utilities in particular has been disappointing in the recent past and the enthusiasm for privatisation of energy utilities has significantly reduced due to lack of popular support and fear by governments of backlash. Bakovic et al. (2003) argue that the main factor behind this was lack of regulatory commitment and balanced decision-making. Although existence of independent regulatory bodies was considered as a pre-requisite for successful privatisation, yet mere existence of such bodies does not ensure success. This is because, Bakovic et al. (2003) argue, in most developing countries regulators face the prospect of increasing the tariffs and the regulators have found it difficult to do so. Instead of adopting a balanced approach to tariff, regulators, they argue, have taken the consumers' side.

Bakovic et al. (2003) argue that regulatory independence is not enough. What is required for a successful privatisation of the energy industries is to limit regulatory discretion through a pre-specified regulatory contract.

Bakovic et al. (2003) provide two definitions of the concept. One states that regulation by contract is regulation without a regulator. In this case, the regulatory contract is similar to any commercial contract and is totally self-contained and self-administered. This essentially eliminates the need to have a separate regulatory entity. The second definition sees the contract as a detailed tariff setting agreement administered by an independent regulator. In this definition, the regulator is not replaced by a contract but the regulatory discretion is reduced by specifying it in a contract.

The main objectives of such a contract are twofold:

- to protect the consumers from monopoly prices and inferior quality supplies;
- to attract private investment by providing an affordable price.

---

[7] See IPART (2001) for further details on these options.

Bakovic et al. (2003) suggest that such an agreement should have the following key features:

- regulatory contract is specified prior to receiving bids for privatisation;
- the contract may be a stand-alone document similar to licence or concession document;
- an independent regulatory body to implement the contract
- pre-committed system of multi-year tariff, with a formula specifying the average tariff or revenue requirement
- regulator will have little discretion over tariff initially; which may be relaxed in future;
- agreement should provide a dispute resolution system;

This alternative approach may come as a surprise to many observers, because the regulatory reform was introduced in many countries to alienate governments from tariff-setting and to help establish cost-based tariffs. To suggest that the regulatory independence may not be such a good thing for achieving privatisation and that it may be better for the governments to retain the tariff-setting authority in the initial years of privatisation amounts to state-centred reform. The sole purpose of this alternative regulation appears to be to promote privatisation at any cost, and to ensure that private investors remain in control, perhaps in collusion with government. This approach does not appear to possess the characteristics of a good regulation.

28.3.1.3 Conduct Regulation

Normally, a regulator can exert control over three elements of an industry: structure, conduct and performance. Structure is traditionally considered as exogenous which is decided by political authorities or is obtained historically. The regulator takes the structure as granted and tries to control performance through incentive or other forms of regulation. Performance regulation focuses on utility outcomes by influencing actions and decisions of the regulated utility but does not try to influence the behaviour of the utility.

Conduct regulation on the other hand focuses on direct supervision of operating and investment actions of the regulated utility. According to Tenenbaum et al. (1992) conduct regulation has taken two forms:

(a) integrated resource planning (IRP) [8] which is a heavily interventionist form of regulation that entails regulatory agency involvement in a company resource planning and selection process taking into account supply and demand side options as well as environmental externalities. The basic premise of IRP is that the utility and its consumers do not appreciate the full cost of the service due to under-priced output and that the utility does not have any incentive to energy

---

[8] We will not discuss IRP in detail in this chapter. Interested readers may refer to Swisher et al. (1997).

28.3 Problems with Traditional Regulatory Approach

conservation. The regulatory solution to this problem is then to require the utility to change its behaviour by forcing it to take environmental costs and demand side options into consideration.

(b) The other form of conduct regulation has been spurred by competition. This has resulted in prescribing code of conducts for various actions. For example, regulators are prescribing detailed guidelines for competitive procurement of capacities. Similar prescriptions of conduct are being made for other aspects of the competitive market such as trading business, access to transmission, etc.

Conduct regulation essentially implies that the regulator forces the utility to do things it would not want to do. In this sense, it represents a failure of regulation (Tenenbaum et al. 1992). When other forms of regulation fail, a regulator resorts to conduct regulation. This is also a difficult type of regulation as the utility may be able to accomplish the same outcome using "a variant of the prohibited behaviour which was not anticipated by the regulator" (Tenenbaum et al. 1992).

Below we discuss price cap, revenue cap and yardstick competition in more detail.

28.4 Price-Cap Regulation

A price-cap, as the name implies, sets the maximum allowed inter-temporal path for price of a product. The price rules are set in advance and depend on non-controllable factors (i.e. the factors that are beyond control of the regulated entity) (King undated). The initial rates under a price-cap regulation are typically set based on traditional rate of return regulation but the subsequent changes are made automatically by using a set formula. These changes are normally made annually, although in principle such changes can be made more frequently as well (Lowry and Kaufmann 1994). The cap remains valid for a fixed period and then it is subjected to a review process and a new price cap is set for the subsequent period. From this perspective, this type of regulation actually results in a multi-year tariff plan.[9]

A typical price cap formula looks like Eq. 28.10 below:

$$\Delta PCI_t = P_t - X_t + Z_t \qquad (28.10)$$

Where, ΔPCI = price cap index growth rate,

P is the inflation factor, X is the X factor introduced to adjust expected productivity improvements, Z is the Z factor introduced to adjust growth in price cap index due to other factors besides inflation and productivity. Two commonly used price indices are retail price index (RPI) and consumer price index (CPI). In the UK, the retail price index was used while CPI is used in Australia (King undated). Accordingly, price cap is commonly associated with RPI-X or CPI-X formulation.

---

[9] See Alexander and Harris (2001) for a discussion on multi year tariff controls.

Equation 28.10 is derived from the economic theory as follows.[10] According to a basic economic theory, the rate of growth of total factor productivity[11] of a firm is equal to the difference between the rate of growth of output prices and that of the input prices. Mathematically, this can be written as

$$dp = dw - dTFP \pm dZ, \qquad (28.11)$$

Where dp is the change in output prices, dw is the change in input prices, dTFP is the change in total factor productivity and dZ is the unit change in costs due to external circumstances. Although Eq. 28.11 looks similar to Eq. 28.10, it is not used in the price cap formula as it looks at the costs and TFP changes in the industry alone. If this is used, the price cap will not have much difference with the rate of return regulation. To provide better incentives, the productivity changes in the regulated industry are compared with the productivity growth in the national economy. An equation similar to (28.11) could be written for the economy as a whole to reflect the changes in TFP. This is shown in Eq. 28.12.

$$dp^N = dw^N - dTFP^N \pm dZ^N, \qquad (28.12)$$

Comparing Eqs. 28.11 and 28.12 we get,

$$dp - dp^N = (dw - dw^N) - [dTFP - dTFP^N] \pm (dZ - dZ^N) \qquad (28.13)$$

Equation 28.13 can be rewritten as

$$dp = dp^N - [(dTFP - dTFP^N) - (dw - dw^N)] \pm (dZ - dZ^N) \qquad (28.14)$$

$$dp = dp^N - X \pm Z \qquad (28.15)$$

Equation 28.15 is same as Eq. 28.10. This suggests that the allowed price change in the regulated firm is given by (Ros 2001):

- the rate of inflation of national output prices;
- less a productivity offset X, which represents the productivity growth differential between the regulated firm and the national economy, adjusted for any differences between the growth of input prices in the regulated firm and the national economy;
- plus exogenous unit cost changes, captured through the differences between changes in exogenous units of the regulated firm and the national economy.

This approach provides an important incentive property and a regulatory property. The incentive property derives from the fact that there is a partial decoupling of current prices from current unit costs. This gives the utility an

---

[10] Here we follow Ros (2001). Also see Bernstein and Sappington (1998).

[11] Factor productivity is a measure of efficiency of conversion of inputs to outputs in the production process. Total factor productivity measures how efficiently a firm transforms all its inputs into aggregate outputs.

incentive to keep the costs below the escalation factor and retain the consequent profits. The regulatory advantage arises as this system is considered as "regulation with a light hand" which is easy and inexpensive from operational point of view (Weyman-Jones 1990).

On the other hand, as the regulated entity reveals a cost saving at end of the capped period, the regulator will attempt to impose a stricter target by passing some of the benefits to the consumers. This reduces the incentive for the firm to make the cost saving and if the benefit received by the firm from the cost saving is less than the cost of achieving such cost reductions, the firm will not engage in any cost saving any further. This ratchet effect destroys the possibility of long term cost reductions (Viehoff 1995). The behaviour of the regulated firm can also change as a result of this weak incentive to cost saving later in the regulatory cycle: a) it may become less concerned to make cost savings; b) it may shift the timing of the actions so as to earn profits early in the cycle.

The price-cap regulation has been used in Britain, Europe and Australia for regulation of privatised entities (Guasch and Spiller 1999, pp 73–74). The type of price cap discussed above constitutes the basic form of price cap. Energy utilities offer a number of services at different prices. In such cases a weighted average price index formula is used instead of a simple, single product price index. The weights used in such a formula can be based on sales volume, cost shares or revenue shares. Similarly, the weights can be based on historic data or forecasts or based on a rule of thumb, or a combination of the above approaches (IPART 2001; Intven 2000). The weighted average price cap requires the operator not to exceed the cap for the weighted average prices. This offers the firm some flexibility in pricing each product. As the prices with heavier weights affect the weighted average more, these prices are likely to be changed less compared to those with light weights (Intven 2000).

Similarly, it is not necessary that all the services have to be subjected to same price caps. Often services are grouped in a basket and the cap applies to the basket of products or services, instead of individual prices. For example, electricity tariffs for residential consumers may be classified by consumption patterns: low, medium and high. Their tariff structure and rates may be different but all these classes may be placed under a single basket and a single escalation rate may be applied.

28.4.1 Choice of Inflation Factor

Generally, the inflation factor used in the formula should correspond to the expected inflation in future. However, in practice historical values are often used for the following reasons (Intven 2000):
- past inflation provides a good indication of the future as well;
- forecasting inflation is complex, time consuming, and may be subject to controversy and manipulation;

- use of forecast values will require error adjustments, which adds to regulatory complexity and regulatory uncertainty.

The disadvantage of using a historical data is that there may be significant variations with the actual results, which will require either frequent revisions or adjustments.

For selecting an inflation factor for the price cap formula, a number of criteria are considered. Some of these are as follows (Intven 2000):

(a) availability from an independent, published source;
(b) timely availability of the index;
(c) easy understandability;
(d) stability in the sense that the same data is not frequently revised retrospectively;
(e) reflective of changes in utilities' costs
(f) consistent with the total factor productivity of the economy.

Commonly used inflation index is the retail price index or consumer price index.

28.4.2 X Factor

As shown above, X factor plays an important role in the price cap regulation. If a high X factor is chosen, the firm does not have much incentive to run the utility efficiently while a low X factor penalises the consumers disproportionately. We have seen that the theory suggests that the X factor should be based on the total factor productivity and not on partial factor productivity (such as labour productivity). Similarly, as the factor productivity can vary significantly from one year to another, it is customary to use long term average factor productivity as opposed to annual TFP.

As can be seen from Eq. 28.11, the X factor is composed of two components: the first part reflects the difference between the economy wide productivity and the regulated industry productivity and can be termed as productivity differential. The second component takes care of the differences in the input prices faced by the regulated industry and the economy, and can be called the input price differential. If the regulated industry is expected to face input price escalations similar to that of the economy as a whole, and the productivity growth would be similar to that of the national economy, then X would be equal to zero. Similarly, if the productivity differential and the input price differential assume same numerical values of opposing sign, the X factor will be zero. On the other hand, if the input prices faced by the regulated industry grows slower (or faster) than that of the overall economy, while the productivity is expected to grow faster (slower) than the overall economy, and when these differentials are not same, the X factor is the sum of the two components. Otherwise, they offset each other and a lower X value will result.

28.4 Price-Cap Regulation

There are two basic approaches for determining the X factor: historical approach and regulatory benchmark approach (Intven 2000). Historical approach, as the name indicates, relies on historical performance of the regulated industry and assumes that the past productivity is a good indicator of the future productivity growth as well. Normally a measure of total factor productivity is preferred to partial factor productivity and from the historical TFP trend, an average productivity rate is determined. Similar productivity growth rates are also required for the entire economy. As Intven (2000) and Ros (2001) indicate this method may be less appropriate for developing countries for the following reasons:

- poor performance of the utilities in the past offers the possibility of greater productivity gains;
- political and economic instability as well as lack of clear regulatory and legal framework may affect productivity gains.
- Data quantity and quality may be an issue. Long series of data may not available for TFP calculations. Similarly, the data is likely to be affected by structural changes in the economy and the sector, which may show significant changes or disruptions in TFP.

The regulatory benchmark approach on the other hand does not believe in relying on the past productivity trend. This may be the case when an industry undergoes drastic changes or when the industry was not performing efficiently earlier. In such a case, the regulator relies on informed judgement, often based on international experiences. This may be the only available approach for developing countries. In general, the regulators will be required to decide on the productivity differential and the input price differential separately.

28.4.3 Z Factor

The Z factor allows adjustment to the price cap due to exogenous factors, i.e. the factors not covered by X factor and inflation and factors which are beyond control of the utility. Regulators can specify a threshold beyond which the Z factor will be applicable. This can be used for a number of purposes (Lowry and Kaufmann 1994; Intven 2000; NER 2003):

- To introduce certain incentives through price cap. For example, to recover a bonus or a penalty designed to promote energy efficiency in the industry, this factor could be used.
- To capture effects of legislative, judicial or administrative actions that have significant impacts on the regulated operator;
- To reflect events which do not represent normal business risks.
- To allow carry forward of the difference between the price actually charged and the price cap (i.e. the correction factor for revenue);
- To allow cost pass through and adjustments for windfall gains.

Inclusion of a Z factor to take care of uncontrollable costs and risk exposure can reduce the incentive property of the price cap as the utility does not have any incentive to hedge the risk (Green and Pardina 1999). However, Intven (2000) suggests that in developing countries where significant exogenous events are more common, the Z factor may be useful for regulatory purposes.

28.4.4 Choice of Form

A price cap regulation in its simplest form is applied on the prices of the regulated services. However, these prices have to be set in the first instance. This brings us to the issue of relationship between regulated revenues and costs. As mentioned in the case of RoR regulation, the prices are set based on the costs. In the case of price cap approach, two alternative approaches are considered: cost-linked form of regulation and cost-unlinked forms of regulation.

In a cost-linked form, individual company's costs and revenues are taken into consideration explicitly. Such an approach relies on calculations similar to a rate base/rate-of return regulation, where different cost elements (operating costs, depreciation, cost of capital, etc.) are considered and reviewed. Total costs are then allocated to different services to decide about the tariffs or prices. The price cap is then applied to the prices so determined. This method then resembles the RoR regulation and requires significant scrutiny on the part of the regulator. As the regulator relies on the information supplied by the utility, the problem of information asymmetry is also present.

In the alternative case, where a cost-unlinked approach is followed, the regulator decides the revenues of the regulated utility based on independent measures of efficiency which are not directly related to the actual or historic costs incurred by the firm. This approach uses what is known as yardstick comparisons or benchmarking techniques for different cost elements and thus attempts to reflect the best practice in the industry. The use of benchmarking in a price cap regulation then makes it far removed from the rate of return regulation.

28.4.5 Advantages and Disadvantages of Price Cap Regulation

Price cap regulation offers several benefits. These include: incentives as mentioned earlier, streamlining of regulatory process, greater price flexibility, reduction in regulatory intervention and micro-management possibility, sharing of gains with consumers, and greater price certainty (Intven 2000).

Some difficulties are also found in the case of price-cap regulations. They are (Guasch and Spiller 1999):

(a) regulatory capture: the choice of X factor leads to considerable debates. Normally, governments want to ensure good return to the privatised entities

initially after the change of ownership. On the other hand consumers criticise regulators for allowing excessive profits to the firms at the cost of consumers. Balancing such a conflicting issue is not easy and often the regulators are accused of biased decisions. If regulators intervene within the period of formula operation, firms resent undue interference.
(b) Cost calculations: Limited cost information is available to the regulators. With this, they are not in a position to determine the feasible unit costs.
(c) Regulatory lag: the cost reductions tend to diminish over time as regulators adjust the price cap index. This implicitly involves an estimation of the profits to be accumulated during the review period. If the regulator over or underestimated, the regulatory approach can appear to be inappropriate. Too frequent adjustments increase regulatory risks while too few adjustments make the regulator ineffective.
(d) Risky return on investment: Under rate of return regulation, the investor was assured of a reasonable return on investments. This is not the case in the price cap mechanism.

28.4.6 Comparison of Price Cap and RoR Regulation

Price cap and rate of return regulation are two commonly used regulatory principles. RoR regulation as applied in the USA and the price cap as applied in the UK both provide excess profits when costs are falling and below normal profits when costs are rising. Both are relatively flexible regulatory designs (Guasch and Spiller 1999, p 83). Beesley and Littlechild (1989) provide a succinct comparison of the two methods as follows:

"RPI-X and rate of return regulation have certain features in common. Both accept the need to secure an adequate return for the company's shareholders in order to induce them to continue to finance the business, without conceding unnecessarily high prices at the expense of customers. Nevertheless, there are significant differences between the two systems, which give RPI-X a potential advantage with respect to incentives and efficiency.

First, RPI-X embodies an exogenously determined risk period between appraisals of prices, whereas rate-of-return regulation makes this period endogenous....

Second, RPI-X is more forward-looking than rate-of-return regulation. The latter tends to be based on historic costs and demands... In contrast, RPI-X embodies forecasts of what productivity improvements can be achieved and what future demands will be and is set on the basis of predicted future cash flows.

Third, there are more degrees of freedom in setting X than are involved in rate-of-return regulation... X is initially set in the context of negotiations about the whole regulatory framework ...In resetting X, the regulator has fewer degrees of freedom but nonetheless can modify (at least at the margins) any aspects of the framework and in practice has done so...

There is an important implication for incentives and efficiency. The exogenous risk period and the forward-looking approach mean that the company is not deterred from making efficiency improvements either by fear of confiscation within the period or by the belief that allowed future prices will simply be an extrapolation of past costs."

It is normally considered that a price cap with a regulatory review procedure brings the regulation closer to RoR regulation because the review tends to focus on the cost of service (Weyman-Jones 1990). Such a review also introduces regulatory uncertainty as regulators may have high level of discretion and may follow inconsistent approaches. Regulatory uncertainty would then reduce the incentive properties of the price cap regulation.

28.4.7 Experience with Price Cap Regulation

As indicated earlier, the price cap regulation was applied to the privatised utilities in the UK. The price cap regulation was first introduced for British Telecommunication in 1984 and since then has been used in many other industries, including electricity and gas industries. The method has been exported from the UK world over—to the USA, Australia, Argentina, and many other developing countries.

The experience from the UK indicates that initially the caps were quite lax and offered high profits to the owners. There is a view that this was an objective of the government itself in order to ensure successful completion of the privatisation process provided generous rents to the private owners. As Beesley and Littlechild (1989) indicate, the initial setting of the price cap for the regulated industry at the time of privatisation was done by the government. The government had a higher degree of freedom in setting the cap. The regulators subsequently had to tighten the caps over time and thereby pass some of the benefits to the consumers. For example, the UK regulator announced new electricity price caps in August 1994 which were supposed to be effective from April 1995. However, in March 1995, the regulator reviewed the price caps and lower caps were set (Guasch and Spiller 1999, p 84). Similar tendencies were noticed in Latin American countries as well.

Another area of concern is that in many countries the prices are not efficiently structured and may be incompatible with competitive outcomes (Green and Pardina 1999, p 29). A price cap imposed on such an existing price structure will continue with the inefficient pricing and may create new problems. Rebalancing the prices is a requirement for better performance but a recent study shows that there is little evidence of price rebalancing in utilities subjected to price cap regulation in the UK and the USA (Giulietti and Price 2005). In developing countries where the pricing problem is more serious, use of price caps without a time-bound rebalancing programme may not yield desired results.

28.5 Revenue Caps[12]

These are a class of regulatory devices used to cap revenues rather than prices. Within this broad category there are a number of alternative forms of schemes, including performance-based regulation (PBR), statistical revenue caps and absolute revenue caps. We will consider PBR separately. Here we consider the generic form of revenue caps.

Revenue caps and revenue per customer caps are commonly suggested as alternatives to price caps. Revenue caps impose an absolute maximum of revenue that could be earned by a regulated firm. This could be applied to each service or each service basket. The regulator does not fix the price to be charged by the utility but only cares about the total revenue earned. The cap would be revised by establishing some links with inflation and other factors. The revenue per customer caps on the other hand allows revenue to grow with the growth in customer base.

This form of regulation has a number of disadvantages: Crew and Kleindorfer (1996) suggest that it provides an incentive to the utility to charge prices above the unregulated profit-maximising monopolist levels. This suggests that some customers (especially the captive customers) may face monopoly prices or even higher when the utility is subjected to a revenue cap. Consequently, the utility will try to restrict the output beyond the level of that of an unregulated monopolist. If additional restrictions are imposed on the utility to avoid such a result, Crew and Kleindorfer (1996) suggest that the regulation will turn into micromanagement of the utility, which will add transaction costs and uncertainty.

Green and Pardina (1999), suggest two additional disadvantages of this type of regulation. First, as the utility will have to set prices first before knowing its revenue, it can never exactly match the regulated revenue in practice. This will then require an adjustment process to correct the differences in revenue earned from the regulated level of revenue. If the utility over-earns, its allowable revenue has to be reduced and vice versa. Second, the utility will have a tendency to expand its business to low-price consumers so as to remain within the revenue cap. This adds volume to the utility without adding much revenue. To prevent such a manipulation, the regulator has to add a number of revenue drivers to the revenue cap formula, which adds complexity to the system.

Crew and Kleindorfer (1996) suggest that revenue caps "shut off the engine of sales" and seek to reduce output and that they actually promote inefficiency.

The National Grid Company (NGC) of the United Kingdom was subjected to revenue caps during 1990–95 for the regulated transmission services. The revenue cap was set based on a revenue requirement estimated along the traditional RoR procedures. NGC provides the transmission services and its level of output is essentially decided by the competitive power generation market and the interaction between generators and demand centres. As the output decision is largely beyond

---

[12] This section is based on Crew and Kleindorfer (1996).

control of the NGC, the revenue cap provided incentive to minimise costs of its services.

28.6 Yardstick Competition[13]

A central problem of regulation lies in the information asymmetry faced by the regulator. Costs faced by the utility are of great importance to regulation but the regulator has to rely on the information furnished by the utility. One of the ways to circumvent this problem is to use yardstick regulation.

Yardstick methods involve a wide range of techniques and applications and can be applied in assessing costs or service quality levels or in assessing future expectations of costs or service quality. Yardstick competition is one of the techniques of yardstick methods and refers to the use of comparisons for setting revenues of a regulated utility on the basis of costs of other companies.

The idea of measurement of productive efficiency was first suggested by Farrel (1957) who used the idea of comparing the observed inputs to the inputs used by an efficient firm to measure the technical efficiency. This is explained in Fig. 28.1 below using a two input case.

SS' represents the various combinations of two inputs X and Y that an efficient firm uses to produce a unit of output.[14] The point P represents the input combinations used by a firm. It can be seen that the firm is using much more inputs compared to an efficient firm. The ratio OQ/OP gives the measure of technical efficiency. But the firm has to use the inputs in a cost effective manner as well. AA' provides the budget constraint. Then cost minimisation requires that the firm produces where the cost function is tangent to the production possibility frontier (i.e. at Q'). The price efficiency is then given by OR/OQ. The overall efficiency is given by the product of technical efficiency and price efficiency [i.e. (OQ/OP) × (OR/OQ) = OR/OP].

The above approach has led to the development of benchmarking. The yardstick regulatory approach uses this concept and was proposed by Shleifer (1985) in his seminal paper. In a competitive market, the market price represents the yardstick and firms that provide goods or services at the least cost prosper. However, in monopoly markets, such direct performance comparisons are not possible in the same market. The yardstick competition uses a set of efficient cost standard determined by other companies' costs and aims at achieving an efficient outcome. Shleifer (1985) concluded that "If the firms are identical or if heterogeneity is accounted for correctly and completely, the equilibrium outcome is efficient".

---

[13] This section relies on Williamson and Toft (2001).

[14] In practice this function is not known and Farrel (1957) suggests that such a function can be constructed from using observed data. The production possibility is then called the best practice frontier.

Fig. 28.1 Measurement of productive efficiency

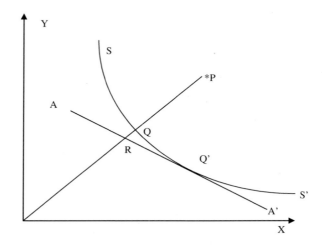

However, in practice it is difficult to ensure identical firms in a regulated market. Different companies face different situations, which need to be taken into account. Yet, it is difficult to account for differences between firms completely and correctly. One solution lies in estimating a regression equation showing how costs vary with exogenous variables and in finding out how each company's costs differ from that predicted by the regression equation. But costs do not vary because of operating efficiency alone and it is difficult to dissociate other factors influencing performance in a regression analysis (Green and Pardina 1999, p 64). Moreover, it should be recognised that there are two sources of productivity gains here: movements towards the frontier and shift of the frontier itself. The relative importance of the two sources of productivity gains varies by industry: some industries are characterised by fast technical changes; for them the shift of the frontier will be important. For industries like transmission and distribution of electricity or natural gas, this possibility may be less important and hence, the role of the second source may be less important (Green and Pardina 1999, pp 65–66). The second is more difficult to identify and predict.

Shleifer (1985) suggests that when the firms are identical, such an efficient outcome requires that the regulator should have the willingness to bankrupt a potentially inefficient firm. When the firms differ, subsidies would be required to avoid bankruptcy, which undermines incentives. These requirements make the practical application of the approach questionable.

Regulators apply alternative methods and techniques for benchmarking. According to one classification, the benchmarks may be linked or unlinked to the actual performance of the individual firm. Another classification is based on best practice frontier or representative or average measure (Jamasb and Pollitt 2001). The average benchmarking may be used with relatively similar costs or when lack of data does not allow frontier methods to be used. Normally statistical, econometric, data envelopment methods or ratios are used in such analyses. These analyses are sensitive to data, model specification, and choice of variables. Williamson and Toft (2001) argue that as normalisation is a very inexact science, the possibility of disputes and mistakes is quite large. This may result in efficient firms being bankrupted or inefficient firms receiving subsidies. As there is no objective way of allocating unexplained cost differences and as it is not possible to attribute the difference to inefficiency alone, this type of regulation results in significant regulatory discretion. Yardstick competition introduces uncertainty and could raise cost of capital.

Williamson and Toft (2001) argue that yardstick competition may harm incentives for the following reasons:

- the assumptions for achieving efficient results are unlikely to be met in practice as they are to demanding or inconsistent with rational behaviour,
- the hypothetical efficiency frontier is not purely based on exogenous frontier;
- the methods of yardstick competition are inherently subjective and discretionary;
- the need to maintain yardstick comparators inhibits future efficiency gains and eases pressures from the management.

The decision to choose individual incentive regulation or yardstick competition is not easy. Chong (2004) suggests the following framework based on static and dynamic efficiency criteria (see Fig. 28.2). Chong suggests that yardstick competition may be appropriate when in a static efficiency framework there is correlation between firms but the public funding is costless. Similarly, in a dynamic set up, when the investment does not lead to macro-economic effects or spill over effects, yardstick competition may be suitable.

28.7 Performance Based Regulation[15]

Performance-based regulation (PBR) is a subset of incentive regulation.[16] As the name suggests, it focuses on the regulated utility's performance characteristics (such as fuel costs, capacity factors, pollution level, resource mix diversity, etc.). Some sort of threshold performance level is set for each characteristic. If the utility beats the target, it is rewarded or escapes penalty.

PBR is "based on the assumption that utility managers will not voluntarily minimize costs under RBR and regulators will lack the complete information to

---

[15] This section is based on Navarro (1996a).

[16] We have considered it separately from price caps and revenue caps considering it a sub-set of incentive regulation. Some authors (such as NARUC 2000 or Comnes et al. 1995) consider PBR as an umbrella concept similar to incentive regulation and discuss price caps, revenue caps and other mechanisms under this. It may be somewhat misleading.

28.7 Performance Based Regulation

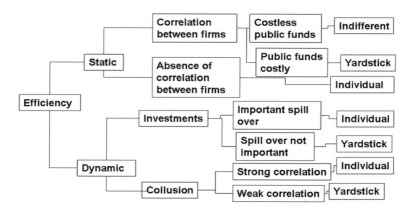

Fig. 28.2 When to use yardstick competition? source: chong (2004)

force them to do so" (Navarro 1996a, p 118). The basic premise of PBR is shown in Fig. 28.3.

The observed average cost is shown by curve AC_{obs} and the minimum cost is AC_{min}. The price is set at AC_{obs} but if the minimum cost were known, the regulator would set the price at the minimum cost. The objective of PBR is to encourage the utility to pursue cost savings by providing a set of incentives so that the utility moves from AC_{obs} to AC_{min}.

The most comprehensive type of PBR is the base rate PBR. The base rate PBR involves three basic steps:

Fig. 28.3 PBR premise

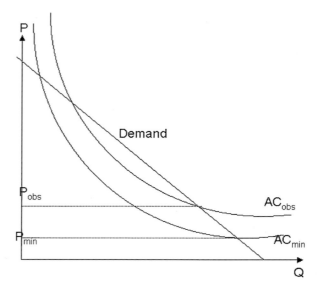

- first, a baseline revenue requirement is to be set. Adjustments to the baseline have to be made for inflation and productivity changes over time.
- Second, the regulator should provide the utility managers with a package of incentives that could lead to lower costs compared to the baseline. This, in other words, implies that a sharing mechanism has to be designed to distribute any cost savings between the utility and the rate payers.
- Finally, there should be a quality control mechanism to ensure that the utility does not achieve the cost reductions by sacrificing quality of service (e.g. reliability, safety, consumer satisfaction, etc.)

28.7.1 Base Revenue Requirement

In setting the baseline revenue requirement, the regulator faces the same problems of gamesmanship, incomplete information and cost revelation as faced in the RoR regulation (Navarro 1996a, p 128). The regulator has two options: follow the traditional method of revenue requirement determination which relies on utility specific information or use a benchmarking approach. The traditional approach will require submissions by the utility, scrutiny of data and claims and finally, a decision about the admissible revenue requirement. The benchmarking approach on the other hand uses the cost structure of a much wider sample of utilities to compare costs. Care is taken to ensure "like with like" comparisons through normalisation as discussed under yardstick competition. This method can be a potentially useful tool subject to the condition that the statistical methods and data are not manipulated for strategic purposes.

One of the concerns regarding the base revenue requirement determination is that the baseline may be inappropriately set, that is it may be either too high or too low. If the base revenue requirement is set at a bloated level, the utility uses the incentives to reduce costs from the inflated level and hence does not make any real cost savings. On the other hand, if the base is set too low (i.e. below the minimum average costs), it will be a punitive measure and the utility will either reduce investment or reduce operating and maintenance costs. Therefore, setting the baseline correctly is very important for PBR regulation.

28.7.2 Sharing Mechanism

The idea of the sharing mechanism is to distribute the benefits of cost savings between the utility and the ratepayers. If the utility is allowed to retain the entire amount of cost saving, it will have the highest motivation of saving costs. But such a sharing mechanism is normally considered as anti-consumer, and will be difficult to implement. Inversely, if the entire cost saving is passed on to the consumers, the utility has no incentive to save.

Navarro (1996a) suggests that the sharing mechanism should be progressive so that the utility receives more benefits with higher levels of cost saving. This is because it will cost the utility more to achieve additional cost saving and it will be uneconomical for the utility to pursue it unless adequately rewarded. As the cost saving potential varies from one utility to another, PBR will be effective when the cost saving can be made at relatively cheaper costs. Where the cost saving possibility is relatively scarce and costly, it may be better not to attempt PBR.

28.7.3 *Quality Control*

This is the third major component of PBR, which establishes a clear link between utility cost savings and its performance. As the objective of the regulator is to achieve minimum cost service subject to the maintenance of certain quality, Navarro (1996a) suggests that the utility should be offered rewards only for cost minimisation and be penalised for violating quality standards. The regulator then faces three separate tasks under this: a) to decide about the quality parameters, b) set threshold limits for each quality parameter and c) to decide about the penalty for violation of quality standards.

Commonly used quality parameters include system reliability, customer service and employee safety. Threshold limits play an important role in PBR and setting them is a difficult task. This is because quality has a direct cost implication. Maintaining the existing level of quality may require bloated costs (thus defeating the PBR objective) while changing the quality levels arbitrarily without analysing the potential implications on reliability, consumer service and safety can lead to criticisms. The non-price dimension of some of these quality parameters makes the decision difficult.

Deciding about the penalties is also difficult in practice. Theoretically, the penalty system should be such that "the marginal penalty from reducing quality below the quality floor is always greater than the utility's marginal benefit" (Navarro 1996a, p 141). However due to non-price characteristics of some quality parameters it is difficult to assign dollar values to some quality reductions. Alternative economic valuation techniques could be used but at a greater regulatory cost, which goes against the PBR objective.

Navarro (1996b) offers the following seven basic rules for the PBR regulator:

- use benchmarking to set the baseline revenue requirement;
- avoid setting the baseline revenue requirement too high;
- use a progressive sharing mechanism to distribute cost savings;
- don't use a regressive sharing mechanism to distribute cost savings;
- use a quality control mechanism;
- do not reward for quality improvements;
- penalties should be commensurate with cost savings.

Thus, PBR may not be a panacea and unless carefully implemented, it may create disincentives than incentives for better performance.

28.8 Conclusion

This chapter has provided a glimpse of traditional and alternative approaches to regulation. Some of the alternative regulations have close similarity with the traditional regulation. Similarly, traditional regulation can be complemented by some form of alternative regulation. Any regulator faces the challenge of information asymmetry and has to decide which option works best in the given context. The regulatory system has to adapt to the changing conditions and emerging challenges of the industry being regulated. There is no guarantee that an approach that works in one place will necessarily work in another environment. Therefore, careful consideration to the environment and the governance structure is essential.

References

Alexander I, Harris C (2001) Incentive regulation and multi-year price controls: choice, constraints and calculations, SAFIR Workshop, New Delhi. http://www.safir.teri.res.in/wkshp/7-8aug2001/ia.pdf

Bakovic T, Tenenbaum B, Woolf F (2003) Regulation by contract: a new way to privatise electricity distribution? Energy and Mining Sector Board Discussion, Paper no. 7, World Bank

Beesley ME, Littlechild SC (1989) The regulation of privatised monopolies in the United Kingdom. Rand J Econ 20(3):454–472

Berg SV (undated) Introduction to the fundamentals of incentive regulation, In: Infrastructure regulation and market reform. Public Utility Research Center, University of Florida

Bernstein JI, Sappington DEM (1998) Setting the X factor in price cap regulation plans, NBER Working paper 6622, National Bureau of Economic Research, Cambridge

Chong E (2004) Yardstick competition vs. individual incentive regulation: what has the theoretical literature to say? ATOM, University of Paris 1, Paris

Comnes GA, Stoft S, Greene N, Hill LJ (1995) Performance-based ratemaking for electric utilities: review of plans and analysis of economic and resource plan issues, vol. 1. Lawrence Berkeley National Laboratory, University of California, California

Crew MA, Kleindorfer PR (1996) Price caps and revenue caps: incentives and disincentives for efficiency. In: Crew MA (ed) Pricing and regulatory innovations under increasing competition. Kluwer Academic Publishers, London

Farrel MJ (1957) The measurement of productive efficiency. J R Stat Soc Ser A Gen Part III 120:253–281

Giulietti M, Price CW (2005) Incentive regulation and efficient pricing. Ann Public Coop Econ 76(1):121–149

Green R, Pardina MR (1999) Resetting price controls for private utilities: a manual for regulators. World Bank, Washington, DC

Guasch JL, Spiller P (1999) Managing the regulatory process: design, concepts, issues and the Latin American and Caribbean story. World Bank, Washington, DC

References

Intven H (ed) (2000) Telecommunication regulation handbook. World Bank, Washington, DC

IPART (1999) Regulation of network service providers, Incentives and principles for regulation, Discussion paper DP32, Independent Pricing and Regulatory Tribunal of New South Wales, Australia . http://www.iprt.net/)

IPART (2001) Form of economic regulation for NSW electricity network charges, Discussion paper DP48, Independent Pricing and Regulatory Tribunal of New South Wales, Australia. http://www.ipart.nsw.gov.au/files/DP48.pdf

Jamasb T, Pollitt M (2001) Benchmarking and regulation: international electricity experience. Utilities Policy 9:107–130

Kahn A (1989) The economics of regulation, principles and institutions. MIT Press, London

King S (undated) Principles of price cap regulation. In: Infrastructure regulation and market reform. Public Utility Research Center, University of Florida. http://bear.cba.ufl.edu/centers/purc/publications/accc/05.pdf

Lowry MN, Kaufmann L (1994) Price cap designer's handbook. Pacific Economics Group, Madison

NARUC (2000) Performance based regulation for distribution utilities, The Regulatory Assistance Project. National Association of Regulatory Utilities Commission

Navarro P (1996a) The simple analytics of Performance-based ratemaking: a guide for the PBR regulator. Yale J Regul 13(105):105–161

Navarro P (1996b) Seven basic rules for the PBR regulator. Elect J April:24–30.

NER (2003) Economic regulation of transmission using incentive based regulation. National Economic Regulator, Pretoria

Newbery DMG (1999) Privatisation, restructuring and regulation of network utilities. The MIT Press, Massachusetts

Parmesano H, Rankin W, Nieto A, Irastorza V (2004) Classification and allocation methods for generation and transmission in Cost-of-Service studies. NERA, National Economic Research Associates, Los Angeles

Ros AJ (2001) Principles and practice of price cap regulation: an application to the Peruvian context, NERA discussion paper, National Economic Research Associates, Washington, DC

Shleifer A (1985) A theory of yardstick competition. Rand J Econ 16(3):319–327 Autumn

Shleifer A (2004) Understanding regulation. Eur Financ Manage 11(4):439–451

Swisher JN, Jannuzzi GM, Redlinger RY (1997) Tools and methods for integrated resource planning: Improving energy efficiency and protecting the environment. UCCEE, Riso

Tenenbaum B, Lock R, Barker J (1992) Electricity privatisation: structural, competitive and regulatory options. Energy Policy, Dec, 1134–1160

Viehoff I (1995) Evaluating RPI-X, NERA topics 17. National Economic Research Associates, London

Viscusi WK, Harrington JE, Vernon JM (2005) Economics of regulation and anti-trust, 4th edn. The MIT Press, Massachusetts

Voll S, Bhattacharyya SC, Juris A (1998) Cost of capital for privatised distribution companies: a working paper for a calculation for India, NERA Working paper, National Economic Research Associates, Washington, DC

Weston F (2000) Charging for distribution utility services: issues in rate design, the regulatory assistance project. http://www.raponline.org/Pubs/General/DistRate.pdf

Weyman-Jones TG (1990) RPI-X price cap regulation: the price controls used in UK electricity. Utilities Policy, Oct:65–77

Williamson B, Toft S (2001) The appropriate role of yardstick methods in regulation, NERA working paper, National Economic Research Associates, Washington, DC

Chapter 29
Reform of the Energy Industry

29.1 Introduction

Since Chile initiated structural reform of the power sector in the 1980s, and similar reforms followed in the UK and Argentina, reform became the normal prescription of the multi-lateral donor agencies like the IMF and the World Bank. Since 1990s, the reform agenda has been pursued vigorously for more than a decade. Various reform models have been formulated and attempted suggesting a wide range of possibilities ranging from minimal structural changes to complete structural disintegration of the activity of the industry. This chapter introduces the rationale behind the reform movement, the steps involved in the reform, the reform options and the sustainability issues related to energy sector reform.

29.2 Government Intervention in Energy Industries

As discussed in the previous chapters, two common approaches used to deal with the natural monopoly problem were regulation of private industries and state-ownership of the industry. Both these approaches suggested government intervention in the energy sector, either through ownership or through regulatory practices. In some cases a further rationale for government intervention has been suggested on the basis of a public good[1] argument. There are a number of alternatives to this argument (Jaccard 1995):

- The choice of primary energy source and technology provides greater societal benefits and costs compared to the private costs. Such an argument has been

---

[1] A public good is characterised by non-exclusivity and non-rivalry. Non-exclusivity implies that it is difficult to exclude others from using the good or service without withholding the good or service. Non-rivalry on the other hand means that consumption by one does not reduce its potential to be used by others.

used to promote nuclear energy, hydroelectric energy, and coal in various countries.
- High share of energy industry investments in the national investment portfolio often implies that the timing and magnitude of the energy sector investment are controlled for macro-economic policy objectives. This could offer broader social benefits.
- The private investors aim at quick recovery of the investment while the public investment takes a long-term view, thereby offering social benefits of investment.

The history of evolution of energy industries[2] around the world is characterised by a number of distinct phases of development:

- Highly competitive markets: initially the activities started under private ownerships with an objective of benefiting from the invention of a new technology or a product or service. The initial phase was marked by cut-throat competition among players, which resulted in price wars. This period also saw duplication of assets, problems of co-ordination and cartelisation in some cases.
- Integrated, monopolistic industry: As the cut-throat competition was not conducive to take advantage of economies of scale and scope, the next phase of development was marked by a movement towards integration and monopolisation of the industry. This has taken two forms: in Europe, nationalisation was the preferred option while in the USA, regulatory approach was followed, where the private ownership continued but the profit was regulated by the government. Both the forms allowed industry consolidation and rapid progress but due to government and regulatory failures as well as developments in technologies, this form of development was questioned.
- The next phase saw the emergence of reforms and restructuring. This continued for more than a decade but new issues emerged in terms of security of supply.
- The fourth phase saw re-emergence of government interventions in the sector.

These four phases could be easily related to a number of energy industries. Box 29.1 provides the example of the electricity industry.

Box 29.1: Historical Review of Developments in the Electricity Supply Industry

The electricity supply industry (ESI) was born at the end of the 19th century as a private enterprise. Electricity supply was available in a few urban centres where there was adequate demand. This sort of development resulted in "electric islands" with lack of compatibility in terms of technical

---

[2] See IEA (1999) for a more detailed account. Also consult EIA (1996); Newbery (1999).

standards and performance (de Oliveira 1997). As better generation and transportation technologies became available, the shortcomings of the isolated, localised, non-integrated systems became apparent. Substantial gains could be achieved through inter-connection of local markets due to exploitation of larger plants with better efficiency, sharing of capacity for better reliability and efficient use of resources. However, reaping these benefits required use of uniform technical standards, centralised decision-making, focussed financial investment in generation and transmission systems, and co-ordination of investment plans. All these proved difficult, given the industry organisation of that time and a profound reform was required.

In the USA, large multi-service utilities were created through mergers and acquisitions. By 1920, 16 large holding companies controlled over 75% of all US generation (EIA 1996). There was also great urban rural disparity at this time. Although state-level regulation was in place since early 1900, federal involvement became important due to higher inter-state transactions through transmission networks and greater reliance on hydro power (a federal government resource). The federal regulation was introduced through the Public Utility Holding Company Act (PUHCA) in 1935. The power industry grew rapidly post-World War II, when the demand grew rapidly. Technological developments, traditional regulation and vertically integrated monopoly structure made supply of reliable, high quality power possible.

In the UK, two stages of public ownership can be identified: a national grid system was established through the creation of Central Electricity Board after the First World War. The CEB completed the national grid in 1933 and by 1938, the overall grid savings from interconnection had risen to 11% of total payments for electricity (Newbery 1999, p. 109). The second phase involved nationalisation of generation at the national level in 1948 through the creation of Central Electricity Generation Board (CEGB). This provided the necessary impetus to rapid development and a well functioning, integrated system resulted over the next two decades.

The post Second World War period was the golden age of the ESI development. Electricity demand grew around 7% per year and the grid was extended to new geographical areas rapidly, often by offering cross-subsidies. ESI development was considered as a source of economic and social development. However, the golden era did not last for ever. In the middle of the 1970s the first signs of strain appeared. The oil shocks ended the era of cheap fuel and led to an era of higher interest rates. Both these factors affected the cost of supply until late 1980s, when these factors started to ease off. But environmental concerns related to electricity supply led to stricter standards, which resulted in higher costs of supply.

On the demand side, the demand for energy intensive appliances in the domestic sector showed signs of saturation. Moreover, energy intensive industries started to relocate to the developing countries, thereby reducing the demand growth to a substantial extent. But utilities started projects based on higher demand growth projections and overcapacity emerged. This overcapacity in turn put pressures on the utilities finances. New technologies emerged on the demand-side and the supply-side, thereby changing the old paradigm. But the utilities paid little attention to these changes and to cost saving, as a result of which the costs started to rise. This situation called for a review of the ESI performance.

Similar developments occurred in developing countries but the situations were quite different. Monolithic, state-owned monopolies emerged with patronage of the World Bank and the IMF. These utilities were able to expand the supply to a large extent but the utilities performed badly financially and operationally due to government interference, high debt levels, inefficient operation of the facilities.

The reform of the sector then promoted competitive supply in segments not infested with natural monopoly characteristics. The wave swept the global electricity industry but a few countries took the process to the logical end while others ended up with a partially deregulated industry. In the developed countries, the drive for sweating out the asset to remove excess capacity led to lower prices. This also promoted use of natural gas where gas was available. But the drive for lower costs also did not promote new capacity in desired technologies as old capacity reached end of life. This along with the tightening of the gas market in the first half of this decade led to security of supply concerns, which in turn prompted demand for interventions. The impending crisis in the electricity sector of the U.K. has forced the regulator to investigate options for ensuring future supply security, including options involving partial or major reversal of the competitive electricity market model.

Source de Oliveira (1997); EIA (1996); Newbery (1999)

29.3 Rationale for Deregulation

The deregulation movement in the energy industry in general and in the networked energy industries in particular was initiated to redress the problems faced by the industry. The main drivers behind such a movement are as follows:

a. *Decline of the natural monopoly rationale* The main economic reason behind public sector ownership or government intervention in the energy industry was the existence of natural monopoly in energy industries. However, in certain areas of the industry this assumption started to be questioned. For example, it is customary to consider three components of the electricity industry: generation, transmission and distribution. Generation component came under scrutiny and attack. It was suggested that the technical gains of employing larger generation plants reached its peak and no further gains could be achieved by increasing the plant size. At the same time, technological innovation made it possible to use small capacities efficiently. This is particularly true of gas turbines operating in the combined cycle mode, which allowed achieving higher efficiencies than conventional thermal plants with much smaller size. The costs of renewable energies, notably wind, have fallen significantly, making them competitive. All these factors undermined the natural monopoly rationale for government intervention (Jaccard 1995).

b. *Regulatory failure* A number of regulatory failures can be identified even in the developed countries. As discussed earlier, the rate of return regulation followed in the USA did not provide enough incentives to the best performing utilities and did not severely penalise the poor performing utilities. Utilities invested in nuclear and other technologies that turned out to be uneconomic ex-post, which the so-called Averch–Johnson effect would justify. Joskow (2000), however, indicates that it is not clear whether these decisions were uneconomic even ex-ante or whether they turned out so due to changed circumstances. In some cases, the old, uneconomic plants were not discontinued under the regulatory regime. The British experience indicates that post reform many old plants were taken out of operation. In the USA, it pays to continue to operate an old plant unless the regulator forces the utility to shut it or imposes heavy penalties (Joskow 2000). There were also cases of politically motivated capacity additions that were uneconomic. Finally, the regulators always lacked adequate information and suffered from the problem of information asymmetry. This affected prudent decision-making. Moreover, there were cases of regulatory capture either by industries or by consumer groups and regulators, who were supposed to be "wise and benevolent" actually acted differently.

c. *Contestable markets* According to Baumol (1982), a contestable market is "one into which entry is absolutely free and exit is absolutely costless". The entrants do not face any cost discrimination and any firm can leave without any obstruction (i.e. while leaving it can recover any costs incurred in the entry process, which in other words means zero sunk costs). "The crucial feature of a contestable market is its vulnerability to hit-and-run entry." An entrant "can go in, and, before prices change, collect his gains and then depart without cost, should the climate grow hostile" (Baumol 1982). According to this theory, a perfectly contestable market never provides more than normal profits—i.e. the economic profit is always zero—because if there is any positive profit, new entrants will hit the market and capture the supernormal profits. Another feature of such a market is that in the long run there is no inefficiency in production, because its existence is an invitation to new entry. Finally, the price in a perfectly contestable market must be equal to the marginal costs as required for Pareto optimality in a first best world.

This new theoretical development has inspired the deregulation debate to a significant degree. The threat of entry or potential competition can be used to discipline the incumbent firm, which as a result will provide the benefits of a competitive market to consumers. However, to be effective a contestable market requires that the regulatory barriers to entry and exit must be eliminated. Normally the exit is unlikely to be costless if the entry involves sunk costs and in that sense, sunk costs can be considered as a barrier to entry.

Most energy industries are characterised by heavy sunk costs, and therefore, it is difficult to ensure free entry and exit. This reduces the possibility of perfectly contestable energy markets but this theory requires policymakers to consider the possibility of potential competition.

d. *Failure of public monopolies* A World Bank (1995) report suggests that the state-owned enterprises (SOE) continue to play important roles in various parts of the world. In the industrialised countries, the share of SOEs in the GDP has started to decline while in the developing world it has remained about constant. The SOEs are an important source of employment and they still absorb a large share of domestic investment. Although the above study is general in nature and is not specific for energy industries, the above characteristics are shared by energy related SOEs. Moreover, their performance has remained unimpressive, characterised by under-priced products, bloated labour forces, major operational inefficiencies, excess demand, and inability to finance investments out of profits (Tenenbaum et al. 1992; Newbery 1999; Bacon and Besant-Jones 2001).

One of the explanations for poor performance of the public sector can be traced from the theoretical construct of the Principal-Agent problem. Following this theory, a public sector enterprise can be considered as an agent of government who is supposed to carry out the objectives assigned to it by the principal (i.e. government). However, the agent may have its own agenda which may be quite different from the assigned objectives. As a result of such a conflict, the enterprise may be ineffectively managed and the performance suffers. This argument thus suggests that public ownership is an ineffective way of mitigating market failure problems.

e. *National debt problem* Tenenbaum et al. (1992) suggest that another important rationale for undertaking restructuring was the problem of financing energy projects through budgetary support, because government or public borrowing adversely affects national debt problem. As de Oliveira and MacKerron (1992) indicate many developing countries borrowed heavily from the capital markets during the period of oil shock when petrodollars were available at relatively low interest rates. However, the situation changed in the 1980s, as the flow of capital stopped and the countries had to devaluate local currencies to overcome macro-economic problems triggered by higher interest rates. The devaluation increased the debt burden of many utilities and made some of them insolvent. Restructuring and privatisation then became an economic imperative.

f. *Decline of the public good rationale* As mentioned earlier, public good rationale was advanced in support of government intervention. However, it appears that the importance of this argument is on the decline. This has happened due to three factors (Jaccard 1995): (1) This argument was mostly used in the case of

29.3 Rationale for Deregulation   689

coal and nuclear energies. However, these two energies have lost shine over the past decades due to environmental and safety concerns. As a result, the need for intervention to promote these resources has declined. (2) The argument of macro-economic benefits of energy investments has also lost much ground on account of the emergence of small, modular technologies. As they are less capital intensive, they impose lower burden on the macro-economy. (3) The importance of oil import dependency has decreased since mid-1980s for many countries, which has reduced the need for intervention in the market to control oil use.

From the above discussion, it is clear that there were significant changes to the economic environment that warrant a change in the way energy business is conducted. While the importance of each driver varies from country to country and from industry to industry, there is a recognition that things need to change. This brings us to ask the following main questions:
What does a reform involve?
Where and how could competition be introduced?
What are the restructuring options?
What are the factors affecting the sustainability of a reform?
These questions are considered in the following sections.

29.4 Reform Process[3]

Reforming a sector or an economy, in institutional terms, involves changing the rules of the game (and hence the institutional environment), changing the organisational structure by abolishing/creating new organisations in line with changed circumstances, modifying the governance mechanism and affecting the process of adaptation of the institutional arrangement in accordance with new rules on the other. The need for changing the rules may arise due to internal or external factors with an objective of improving performance. The institutional endowment shapes the first while the organisation structure and mode of transaction would influence the adaptability.

29.4.1 Changing the Rules Requires Stability of Rule Makers

In general, any reform process attempts to introduce changes in the ownership, industry structure, governance mechanisms and the nature of transactions. The extent and depth of such changes are decided through a continuous process of bargaining among various political, social and economic stakeholders and

---

[3] This section is based on Bhattacharyya (2007a).

organisations. Any major reform being a political process, the rules governing legislating and executive bodies and their stability would normally limit the opportunistic behaviour of the political institutions and thereby reducing uncertainty of the institutional environment. This in turn would increase confidence in contracting and would provide credible commitment towards the reform process. Thus when a political party controls both the legislative and executive bodies and there is weak opposition to it, far-reaching reform could be undertaken. Low political stability could also lead to reform when informal rules, ideology and social aspects prevent the political institutions to adopt opportunistic behaviour. However, when the political system is characterised by highly opportunistic behaviour, irrespective of political stability, reform is quite unlikely. Similarly, the depth and complexity of the reform process is expected to be dependent on the political stability and the commitment to reform. Highly complex and time consuming reform is unlikely to be undertaken during the periods of political instability, although minor, less complex reforms could move forward.

Economic prospect of a country is often linked to the stability of the political systems, and the outlook of the government. In many cases, the economic performance of a government is not the deciding factor in the political selection process, although it influences the selection. In these circumstances, political systems often adopt short-term policies ignoring/overlooking long-term perspectives and objectives. In a stable political environment with uncertainty about long-term political perspectives (as is common in many developing countries with democratic systems), the reform is likely to look for quick-fixes that yield immediate results, often compromising long-term prospects of the economy. Major reforms are possible only where there is political stability with good long-term political prospects for the ruling party. In a case where the political system is unstable and the prospect for the ruling party is unclear or short-term in nature, reforms would lead to opportunistic/rent-seeking behaviour. This introduces uncertainty in the business environment and only opportunistic firms and corrupt practices are expected to gain from such a process. This is reflected in the economic perspective of the country and in the economic performance. In an unstable political environment, no long-term and highly involved reform is unlikely to be undertaken (see Fig. 29.1).

29.4.2 Danger of Derailment at Every Stage of the Reform Process

Bargaining and trade-offs accompany the entire process of reform, the generic elements of which is shown in Fig. 29.2. An appraisal study is generally the first step, where consultants, funding agencies, the executive and the bureaucracy are involved to initiate and conceptualise the course to be undertaken. Each player will have its own self-interest, motive and perceptions about the reality. It is likely that

29.4 Reform Process

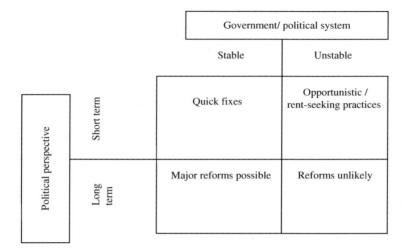

Fig. 29.1 Importance of rule makers' stability in reform

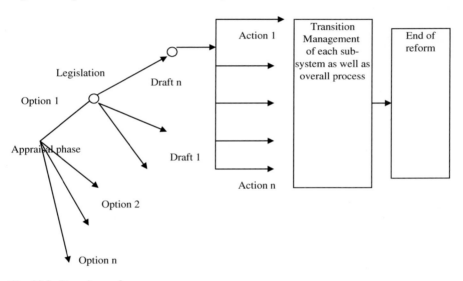

Fig. 29.2 Steps in a reform process

there is information asymmetry and differences in cognitive capacities among the actors on one hand and conflicts of motives, interest and perception on the other. Consequently, the suggested or recommended option is often obtained through a first round of bargaining process and may not be the most efficient outcome in economic sense.

The next major step in the reform cycle is the reform-related legislations. This process has its own process cycle and involves more broad-based stakeholder participation including political opponents, employee trade unions, NGOs and general public. Drafting, debate, revision and finally legislative approval are the four steps of this process and the approved legislation is a manifestation of a socio-political bargaining exercise. The legislation that provides the framework for reform process can thus be constrained and compromised by the approval process itself. The legislation process introduces path dependence and irreversibility to the process of reform.

Undertaking enabling reform actions constitute the next step where the Executive has to carry out its duties and responsibilities assigned to it by the legislation. This could involve a wide range of tasks, including framing of subordinate rules and regulations, creation or abolition of organisations, vesting of assets and liabilities from and to organisations, transferring personnel, etc. The enabling acts may be one time acts or phased over time. Although the possibility of bargaining may appear to be less at this stage, in practice this stage involves detailed design of the reform process and the work starts on the ground. Therefore, the intensity of conflicts tends to be more here between the immediate stakeholders and can result in compromises that could affect the implementation of the reform and its performance.

If the reform unveils in a phased manner, initiation of the enabling actions leads to a transition period through which the reform process is expected to reach its final shape. The process may reach deadlock at any of the stages of the reform due to conflicts, lack of compromise solutions, lack of adequate political support, etc.

29.4.3 Importance of Overall Acceptance of Changed Rules

Acceptance of the changes to the rules of the game by various stakeholders is the key to the success to any reform process. This is the level where each stakeholder forms its own opinion about the cost-benefit of the reform. This process is highly influenced by a number of factors—economic, social, cultural, history, past experience, level of education and awareness, etc. While the adoption of the changed rules depends on the political system and process of a country and hence is a politically motivated process, its success depends on creating champions of reform who themselves would believe in the process and would influence others to become believers. As the reform process often creates different sets of winners and losers, it is not always easy to make everyone happy. If the unhappier section is more vocal and organised than the winners, and if they could influence the general opinion in their favour, the obstacles to reform implementation and progress would increase. It is quite logical for each stakeholder to try to protect its own self-interest and in doing so to try to garner support of other stakeholders. The losers are quite likely to form a strong opposition and to promote anti-reform sentiments.

Often the top-down approach to reform fuels such opposition as well. As discussed in the stages of reform, bargaining continues at each stage. But once the process is politically decided, it is imposed on all the stakeholders, many of whom may still perceive the process as unfavourable to them. Such stakeholders or at least those who are more assertive would try to put spanners in all possible manners. As the perception is about the expected future gain or loss, the inherent uncertainties about the key variables (economic and others) are exploited by all concerned to fashion their views. Thus favourable projections of key variables would create a rosy picture that would form the basis of pro-reform propaganda. Similarly, the pessimistic projections would be used by the opponents to portray a gloomy picture. The reality may be either of these extremes or something in between or something quite different depending on how the reality is actually accurately envisaged at the time of undertaking the reform.

Changes in the rules of the game of a reform process modify the contractual arrangements at several levels, namely government-firm, government-regulator, firm-regulator, firm-consumer, firm-firm, government-consumer, and regulator-consumer interactions. At the pre-reform stage, one or more of these levels may not be explicit due to lack of clarity and functional separation. Post-reform performance is likely to depend on the credible commitment of these contractual arrangements, their enforcement and stability. Once again institutional endowment plays an important role.

Changes in institutional environment naturally affect the institutional arrangements, the nature of transactions and hence the way transactions would be carried out. Although asset specificity is not affected as such, restructuring and reform change the frequency of transactions, risk involved in them and adaptability of the organisations.

Evidently, uncertainty and the asset specificity would play an important role in deciding the type of relational arrangement that would develop. Following Williamson's classification of asset specificity, we consider three types of assets—non-specific, moderately specific and highly specific. When uncertainty is low and the asset is non-specific, markets will perform better but in case of highly uncertain environment, entry to the market is expected to be limited and the risk premium will be high. In case of highly specific assets, low uncertainty is desirable for market development. Quasi-integration and long-term contracts are expected here. But if uncertainty is high, new investment is unlikely to materialise, as even with long-term contracts, investors will not be sure to recover their investments.

29.4.4 Adaptation to the New Environment

Adaptation of the organisation/individuals to the changed environment is the third important aspect in any reform. The changed rules of the game require a new/different set of actions/transactions from different players—existing and new. The success of the reform process partly depends on how fast and how appropriately the players have conformed themselves to the new requirements. Evidently, when the changes introduce a higher degree of randomness in the system (i.e. higher level of entropy), a widespread adaptation is required. The ability to adapt varies among players.

29.4.5 Transition Management

No reform is instantaneous. Once the reform is initiated, it takes some time to move from the existing state to the new state and the intervening period is called the transition period. During this period all enabling actions are taken to change the economic and transaction environment as well as changes related to structure and organisation of the sector. This period also involves learning-by-doing for various stakeholders in their attempt to adapt to the new environment.

The importance of the transition period arises due to the fact that during this period the players continue to play the game in an evolving environment where the rules of the game change even during the play. The dynamics of the events, the reaction of the players to the evolving environment and their ability to adapt to the changes shape the outcome of the events. In order to obtain the desired outcome of the reform process, the dynamics, players' reaction and the adaptation process have to be managed along the lines envisaged in the reform. There lies the importance of transition period management.

Evidently, this is not an easy task and depends largely on the local environment, and socio-political considerations. Experience from other places often does not help because of case specific nature of the issue. Lack of proper transition management policies often leads to disastrous results.

From the above discussion, the key areas of any reform process that affect the outcome significantly are

a. stability of the political decision-making system;
b. overall acceptance of the rules of the game;
c. ability to adapt to the changed environment; and
d. proper transition period management.

The devil lies in the detail and here the secondary legislations (rules, regulations and procedures) play an important role. The nature of these secondary legislations decides the incentive structures that the new system is going to provide. It is possible to miss the opportunity and remain in the old mental set-up to defeat the real purpose of the reform. It is also possible to follow progressive policies and procedures, but that may be more difficult to distil through. Here the capabilities of the key organisations and actors influence the decisions to a large extent. As in most of the cases, the law provides a broad guideline and parameters within which the actors should perform. Detail design often controls the fate of the reform.

29.5 Options for Introducing Competition

Although the question of introduction of competition depends on the technical and cost characteristics of a business and therefore it requires industry-specific focus, it is possible to identify a number of options such as the following (Klein 1996):

29.5.1 *Competition for the Market*

This option allows introduction of competition through bidding for the monopoly franchise. For the distribution function, it is quite common to grant a monopoly franchise over a service territory, which often implies an exclusive right over the territory with an obligation to serve. Consumers do not have any access to other service providers (Tenenbaum et al. 1992). The service would remain a monopoly but to exert competitive pressures such a business should be delineated and the franchise should be auctioned off to the bidder requiring the lowest price from the consumers. The price and other terms of the franchise or concession have to be adjusted over time—either through rebidding or through regulation. The rebidding approach is preferable if one wants to escape from the regulatory problems.

A typical franchise has the following features (Guasch and Spiller 1999):

- the franchising authority and franchise holder share a contractual relationship;
- the contract remains valid for a limited period and may be renewable;
- the franchise holder enjoys the right to use the assets without assuming ownership rights;
- the concessionaire is responsible for developing all new facilities required to carry out the business and retains temporary ownership of these assets until they are handed over to the franchising authority;
- the concessionaire assumes the risk for ensuring appropriate service and for maintaining and operating the facilities in good condition;
- the concessionaire is remunerated according to a contractually agreed tariff, collected from the customers.

The advantages of franchising natural monopoly activities are as follows (Guasch and Spiller, 1999):

- it reduces the possibility of regulatory capture and political interference in management;
- franchising encourages cost efficiency as the franchise contracts specify maximum prices for set quality standards and investment requirements;
- encourages productive efficiency by ensuring low cost as well as a reasonable return to the concessionaire;
- competition for franchise bids ensures efficient pricing as well.

But this is not devoid of problems either. When sunk cost is involved, the concessionaire has to be allowed to recoup costs at the time of rebidding. This requires valuation of the assets which can be fixed by a regulatory agency and the bidder offering lowest price to consumers could be selected. The other option is to bid for the asset value, for which expected future prices have to be specified by a regulatory agency. The valuation exercise has similarities with the rate case in a traditional regulation and without proper incentives, the incumbent will not be interested to maintain the assets while a high asset value will raise consumer prices. Despite these problems governments may be able to subject franchisees to challenges by putting time limits on franchises and requiring some form of rebidding for renewal.

However, the idea of franchise bidding is not free from disputes. It has been pointed out that the bidding may suffer from lack of competition or participation of colluded parties. Contracting and monitoring of contracts are also difficult for complex services. Finally, there are transactional difficulties as well (World Bank 2004, p. 38).

29.5.2 Competition in the Market

Within the existing network competition can take different forms such as open access, pooling and timetabling.

29.5.2.1 Open Access

Open access implies opening up the access to monopoly network so that competitive suppliers can use any available network capacity for providing the service. This is possible in electricity transmission and distribution systems as well as in oil or gas pipelines. Open access can relate either to the transmission or wholesale level or to the retail or distribution level or to both the levels.

As the marginal cost of such capacity is close to zero, the network owner always earns by allowing additional access to the grid, if the capacity is available. If the capacity is constrained, some sort of rationing is required, which can be done through access pricing. If the network owner could charge the difference between the consumers' willingness to pay and the producers' marginal cost, the pricing would be efficient. This would imply that the lowest cost producers will supply to the highest paying consumers. However, this situation leads to monopoly profits to the network owner, which then requires regulatory intervention.

One of the concerns in the case of open access is that if the network owner has interests in the upstream or downstream operations, there can be conflict of interests and the network owner may discriminate or resort to predatory pricing strategies to drive out competitors. In order to ensure non-discriminatory access, regulators often impose certain obligations. Generally, network segment is separated from the other components of the business and is held under different ownership and limits to vertical integration are imposed.

In some cases, users of network may buy rights to use the grid capacity and such rights may be transferable by resale in a secondary market. This then results in a contract carriage. Resale of capacity rights allows users without contracted capacity to obtain access to the grid through purchase of rights in the retail market. If the retail market price is not regulated, then the prices would be same as that of a monopoly selling directly. Therefore, prices cannot be set in a free market under such a condition.

Open access system faces significant implementation challenges but it is essential for ensuring competition in the competitive segments of the market.

29.5.2.2 Pooling/Timetabling

It may be difficult to define, adjust and enforce the capacity rights of an open access regime that relies on sale of capacity rights on a non-discriminatory basis to multiple users. This may be particularly true of the electric power industry, where capacity use or unused depends on the physical flow in the entire system. In such a case, an alternative is to rely on a central dispatch system that allocates capacity based on a system optimisation approach. The suppliers who are selected to dispatch will automatically earn the right to use the capacity and thus have open access to the grid.

The pool system was used in the initial phase of the electricity industry reform in England and Wales. The English pool used day-ahead auctions for every half-hour period of any given day. Bids were invited from generators on the day ahead of the operation. Every morning generators were required to declare the generating sets available for the next day and announce prices for each generating set. The system operator then ranks the bids in terms of price and the system marginal price is the bid price clearing the expected market demand for the every half hour period of the day ahead. Figure 29.3 indicates the mechanism. Thus the pool price was decided for every half an hour period on a day-ahead basis and settlement was made based on the pool price.

In the pool system, the transmission system remains a natural monopoly and hence it has to be regulated. The charge for using the transmission system has to be decided by the regulator.

29.5.2.3 Competition for New Capacity Expansion

This can apply to both the competitive segment as well as the monopoly segment of the network industry. In the competitive segment (or potentially competitive segment) such a competition can be introduced through a competitive capacity procurement process by inviting bids for a specified capacity. The most favourable bidder is normally chosen to provide the capacity. If the lowest price of the output is the criterion for selecting the bids, the price is directly obtained from the competitive process and can be considered as an efficient outcome, if the bidding is competitive and there is no collusion among bidders. Such a competitive process can be grafted even in a vertically integrated industry structure or in other forms of industry structure.

Competition in network capacity expansion can also be introduced. The network capacity should be expanded where there is capacity shortage. If the capacity is short in a particular segment during some periods, the price for network usage should provide the signal (which requires a pricing system differentiated by

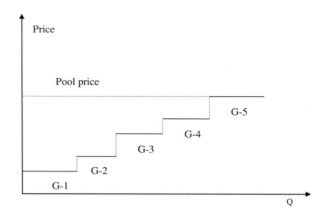

Fig. 29.3 Pool price mechanism

location and by time of use) to the users and developers. New capacity is expected to be developed to relieve congestion or capacity shortage. However, when the network is granted a monopoly status, either new entry has to be permitted to develop such capacity expansion or the additional capacity may have to be procured through a competitive process, where the bidders builds and transfers the capacity to the monopoly franchisee.

29.5.2.4 Competition Among Multiple Networks

The reason for initiating such a competition is that the extent of natural monopoly of a system is not known ex ante. By introducing competition, it could be verified whether the market is a natural monopoly where only one firm survives. Competition would result in some duplicate capacity, but it is argued that redundant capacity is required to try out new things and to check monopoly behaviour (Klein 1996). Moreover, competition allows experiments with new ways of doing business and can bring more welfare benefits than the cost of introducing competition.

In energy industries such competition can be introduced by allowing multiple franchises for serving a particular area and allowing multiple networks (either for the entire area or for selected areas). The problem of scheduling arises in the case of electric networks but when congestion arises due to capacity shortage, alternative routes for certain segments could become viable.

29.6 Restructuring Options

Reforming the power sector involves moving away from the vertically integrated monopoly structure of the industry. A wide range of options are available for such changes—ranging from a minimal modification to the structure to the radical,

29.6 Restructuring Options

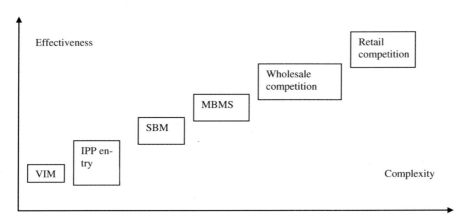

Fig. 29.4 Alternative reform options in the power sector

comprehensive changes.[4] Allowing entry to generation segment without modifying the rest of the sector falls under the minimal reform. Complete unbundling of generation, transmission and distribution and allowing choice to all consumers forms the most radical reform of the industry. In between there is a plethora of options, mostly of transitory nature. Figure 29.4 presents a view of these options in terms of complexity of their implementation and degree of consumers' choice.

In what follows, the following models will be considered:

- Vertically-integrated monopoly
- Monopsony (or single buyer) model
- Transition models
- Price-based pool, wholesale competition
- Cost-based pool, wholesale competition
- Open access, wholesale competition
- Full customer choice

We will discuss each model and briefly present their features, advantages and disadvantages.

29.6.1 Vertically Integrated Monopoly Model (VIM)

Vertically integrated monopoly is the traditional model of the sector. Here, a single entity undertakes all the activities of the sector, namely generation, transmission and distribution. In many places, small vertically integrated entities were

---

[4] See Tenenbaum et al. (1992); Hunt and Shuttleworth (1996); Newbery (1999); World Bank (2004); IEA (1999); IEA (2001) for further details.

developed initially, serving small areas of a region (often urban). Most of these were privately owned. As the technology improved and demand for electricity grew, utilities expanded their services and often merged with other utilities to benefit from economies of scale. In many countries (developing as well as European), this was done through nationalisation of private assets. Customers do not have choice of suppliers in this model.

Until late 1970s, this model was prevalent in the world and it is still being followed as a basic model in many countries around the world. The rationale behind vertical and horizontal integration in the electricity industry is summarized as follows:

Until recently it was believed that the power supply industry is a natural monopoly and even now, it is believed that transmission and distribution activities are natural monopolies. This implies that:

- economies of scale with increasing plant size;
- integrated planning and operation can provide significant benefits
- single ownership eases coordination of investments, planning and operation of systems, thus saving costs

Marginal pricing for a natural monopoly could lead to losses, if the average cost is higher than the marginal cost. One way to avoid the problem is to provide subsidy to the utility, which is easier if the ownership lies with the government. So, public ownership was promoted. Other factors favouring public ownership are:

- long gestation period of electricity projects;
- lack of market demand and lack of viability in initial days;
- capital intensiveness of projects, etc.

According to the Transaction Cost theory, Williamson (1995) categorised the governance structure of the firms by the asset specificities and the nature of transactions as shown in Table 29.1.

For non-specific assets, there is no need for fancy contracts. They are easily re-deployable and competition will prevent any undue advantage. When assets are intermediately specific, relational contracts come into place. When the specificity is high, the costs of contracting rise. Each party has more at risk and must be involved in pre-contract planning, monitoring and enforcement of contracts. Unified governance where one party buys out the other and takes charge of all the transactions becomes cost effective, if law permits. This is relevant in the case of electricity industry. The coordination between generation and transmission to meet

Table 29.1 Contracts and nature of transactions and asset specificity

Frequency of transactions	Degree of asset specificity		
	Non-specific	Intermediate	Highly specific
Rare	Classical contracting	Trilateral contracts	Trilateral or unified
Frequent	Classical	Bilateral contracts	Unified contract

Source: Williamson (1995)

the load is complex so that one system operator who can make decision on starting up and shutting down a generation unit is needed. The transaction involved in negotiating generation sequence would be extremely complex if the system operator is not empowered to make the decision. The transaction costs would be too high if the relevant entities are all independent and the costs must be added at each activity.

29.6.1.1 Advantages

This model was successful in ensuring a steady growth of the industry and in providing electricity in a reliable and safe manner for more than three decades. The electric industry grew significantly in size during this period and co-ordination of different activities as well as investment and operating decisions did not pose any problem. Single ownership coupled with regulated environment ensured that consumers receive power at a reasonable rate. Implementation of state policies (such as rural electrification, increasing access to electricity, fuel choice, technology choice, etc.) was relatively easy. However, implementation of social policies through electric utilities became a source of problem in many developing countries, as the utilities did not get properly remunerated for these activities.

29.6.1.2 Disadvantages

As indicated in earlier lectures, a natural monopoly requires a balancing act: to prevent the misuse of monopoly power while ensuring adequate revenue to firm. Whether it is a regulated independently or implicitly by a government agency, this is not an easy task due to conflicting nature of stakes involved. Consequently, in many cases the tariff got divorced from the cost structure under the influence of state intervention. Subsidies and cross-subsidies were deliberately used to manipulate tariffs for specific advantage. The utility does not have any incentive to improve performance when it knows that government will support it in any case. There is a tendency to over-invest in capacity, which provides return in a rate of return regulation regime.

29.6.2 Entry of Independent Power Producers (IPP)

This is the simplest form of reform where entry to generation is allowed through entry of independent power producers, who may be private or public or joint-sector entities. The vertically integrated utility continues to exist and operate as before but new generating entities are allowed to produce electricity and sell to the utility. In some cases, IPPs may be allowed to sell directly to consumers as well, which requires further opening of the market and open access to the grid.

Normally, new generation capacity procurement is initially done through this approach, while the incumbent utility continues to produce in its own generating facilities. However, it is also possible to divest a part of the generating assets to create more IPPs. Capacity procurement may be through a competitive process or through a negotiated route. Competitive procurement is generally considered to be more efficient than negotiated deals.

IPPs enter into a long-term power sale agreement with the vertically integrated utility. Often, the IPP is required to sell all of its output to the utility but it may be allowed to sell power to others (third parties) in case the utility fails to respect its obligations under the contract. In reality however, it may be difficult to implement such a provision as it may be in conflict with other laws or regulations governing the sector. For example, if the law governing the electricity sector does not require the transmission grid to offer services to others, an IPP may have little option than to sell it to the vertically integrated utility. The issue of credible commitment on the part of the utility arises in many developing countries as the cost of power purchased from the IPP may constitute a significant component of the utility revenue. In a vertically integrated structure, the utility had more flexibility in managing its limited financial resources. Issues related to IPP entry are indicated in Box 29.2.[5]

Box 29.2: Issues Related to IPP Entry

Objective: to ensure adequate investment in generation at reasonable (competitive) rates.

Risks: Two types of risks are identified, namely risk of contracting and contractual risks. Risk of contracting is affected by the institutional environment, the judiciary system, existence of contract enforcing mechanisms, and the reputation/practice of honouring contracts. These are more related with the institutional endowment of a country. Contractual risks on the other hand are related to business transactions and economic environment that affect the financial performance of the IPP. Factors affecting this type of risk are changes in sales volume (volume risk), changes in the rate of sales (rate risk), possibility of non-payment by the contractor (payment risk), possibility of changes in the fuel prices and fuel supply (fuel supply risk), possible fluctuations in the exchange rates (exchange rate risk), etc.

IPP entry may lead to the following conflicts:

i. dispatching conflicts – IPP contract may require preferential dispatching of IPP plants;

---

[5] See also RAP (2000).

ii. Tariff conflict—If the utility purchasing IPP power is not allowed to pass through the cost, the utility may face revenue problems, which in turn increases the payment risk.
iii. Billing dispute between the IPP and the utility;
iv. Technical dispute regarding technical performance of the IPP plants;
v. Other disputes related to interpretation of the contract.

Transaction cost: The cost of contracting, contract monitoring and enforcement will come under transaction cost. As the number of contracts increases, this cost can assume significant proportions for the utility. Cost of risk management due to exposure to various risks would also appear, which may not be recognised by the utility as a separate cost.

Precautions: The following precautions are required while allowing IPP entry:

- The government should not distort the incentive schemes to promote a particular technology/fuel choice for short term gains.
- The government should not provide undue advantage to IPPs and place undue burden on the utility. The policy should provide level playing fields for all.
- Contracts should be carefully negotiated upon proper analysis of all terms and conditions, without extraneous influence.
- Transparent process – selection of IPPs based on a transparent process with specified selection criteria would ensure reasonable rates for IPP power.

29.6.3 Single Buyer Model

This model introduces a single purchasing agency at the wholesale level. This single entity, which is often state-owned, performs the transmission and wholesale supply functions. Vertical separation of generation, transmission and distribution functions of the electric utility is often undertaken but is not a pre-requisite. When no functional separation is undertaken, the model ends up as an IPP entry model discussed before. Similarly, when both generation and distribution functions are separated, the term single buyer becomes more appropriate. In what follows, we assume such a configuration.

As shown in Fig. 29.5, there are multiple generators for producing power. Similarly, there are a few regional or local monopolies for power distribution and retail supply. Generators enter into power purchase agreements with the single buyer who in turn performs the trading function (i.e. selling power to distribution and retail supply companies). As the single buyer and the distribution companies are monopolies, a regulatory body oversees their functioning and performance and fixes tariffs.

Fig. 29.5 Schematic of single buyer model

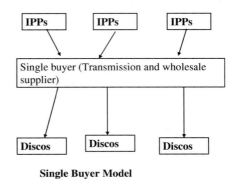

Single Buyer Model

29.6.3.1 Advantages

The model introduces competition for new generation capacities Vertical separation of activities can also provide better cost information and scope for improvement. As the single buyer purchases power through PPA, there is certainty of power purchase price and the chance of surprises is remote. As the single buyer is often government controlled, it is easy to take social and other considerations of the government into account. The skill requirement is not high as the changes are incremental only. It is easy to maintain uniform retail tariffs and even discriminatory tariffs.

29.6.3.2 Disadvantages

As competition is introduced only at the level of new generation capacities, the expected benefits from competition is rather low here. The single buyer assumes the responsibility for the entire power generated although it is not responsible for final sale. As the revenue flows backwards from the distributing companies to single buyer to generators, any default by distribution companies either due to poor tariff rates or inefficiency or a combination of different factors, the single buyer is likely to default on its dues to generators and the entire chain gets affected. Although this is mainly a contractual problem, events in certain parts of the world have highlighted the risk assumed by the single buyer. The single buyer makes decisions about capacity additions and generation system expansion. Such decisions are easily influenced by political or other considerations due to state control. Distribution companies have less say in such decisions although they are the likely to use the system and pay for the power generated. Where rate of return regulation is used, utilities have tendency of higher capital expenditure to earn better return. This is perceived to be a problem of in this model as well.

Although this model is very popular with many developing countries facing pressure for reform, it was considered as a dangerous option at some point of time (Lovei 2000), because of the path dependence introduced by such reforms. Once the single buyer enters into long-term contracts with the generators, it becomes difficult to undertake further reform to introduce further competition in generation. However, with limited reform progress in developing countries, this option is returning to the agenda once again (see Arizu et al. 2006).

29.6.4 Transitional Models

As the name indicates, these models are intermediary arrangements for passage from one structure to another in the reform process. These arrangements have been advocated for smaller systems after the California crisis because the risks of directly moving to spot markets may outweigh the benefits of spot markets. The idea behind these models is to adopt an intermediate structure for a transition period at the end of which the industry moves to a final structure. The path dependence of the restructuring process requires that the intermediate structure has to be appropriate and compatible with the final structure; otherwise the intermediate structure would create hindrance to reach the final goal.

There can be various alternative transitional structures incorporating one or another feature of the single buyer, VIM or wholesale competition models. Two such structures are discussed below: a) Multi- buyer multi–seller model without retail competition (model A, Fig. 29.6 and b) multi-buyer multi-seller with limited retail competition (model B, Fig. 29.7).

In both the models, generation is unbundled from the transmission service and a number of generating companies would operate in the system. Similarly, there would be a number of distribution companies, which may be regulated monopolies. The supply market would have two components: a competitive segment and a non-competitive segment. The non-competitive segment acts like a single buyer model, where the generating companies would sell power through PPA to the balancing buyer/seller. But some energy would be available for trading competitively. This can be achieved by defining the proportion of generation capacity to be released from each generator's PPA to balancing buyer–seller. Initially, the

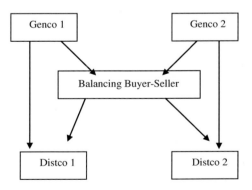

Fig. 29.6 Model A: Multi-Buyer with Balancing Buyer–Seller

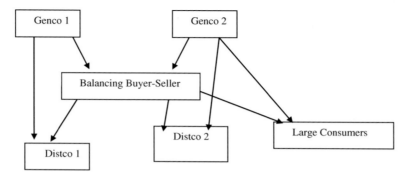

Fig. 29.7 Model B: Multi-buyer multi-seller with some retail competition

proportion of available competitive energy would be relatively small but this could be increased over time. Generators would compete with each other for selling their competitive portion of power to distribution companies using bilateral contracts for physical delivery. The balancing buyer/seller would sell power to distribution companies under a regulated bulk supply tariff. The balancing buyer could be the transmission company, which is often a state-owned company.

The second model (see Fig. 29.7) introduces limited amount of retail competition by allowing large consumers to purchase power from the generators directly through bilateral contracts. Competition arises at the retail level as eligible customers would have the choice to take supply from the distribution companies or the generators directly. Large consumers could be selected based on some criteria such as connection voltage (for example, direct connection to the transmission system) or some measure of size (for example, energy consumption). Other features of model A remain unchanged. Allowing large consumers to participate in the bilateral trade introduces competitive pressures on the system.

These models require detailed procedural arrangements for dispatch of energy traded under bilateral contracts. The balancing buyer/seller would have to take responsibility for balancing the system in real time. It would also determine energy volumes for financial settlement. Since it is not be possible to distinguish between generation to fulfill a bilateral contract (between generator and distribution/retail company) and generation for sale under a PPA, rules would be required to allocate generation between bilateral contracts and PPAs. Settlement mechanisms for energy imbalances between contracted and metered generation would have to be developed. Similarly, settlement mechanism would be required for energy imbalances between contracted and metered demand.

29.6.4.1 Advantages

Model A allows some wholesale competition and provides incentives on generators to improve productive efficiency. Frequent price determination would allow reflection of generation costs more accurately and shorter trading periods would encourage the development of shorter-term bilateral contracts to manage the risks faced by market participants.

In the early years of market operation initial contracts could be regulated to provide the necessary degree of certainty required by potential investors. At the same time, it avoids a major problem of single buyer model by allowing risks to be shared by market participants.

It provides a foundation for introducing Competition. The balancing market could provide an opportunity to simulate the spot market operations. At the same time, the model can be easily implemented without significant efforts.

This model permits implementation of certain policy objectives through the balancing buyer and the transmission company. For example, a uniform retail tariff regime throughout the country can be achieved in this model.

Model B retains the above advantages and provides more incentive to improve productive and allocative efficiency.

29.6.4.2 Disadvantages

Model A has the following disadvantages:

From system operation point of view, this model is more complex than the single buyer model or VIM. It would require establishment of procedures for dispatching, accounting and apportioning.

As distribution companies would be purchasing power from the generating companies, existing PPA would require some renegotiation to allow for competition and assignment to discos. The basis for such assignment has to be developed.

Both generating companies and distribution companies would be exposed to risk, as the financial viability of both would be inter-dependent. Consequently, there may be some tendency of re-integration of these business segments.

Appropriate metering and accounting systems have to be in place prior to implementation of this model.

Model B has similar disadvantages. Additionally, it reduces the flexibility of cross-subsidisation through retail tariff because some of the traditional subsidizing categories would be eligible customers for retail choice and hence their tariff would be decided competitively. Where cross-subsidy is a serious issue, introduction of limited retail competition would threat the viability of the distribution companies.

29.6.5 Wholesale Competition: Price-Based Power Pool Model

In a competitive wholesale market retailers and other bulk customers have choice of supplier, as multiple sellers for bulk power compete with each other to sell their power. The most common form of wholesale competition is the "power pool" model and three variants have emerged:

a. the initial English system of compulsory pool, which is also known as "gross" pool or price-based pool;
b. the voluntary trading arrangement with a balancing pool also known as "net" pool; and
c. the variants on the Argentine system where pool price is based on costs, also known as cost-based pools. This section covers price-based pools and the two subsequent sections are devoted to "net" pools and cost-based pools.

In a price-based pool (see Fig. 29.8), the price at which generators sell electricity is decided through a competitive bidding process. Generators submit price bids indicating how much they can generate and the price they want. This is done at a pre-determined time (say one day ahead of delivery). Based on the price bids received, the regulator (or the system operator) establishes the merit order dispatch schedule paying due attention to the expected demand from consumers (as submitted by the suppliers). A list of successful bidders is prepared and these generators are chosen for supplying the power. The price of the most expensive unit scheduled to be dispatched to meet forecast demand sets the system marginal price. A capacity payment component is normally added to provide incentive to generators to maintain adequate reserve margins and also to recover their full costs. The system marginal price and the capacity payment constitute the Pool Purchase Price (PPP) for any settlement period and all generators receive this price irrespective of their bid. Generators and retailers can enter into agreements outside the pool but in a gross pool, physical delivery has to be settled in the pool and these agreements become financial instruments for hedging risks.

29.6.5.1 Advantages

A wholesale competition model introduces choice to retailers and eligible customers at wholesale level and avoids some of the pitfalls of the single buyer model or regulated MBMS models. The pool provides efficient economic signals to

Fig. 29.8 Price-based wholesale competition model

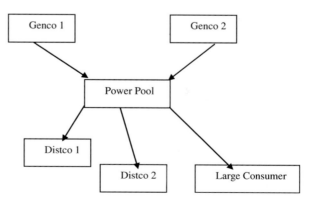

generators and retailers. Credit risk is taken on by all participants in the pool and not by any single entity. As individual distcos or large customers become responsible for decisions relating to investment in capacity, investment decisions are likely to be better. Mandatory participation of all players of the game ensures liquidity in the new market. It could also encourage new participants to enter the market, for example captive generation or large consumers. As the system operator dispatches all energy, real time system balancing remains relatively simple.

The model should be less prone to the adverse effect of vested interests, nonprofit incentives of management, and poor purchasing/contracting capability. The model could meet the national objectives of meeting demand security and reliable supply. Through some design of rules, diversity of energy sources could be encouraged and volatility in prices can be minimized. The model offers easy transition to full customer choice model.

29.6.5.2 Disadvantages

Compared to the previous models, this is a more complex reform and its implementation process is difficult. This model requires systems and processes for bid handling, price determination, metering energy sales, energy accounting and settlement. The price-based English pool was criticized for not being transparent, for being a price-setting mechanism rather than a true market, and for distorting the market to the disadvantage of flexible plants including coal plants. The abuse of market power by generators is a serious problem in many pools. The UK pool system has been criticized as a "generators' club". Experience indicates that generators with significant market power have succeeded in manipulating market prices.

The overall financial health of the sector is a pre-requisite for this model to be effective. Distcos may face difficulties in managing price and volume risk and could be affected financially. It would be difficult to maintain uniform retail tariffs under such a system without explicit subsidies. Wholesale price volatility has been a problem in many jurisdictions and could lead to unviable distribution businesses.

29.6.6 Wholesale Competition: Net Pool

Net pool (see Fig. 29.9) aims to remedy the problems faced in a gross pool. Here participants strike contracts for physical delivery of power (as opposed to financial contracts used alongside mandatory pools). Participants are responsible for dispatching themselves to their contracted positions. The participation in the pool is voluntary. The pool acts as the balancing market to balance the system in real time.

Net pool provides the real choice to retailers and large customers, as the generators have to compete with each other to seek out customers. In contrast with gross pool, where generators' bids decided the system marginal price and the pool

Fig. 29.9 Schematic presentation of net pool model

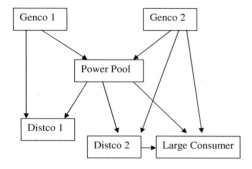

power price, here buyer side of the market is fully involved in the setting of price and thus faces stronger incentives to be efficient. It provides incentives for both investments in new generation and operation of existing plant as the full market could become eligible for competitive supply and price signals would be given over short time periods.

There is no single system marginal price for electricity and so less potential for manipulation of prices. However, any market power on the part of generators or discos will tend to be manifested in the terms of negotiated contracts. Regulation of contracting is a possible solution to this problem but will result in distorted economic signals.

29.6.6.1 Advantages

The main advantage of this model stems from its economic efficiency. It provides efficient signals to consumers and generators and involves a deeper participation of the market players. It retains the advantages of price-based pool discussed above and sharpens the incentive component for better performance.

29.6.6.2 Disadvantages

It is the most sophisticated market and the most complex in both implementation and operation. This model is more complex than the price-based pool discussed above. For a developing country without much experience in similar markets, the net pool model does not appear appropriate. Even in developed countries like UK, it has been implemented as an evolutionary step from previous forms of competition. The main complexities arise from "self-dispatch" by generators to meet their contractual positions and the balancing mechanism, which often takes the form of a near real time auction process. Such systems can add huge complexity and cost that may not be feasible or justified in a newly liberalizing market.

As the market is voluntary, liquidity may be a problem – particularly where there are long-term contracts for a large proportion of load. There is less transparency, as the Market Operator is not aware of the commercial terms of contracts struck between participants. This model would require sophisticated metering, technology and skills to operate the system.

It would be difficult to maintain uniform retail tariffs under such a system without explicit subsidies. Wholesale price volatility has been a problem in many jurisdictions and could lead to unviable distribution businesses.

29.6.7 Wholesale Competition: Cost-based Pool

This is a variation of the other pool models discussed above where the pool pricing method is based on costs and not on bids. As opposed to price-based bids where bidders submit non-verifiable prices indicating their willingness to supply under various conditions, cost based bids reflect actual fuel and O&M costs of generation. This requires systematic auditing or verification of actual costs.

The exact details of how prices are set can vary from one system to another. In Chile, short run marginal costs (SRMC) are determined based on formulas. The short-run marginal cost takes into account, inter alia, reservoir levels, plant availability, operating costs and power rationing requirements, and is optimized over a 12 or 48 month horizon. The spot price is a combination of SRMC and a capacity component based on the costs of a 50 MW gas turbine plant. Load dispatch is done on strict merit-order basis. However, on the supply side, the pool is open only to generators.

In Argentina, CAMMESA, the market operator, used to determine the hourly spot prices based on the short-run marginal cost principles. Fuel prices declared by the market participants (subject to a cap and which can be varied every 6 months), technical characteristics of generating stations and the hourly demand on the system were considered for this purpose. The spot price was the price of the last generator dispatched. In addition, the suppliers receive a fixed payment for capacity of US$ 10/MW for each peak hour and a variable capacity charge to ensure system reliability. Load dispatching was strictly based on merit order. The supply side was not a "generators' club" but is open to other participants.

29.6.7.1 Advantage

A cost-based pool allows verification of costs of generators through independent audits and constrains gaming behaviour of participants. Therefore, it curbs the market power of the participants or rather prevents them to exploit their dominant position. Thus a cost-based pool is considered to be a better option than a price-based system.

A cost-based system can also boil down to an administrative mechanism of price-setting where specific characteristics of a system can be taken into account. The Chilean system is essentially such a system, designed to take care of the dominant hydro position of the country. Although such an administrative system can lead to loss of economic efficiency, such mechanisms may improve initial acceptability of a pool to developing countries.

The price volatility is less in a cost-based system due to rigidity imposed on the price variability through caps, limitations on price change and verification of costs. It also ensures that the generators are adequately remunerated.

As cost is the pricing criteria, it would allow development of cheaper regional power and participation through pool.

29.6.7.2 Disadvantage

Cost-based pools also have the same disadvantages as other pools discussed above. In addition, this form has the following disadvantages:

1. higher level of regulatory supervision and control for cost verification;
2. could be inefficient if price setting becomes an administrative mechanism;
3. the price-setting mechanism could be less transparent and obscure.

29.6.8 Wholesale Competition through Open Access

Open access essentially implies liberalizing entry to the market by providing access to the transmission and distribution network. Open access offers the possibility for electricity suppliers, especially the Independent Power Producers, trading companies and/or customers, to make use of electricity grids they do not own or control in order to sell or buy electricity. Third Party Access (TPA) is concerned with the "transportation" of electricity over network facilities at transmission and distribution level. Without such access to the grid, no liberalization process can work.

Energy Policy Act of 1992 introduced the open access philosophy in the USA. This introduces contestability to the market without requiring radical restructuring. Third party access is one of the models allowed by the European Commission for liberalization of the electricity market in European Union.

Normally the regulatory framework would ensure open access to the wires. For example, in a single buyer model, the license or concession agreement with the transmission company would require the licensee or the concessionaire to comply with the open access criterion. Access to the wires could be refused when the system is capacity constrained or would be illegal or in contradiction with other provisions of the law.

As discussed above, open access need not involve any structural change of the supply industry. It can be introduced while retaining monopoly over transmission and distribution activities and public ownership of the utilities. Both traditional VIM and SBM would be able to implement such a policy.

29.6.8.1 Advantages

The attractiveness of this model derives from the minimal restructuring requirement required to implement it. It is suitable for those countries who are willing to maintain state control over the electricity sector but nonetheless would like to liberalize their electricity industry to some extent either to comply with some regional requirements or for any other reasons.

When open access is implemented in a vertically integrated model or in a single buyer model, advantages mentioned for the respective model would also apply here.

29.6.8.2 Disadvantages

While open access is a pre-requisite for success of any liberalization process, only open access is not a sufficient condition for achieving the objectives of any reform process. Traditional utilities may be able to develop approaches to discourage or deny access through a set of restrictive practices. This could easily be the case in developing countries where there are transmission or distribution network constraints, problems of cross-subsidies and other constraints. TPA could lead to intrusive and burdensome regulatory oversight.

Open access may not lead to the full benefits of reform as the structure of the industry undergoes little change. Only larger consumers who are eligible to participate would benefit and this could deteriorate the situation of other consumers.

29.6.9 Full Customer Choice: Retail Competition Model

This is the ultimate model in the electricity restructuring process where all customers have access to competing generators and the option of selecting their supplier. Retail electricity competition would allow residential, commercial, industrial and any other consumer to choose their supplier rather than automatically taking supply from the supplier in the geographic region of the consumer (see Fig. 29.10). Consumers may or may not decide to change the supplier but they would have the choice to do so. The difference of retail competition from the wholesale competition is that in wholesale competition the choice is open to eligible consumers and retailers, while in retail competition, all consumers have the choice. Introduction of retail competition to all sections of consumers may not take place simultaneously but may take place in a phased manner.

Retail competition involves complete separation of generation and retailing from the wires business. Moreover, in order to reap all the benefits of retail competition, generation, wholesale supply and retail supply should have to be competitive and transmission and distribution wires should provide open access to all participants. It implies that transition to retail competition is not easy even from

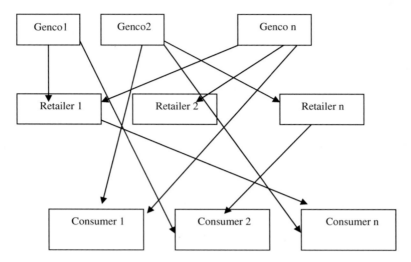

Fig. 29.10 Schematic diagram of retail competition

certain models discussed above because of the path dependence of the reform process itself and the prerequisites of retail competition.

Retail competition is a recent development and there are a handful of examples where the principle has been applied. In England and Wales, at the time of privatization, 5000 consumers with more than 1 MW demand were free to contract with any supplier. In 1994, consumers with more than 100 kW were allowed to participate in retail competition (which involved another 45000 consumers). The remaining 22 million consumers were granted the right to choose their supplier from late 1998. In USA, California was the first state to introduce retail competition in 1996. But since the recent crisis in California, the progress in implementation of retail competition programs has been slow.

29.6.9.1 Advantages

An effective retail competition is expected to result in lower prices, better quality and better innovation than was possible under regulated regime. It would result in efficient production, investment and consumption and thus would improve allocative and productive efficiency. Competition at the wholesale level would provide incentives for efficient production in the short- and long-run. Similarly competitive retail supply would also provide incentive for efficient consumption. Consequently, there could be environmental benefits in terms of reduced pollution from efficient production and consumption of electricity.

The presence of a wholesale spot market would indicate how much it is worth investing to make a plant available. The pricing of wholesale power would provide the right incentive to take correct decisions about building, maintaining, running and closing power plants. Competition would facilitate proper technology choice and free capital for investment in other productive and developmental activities.

In a full customer choice model, consumers would respond to the market prices more effectively and would help manage demand better than in other models discussed above. The consumers have the right to choose their supplier according to their preference and could select: the nearest one, or the cleanest one, or the cheapest one or the best service provider.

Better productive and allocative efficiency would have macro-economic effects in terms of lower prices of products, and better terms of trade.

29.6.9.2 Disadvantages

As can be imagined, this is the most difficult to implement of all the models discussed here. It requires technical and managerial skills to introduce this system.

Metering becomes a major problem in retail competition. All consumers must have appropriate meters to allow recording of how much each consumer has consumed from competing retailers in each settlement period. In the UK, metering proved to be a logistical problem and in terms of consumer comprehension.

There can be no obligation to supply in retail competition model, as monopoly supply would cease to exist. This could affect the policy objectives of the government of a developing country where rural electrification is an issue.

Implementation of social policy programs through life-line rates would not be possible any more because the retailers would not be able to offer different rates for different consumers. Governments would have to explore direct subsidy mechanism for such social programs.

Special promotion of certain technologies such as renewable technologies would be difficult in this model and as in social policy programs, here also, explicit direct subsidy systems have to be developed.

As mentioned in wholesale competition, moving to retail competition is a big decision and such a decision has to be taken very carefully. For developing countries with a number of specificites, retail competition should be aimed only at the end of transition period of 5–10 years.

29.7 Reform Sustainability: A Framework for Analysis[6]

Von der Fehr and Millan (2001) suggest that the objective of the power sector remains unchanged before and after reform—to deliver electricity to the population in a financially, economically, socially, politically and environmentally sustainable manner. Although they suggest a three-stage analysis framework,[7] they do not elaborate on the concept of sustainability by elaborating each element. Similarly, Benavides (2003) who believes that reforms should not be viewed only as a process of changing initial conditions, suggests some basic principles[8] to make reform work but does not discuss reform sustainability as such. In order to bridge the gap, first I elaborate on what could be considered as a sustainable reform.

The framework is presented in Fig. 29.11. A reform can be considered as sustainable if it is politically acceptable, financially viable, economically efficient, socially desirable, environmentally benign and implementable as a project. The project management aspect of a reform has not been considered earlier in the literature but given its importance, it has been included here.

Any reform being a political process, its acceptability depends on three conditions: reforms have to be politically desirable, feasible and credible (World Bank 1995). A reform can become politically desirable when a country faces an economic crisis due to domestic or international factors or faces political changes (e.g. regime changes or changes in governing coalition), bringing ideological shifts in policymaking or governance. Political desirability of any reform does not necessarily ensure its feasibility, which requires garnering necessary political support to implement the reform policies and effectively withstand opposition to reforms. Credible commitments on the other hand are judged whether the promises are kept and whether national/international restraints to policy reversals are at work (World Bank 1995).

Financial viability implies that the power sector runs on commercial principles and is able to cater to its financial needs without state support. In a regulatory environment, this amounts to

a. Cost minimisation through improved operating performance, prudent investment decisions and efficient commercial/contractual performance.
b. Adequate revenue generation through properly set tariffs for services and efficient commercial operation.
c. Reduction in state support in the form of revenue subsidy and capital support.

Improving economic efficiency through introduction of competition and choice to the consumers has been the predominant logic for reforms. The economic efficiency would aim for (1) effective competition for investment in new plant capacity, (2) effective competition among generators in the operation of existing plant, (3) effective wholesale competition to supply distribution/retail companies, and (4) effective retail competition to supply final customers (Bhattacharyya and Dey (2003)). Moreover, ensuring an adequate system to avoid shortage of capacity (which may be in generation, transmission or distribution), managing the system

---

[7] This consists of a. identifying the objectives and constraints of reform, b. analysing how these are taken care of and identifying the threat to reform and c. to suggest improvements/modifications to remove the threats.

[8] These include credibility of the regulator, use of simple instruments, avoiding problems of euphoria or pessimism, etc.

29.7 Reform Sustainability: A Framework for Analysis

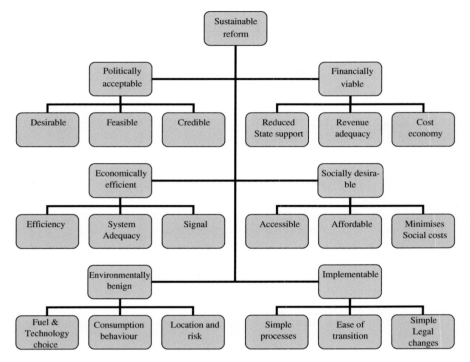

Fig. 29.11 Sustainability conditions for power sector reforms

using economic principles and finally, providing correct signals to different actors of the game are also aimed through economic efficiency.

Three main social aspects of power sector reform have attracted attention as well (see Dubash (2002) and Wamukonya (2003)): access to power, affordable prices and minimising other social costs (such as job reduction). Reformed markets with private participants are unlikely to have necessary incentives to undertake financially non-remunerative activities to provide access to the poor unless transparent and direct compensation is provided (Dubash (2002) and Heller and Victor (2004)). Similarly, affordable price for electricity is a related social concern especially in developing countries where the prices in the pre-reform period were subsidised for many consumers and price reduction is not expected through reforms. Moreover, as commercial energies have to compete with traditional energies for most important demands and given the lack of regular flow of money income for the poor due to existence of in-kind payments for certain services and participation of the poor in informal sector activities, the possibility of substitution by dirty fuels exists. In addition, barriers like access cost, inability to pay for access and taxes and duties affecting costs (Barnes and Halpern 2000) may make subsidies inevitable in the short run but subsidies tend to distort economic decisions, run counter to reform arguments and manifest in politicisation of energy supply issues.

Similarly, reforms may introduce other social costs due to macro-economic impacts of abandoning uneconomic assets, impact of these assets on employment, fuel supply, etc. Unemployment or job loss can be a major problem. These social costs need to be minimised to make a reform socially desirable.

Attention has been drawn to the effects of competition and restructuring on the environment, as reform influences fuel and technology choice, consumption behaviour and investment decisions. Two opposing changes in the power sector reform, namely the possibility of higher emission due to lower price and consequent higher demand on one hand (Vrolijk 2002) and the changes in the decision-making perspective for fuel and technology choice on the other hand could positively or negatively affect the environment. Fuel switching, efficient technology use, opportunities for alternative fuels and clean technologies (such as cogeneration) can accompany a power sector reform, benefiting the environment. But neglect of long term interest such as renewable energies and energy conservation for short-term profit motives may affect the environment in the long-run (Byrne and Mun 2003). The net effect of these opposing forces is not obvious (Brennan et al. 2002).

Finally, any reform is a project that has to be implemented and hence it must be a manageable project. This in turn implies that the reform is not unduly complex or does not require unduly complex and time-consuming legal implementation and allows transition to more competitive models relatively easily (Bhattacharyya and Dey 2003).

Any power sector reform that satisfies all these conditions simultaneously can be considered to be sustainable. As indicated earlier, some of the conditions can be in conflict with other conditions and can be a source of concern for sustainability. The time dimension has not been explicitly considered above.

29.8 Experience with Energy Sector Reform

Since Chile initiated structural reform of the power sector in the 1980s, and similar reforms followed in the UK and Argentina, reform became the normal prescription of the multi-lateral donor agencies like the IMF and the World Bank. Since 1990s, the reform agenda has been pursued vigorously for more than a decade (Bacon and Besant-Jones (2001), World Bank (2004)). Recent reviews, however, indicate that not much progress has been made in many parts of the world (Bacon and Besant-Jones 2001; WEC 1998; World Bank 2003). While most of the countries have initiated the reform process by introducing minimal changes, major reforms did not take place in most of the countries.

The progress has slowed down considerably after the California crisis. With high oil prices and heightened supply security concerns, the reform is not on the top of the government agenda.

A number of factors explain the predicament of the sector reform initiatives:

- *Small size of the industry* The developing world, with a small sized power sector, may gain little from the unbundling exercise.

The World Bank (2004) report indicates that "60 developing countries have peak system loads below 150 MW, another 30 between 150 and 500 MW, and possibly another 20 between 500 and 1000 MW. Even a 1000 MW system is small for introducing competition.

There is no standard template to follow in this regard.

- *Investment issue* It is difficult to ensure investment in expansion of network infrastructure in an unbundled industry.

Developed countries are experiencing this problem faced with capacity renewal challenge.

Developing countries, with inadequate capacity and high demand growth, face more challenge.

- *Lack of regulatory capacity* The regulatory capacity for ensuring proper functioning of the unbundled system is lacking in many countries. The regulatory task becomes challenging in an unbundled environment and with low regulatory capability. It took developed countries a long time to build and develop expertise in regulatory capabilities. Developing countries with fewer resources will find it more challenging to develop required regulatory capabilities.
- *Strong path dependence* Experience shows that it is difficult to make changes to structural choices once they are introduced. Thus if privatisation precedes restructuring, it is more difficult to restructure the industry as such change requires co-operation and participation of the private parties, which is difficult compared to restructuring state-owned entities. It is often suggested that restructuring should be done first followed by privatisation. In other words proper sequencing is essential to the success of any privatisation programme.

In terms of effects of reform undertaken so far, the World Bank (2004) report suggests the following:

- more than $750 billion has been invested in infrastructure projects in the developing world between 1990 and 2001. However, a large share of this investment has gone to Latin America and East Asia, while Sub-Saharan Africa, Middle East and North Africa received a miniscule amount of this investment.
- Annual investment peaked in 1997 to $130 billion and started to decline since then. By 2001, the volume has reached $60 billion. Utilities are finding it difficult to ensure system expansion.

"Privatisation and deregulation have significantly improved physical performance, service quality and other efficiencies in many developing countries."

Although it is normally expected that price reform will accompany privatisation, the progress in this respect has been slow in many developing countries. The prices still cover only a small fraction of the long run marginal costs.

There can be some impacts of the reform on the poor as prices are reformed and cross-subsidies are removed. The improvements in service quality, increased access and less stress on the public finance should mitigate these distributional effects to some extent.

There is thus recognition that reform and restructuring options have to be carefully analysed, designed and implemented in order to produce benefits for the society.

29.9 Conclusions

This chapter has presented an overview of energy sector reforms, with a special focus on the electricity industry. The energy sector reform is one of the most discussed issues of the energy sector since the 1990s. It is essential to have some understanding of the concepts and issues related to these reforms. Given that the electricity sector was most commonly targeted, this chapter has considered the changes that took place in the electricity industry. The chapter has presented the reasons for reforming the energy sector and the alternative options that were widely used around the world. It ends with a brief review of the experience related to the reform and suggests that despite all the attention received, the reforms did not really deliver in most cases. As new challenges (such as the climate change issue and the security of supply issue) emerged, a new debate has started about the need for intervention in the market. This has undermined the reform drivers to some extent.

References

Arizu B, Gencer D, Maurer L (2006) Centralised purchasing arrangements: international practices and lessons learnt on variations to the single buyer model, Discussion Paper 16, Energy and Mining Sector Board, World Bank (see World Bank website)

Bacon R, Besant-Jones W (2001) Global electric power reform, privatisation and liberalisation of the electric power industry in developing countries. Annu Rev Energy Environ 26:331–359

Barnes DF, Halpern J, (2000) The role of energy subsidies, Chapter 7, in energy services for the world's poor. ESMAP, World Bank

Baumol WJ (1982) Contestable markets: an uprising in the theory of industry structure. Am Econ Rev 72(1):1–15

Benavides JM (2003) Can reforms be made sustainable? Analysis and design considerations for the electricity sector. Technical Paper, Inter-American Development Bank, Washington, DC

Bhattacharyya SC (2007a) Power sector reform in South Asia: Why slow and limited so far? Energy Policy 35(1):317–332

Bhattacharyya SC (2007b) Sustainability of power sector reform in India: What does recent experience suggest. J Clean Prod 15(2):235–246 Special issue

Bhattacharyya SC, Dey P (2003) Selection of power market structure using the analytical hierarchy process. Int J Glob Energy Issues 20(1):36–57

References

Brennan TJ, Palmer KL, Martinez SA (2002) Alternating currents: electricity markets and public policy, resources for the Future, Chapter 15, restructuring and environment protection, Washington, DC (See http://www.rff.org/rff/Documents/RFF-Bk-AltCur-ch15.pdf)

Byrne J, Mun Y (2003) Rethinking reform in electricity sector: power liberalisation or energy transformation? In: Wamukonya N (ed) Electricity reform: Social and environmental challenges. UNEP, Riso, Denmark

de Oliveira A (1997) Electricity system reform: World Bank approach and Latin American reality. Energy Sustain Dev 3(6):27–35

de Oliveira A, Mackerron G (1992) Is the World Bank approach to structural reform supported by experience of electricity privatisation in the UK? Energy Policy, February 153–162

Dubash N (ed) (2002) Power politics: equity and environment in electricity reform, World Resources Institute, Washington, DC

EIA (1996) The changing structure of the electric power industry: An update. Energy Information Administration, Washington, DC

Guasch JL, Spiller P (1999) Managing the regulatory process: design, concepts, issues and the Latin American and Caribbean story. World Bank, Washington, DC

Heller TC, Victor D (2004) A political economy of electric power market restructuring: introduction to issues and expectations, Working paper 1, Program on Energy and Sustainable Development. Stanford University, Stanford

Hunt S, Shuttleworth G (1996) Competition and Choice in electricity. Wiley, London

IEA (1999) Electricity market reform: An IEA Handbook. International Energy Agency, Paris

IEA (2001) Competition in electricity markets. International Energy Agency, Paris

Jaccard M (1995) Oscillating currents: the changing rationale for government intervention in the electricity industry. Energy Policy 23(7):579–592

Joskow P (2000) Deregulation and regulatory reform in the US electric power sector, MIT Discussion paper (See http://www.iasa.ca/ED_documents_various/joskow03.pdf)

Klein M (1996) Competition in network industries. The World Bank, Washington, DC

Lovei L (2000) The single buyer model: a dangerous path toward competitive electricity markets, Public Policy for the Private Sector, World Bank, Note 225, December

Newbery DMG (1999) Privatisation, restructuring and regulation of network utilities. MIT Press, Mass

RAP (2000) Best practices guide: implementing power sector reform, Regulatory Assistance Project, Vermont, USA

Tenenbaum B, Lock R, Barker J (1992) Electricity privatization: structural, competitive and regulatory options. Energy Policy, pp 1134–1160 December

Von der Fehr N-HM, Milan JJ (2001) Sustainability of power sector reform in Latin America: an analytical framework, Working paper, Inter-American Development Bank, Washington, DC

Vrolijk C (ed) (2002) Climate change and power: economic instruments for European electricity. Earthscan, London

Wamukonya N (ed) (2003) Electricity Reform: social and environmental challenges. United Nations Environment Programme, UNEP Riso Centre, Denmark

WEC (1998) The benefits and deficiencies of energy sector liberalization. World Energy Council, London

Williamson OE (1995) The institutions and governance of economic development and reform. In: Proceedings of the World Bank Annual Conference on Development Economics 1994, World Bank, pp 171–197

World Bank (1995) Bureaucrats in business: the economics and politics of government ownership. World Bank, Washington, DC

World Bank (2003) Power for development, a review of the World Bank group's experience with private participation in the electricity sector. World Bank, Washington, DC

World Bank (2004) Reforming infrastructure: privatisation, regulation and competition, World Bank